RELIABILITY TECHNOLOGY

Reliability Technology

A. E. GREEN
and
A. J. BOURNE

Systems Reliability Service,
United Kingdom Atomic Energy Authority,
Risley, Warrington, Lancashire

A Wiley–Interscience Publication

JOHN WILEY & SONS

Chichester · New York · Brisbane · Toronto

Copyright © 1972 John Wiley & Sons Ltd. All Rights Reserved. No part of this publication may be reproduced, stored in a retrieval system, or transmitted, in any form or by any means, electronic, mechanical photo-copying, recording or otherwise, without the prior written permission of the Copyright owner.

Library of Congress catalog card number 73-161691

ISBN 0 471 32480 9

Reprinted March 1977
Reprinted October 1978

Printed by Unwin Brothers Limited
The Gresham Press, Old Woking, Surrey
A member of Staples Printing Group

Preface

Interest in the characteristic of reliability ranges from the user of the smallest domestic appliance to those responsible for the management of the largest industrial concern or technological project. Time and time again the question 'Is it reliable?' is asked about the item concerned. The question may be posed in different forms such as, 'Can it actually do the job intended?' or, 'Can it do it with an acceptable cost?' or, 'Can it do it both when and where required?'. Behind all these questions is the basic human desire to obtain value for money and to avoid the frustrations, penalties and possible dangers associated with untrustworthy products. The points made by the questions are perfectly understandable, but is there some general method by which they can be satisfactorily answered? In addition, is there some way in which any answer produced can be shown to be correct? It is the purpose of this book to examine these problems, to describe the techniques for solving them and to show the application over a range of technological products.

It can be said generally that reliability costs money, but the achievement of reliability may save money, may have sociological advantages and may prevent the loss of human life. As with all practical problems of this type, there is a need to preserve an economic balance between the cost of reliability and the advantages which accrue from its achievement. The simple question 'Is it reliable?' is not, therefore, sufficiently searching and should be replaced by a question of the form 'Is it reliable enough?'. The introduction of the word 'enough' implies a quantification of the reliability concept. In other words, if some economic balance is to be obtained, reliability must be defined and used in such a way that it becomes a measurable quantity. The opening chapter of the book formulates such a definition and the subsequent chapters show how reliability criteria based on the definition can be applied to items in many branches of technology.

The actual evaluation of reliability in a particular application may be derived by practical or theoretical means. Practical evaluation involves a knowledge of past history of reliability performance or the setting up of specific tests designed to reveal a particular reliability characteristic. In such cases special techniques are required to process the data collected and to interpret the information in a meaningful fashion. These techniques, which are described in the book, are based upon the statistical concepts of 'goodness of fit', 'parameter estimation' and 'confidence limits'. In the case of theoretical evaluation, there is a need to formulate a reliability mathematical model which is an adequate representation of the item being considered. This is a process of synthesis which involves the techniques of probability theory, logical flow analysis, system modelling and various specialized mathematical concepts. The interdisciplinary use of these techniques is described and developed in the book together with the ways in which the appropriate input data may be obtained.

The practical evaluation of reliability generally involves the availability of large samples for test purposes or the availability of time in which to collect appropriate data from field experience. In the past, the reliability and capability of many items, such as the wheel or lever, have been established by trial and error methods over long periods of time. The rate of technological development today is such that the luxury of time or the cost of full sample testing can rarely be afforded. This is particularly true in high risk systems where the consequences of unreliability can lead to large capital loss and hazard to human life.

The aircraft industry, from the first powered flights of a few seconds' duration by the Wright Brothers in 1903, has progressed to aircraft carrying large numbers of passengers round the world. In the 1960's, man-carrying space vehicles became an acccpted part of our lives. The speed and carrying capacity of all forms of transportation is steadily mounting. Chemical and other industrial plants have rapidly increased in size, complexity and potential risk. Electrical generation is carried out by the use of nuclear reactors, yet a nuclear reactor did not exist prior to 1942. Such happenings have one thing in common, they are taxing our skills to produce a reliable solution. In many cases the only way of evaluating the reliability is by a theoretical process of reliability synthesis as described in this book.

In addition, as the consequences of unreliability become more severe, it may be necessary not only to evaluate the reliability but also to substantiate that the evaluated reliability is correct and can be achieved. It is from this need to cross-check on the reliability performance of a system that a process of independent reliability assessment has evolved. This process is directed at substantiating the adequacy of a system, as opposed to just saying what one would do if one were designing it.

To understand the assessment process it is necessary to have a knowledge of how a problem may be logically solved. The setting up of the problem in terms of the reliability and general performance requirement is discussed. The techniques used to assess the actual performance of the system and to substantiate that the performance meets the requirement are described. A method of organizing and carrying out this type of work is presented together with some of the benefits which accrue from establishing a logical approach to reliability evaluation.

PREFACE

Reliability technology may be defined as the scientific study of the trustworthy nature of devices or systems in practical and industrial situations. The scientific study involves not only evaluating reliability in a quantitative manner, but also substantiating the validity of the evaluation. There has been a lack of a comprehensive treatment of these aspects of reliability assessment and this book has been written to fulfil this need. The approaches and the techniques are made available as a reference for those who are engaged in reliability assessment or called upon to make such an assessment. The design engineer and user, particularly of systems which involve risks, obviously have a great interest in the question 'Is it reliable?' or 'Is it reliable enough?'. Similarly, students engaged in the study of any form of system or equipment engineering will have a need to understand how to substantiate the reliability of a design if dependence is to be placed upon it.

Contents

CHAPTER 1 RELIABILITY CONCEPTS

Introduction	1
An Historical Review	2
Some High Reliability Systems	5
Aircraft systems	5
Power plants	11
Chemical plants	14
Electrical supply systems	16
Systems in general	18
A Probabilistic Definition	21
Some Probability Concepts	26
Reliability in the Time Domain	31
Technological System Applications	36
Conclusions	43
Questions	44

CHAPTER 2 AN APPROACH TO RELIABILITY ASSESSMENT

Concept of Overall Reliability	47
Some Further Probability Concepts	50
System Reliability	57
Capability of Performance	60
Variability of Performance	62
Required Performance	72
The Overall Reliability Function	75
Questions	80

CHAPTER 3 THE PERFORMANCE REQUIREMENT

Requirement Concepts	83
Basic Requirement	86
Requirement Conditions	91
Requirer conditions	92
Space conditions	93
Time conditions	97

Input conditions	99
Output conditions	105
Overall Requirement	108
Requirement Limits	112
Financial Requirements	115
Requirement Variations	118
Questions	120

CHAPTER 4 THE PERFORMANCE ACHIEVEMENT

Achievement Concepts	122
Introduction	122
The interface problem	122
System definition	125
Capability and variability	129
Functional Capability	130
General concepts	130
Logistics	131
Logic	132
Boolean Algebra	136
Truth tables	143
Logic Diagrams	145
Matrix Representation of a Network	149
Functional Diagrams	152
Generalized Systems	155
Questions	158

CHAPTER 5 VARIATIONS IN THE PERFORMANCE ACHIEVEMENT

General Concepts	161
Achievement Variations	162
The Transfer Function Approach	165
Systematic Variations	169
Random Variations	174
Continuous distributions	175
Discrete distributions	181
Combinations of continuous and discrete distributions	188
Achieved Performance	190
Questions	191

CHAPTER 6 THE TRANSFER CHARACTERISTIC

Transfer Characteristic Concepts	193
The Laplace Transform	194

CONTENTS

Transforms of Particular Functions 195
    Step function 195
    Impulse function 197
    Ramp function 198
    Exponential functions 200
Transforms of Particular Operations 201
    Differentiation and integration 201
    Multiplication by an exponential 204
    Change of variable 205
    Convolution 206
Transfer Function 209
Effects of Particular Transfer Functions 216
    First-order system 216
    Second-order system 220
Transfer Function Algebra 225
The Transfer Matrix 229
    Questions 234

CHAPTER 7 PROPERTIES OF DISTRIBUTIONS
The Distribution Function 238
Moments of Distributions 242
The Cumulative Distribution Function 257
Conditional Distribution Functions 262
Hazard Distribution Functions 269
    Questions 272

CHAPTER 8 COMBINATIONS OF DISTRIBUTIONS
The Need for Combinations 276
Sums and Differences of Distributions 277
Products of Distributions 290
Quotients of Distributions 296
General Combinations of Small Variations 299
Selected Combinations of Distributions 300
    Questions 309

CHAPTER 9 SAMPLING, ESTIMATION AND CONFIDENCE
Sampling 313
Estimation and Confidence 315
    General concepts 315
    Sample means 320
    Sample variances 328
    Sampling from the normal distribution 334

Sampling from the exponential distribution	340
Sampling from the Poisson distribution	346
Sampling from the binomial distribution	350
Weighted combination of estimates	353
Goodness of Fit	359
General	359
Graphical methods	359
Analytical methods	363
Sample Testing	370
Questions	382

CHAPTER 10 RELIABILITY CONSIDERATIONS FOR SYSTEMS

Overall Concepts	386
The Correlation Process	387
Simple correlations	387
Multiple correlations	394
Dependent correlations	397
Correlations associated with discrete states	402
Changes of State	404
Practical implications	404
Irreversible changes of state	405
Reversible changes of state	417
Reliability Parameters of Interest	424
Questions	425

CHAPTER 11 SYNTHESIS OF SYSTEM RELIABILITY

The Need for Synthesis	430
The Two-element System	431
Non-redundant Elements	432
Change of state representation	432
Irreversible changes of state	437
Reversible changes of state	445
System renewals	450
Variational aspects	454
Parallel Redundant Elements	455
Change of state representation	455
Irreversible changes of state	458
Reversible changes of state	465
System renewals	468
Variational aspects	469
Standby Redundant Elements	470

CONTENTS xiii

    Change of state representation 470
    Irreversible changes of state 471
    Reversible changes of state 474
    Variational aspects 478
    Questions 478

CHAPTER 12 SYNTHESIS OF COMPLEX SYSTEMS

    Types of Complex Systems 483
    Exclusive Elements 484
    Simple Common Elements 490
    Cross-linked Common Elements 494
    Majority-vote Elements 497
        Change of state representation 497
        Irreversible changes of state 502
        Reversible changes of state 512
        Variational aspects 516
    General Procedures for Synthesis 516
        Questions 522

CHAPTER 13 THE APPLICATION OF RELIABILITY ASSESSMENT

    General Review 527
    Reliability Data 530
        General aspects 530
        Acquisition of data 531
        The use of data 535
    The Mathematical Model 541
        Characteristics and type 541
        Typical assumptions 544
    The Perspective of Reliability Assessment 549

APPENDIX 554

ANSWERS TO QUESTIONS 600

LIST OF SYMBOLS 613

GLOSSARY OF TERMS 622

SUGGESTED REFERENCES FOR FURTHER READING . 628

INDEX 631

CHAPTER 1
Reliability Concepts

Introduction

Reliability, as a human attribute, has been sought after and praised since time immemorial. Few would disparage the reliable man or decry his worth in industry, commerce or society generally. Trustworthy, dependable and consistent are almost synonymous adjectives which may be applied to those on whom reliance can be placed and these give an indication of why the characteristic of reliability is so valued. Inherently, man always finds comfort where there is trust and craves for those things which are consistent and predictable.

Although reliability in human behaviour is valued, it is difficult to define the characteristic precisely or to measure its exact worth. On the other hand, there are obviously degrees of reliability and no definite line can be drawn between the man who is reliable and the man who is not. It is nearly always possible to adjudicate on a comparative basis and decide whether one particular individual is more reliable than another. This judgement becomes easier when it is related to some definite human function. For instance, the degree of punctuality of individuals in arriving at work or attending meetings could be used as a measure of their reliability in performing this particular function.

It is interesting to note, in fact, that the characteristic of reliability is usually used to describe some function or task. In the widest sense, it may be said to be a measure of performance. The man, who, as an emissary, always says the right things, may be described as reliable because he is good at this diplomatic task. The man who always completes his work in the scheduled time may be said to be reliable because he succeeds in this type of activity. The man who is always in the right place at the right time may be called reliable because he is functioning in some particular desired manner. Generally, then, the quality of a man's performance and also the time at which or in which he performs may be a measure of his reliability in association with any particular task.

It is easy to see, with this general concept of reliability, how man has applied the term 'reliability' not only to human activity but also to the performance of functional objects of his own make or invention. Just as a man may feel let down by his fellow men, so he may feel frustrated or disappointed if the functional objects with which he deals do not perform in the manner desired. With objects, perhaps more than with people, the lack of reliability may lead to more than just a feeling of disenchantment. Unreliability of functional objects can waste man's time, cost him money or even endanger his life. As the consequences of this type of unreliable behaviour become more severe, so man's interest in reliability and his desire for reliable products becomes more acute.

At one end of the scale, for instance, a desire will exist, say, for a reliable television set. If the set breaks down, the user may be annoyed, he may feel frustrated and it may cost him some money to have it repaired. An acceptable standard for the reliability of a television set will obviously depend upon what value the user places on his freedom from annoyance and frustration and on the size of his maintenance bills. Higher up the scale of consequences, there will be a desire on the part of an airline operator to have reliable aircraft. If an aircraft crashes it will do considerably more than annoy the operator. There will be all the implications arising out of the loss or the maiming of human life and it will certainly cost the operator a considerable amount of money. Obviously, the standard of acceptable reliability for the aircraft is going to be considerably different from that for the television set.

As the previous two examples illustrate, the characteristic of reliability is of interest to the user of any functional object, to the management of any industrial or commercial concern and to those responsible for any technological project. The degree of interest and the standard of reliability to be achieved are obviously coupled to the consequences of unreliable behaviour. Improving reliability will generally cost money, but the achievement of reliability may save money, may have sociological advantages and may prevent the loss of human life. As with all practical problems of this type, there is a need to preserve an economic balance between the cost of reliability and the advantages which accrue from its achievement. For any functional item, therefore, it is not only of interest to know whether it is reliable but, more particularly, whether it is reliable enough. The introduction of the word 'enough' implies a quantification of the reliability concept. In other words, if some economic balance is to be obtained, reliability must be defined and used in such a way that it becomes a measurable quantity.

An Historical Review

Any functional item will have a certain desired or required performance specification. If, under the appropriate conditions, the item achieves the required performance or continues to achieve it, then the item may, in a qualitative sense, be termed reliable. If, on the other hand, the item fails to

RELIABILITY CONCEPTS

perform in the desired manner or nearly always fails so to perform then, again in a qualitative way, it may be described as unreliable. Reliability, then, is a measure of the degree of successful performance of a functional object under the required conditions of operation. The quantification of reliability, therefore, resolves itself into how success may be measured or, conversely, what measure can be attached to failure. It is perhaps useful at this stage to see how this idea of the measurement of success or failure in connexion with reliability has developed over the past few decades.

In the expansion of the aircraft industry after the First World War, the fact that an aircraft engine might fail was partly instrumental in the development of multi-engined aircraft. Comparisons were made between one and two and between two and four engined aircraft from the point of view of successful flights. The comparisons tended to be purely qualitative at this stage and little attempt was made to express the reliability in terms, say, of the proportion of flights which were deemed successful for the different engine configurations. As more aircraft took to the air information was gradually collected on the number of aircraft system failures which took place in a given number of aircraft over a given length of time. This led, in the 1930's, to the concept of expressing reliability or unreliability in the form of an average number of failures or as a mean failure-rate for aircraft. From these beginnings, as would be expected, various inferences were made as to what should be the reliability criteria for aircraft and the necessary safety levels began to be expressed in terms of maximum permissible failure-rates. In the 1940's requirements were being given for aircraft in such terms as the accident rate should not exceed, on average, 1 per 100,000 hours of flying time. Other similar types of numerical expressions for some facet of reliability began to appear. For instance, during the 1960 decade it may be deduced that an aircraft has been involved in a fatal landing approximately once in every million landings that have taken place. At the start of the automatic landing system development, the degree of reliability required from the system was specified in terms of the fatal landing risk not being greater than 1 per 10^7 landings.

A further field of interest is the work which was done on the V1 missile in Germany during the Second World War. The development of the V1 started in 1942 and the original concept for its reliability was that a chain cannot be made stronger than its weakest link. Although efforts were directed along this line, there was 100% failure of all missions in spite of all the resources available under wartime conditions. Eventually, it was realized that a large number of fairly strong 'links' can be more unreliable than a single 'weak link' if reliance is being placed on them all. Work based on this idea gave rise to a great improvement in the reliability of the V1. It is interesting to note that the reliability ultimately achieved was that 60% of all flights were successful and a similar figure also applied to the V2 missile. The successes which have been achieved with unmanned earth satellites suggest that the reliability of such devices is still in the region of 60% to 70% successful flights. Here it appears that although there has been a basic improvement in the component part reliability, this has been offset by the increase in complexity.

The armed forces have developed a keener interest in reliability with the advent of more complex military equipment. Even in the Korean War the unreliability of electronic equipment called for some strong measures in the U.S.A. About this time the U.S. Armed Services reported that $2 per year was necessary to maintain every dollar's worth of electronic equipment. In the case of the British Forces it has been stated that lack of reliability and maintainability in the equipment that it buys forces the Royal Air Force to spend about half of its resources on maintenance rather than operations. The Army from time to time has carried out surveys on the reliability of its fleets of wheeled vehicles. Numerical evaluations of the initial costs, the maintenance costs and the degree of availability achieved have enabled them to plan the most economical maintenance and replacement intervals. Similar reliability calculations have been carried out by industrial and commercial users of vehicles. These examples emphasize the importance of reliability from the cost and maintenance point of view. A number of Service departments have actively considered in recent years the placing of contracts for military equipment in which reliability is quantitatively expressed as a performance parameter. The contracts carry financial penalty and bonus clauses depending upon the degree of reliability ultimately achieved. Similar numerical specifications for reliability have already made their appearance in connexion with various component parts used in the electronics industry.

From the 1950's the nuclear industry has grown up in an atmosphere where reliability has been of paramount importance from the safety point of view. Special reliability assessments have been made of all aspects of nuclear power reactor construction and operation but, initially, the actual quantification of reliability performance began with the reactor's automatic protective system. This is the system which is designed to safeguard the reactor and keep it in a safe state should any abnormal conditions arise. Assessments have shown that the chance of such a system failing to perform its required function on demand is typically about once in every 10,000 occasions. Since the demands for such action are generally infrequent the overall chance of any untoward occurrence may be expressed quantitatively as one in several million. More recently, other aspects of nuclear power reactors have been expressed in numerical reliability forms. Considerations have been given to the reliability of turbines, circulators, boiler feed pumps and standby electrical supply systems. The electrical supply industry, generally, has used reliability evaluation techniques to determine costs of supply configurations and availability of supplies to consumers. The availability of supplies to a particular consumer, for instance, might be expressed in terms of an average of 30 minutes loss of supply per year of usage.

The nuclear industry is not the only one where hazards may arise due to the unreliable operation of plant or equipment. This occurs to a greater or lesser extent in all industrial and commercial undertakings. Dangers to people can arise from collapsing structures, flying missiles, crashes, excessive temperatures, poisonous gases and explosions. Where such dangers are appreciable a need has arisen for the reliability evaluation of the plant or equipment which provides the necessary protection. Gradually, therefore, the techniques of reliability evaluation are finding their way into the process industry, the chemical industry and the transportation undertakings.

This brief review of the way in which the interest in reliability technology has developed leads to a number of conclusions. First, the unreliability of any functional item or technological project can cause financial loss, can cause human frustration and dissatisfaction and, in the ultimate, can cause death. Secondly, the degree of required or acceptable reliability depends on the consequences of failure. Thirdly, there is a need for a scientific method of reliability evaluation so that the benefits of reliability achievement may be balanced in some way against all the risks and consequences of failure. Fourthly, the precision of the scientific method and the extent of its application need to increase as the risks associated with any particular technological project become more severe.

Some High Reliability Systems

It would be useful at this stage to consider a few technological systems where there is a need for high reliability. As suggested earlier, such applications will become apparent in high risk situations where there is a large amount of capital at stake or danger to human life.

Aircraft systems

In the operation of aircraft there is obviously a safeguard required for the capital investment in the aircraft which could be of the order of a few million pounds per aircraft. Additionally, if a crash takes place subsequent damage to other property could result and, of course, there is the possibility of the loss of human life. Taking accidents which could involve fatalities during the landing phase, an analysis of a sample of airline flying, prior to automatic landing systems, showed a rate of about 0.65×10^{-6} per landing. Interpreting this figure as applied to the world's fleet of large passenger-carrying aircraft at that time, it showed that there would be a landing accident on average about once per year on the basis of two landings per day by each aircraft.

It is also interesting to note that the analysis of aircraft accidents showed about 40% of accidents to civilian aircraft occurred during the landing approach. Obviously, from the operator's point of view, there was and still remains a great economic incentive for the ability to fly and land in zero visibility as well as trying to decrease the number of fatal landings.

This gives some background to the development of automatic landing systems and, as stated earlier, the British Air Registration Board laid down that a landing system should not contribute a rate greater than 10^{-7} fatal accidents per landing. Figure 1.1 shows a typical automatic approach and landing arrangement. Such a system may utilize two radio aids in the instrument landing system (I.L.S.), a leader cable system and a radio altimeter. Signals are fed from the radio systems into autopilots which control such items as rudder, aileron, elevator and throttle by means of servomotors. Other items may require control, for example pitch, which may involve radio guidance from the ground.

The need to meet the reliability requirement has led in practice to at least a complete duplication of the channels of equipment involved. This is the well-known principle of equipment redundancy which will be found in many highly reliable systems. Such redundancy implies that the system will generally survive a single equipment fault and continue with its intended operational function.

An arrangement of equipment for a typical automatic landing system is shown in Figure 1.2. If one of the control axes is considered, then there will be a duplication of sub-systems as a minimum and one method is to have one sub-system in operation and the other on standby. If failure of the operating sub-system takes place then automatic change-over to the standby sub-system immediately takes place. Clearly, this leads to the use of high integrity monitoring techniques. Figure 1.3 gives an example of a monitored duplicated system where two automatic change-over positions are used; one is for the two ground radio installations which are monitored and the other for the autopilot outputs.

Another type of redundancy that can be used for control is that known as a triplex system which consists of three sub-systems. The output of the three sub-systems is compared so that in the event of a fault occurring in one sub-system, it can be located and the faulty sub-system disconnected. An example of this type of control system is shown in Figure 1.4, which is for an elevator control function.

In the development of such aircraft systems the important fail-safe principle is used in the basic design. This principle is often used in systems dealing with high risk situations. In the context of these aircraft systems it means that if a single failure takes place in one sub-system, there is a capability of automatic disconnexion without alteration to the flight path of the aircraft.

RELIABILITY CONCEPTS

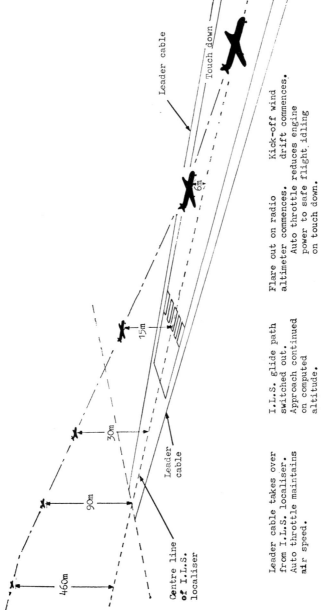

Figure 1.1 Automatic approach and landing system

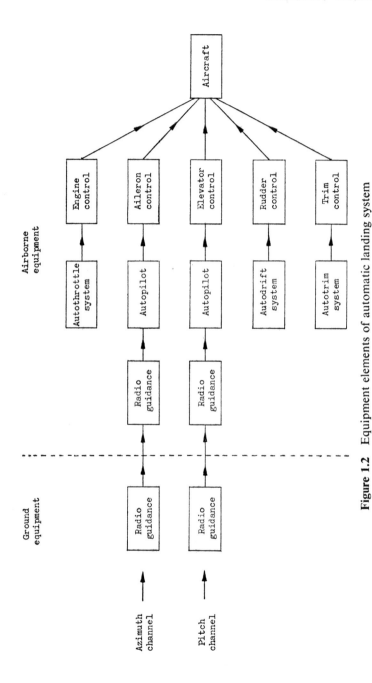

Figure 1.2 Equipment elements of automatic landing system

RELIABILITY CONCEPTS

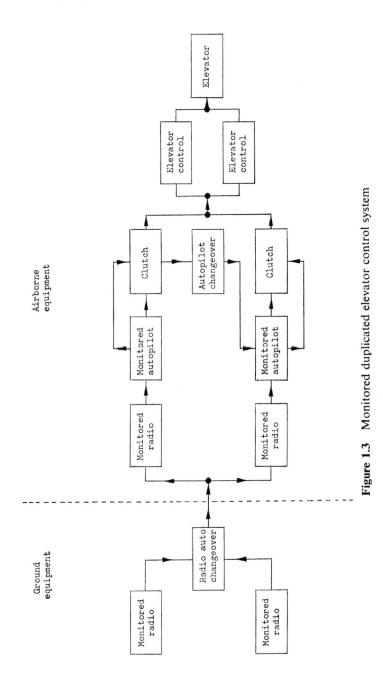

Figure 1.3 Monitored duplicated elevator control system

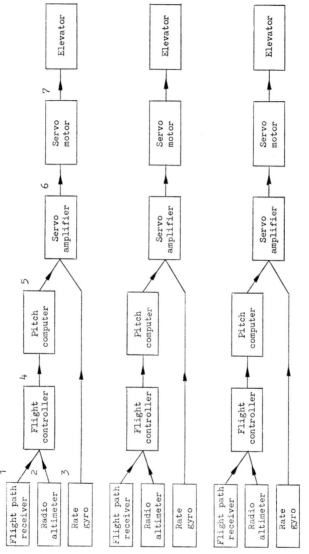

Figure 1.4 Triplex elevator control system

RELIABILITY CONCEPTS

Adjudication between different redundancy techniques and system configurations for particular applications requires a numerical evaluation of reliability and performance and this will be dealt with later.

Power plants

The need for high reliability occurs in many forms of process plant. In the case of a nuclear power plant lack of availability may cause it to be uneconomic. If such a plant has an electrical power output of 1000MW then the loss of revenue due to unavailability becomes £2000 per hour assuming an electricity selling price of £0.002 per unit. This lack of availability can be caused by failures in the plant. The capital investment could be £100M and such investment requires safeguarding. Also, there may be the possibility of risk to human life if large quantities of radioactive materials were released due to some remote large-scale failure. This poses for the operator the need to balance off his operational and safety requirements. Furthermore, unreliability could lead to a high risk situation which he would not readily tolerate. An example of a high reliability system from this application would be that of an automatic protective system for shutting down the nuclear reactor should particular fault conditions arise. The basic elements of such an automatic protective system are shown in the block diagram of Figure 1.5.

The 'protective channels' are designed to sense by means of transducers and amplifiers any abnormal changes in certain reactor parameters. The parameters chosen are generally those which show the greatest or most reliable change in value during those relevant reactor fault conditions which need to be met by automatic protective action. Not all the protective channels will respond to any particular reactor fault condition but it is normally arranged so that at least two initiate protection for any one relevant fault. This gives a measure of diversity in reactor protection and ensures a higher reliability against any common fault in the protective system. As an illustration of this diversity the case may be considered of a reactor fault where the drive is lost to the coolant circulators. This condition will be sensed by a 'circulator supply' protective channel and a short time later by the 'fuel temperature' protective channel. Both these channels will initiate a reactor shutdown independently so that a failure of either of them will not inhibit the overall reactor protection. This principle of diversity is an important principle in dealing with common mode failures which could prevent an ordinary redundant system from functioning.

Apart from the diversity between channels, it is normal to at least duplicate the measurement within any one channel since no equipment can be made 100% reliable. A temperature measurement at one point in the reactor may

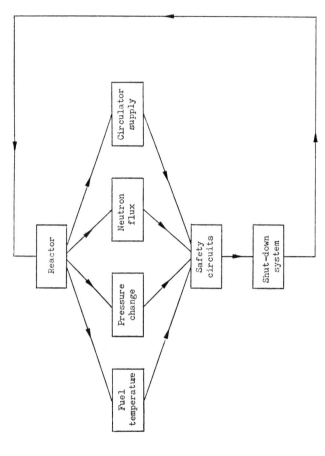

Figure 1.5 Block diagram of a reactor protective system

RELIABILITY CONCEPTS

be carried out by the use of duplicate thermocouples and the measurement transmitted through duplicate cables to duplicate temperature trip amplifiers.

The reliability of the system is normally affected by two main types of equipment faults. These are usually designated as 'fail-dangerous' and 'fail-safe'. The former describes an equipment fault which would inhibit or delay automatic protective action should a plant condition arise demanding protection. The latter describes an equipment fault which, irrespective of input, causes the protective system to move nearer to the trip point or to initiate trip or shutdown action. Both types of faults have adverse effects. The fail-dangerous fault has a direct and detrimental effect on safety. The fail-safe fault may lead to an undesirable frequency of shutdowns which, in a power-producing plant, would be costly and bring the automatic protective system into disrepute.

Considerations of fail-dangerous and fail-safe equipment faults generally lead to a compromise in which simple duplication is replaced by a number of redundant system lines working on a majority-vote principle. The simplest majority-vote scheme requires three redundant instrument lines from which the coincident output of any two is taken as truth, often known as a 2-out-of-3 logic. This processing of redundant information on a coincident or majority basis usually takes place in the safety circuits. Other forms of logic may be used depending upon the required degree of reliability and upon the type of maintenance and service which is applied to each protective channel. This use of redundancy is similar to that of the triplex control system previously described under aircraft systems.

Another point of interest is that the safety circuits usually represent the focal point of the whole protective system and it is essential to minimize the effects of faults at this stage. Ideally, any common point of the system should occur functionally as late in the system as possible, thereby maintaining the maximum safeguard against a common fault. The shutdown system is normally divided into a number of sub-systems or individual elements. The total number is chosen to cater for the likely failure of individual sub-systems and still leave an adequate margin in hand for shutting down the reactor. The forces used for shutting down the reactor are generally those which can be guaranteed to exist. For example, the insertion of absorbers into the core makes use of gravity if at all possible. On the other hand, the forces opposing insertion have also to be carefully considered, such as the hydraulic forces of coolant flow or the frictional forces due to inadequate clearances.

Ideally, such a protective system should always shut down a plant when called upon to do so but should never shut it down when not called upon to do so. In practice, the probability of failing to shut down the reactor when called upon to do so may be better than 1 in 10,000 for a typical system and the spurious shutdown rate would not be expected to exceed more than once a year.

Chemical plants

In the operation of chemical processing plant there is a strong economic incentive to maintain throughput as per the flow sheet. However, in the search to produce more of the less easily obtained chemicals and to improve efficiency of plant operation, there has been a trend towards chemical plants becoming more complex and being of an increased capital cost. Some of the processes are conducted at higher pressures and temperatures with higher concentrations of reactive chemicals than previously. There is obviously more energy available giving rise to the possibility of greater plant damage under fault conditions. Not only can the damage to the plant give rise directly to economic loss, but indirect loss due to the loss of future markets. In addition, human life can be at stake and persons working outside the confines of the factory may, under certain conditions, suffer harm as a result of the emission of poisonous materials in some processes.

Some initiating causes of such emergencies may occur more frequently than others and may arise due to loss of services such as power supplies, cooling water, steam, air and inert gas. These could give dislocation of plant operation by means of fire, explosion, spillage and runaway reaction in the process involved. In some processes certain conditions may be difficult and uneconomic to contain, for example, detonation, and great reliance is placed upon a high integrity shutdown system and the use of alarm systems to warn the operator of the onset of unacceptable plant operating conditions. Figure 1.6 shows a simplified flow diagram for a typical chemical process plant.

In this typical plant both air and reagent gas are passed continuously into a reactor vessel in which a reaction is carried out at a temperature of approximately 1000°C and atmospheric pressure. The reactor outlet gases are taken through a cooler which allows the unwanted gases to be condensed and drained off, the product gas then being pressurized to about 800 kN/m^2 before passing to stock tanks for subsequent usage in other plants.

Explosive conditions could arise in the plant—and also in those plants using the product gas—if:
(1) the temperature in the reactor fell,
(2) the ratio of the two gases was not maintained within the required limits,
(3) the reaction was incomplete.

To safeguard against these hazards an automatic shutdown system is normally installed. This would cater for a divergence from safe to hazardous conditions which may take place within a short time. Consequently little reliance can be placed upon the operator even if the plant were constantly

RELIABILITY CONCEPTS

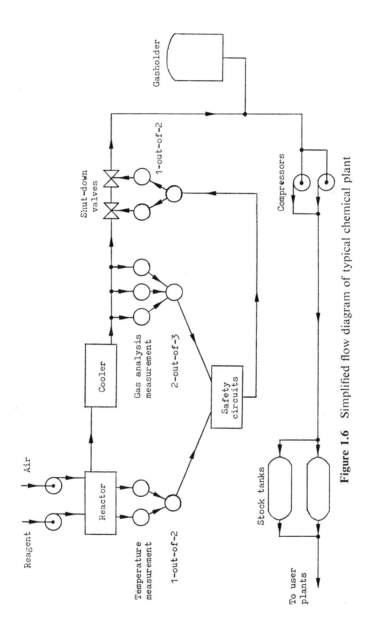

Figure 1.6 Simplified flow diagram of typical chemical plant

attended. The basic elements of the shutdown system are shown in Figure 1.6, the temperature in the reactor is monitored by two measurement lines arranged in a 1-out-of-2 logic and a shutdown signal is given on a fall in temperature. The composition of the product gas is also monitored on a 2-out-of-3 basis. When unacceptable deviations occur in either of these process variables then the plant is automatically shut down by means of signals from the safety circuits operating the shutdown valves on a 1-out-of-2 basis. This type of system is very similar to the one shown (see Figure 1.5) for a nuclear reactor and, in practice, similar reliability requirements may be placed upon the system.

Electrical supply systems

An electrical power system consists of generating capacity which is linked by a transmission system to loads to be supplied. The distribution network is normally at the national level and a principal criterion of interest is the resulting over-all reliability of the system. Obviously, if a service is being offered to a customer then the over-all cost of a power system and its availability are of prime importance. As power stations tend to become larger, then often there is a tendency for them to be not always sited at the points of bulk load and transmission costs and associated reliability can be progressively more important. Large capital expenditure is involved in such electrical supply systems and the loss of electrical supplies to many consumers can be very much more than just an economic embarrassment. In some applications it can also lead to the risk to human life. Once again there arises the basic ingredients for high risk situations and the need for highly reliable systems.

Figure 1.7 shows a simplified electrical supply system which could be found at some fairly high power application. The loads which are to be supplied, X and Y, represent 50% load each and the source of the supply is taken from source A and source B either of which could supply the load. A and B feed into two bus-bars, the main bus-bar and the reserve bus-bar. These bus-bars can be coupled together if need be by the bus-coupler or, in the case of fault conditions, the main bus-bar can be split into two by means of the inter bus-bar connector. Parallel feeds are taken to boards 1 and 2 which, once again, have the facility of linking via inter bus-bar connectors and then they are fed to boards 3 and 4 with similar facilities for interconnecting if required. In this system redundancy is quite evident both in the case of the main and reserve bus-bars and the various parallel feeds so that a failure of one line would not prevent supply being maintained to one of the loads X or Y. Diversity is also represented because source A and source B could be from different power stations and the transmission cables routed differently so that they would not, in general, be affected by the same types of faults or,

RELIABILITY CONCEPTS

perhaps, weather conditions. The circuit breakers enable the system to be disconnected if required and the transformers which are shown enable voltage reductions to be undertaken appropriate to the particular supply requirements.

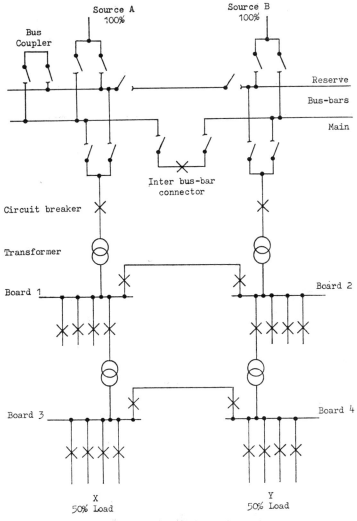

Figure 1.7 Electrical supply system

In the design of the system it may be necessary to provide protective devices to prevent damage to expensive items of equipment. This requires careful consideration to be given to preferred modes of failure. For example, the circuit breakers may be designed to open in the event of system fault conditions arising. However, with reference to the consumer, a loss of electrical supply may lead to operational and safety embarrassment. This means that a balance may need to be sought between the fail-safe requirements of the producer and consumer. In applications where a high guarantee of electrical supply is required the general principles of redundancy and diversity are utilized.

Systems in general

In discussing some systems where reliability is at a premium, it is seen that certain basic principles of design have emerged. These are the use of diversity,

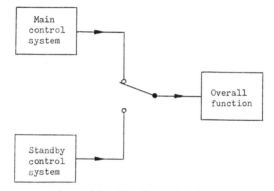

Figure 1.8 Standby redundancy

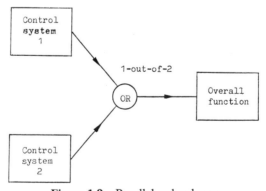

Figure 1.9 Parallel redundancy

redundancy and designing with a preferred mode of failure in mind such as always endeavouring to produce a fail-safe condition.

The systems bring out two forms of redundancy:
(1) standby redundancy, which is shown in Figure 1.8, and
(2) parallel redundancy, which is shown in Figure 1.9.

RELIABILITY CONCEPTS

Clearly, in each case one can use multiple standby systems or multiple parallel systems. This could be further illustrated by considering an example of a structure in which redundancy is used, as shown in Figure 1.10, where the ideas are well known.

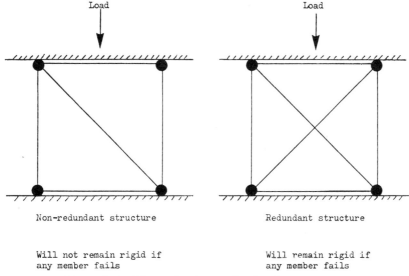

Figure 1.10 Non-redundant and redundant structures

Another example could be the bearing assemblies as shown in Figure 1.11 and 1.12 where the load can be carried by a single race hence surviving under certain conditions a failure on the other race. In the standby case, a combination of a plain bearing and a ball bearing allows one to fail but for operation to continue on the other type of bearing. This is an interesting example because it also brings in the idea of diversity which is often used to overcome a common fault condition. In the case of the parallel redundant bearing assembly, this would be dependent upon checks being made at maintenance, whereas in the standby arrangement it would be useful to have a monitoring system to sense when the plain bearing is working.

Illustrations of fail-safe conditions may be found in various applications, a typical example is that of the 'dead-man's' handle on the driving control of certain railway trains. Here it requires the engine driver to maintain a pressure on the driving control handle. If for any reason the engine driver becomes incapacitated, then the control handle returns to a position so that the speed of the train is reduced to zero.

Obviously, it is necessary to adjudicate between the various approaches and their particular applications on the grounds of reliability. Some designers

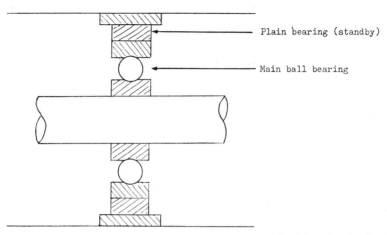

Figure 1.11 Standby redundancy in a bearing assembly

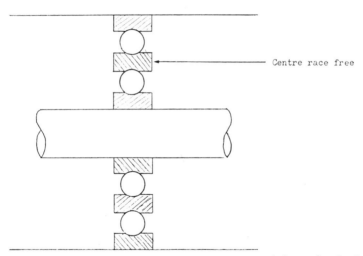

Figure 1.12 Parallel redundancy in a bearing assembly

RELIABILITY CONCEPTS 21

may require to investigate various lines of action to improve reliability, others may wish to show whether a particular system is adequately reliable or not. For this to be approached in a quantitative and logical fashion the definition of reliability based on the idea of probability requires to be developed.

A Probabilistic Definition

So far it has been seen that there is a need for a scientific definition of reliability and for a method of reliability evaluation which can be applied to a wide range of technological projects. As a start, therefore, Reliability Technology, which is the subject of this book, may be defined as the scientific study of the trustworthy nature of devices or systems in practical and industrial situations. The quality of reliability itself has already been shown to be some measure of the success of an item in performing some desired function. A number of expressions for success or failure were mentioned in the historical review and in the discussion on high reliability systems. For instance, the number of failures which occurred in a given number of items over a given period of time were mentioned in connexion with aircraft engines. Mention was also made of the fact that failures occurring over a time interval could be expressed as a mean failure-rate. Some reliability statements appeared in the form of one failure for so many trials or events. In other instances this latter form of statement was expressed in terms of the proportionate success or percentage of successful events in the total number of trials. Sometimes such a proportionate idea was applied to times giving a ratio of, say, out-of-service time to total time and the word 'availability' was used in this context. In other cases use was made of the word 'chance' and this chance was then expressed as a ratio or fractional number.

Before defining reliability more specifically it is worth while to stop and examine these various terms, words and numerical expressions in order to get a clear idea of their inter-relationships and the fundamentals on which they are based. It is apparent, of course, as has been stated before, that they all spring from the concept of success or failure. If failure is briefly defined as that state which prevents the item of interest from performing in the manner desired, then all man-made devices are prone to or subject to some type of failure during their working life. Obviously the total number of failures which have occurred or are likely to occur in a particular device or system are of interest and could be used as some index of reliability. A quantity, N, can therefore be defined as:

$$N = \text{total number of failures} \qquad (1.1)$$

Such an index of reliability could be important and useful. For instance, it determines the number of times that a system or device may be out for repair. It is a factor determining the number of maintenance staff to be employed and also the stock level of spare parts to be carried. If the number of failures is high, it may undermine everybody's confidence in the viability of the system. However, the number of failures occurring is not a sufficient index in itself because it lacks a base for comparison. The operator of a process plant, for instance, would want to know not only that there had been, say, 10 failures on his plant but also whether these 10 failures occurred over an operating period of, say, 1 year or 10 years. He might also want to know how the failures were distributed, whether they were bunched together in one particular period or whether they were evenly spread out. Interest would also be aroused as to the effect of each failure, how long it took to repair the failures and what it cost. Here the number of failures occurring and the repair of the failures may be judged against some base of time, but failures in the time domain are not always the criterion of interest. For instance, the management of a transportation system may be more interested in the number of failures related to the total number of vehicles employed or to the total number of journeys made. An airline operator may want to judge his reliability in terms of the proportion of successful flights; a railway executive may be interested in the availability of engines at the start of each journey, or an industrial inspectorate may be concerned with the number of rejects compared with the total number of items produced. In all these cases the number of failures is being related to the total number of relevant events in which a failure can occur.

Failures then, or for that matter successes, may be related to either space or time domains or to some combination of the two. Considering first the space domain, the case may be taken where failures are compared with some total number of events or operations which have or which are likely to take place. Let:

$$n = \text{total number of events} \qquad (1.2)$$

Each event in the quantity n can be considered as having one of two outcomes, either success or failure. In the simplest case the outcome for each event may be taken as being time independent. If, as before, the total number of failures is N then:

$$n - N = \text{total number of successes} \qquad (1.3)$$

Two proportionate quantities may now be derived. First a quantity, $\hat{P}_f$, which represents the proportionate number of failures where:

$$\hat{P}_f = \frac{N}{n} \qquad (1.4)$$

and, secondly, a quantity, $\hat{P}_s$, which represents the proportionate number of

RELIABILITY CONCEPTS 23

successes where:

$$\hat{P}_s = \frac{n-N}{n} \qquad (1.5)$$

Both the quantities $\hat{P}_f$ and $\hat{P}_s$ could be used as a reliability index. They are each, of course, fractional numbers which can range in value from zero to unity. These numbers, multiplied by 100, will give the proportionate failure or success in the form of a percentage as was used earlier when discussing successes of missile flights.

For any particular series of n events the quantity $\hat{P}_f$ may well be a factual expression of what has occurred in the way of failures, but the question arises as to what general meaning can be attached to $\hat{P}_f$. For instance, the series of n events may have yielded N_1 failures but if the series was repeated under the same conditions then it is likely that some new number of failures, N_2, would have occurred. Obviously the value of $\hat{P}_f$ is not a constant but for a large number of series of events it would seem legitimate to take a reliability index based on the mean value of all the individual estimates of $\hat{P}_f$. This mean value of $\hat{P}_f$ would then appear to represent a reasonable expectation of what might be the outcome from some future series of events under the same conditions.

The quantity $\hat{P}_f$, then, can be used as an estimate of future behaviour based on some sample of the outcomes from a previous series of events. It may be intuitively argued that $\hat{P}_f$ will approach some true and constant value as the sample of events from which it is estimated becomes infinitely large. A quantity P_f is now introduced which represents this limiting value of $\hat{P}_f$ such that:

$$P_f = \lim_{n \to \infty} \frac{N}{n} \qquad (1.6)$$

and, expressed in this way, P_f is defined as the probability of occurrence of a failure at each event.

Example 1.1

A factory producing fuses has an average output of 15 million fuses per year. Over a 10-year period it has been found that 30,000 fuses have been rejected during final test and inspection. What is the probability that any fuse being inspected will be rejected? If the inspection department only detects 95% of all sub-standard fuses, what is the chance that any fuse a customer receives will be a reject?

The total number of fuses, n, produced in the 10-year period is:

$$n = 10 \times 15 \times 10^6$$
$$= 15 \times 10^7$$

The number of rejects or failure events, N, over the same period is:

$$N = 3 \times 10^4$$

Therefore, from Equation (1.6), the estimated probability, $\hat{P}$, that any fuse will be a reject is:

$$\hat{P} = \frac{N}{n}$$

$$= \frac{3 \times 10^4}{15 \times 10^7}$$

$$= 2 \times 10^{-4}$$

that is to say, 1 chance in 5000.

If the 30,000 fuses represent 95% of all the sub-standard fuses produced, then the number of rejects, N', which got through inspection was:

$$N' = 30{,}000 \times \tfrac{5}{95}$$

$$= 1579$$

Therefore, the estimated probability that a customer will receive a reject fuse is:

$$\hat{P} = \frac{1579}{15 \times 10^7}$$

$$\simeq 10^{-5}$$

or, approximately, 1 chance in 100,000.

This has introduced the concept of probability and it is seen that it has a close relationship with a reliability index which is expressed in the form of proportionate failure or success. The practical meaning of P_f is that in a large number of repetitions of the relevant event the proportionate frequency for the occurrence of failure will be approximately equal to P_f. A similar meaning may be attached to the probability of success, when it is defined as:

$$P_s = \lim_{n \to \infty} \frac{n - N}{n} \qquad (1.7)$$

Thus defined, probability generally, whether it be of success or failure, is a number between zero and unity, that is:

$$0 \leqslant P \leqslant 1 \qquad (1.8)$$

Also, from Equations (1.6) and (1.7), it is seen that:

$$P_f + P_s = 1 \qquad (1.9)$$

RELIABILITY CONCEPTS 25

which may be written in the more general form:

$$P + \bar{P} = 1 \qquad (1.10)$$

Equation (1.10) states that the probability of a particular outcome from an event occurring plus the probability of this same outcome not occurring is equal to unity, where $\bar{P}$ is the symbol used to represent the probability of the outcome not occurring.

Example 1.2

For a missile about to be launched an estimate is given, on the basis of past performance, that its chance of success will be 0.68. If a total of 25 missiles of the same type have been previously launched under the same conditions, what is the number of unsuccessful missions that have already occurred?

The estimated probability of success, $\hat{P}_s$, is:

$$\hat{P}_s = 0.68$$

Therefore, from Equation (1.9), the estimated probability of failure is:

$$\hat{P}_f = 1 - \hat{P}_s$$
$$= 0.32$$

If this is estimated on the basis of a total of 25 events, then the number of failures, N, from Equation (1.4) is:

$$N = 25 \times 0.32$$
$$= 8$$

These initial ideas of probability together with the associated concepts of mean values, expectations and estimations are enlarged upon in subsequent chapters of this book. For the present it may be seen that very often reliability and probability are synonymous. Some of the fractional numbers and statements of chance used in the historical review and later sections were, in fact, forms of probability expressions. From this point of view it is possible now to put forward a tentative definition of reliability couched in probabilistic terms:

Reliability is defined as that characteristic of an item expressed by the probability that it will perform its required function in the desired manner under all the relevant conditions and on the occasions or during the time intervals when it is required so to perform.

The probability of performing functions on specific occasions which are time independent has already been briefly mentioned. The other aspect, which is brought out again in the preceding reliability definition, is to see how the same probabilistic concepts may be applied to events occurring in the time domain.

Some Probability Concepts

It is necessary, first of all, to review one or two basic relationships which are fundamental to the way in which probabilities may be combined. Consider two entirely different series of events. In the first series of events it is observed whether a particular outcome of an event, designated as A, does or does not take place. Similarly, it is observed in the second series of events whether an outcome of these events, designated as B, does or does not take place. The outcomes A and B are completely independent, that is the occurrence of A does not depend upon the occurrence of B and vice versa.

Suppose that in the first n occasions of the first series of events A occurs N_a times. Suppose also that in the first n occasions of the second series of events B occurs N_b times. From the definition of probability given by Equation (1.6) it can be seen that the probability of occurrence of the outcome A, $P(A)$, is given by:

$$P(A) = \lim_{n \to \infty} \frac{N_a}{n} \qquad (1.11)$$

and, similarly:

$$P(B) = \lim_{n \to \infty} \frac{N_b}{n} \qquad (1.12)$$

Now consider the probability of both outcomes A and B occurring. A condition for this combined event is that A must definitely occur and this occurrence takes place N_a times in n performances. In the N_a times that A occurs the outcome B occurs, on average, $N_a N_b/n$ times. Hence, the average occurrence of the combined event, N_{ab}, is:

$$N_{ab} = \frac{N_a N_b}{n} \qquad (1.13)$$

If the symbol $P(AB)$ represents the probability of the combined event, that is the probability of A AND B occurring, then, by definition:

$$\begin{aligned} P(AB) &= \lim_{n \to \infty} \frac{N_{ab}}{n} \\ &= \lim_{n \to \infty} \frac{N_a N_b}{n^2} \\ &= \left(\lim_{n \to \infty} \frac{N_a}{n} \right) \times \left(\lim_{n \to \infty} \frac{N_b}{n} \right) \end{aligned}$$

Hence, from Equations (1.11) and (1.12):

$$P(AB) = P(A)P(B) \qquad (1.14)$$

From which it follows generally for k independent outcomes where the overall probability is required for the probability of occurrence of outcome A_1 AND outcome A_2 AND outcome A_3 etc., that:

$$P(A_1 A_2 \ldots A_k) = P(A_1)P(A_2) \ldots P(A_k) \qquad (1.15)$$

In the case where two outcomes or events are completely dependent upon each other, that is, outcome A always leads to outcome B and vice versa, then:

$$P(A) = P(B)$$

and Equation (1.14) becomes:

$$P(AB) = P(A) = P(B) \qquad (1.16)$$

Equation (1.15) can be modified in a similar manner where cases of complete dependence exist.

Example 1.3

A domestic radio receiver contains 100 individual electronic components and it is found that the probability of successful operation of each component can be taken on average to be 0.99 over a typical 10-year period of usage of the set. If individual component failures are assumed to be independent of each other, what is the probability of any one radio receiver completing a normal 10-year period without failure?

Applying the general formula of Equation (1.15), the overall probability of successful operation without any component failure is:

$$P_s = (0.99)^{100}$$
$$\simeq 0.366$$

There is, therefore, approximately a 37% chance that such a radio receiver would complete a typical 10-year operating period without failure.

Another combination of interest is the probability of occurrence of either outcome A or outcome B or both. This may be stated logically as the probability of outcome A OR the probability of outcome B and written symbolically as $P(A+B)$. The combined outcome is made up of three possible states and hence is given by:

$$\begin{aligned} P(A+B) &= P(A)P(\bar{B}) + P(B)P(\bar{A}) + P(A)P(B) \\ &= P(A)[1-P(B)] + P(B)[1-P(A)] + P(A)P(B) \\ &= P(A) + P(B) - P(AB) \end{aligned} \qquad (1.17)$$

or alternatively:
$$P(A+B) = 1 - P(\bar{A}\bar{B}) \tag{1.18}$$

provided that A and B are independent and not mutually exclusive.

Example 1.4

A particular type of aircraft engine fitted to aircraft which operate a regular service over the same route has been found to have failed in flight on two occasions during an operating period in which 4000 aircraft journeys were made. If the aircraft in question are fitted with two such engines, what is the probability that both engines will fail during the course of any one journey over the scheduled route? What is the chance, expressed as a percentage, that a flight will be completed without either engine failing? The assumption is made in both cases that the failure of one engine does not in any way affect the performance of the other.

The estimated probability of failure per engine per flight is:

$$P_f = \tfrac{2}{4000}$$
$$= 5 \times 10^{-4}$$

Therefore, the probability of failure of both engines per flight is, from Equation (1.14):

$$P_f = 5 \times 10^{-4} \times 5 \times 10^{-4}$$
$$= 2.5 \times 10^{-7}$$

that is, a chance of 1 in 4 million.

The probability that either one or both engines fail can be derived from Equation (1.17), whence:

$$P_f = 5 \times 10^{-4} + 5 \times 10^{-4} - 5 \times 10^{-4} \times 5 \times 10^{-4}$$
$$\simeq 10^{-3}$$

and thus the chance of success, P_s, that neither engine fails is:

$$P_s = 1 - P_f$$
$$= 1 - 10^{-3}$$
$$= 0.999$$

or, expressed as a percentage:

$$P_s = 99.9\%$$

This same result could have been derived another way by first of all obtaining

RELIABILITY CONCEPTS

the probability of success per engine per flight, that is:

$$P_s = 1 - P_f$$
$$= 1 - 5 \times 10^{-4}$$
$$= 0.9995$$

and then P_s is obtained by applying the product rule of Equation (1.14):

$$P_s = 0.9995 \times 0.9995$$
$$\simeq 0.999$$

as before.

If A and B are completely dependent, then Equation (1.17) becomes:

$$P(A+B) = P(A) = P(B) \tag{1.19}$$

If A and B are mutually exclusive, that is the outcome A completely precludes the outcome B and vice versa, then the product term $P(AB)$ in Equation (1.17) is zero and the equation reduces to:

$$P(A+B) = P(A) + P(B) \tag{1.20}$$

Generally, then, for independent events which are not mutually exclusive:

$$P(A_1 + A_2 + \ldots + A_k) = 1 - P(\bar{A}_1 \bar{A}_2 \ldots \bar{A}_k) \tag{1.21}$$

and for events which are mutually exclusive:

$$P(A_1 + A_2 + \ldots + A_k) = P(A_1) + P(A_2) + \ldots + P(A_k) \tag{1.22}$$

Example 1.5

A firm manufacturing a nominal 50 kW motor car engine produced in one year 10,000 such engines. The performance of each engine produced

Table 1.1

kW achieved	No. of engines
47.0–47.9	84
48.0–48.9	761
49.0–49.9	6852
50.0–50.9	2148
51.0–51.9	139
52.0–52.9	16
Total:	10,000

was measured and classified according to the power output in kW that was achieved. The number of engines falling into the various kW classifications to one decimal place of measurement were found to be as shown in Table 1.1.

If the same standard of production is achieved in the future, what is the probability of (a) having an engine greater than or equal to 50 kW? (b) having an engine whose power output lies between 48.9 and 51 kW? and (c) having an engine which is either less than 48 kW or more than 51.9 kW.

Each of the kW ranges may be considered as an outcome of the total series of events which is the production of engines. The probability for each range, therefore, is estimated from the number of engines in each range divided by the total number of engines. Hence, a table of probabilities can be constructed as in Table 1.2. Each probability represents a mutually

Table 1.2

Range	kW achieved	Probability
1	47.0–47.9	0.0084
2	48.0–48.9	0.0761
3	49.0–49.9	0.6852
4	50.0–50.9	0.2148
5	51.0–51.9	0.0139
6	52.0–52.9	0.0016
	Total:	1.0000

exclusive event since a particular outcome can only fall in one particular range. Any combination of the probabilities, therefore, will follow the summation rule of Equation (1.22). In particular, the summation of all the probabilities for each range is equal to unity as would be expected.

The probability of having an engine greater than or equal to 50 kW is given by the summation of the probabilities for ranges 4, 5 and 6:

$$P = 0.2148 + 0.0139 + 0.0016$$
$$= 0.2303$$

The probability of having an engine whose power output lies between 48.9 and 51 kW is given by the summation of the probabilities for ranges 3 and 4:

$$P = 0.6852 + 0.2148$$
$$= 0.9000$$

and the probability of having an engine which is either less than 48 kW or more than 51.9 kW is obtained from the sum of the probabilities for ranges 1 and 6:

$$P = 0.0084 + 0.0016$$
$$= 0.01$$

Reliability in the Time Domain

If, in a time domain application, N failures occur in some total time of interest T then, to keep in line with the definition of reliability already given, it is necessary to know the probability of failure as some function of N and the time T.

The total time T may be considered as being made up of n equal increments of time δt. The time increment δt needs to be chosen so that any increment contains either no failures or not more than one failure. Each time increment δt may then be looked upon as being either a successful or a failed event. Following the line of argument used in deriving Equation (1.6), the probability of any time increment δt being classified as a failure event is:

$$P_f = \lim_{n \to \infty} \frac{N}{n} \qquad (1.23)$$

Since both N and T have been taken as finite, then the value of P_f tends to zero as n tends to infinity or, which amounts to the same thing, as δt tends to zero. This is a result which would be expected since it is saying that there is a zero probability of a failure occurring in an infinitesimal period of time. However, there is obviously some finite probability of failure related to the total time T and this may be deduced by considering the limiting combination of the probabilities of failure for each time increment δt.

The combined probability $P(T)$ of at least one failure occurring in the total time T is made up of the probability of a failure occurring in any one or more of the increments of time δt. Hence, from Equations (1.21) and (1.23):

$$P(T) = 1 - \lim_{n \to \infty} \left(1 - \frac{N}{n}\right)^n \qquad (1.24)$$

provided that the value of P_f and hence the value of N can be taken as being constant and time independent relative to the position of any time increment δt in the complete time range T.

It will be recognized, by applying the exponential limit theorem, that Equation (1.24) reduces to the form:

$$P(T) = 1 - e^{-N} \qquad (1.25)$$

Since the assumption has been made that the value of N applies to any part of the overall time scale of interest and that there is no systematic spreading or bunching of failures in the time domain, then it is possible to define a mean failure-rate parameter, θ, such that:

$$\theta = \frac{N}{T} \qquad (1.26)$$

Using this relationship, Equation (1.25) can be written in the form:

$$p(t) = 1 - e^{-\theta t} \qquad (1.27)$$

where t is now any time interval of interest over which the value of θ may be taken as being time independent or constant and $p(t)$ is the probability of one or more failures occurring in this time interval when no failure was known to exist at time zero.

Equation (1.27) is the exponential probability function, which together with other functions of this type, will be discussed extensively in the later chapters. It is interesting at this stage to note that the function reduces to the approximate form:

$$p(t) \simeq \theta t \qquad (1.28)$$

when $\theta t \ll 1$.

Under certain circumstances, therefore, the reliability performance of a device or system in the time domain may be characterized by a failure-rate or by a probability of failure over some time interval of interest.

Example 1.6

A taxi operator owns a fleet of 10 cars. The past records show that the average number of times each car breaks down under normal running conditions is once in every 2 years and this mean value remains sensibly constant over long periods of time. What are likely to be the average number of calls on the breakdown service in a year? What is the probability of at least one breakdown in any 3-monthly period?

A mean failure-rate, θ, can be ascribed to each car following the definition of Equation (1.26) and taking the figures of 1 fault in a 2-year period:

$$\theta = \tfrac{1}{2}$$

$$= 0.5 \text{ breakdowns/year}$$

For a fleet of 10 cars, the overall failure-rate for the fleet, Θ, is:

$$\Theta = 10\theta$$

$$= 5 \text{ breakdowns/year}$$

Therefore, there are likely to be an average of 5 calls on the breakdown service each year.

The probability of at least one breakdown in any 3-monthly period may be calculated from Equation (1.27) where:

$$\Theta = 5 \text{ breakdowns/year}$$

$$t = \tfrac{1}{4} \text{ year (3 months)}$$

RELIABILITY CONCEPTS

Hence:

$$p(t) = 1 - e^{-(5 \times \frac{1}{4})}$$
$$= 1 - e^{-1.25}$$
$$= 0.714$$

giving a 71.4% chance of at least one breakdown in any 3-monthly period.

Other aspects of reliability performance in the time domain arise, however, when there is a need to ascribe some measure to the availability, or unavailability, of a functional object. The concept of availability becomes important when failures lead to some finite length of time associated with repairing, restoring or replacing the failed item. In the simplest case it may be assumed that, when an item fails, it is out of action for some fixed length of time τ_r which represents the repair time or its equivalent. If, using the previous symbolism, it is taken that N failures occur in a total time T then the total 'down' time or 'dead' time associated with the interval T is:

$$T_d = N\tau_r \tag{1.29}$$

and, similarly, the total available time, T_a, is given by:

$$T_a = T - N\tau_r \tag{1.30}$$

Although both Equations (1.29) and (1.30) represent indices of reliability, it is often more useful to express these times as proportions or fractions of the total time. A quantity, mean fractional dead time, $\hat{D}$, may, therefore, be defined which represents the mean proportion of dead time over the total time of interest:

$$\hat{D} = \frac{\text{mean time in the failed state}}{\text{total time}} \tag{1.31}$$

In the case of the simple failure and repair model being discussed it is seen that:

$$\hat{D} = \frac{N\tau_r}{T} \tag{1.32}$$

and a limiting case of this mean fractional dead time may be considered where:

$$D = \lim_{T \to \infty} \frac{N\tau_r}{T} \tag{1.33}$$

from which it is seen that the quantity D has characteristics similar to those of a probability as defined by Equation (1.6).

Mean availability, $\hat{A}$, may be defined in a similar way where:

$$\hat{A} = \frac{\text{mean time in the working state}}{\text{total time}} \qquad (1.34)$$

and hence, for the simple failure and repair model:

$$A = \lim_{T \to \infty} \frac{T - N\tau_r}{T}$$

$$= 1 - \lim_{T \to \infty} \frac{N\tau_r}{T} \qquad (1.35)$$

It then follows from Equations (1.33) and (1.35) that:

$$A + D = 1 \qquad (1.36)$$

Provided that the ratio N/T can be taken as being time invariant then the number of failures which occur may be expressed in terms of a mean failure-rate. In the case of the simple repair model being considered the failure-rate characteristic, θ, refers to the actual working time T_a and not to the total time T, hence:

$$N = \theta T_a \qquad (1.37)$$

Combining Equations (1.30) and (1.37) gives:

$$N = \frac{\theta T}{1 + \theta \tau_r} \qquad (1.38)$$

and then, from Equation (1.33):

$$D = \frac{\theta \tau_r}{1 + \theta \tau_r} \qquad (1.39)$$

or:

$$D \simeq \theta \tau_r \quad \text{if} \quad \theta \tau_r \ll 1 \qquad (1.40)$$

Similarly, from Equations (1.35) and (1.38):

$$A = \frac{1}{1 + \theta \tau_r} \qquad (1.41)$$

These simple expressions for mean fractional dead time and mean availability will be discussed further in the later chapters for the cases where both the failure-rate characteristic and the repair processes may themselves be functions of time and for where different methods of detecting failures and carrying out repairs are taken into account.

RELIABILITY CONCEPTS

Example 1.7

An electrical supply system is subject to failure which causes loss of supply to the consumer. The mean time between such failures is known to be 398 h and the mean time to repair the failures and restore the supply is known to be 2 h. What is the average value of the availability of the supply to the consumer over a long period of time?

For a mean time between failures of 398 h which is time independent the corresponding value of the failure-rate, θ, is:

$$\theta = \tfrac{1}{398} \text{ failures/h}$$

and for a repair time, τ_r, of 2 h it is seen from Equation (1.41) that the average availability is given by:

$$A = \frac{1}{1 + \theta \tau_r}$$

$$= \tfrac{398}{400}$$

$$= 0.995 \quad \text{or} \quad 99.5\%$$

Example 1.8

A fog lamp on a car, which is not normally switched on, is subject to failures of the bulb and wiring which would cause the fog lamp to fail to light when required to do so. Such failures, which occur at a mean rate of 1 failure every 38 weeks, are not directly revealed but stay, on average, in existence for a mean period of 2 weeks until they are rectified. If over some period of time the chance of the driver of the car running into fog is 1 in 100, what is the chance that the fog light will not come on at the time it is required?

The problem here is to calculate the combined probability, P, for the presence of fog and for the fog lamp being in the failed state. From Equation (1.14):

P = (probability of fog) $\times$ (probability of fog lamp being in failed state)

The probability of the fog lamp being in the failed state is the mean fractional dead time, D, which, from Equation (1.39), is given by:

$$D = \frac{\theta \tau_r}{1 + \theta \tau_r}$$

where, in this case:

$$\theta = \tfrac{1}{38} \text{ failures/week}$$

$$\tau_r = 2 \text{ weeks}$$

Hence:
$$D = \tfrac{1}{20}$$
$$= 0.05$$

and, the probability of fog, P_f, is:
$$P_f = 0.01$$

Therefore:
$$P = P_f \times D$$
$$= 0.01 \times 0.05$$
$$= 0.0005$$

or a chance of the fog lamp failing on demand of 1 in 2000.

Technological System Applications

In order to illustrate the reliability concepts discussed so far and at the same time introduce the sort of reliability problems which may occur in practice, a number of simple and typical systems are now considered. Some background to these systems has already been described earlier in this chapter.

Example 1.9

A chemical plant with a certain through-put of process gases is subject to an explosion hazard if the gas concentrations deviate outside certain specified limits. The plant is protected by an automatic shutdown system which measures both flow ratios and gas concentrations and operates shutdown valves if either of these measurements moves outside the required limits. A block schematic of the shutdown system showing the logical flow and combination of information is given in Figure 1.13. The logical OR boxes mean that the correct shutdown information is passed on if any of the inputs have a signal requiring shutdown. The gas analysis measurement, the flow ratio measurement and the shutdown valves are all duplicated so that, in each case, both items require to fail before that particular function is invalidated. For simplicity, it is assumed that the significant failures that can occur in each block and prevent it operating can be represented by some mean failure-rate, θ, and that such failures are independent and not directly detected but have associated with them some mean time in the failed state, τ_r. If, over the operating life of the plant, there is a 1 in 10 chance that a plant condition will arise demanding protection by the automatic shutdown system, what is the corresponding chance that an explosion hazard will occur? Typical values of θ and τ_r are assumed to be as shown in Table 1.3.

RELIABILITY CONCEPTS

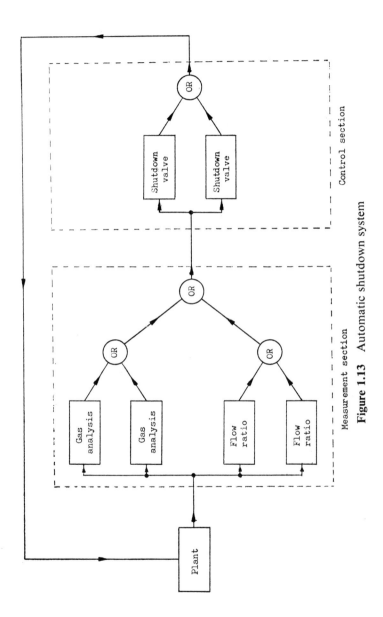

Figure 1.13 Automatic shutdown system

Table 1.3

Functional block	θ (faults/year)	τ_r (years)
Gas analysis	3.0	0.04
Flow ratio	2.0	0.04
Shutdown valves	0.2	0.05

The problem here is to calculate the combined probability, P, for the occurrence of a demand on the shutdown system, p_d, together with the probability of the shutdown system being in a failed state at the time of the demand. This latter probability can be taken as the mean fractional dead time of the system, D_s. Therefore:

$$P = p_d \times D_s$$

The quantity D_s is made up of the mean fractional dead times for the various functional parts of the system.

For a single gas analysis unit, from Equation (1.40), the value D is given approximately by:

$$D = \theta \tau_r$$
$$= 3.0 \times 0.04$$
$$= 1.2 \times 10^{-1}$$

For the complete gas analysis section, the combined probability of it being in the failed state, D_g, is, from the product rule of Equation (1.14):

$$D_g = D^2$$
$$= 1.44 \times 10^{-2}$$

Similarly, for the flow ratio section, the complete mean fractional dead time, D_f, is:

$$D_f = (2.0 \times 0.04)^2$$
$$= 6.4 \times 10^{-3}$$

The values of D_g and D_f are then combined in another OR function to make up the complete measurement section for which the mean fractional dead time, D_m, is seen to be given by:

$$D_m = D_g \times D_f$$
$$= 9.216 \times 10^{-5}$$

In the case of the control section, the mean fractional dead time, D_c, is given by the product of the mean fractional dead times for each shutdown

RELIABILITY CONCEPTS

valve, that is:
$$D_c = (0.2 \times 0.05)^2$$
$$= 10^{-4}$$

A failure of either the measurement section or the control section will inhibit the operation of the automatic shutdown system. Therefore, the probabilities of either or both being in the failed state is computed from the relationship of Equation (1.17) which gives:

$$D_s = D_m + D_c - D_m D_c$$
$$\simeq 1.9 \times 10^{-4}$$

Since the probability of a demand during the plant life is 1 in 10 then from the relationship:

$$P = p_d \times D_s$$
$$= 0.1 \times 1.9 \times 10^{-4}$$
$$= 1.9 \times 10^{-5}$$

it is seen that the probability of an explosion hazard during the plant life works out to be 1.9×10^{-5} or a chance of about 1 in 50,000.

Example 1.10

A particular type of aircraft is fitted with a triplex autopilot system. This comprises three sensibly independent autopilot sub-systems each of which is capable of achieving its purpose alone but all of which normally work together. The complete autopilot provides automatic control for the aircraft elevator, aileron and rudder systems. A block schematic of the triplex elevator control system is shown in Figure 1.4. It has been found that the failure characteristics of each block in this system can be represented by a mean failure-rate function, θ, and that such failures are reasonably independent of each other. Also, the failure of any one block results in a failure of the sub-system of which it is a part. The elevator control system is checked for complete working immediately prior to a 2-h flight. What is the probability that no automatic elevator control will be available by the end of the flight if the assumed typical values for θ are as shown in Table 1.4?

The probability that an item with a failure-rate characteristic of θ will have failed by a time t when it was known to be working at time zero is given by Equation (1.27), that is:

$$p(t) = 1 - e^{-\theta t}$$

For the items in the present example, θ always lies between 10^{-3} and 10^{-4} faults/h and t has a value of 2 h. Hence θt is always very much less than unity so that:

$$p(t) \simeq \theta t$$

The probabilities of failure for each item in the elevator control system during the 2-h period, therefore, work out to be as shown in Table 1.5. Any one

Table 1.4

Item No.	Item description	θ (faults/h)
1	Flight path receiver	10.0×10^{-4}
2	Radio altimeter	6.0×10^{-4}
3	Rate gyro	2.0×10^{-4}
4	Flight controller	3.5×10^{-4}
5	Pitch computer	5.5×10^{-4}
6	Servo amplifier	2.0×10^{-4}
7	Servo motor	1.0×10^{-4}

Table 1.5

Item No.	$p(t)$
1	2.0×10^{-3}
2	1.2×10^{-3}
3	0.4×10^{-3}
4	0.7×10^{-3}
5	1.1×10^{-3}
6	0.4×10^{-3}
7	0.2×10^{-3}

or more of the items failing constitutes a sub-system failure. Therefore, if p represents the probability of a sub-system failure, then from Equation (1.21):

$$p = 1 - [\overline{p_1(t)}\,\overline{p_2(t)}\,\overline{p_3(t)} \ldots \overline{p_7(t)}]$$

but since the product terms of $p(t)$ are small, this can be approximated to:

$$p = p_1(t) + p_2(t) + p_3(t) + \ldots + p_7(t)$$
$$= (2.0 + 1.2 + 0.4 + 0.7 + 1.1 + 0.4 + 0.2) \times 10^{-3}$$
$$= 6.0 \times 10^{-3}$$

All sub-systems require to fail before a complete failure of the elevator control is produced. Hence, if P is the probability of complete system failure, then, from Equation (1.15):

$$P = p^3$$
$$= (6.0 \times 10^{-3})^3$$
$$= 2.16 \times 10^{-7}$$

RELIABILITY CONCEPTS

Example 1.11

A small nuclear research reactor has three absorber rods which are suspended above the reactor and are designed to drop into the reactor core and shut the reactor down in the event of any untoward incident. The three rods are designated A, B and C and it has been found that the probability of each of these rods failing to drop on demand is $p_a = 0.005$ $p_b = 0.01$ and $p_c = 0.001$. If it is known that any two or more of three rods entering the reactor core will safely shut the reactor down, what is the probability of failing to shut the reactor down when called upon to do so?

One condition of failure would be if, say, rod C entered the core but rods A and B failed to do so. The probability of this combined condition is seen, from Equation (1.15), to be:

$$p = p_a \times p_b \times (1-p_c)$$

or:

$$p = 0.005 \times 0.01 \times 0.999$$

$$\simeq 5 \times 10^{-5}$$

With a total of three rods there are, altogether, eight possible conditions some of which represent failure and some of which represent success. If the symbols F and S are used to represent failure and success respectively, then all these conditions can be tabulated together with their respective probabilities of occurrence as shown in Table 1.6. Each of the conditions represents

Table 1.6

Condition	Rod A	Rod B	Rod C	System	Probability
1	S	S	S	S	$(1-p_a)(1-p_b)(1-p_c)$
2	S	S	F	S	$(1-p_a)(1-p_b)p_c$
3	S	F	S	S	$(1-p_a)p_b(1-p_c)$
4	S	F	F	F	$(1-p_a)p_b p_c$
5	F	S	S	S	$p_a(1-p_b)(1-p_c)$
6	F	S	F	F	$p_a(1-p_b)p_c$
7	F	F	S	F	$p_a p_b(1-p_c)$
8	F	F	F	F	$p_a p_b p_c$

a mutually exclusive event and, therefore, any combination of the probabilities will follow the summation relationship of Equation (1.22). In particular, the sum of all the probabilities for each condition is unity as would be expected and as can be readily checked.

The probability of system failure, P, is given by the sum of the probabilities of occurrence for conditions 4, 6, 7 and 8 as can be seen by reference to

Table 1.6. Therefore:

$$P = (1-p_a)p_b p_c + p_a(1-p_b)p_c + p_a p_b(1-p_c) + p_a p_b p_c$$
$$= p_a p_b + p_b p_c + p_c p_a - 2p_a p_b p_c$$

and putting in the numerical values for p_a, p_b and p_c:

$$P = 5 \times 10^{-5} + 10^{-5} + 5 \times 10^{-6} - 5 \times 10^{-8}$$
$$= 6.5 \times 10^{-5}$$

Example 1.12

It is simply assumed that an accident may occur at automatic level crossings as a result of two possible sequences of events. These are, (a) the complete failure of the automatic mechanism for sounding the alarm and closing the barriers, in which case a train may hit a vehicle which is in the process of negotiating the crossing, and (b) the stalling of a vehicle on the crossing just prior to the automatic closure of the barriers and the inability to move the vehicle before the passing of the train.

On a particular railway system 100 automatic crossings are in existence and both the arrival of trains and the crossing of road vehicles takes place with an equal likelihood at any time. The arrival of trains takes place with a mean rate of one train every 2 h and the crossing of vehicles at a mean rate of 100 vehicles/h.

The transit time of road vehicles, during which time they may be at risk, has a mean value of 3 s but if a vehicle stalls then its mean risk time rises to 15 s. One in every million vehicles using the crossing stalls on the crossing and the probability of complete failure of the automatic system is also one in a million per demand for operation.

If all events are assumed to be independent and the same conditions and parameters pertain over a 10-year period, how many accidents would be expected over this period of time?

Both postulated accident conditions are a function of the number of pertinent train arrivals over the 10-year period multiplied by the relevant probability of a vehicle being at risk on a crossing. With a train frequency of one per 2 h, then the total number of train arrivals, N_t, for 100 crossings over 10 years is:

$$N_t = \tfrac{1}{2} \times 100 \times 10 \times 8760$$

taking 8760 h in a year. Thus:

$$N_t = 4.38 \times 10^6$$

There is a probability for each of these arrivals of 10^{-6} that the train fails to operate the warning signals and the crossing barriers. Therefore, the average

RELIABILITY CONCEPTS

number of unheralded train arrivals of this type, N_t', is:

$$N_t' = 4.38 \times 10^6 \times 10^{-6}$$
$$= 4.38$$

The normal probability of a vehicle being on a crossing is the mean rate of vehicle crossings multiplied by the transit time at risk (compare Equation 1.40). This probability, P, is thus given by:

$$P = 100 \times \tfrac{3}{3600}$$
$$= \tfrac{1}{12}$$

The probability of a stalled vehicle being on the crossing, P', is the mean rate of stalled vehicles multiplied by the corresponding transit time at risk for stalled vehicles, that is:

$$P' = 100 \times 10^{-6} \times \tfrac{15}{3600}$$
$$= \tfrac{5}{12} \times 10^{-6}$$

Accident condition (a) is produced by the number of unheralded train arrivals N_t' together with the normal probability, P, of a vehicle being on the crossing:

$$N_a = N_t' \cdot P$$
$$= 4.38 \times \tfrac{1}{12}$$
$$= 0.365$$

Accident condition (b) is produced by the number of normal train arrivals N_t, together with the probability, P', of a stalled vehicle on the crossing:

$$N_b = N_t \cdot P'$$
$$= 4.38 \times 10^6 \times \tfrac{5}{12} \times 10^{-6}$$
$$= 1.825$$

Therefore, the total number of expected accidents over the 10-year period is:

$$N = N_a + N_b$$
$$= 0.365 + 1.825$$
$$= 2.19$$

that is, approximately two accidents would be expected.

Conclusions

This chapter has reviewed the way in which there is a need for a scientific study and definition of reliability and has introduced the probabilistic nature of a typical reliability definition. Some of the simple rules which arise from this type of definition have been explained and the corresponding methods of calculating reliability have been illustrated by a number of practical examples.

In the basic concepts so far discussed, a number of simplifying assumptions have been made, some of which have been mentioned and some of which have been inherent in the method of treatment. The making of these assumptions has been deliberate in order to simplify and minimize the work undertaken in this introductory chapter.

Questions, of course, need to be raised as to the validity of any assumptions in any particular application and as to the trustworthy nature of the answers produced by the various basic mathematical models that have been used. The later chapters of this book examine these types of points in detail and deal with the more complex mathematical models that some of the assumptions may demand. The whole question of independence, practical limitations, estimation and confidence is dealt with in subsequent chapters as and when the points arise.

Questions
1. Discuss what kind of quantitative meanings could be attached to the word 'reliability' in connexion with, say, the usage of a motor car. Enumerate any different aspects that might arise between the private user of a single vehicle and a public company owning a large fleet of vehicles.
2. Explain why an interest in reliability, particularly from the point of view of quantification, has grown more rapidly in some industrial activities and technological projects rather than in others. Choose a particular activity or project and discuss any needs that may at present exist for a scientific evaluation of reliability.
3. Describe in what way reliability may be considered as being a measurement of performance with typical technological illustrations. From this discussion outline how reliability may be expressed in terms of probability.
4. Explain what is meant by the probability of a particular outcome from an event. If the event has three possible outcomes derive, from first principles, the probabilistic relationships between the three outcomes when all outcomes are mutually exclusive. If the probability of one outcome is 0.2 and another is 0.45, what is the probability of the third outcome?
5. Derive the formula for the probability of occurrence of either or both of two independent outcomes in terms of the individual probabilities of occurrence of these outcomes. Deduce a similar expression in terms of the individual probabilities that the outcomes do not occur and prove that $P(A+B)+P(\bar{A}\bar{B}) = 1$.

RELIABILITY CONCEPTS   45

6. A company producing electric light bulbs has an annual inspected output of 7.8 million bulbs and its inspection department is assessed as having a reliability of 0.9. A particular customer buys a batch of 4500 light bulbs from this company in which he finds that 9 are faulty. On the basis of these data, what is the estimate of the average number of bulbs which the company rejects each year in the inspection department?

7. A solid fuel booster engine has been test fired 2760 times. On 414 occasions the engine failed to ignite. If a projectile is fitted with three identical and independent booster engines of this type, what is the chance on launching of the projectile that (a) all three engines fail to ignite, (b) at least one of the engines fails to ignite?

8. A chain made up of n identical links is placed under a certain test stress and it is found that a link has failed 12,312 times in a total of 3×10^6 tests. If the probability of a single link failure under the same test conditions is 7.6×10^{-5}, what is the estimate of the total number of links, n, in the chain?

9. A railway executive operates an identical train service between place A and place B 20 times a day for 250 days each year. The scheduled time for the journey is 2 h 45 min. Over a two year operating period it has been found that journey time has ranged, to the nearest minute, from 2 h 30 min to 2 h 39 min on 746 occasions, from 2 h 40 min to 2 h 49 min on 2423 occasions, from 2 h 50 min to 2 h 59 min on 5564 occasions and that the train has taken 3 h or more on the remaining occasions. If a traveller sets out on one of these trains from place A, what is the probability that (a) his journey time will be outside the range 2 h 40 min to 2 h 49 min, (b) that a friend meeting him at place B will be kept waiting a quarter of an hour or more for his arrival, (c) that the train will be at least 5 min early?

10. It has been found in practice that a range of domestic washing machines have a mean failure-rate of 0.2 faults per year which remains sensibly constant and time independent over a 10-year period of normal usage. If a householder buys a new machine of this type, what is the probability that at least one failure will have occurred by the end of (a) the first year of use, (b) the fifth year of use and (c) the tenth year of use?

11. An industrial process plant is controlled by two identical on-line computers in such a way that either of the computers is capable of controlling the plant on its own and faults on one computer do not affect the operation of the other. Each computer consists of 10,000 electronic components where the failure of any one component results in a computer failure. Each component has a mean constant failure-rate of one fault per 10^7 h. Assuming the computers to be working correctly at the start of a process run and assuming that computer faults cannot be repaired during the

run, what is the probability of loss of plant control for a process run of 24 h?

12. Two diesel generator sets which are independent and designed to operate continuously are each capable of supplying the full load requirements in a particular situation. If one diesel generator has a mean constant failure-rate of 9 faults per 10^4 h and on the occurrence of each fault is out of service for a fixed time of 50 h, and the other diesel generator has a corresponding failure-rate of 15 faults per 10^4 h and an out-of-service time per fault of 20 h, what is the mean availability, expressed as a percentage, of full load supplied?

13. A large office block has a fire detection and alarm system which is subject to a mean constant failure-rate of two failures per year (1 year = 8760 h) and each failure that occurs takes, on average, 4 h to detect and repair. The system is also subject to a quarterly routine inspection and test on which occasions it is out of action for a fixed time of 1 h. If the expected probability of fire occurrence in the building over a period of time is 0.073, what is the probability of an undetected fire by the alarm system over the same period of time?

14. A nuclear reactor protective system has three independent and separate measurement channels each of which is capable of detecting certain relevant reactor fault conditions. The measurement channels detect temperature, coolant flow and neutron flux and their respective mean availabilities are 0.975, 0.96 and 0.96152. Protective signals from any of the measurement channels are designed to operate four shutdown mechanisms, any two of which are capable of shutting down the reactor. The shutdown mechanisms are completely independent and each has a mean availability for correct operation of 0.98. If the probability of a relevant reactor fault condition arising in a certain period is 3×10^{-3}, what is the probability that the reactor will not be safely shut down by the protective system over the same period of time?

CHAPTER 2
An Approach to Reliability Assessment

Concept of Overall Reliability

The general problem posed by Chapter 1 was that of assessing the reliability of a complete technological system. Some simple approaches to this problem have already been mentioned and an indication has been given as to how some reliability parameters may be calculated. However, reliability parameters such as failure-rate, repair time, availability or fractional dead time do not represent the complete picture. Overall reliability, in its widest sense, is a measure of the relationship between the complete achieved performance spectrum of the system against the corresponding required performance under all the relevant conditions of space and time. The definition of reliability given in Chapter 1 implied this overall concept and it may be further illustrated by the block diagram of Figure 2.1.

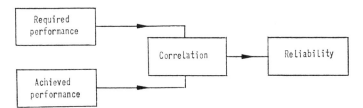

Figure 2.1 Overall reliability concept

The diagram shows that reliability is determined by two main factors. These are the 'Required performance' and the 'Achieved performance'. Each factor is dependent upon the answer to a simple question as follows:
  Required performance—What is the system supposed to do under all the relevant conditions?
  Achieved performance—What is the system likely to do under all the relevant conditions?
  Although the questions themselves are simple the determination of the answers may be complex. System performance, for instance, may be defined by the interrelationship of a large number of individual performance parameters. Such overall performance may then in itself be dependent upon the

inputs to the system, the environmental conditions of the system and the time and place in which the system operates. In addition, both the description of required performance and of achieved performance may be probabilistic in nature.

Under any of the relevant conditions, including any point in space and time, the reliability of a system is the chance of the achieved performance falling within the bounds of the required performance. This is the procedure denoted by the 'Correlation' block in Figure 2.1 and the output of this block leads directly to the measure of reliability. In this way, the reliability resolves itself into the answer to the question:

Reliability—What is the probability, under all the relevant conditions, of the achieved performance meeting the required performance?

which is in accord with the definition of reliability previously given in Chapter 1.

It is now clear that every block in Figure 2.1 may involve the manipulation of the relevant characteristics in probabilistic terms. The reader has already been introduced in the previous chapter to some elements of probability theory. On this basis, the overall reliability concept may be illustrated in a simplified form. Suppose that under a certain set of conditions the requirement for one of the performance characteristics, x, of a system is either that the characteristic should be greater than x_{q1} or that it should be greater than x_{q2}, x_{q1} being less than x_{q2}. The probability that this requirement depends on the performance being greater than x_{q1} is $f(x_{q1})$ and the corresponding probability for x_{q2} is $f(x_{q2})$. If no other requirement is possible and the requirements are mutually exclusive, then:

$$f(x_{q1}) + f(x_{q2}) = 1$$

Under the same conditions, the achieved performance of the system in relationship to the characteristic x is either x_{h1} or x_{h2} where:

$$x_{q1} < x_{h1} \leqslant x_{q2} \quad \text{and} \quad x_{q2} < x_{h2}$$

The probabilities associated with the two possible achieved performance values are $f(x_{h1})$ and $f(x_{h2})$ and these are also assumed to be complete and mutually exclusive so that:

$$f(x_{h1}) + f(x_{h2}) = 1$$

The situation may be represented in diagrammatic form as shown in Figure 2.2.

For requirement x_{q1}, either achievement value leads to success, but for requirement x_{q2}, only achievement value x_{h2} gives success. The probability equation for the reliability, R, of the system is therefore:

$$R = f(x_{q1}) + f(x_{q2})f(x_{h2}) \tag{2.1}$$

making use of the laws for combining probabilities developed in Chapter 1.

AN APPROACH TO RELIABILITY ASSESSMENT

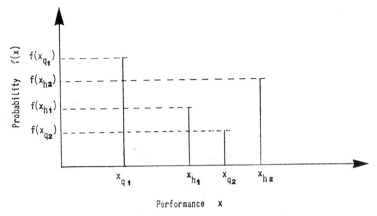

Figure 2.2 Simple requirement–achievement diagram

The same result can also be derived by the truth table method described in the previous chapter in connexion with Table 1.6. In the present case, four conditions are possible and these are illustrated in Table 2.1.

Table 2.1 Simple requirement–achievement conditions

Condition	Requirement (x_q)	Achievement (x_h)	System state
1	1	1	Success
2	1	2	Success
3	2	1	Failure
4	2	2	Success

Only condition 3 is a failure, therefore:

$$R = 1 - \text{(probability of condition 3)}$$
$$= 1 - f(x_{q2})f(x_{h1})$$
$$= f(x_{q1}) + f(x_{q2})f(x_{h2})$$

as before.

Obviously, the summation of the probabilities for conditions 1, 2 and 4 will again lead to the same result.

Example 2.1

A requirement exists for an aircraft to land smoothly at a particular airport. Under normal conditions this requirement can be met by either manual or automatic landing control. Under sudden gusty conditions, however, the requirement can only be met by manual control. If the pilot has a 1 in 10 chance of selecting manual control and there is also a 1 in 10 chance of gusty conditions suddenly developing, what is the reliability of a smooth landing?

In the terminology of Figure 2.2:

$$f(x_{q1}) = 0.9$$

therefore:

$$f(x_{q2}) = 0.1$$

and:

$$f(x_{h1}) = 0.9$$

therefore:

$$f(x_{h2}) = 0.1$$

Hence, from Equation (2.1):

$$R = 0.9 + 0.1 \times 0.1$$
$$= 0.91$$

The simple type of requirement–achievement problem just discussed depends on the representation of the required and achieved performances by a small series of discrete values. In many cases, however, the discrete values may be very large in number or they may merge to form a continuum of all possible values between certain limits. The reliability assessment now depends on how such a continuous distribution of required and achieved performance values can be represented and combined in a probabilistic way. It is necessary, at this stage therefore, to divert slightly from the main theme in order to consider this point.

Some Further Probability Concepts

Aspects of system performance can only rarely be defined by the possibility of just one or two discrete values as illustrated in Figure 2.2. More generally, a performance parameter may, within certain limits, take on any value. This spread in likely performance is sometimes called variability. Consider, for example, a system which is designed to respond in a time, say, of 1.5 s. It would be very surprising if, each time the system response time was measured, it came out to be exactly 1.5 s. Variability in both the time response of the system and in the measurement of such times is the normal thing to be expected. A typical series of results which might be experienced in practice is shown in Table 2.2.

This table gives the response time values for a series of 85 different measurements. If the inaccuracy of measurement is assumed to be small compared

AN APPROACH TO RELIABILITY ASSESSMENT

with the values quoted, then the differences in reading are some measure of the variability in the system performance. The immediate problem posed by such results is that of how much variability exists and how can it be measured. It takes some time to find the lowest response value which is 1.25 s or the highest which is 1.80 s. It would take considerably longer to evaluate the average or mean value which happens to be 1.51 s. What is required is some method of summarizing information of the type contained in Table 2.2.

Table 2.2 Series of measurements of response time in seconds

1.48	1.46	1.49	1.42	1.35
1.34	1.42	1.70	1.56	1.58
1.59	1.59	1.61	1.25	1.31
1.66	1.58	1.43	1.80	1.32
1.55	1.60	1.29	1.51	1.48
1.61	1.67	1.36	1.50	1.47
1.52	1.37	1.66	1.44	1.29
1.80	1.55	1.46	1.62	1.48
1.64	1.55	1.65	1.54	1.53
1.46	1.57	1.65	1.59	1.47
1.38	1.66	1.59	1.46	1.61
1.56	1.38	1.57	1.48	1.39
1.62	1.49	1.26	1.53	1.43
1.30	1.58	1.43	1.33	1.39
1.56	1.48	1.53	1.59	1.40
1.27	1.30	1.72	1.48	1.66
1.37	1.68	1.77	1.62	1.33

In order to summarize, it may be decided that a difference in response time of, say, 0.1 s is insignificant compared with the overall spread of 0.55 s. In this case the complete range may be covered by groups of response times 0.1 s wide and the numbers of response times falling within each group recorded. The result of such an exercise is the frequency table shown in Table 2.3. Not only does this table contain all the relevant information from Table 2.2, but it gives a much better picture of the variability. Note that the group intervals are defined to three decimal places so that there is no ambiguity as to the interval in which any measurement falls.

Obviously nothing that has been said is unique to system response time measurement. It would apply generally to any measured quantity. In general, this quantity when used in the statistical sense is called the variate. A variate, such as response time, is described as a continuous variate since it can take on any value within a certain range. Not all variates fall into this category. Some, such as the number of forced plant shut-downs in a year or the number of system faults occurring in a given length of time, are discrete. Broadly speaking, both continuous and discrete variates are dealt with and classified in similar ways.

Table 2.3 Frequency table for response time measurements

Group interval	Frequency of response times
1.205–1.305	7
1.305–1.405	14
1.405–1.505	21
1.505–1.605	23
1.605–1.705	16
1.705–1.805	4
Total	85

It is clearly feasible to exhibit the frequency table pictorially. One way of doing this is the Frequency Polygon, in which the frequency in a group is plotted against the mid-point value of that group. But a more useful way, perhaps, is to form a histogram. For the continuous variate, the histogram

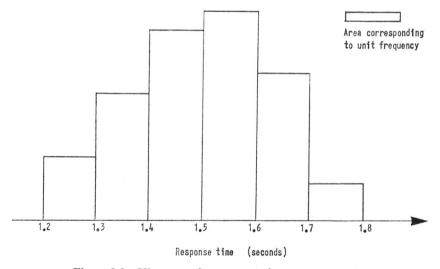

Figure 2.3 Histogram for response time measurements

in which area is equivalent to frequency maintains the continuous characteristic of the variate. Such a histogram, for the case of the response time measurements of Table 2.3, is shown in Figure 2.3. There is again nothing special about response time values and the ideas embodied in this histogram could be applied to any continuous variate. For discrete variates, the rectangles would be replaced by ordinates.

It is not difficult to appreciate that the histogram may be considered as representing proportionate frequency (i.e. the ratio of the number of events of interest to the total number of events) instead of actual frequency. In this

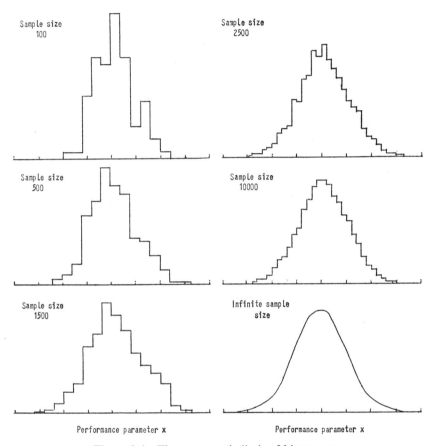

Figure 2.4 The asymptotic limit of histograms

case the areas of the constituent rectangles can be considered as measuring probability in the way in which it has been defined in Chapter 1. If, in this case, larger and larger samples of the variate were obtained and the histograms of proportionate frequency were plotted, a point would be reached where further information changed the outline of the histogram negligibly. Continuing in this way it can be seen, from Figure 2.4, that the outline of the histogram approaches some smooth curve.

It is a fundamental assumption in statistics that such a limiting curve exists. The curve is essentially the asymptotic form of the histogram which contains all possible observations or values. In many cases the total possible observations are infinite in number. In other cases they are finite and can be either large or small. The general statistical word used for this total possible aggregate is 'population'. The population may be real or imagined, existing or hypothetical. In reliability technology, it may represent overall systems of which only a few will actually exist or it may represent component parts which are manufactured and produced by the million.

A limiting curve of the proportionate frequency histogram of the type depicted in Figure 2.4 is called a probability density function (p.d.f.) when the area is used as a measure of probability. In this limiting case, the population, n, tends to infinity and the chance that any specified member of the population has a value in a range m is given by:

$$\lim_{n \to \infty} \frac{m}{n}$$

which is in accord with the definition of probability given in the previous chapter.

In general usage, the notation is that some variable such as x represents both the variate and a possible value of the variate and the symbol $f(x)$ is used to denote the p.d.f. From the way in which $f(x)$ has been defined, it will be seen that:

$f(x)\,dx$ is the probability that $(x - \tfrac{1}{2}dx) \leqslant x \leqslant (x + \tfrac{1}{2}dx)$

and:

$$\int_a^b f(x)\,dx \text{ is the probability that } a \leqslant x \leqslant b$$

If this latter integral is extended to cover the entire range of the variate then its value must be unity. Often this means that:

$$\int_{-\infty}^{\infty} f(x)\,dx = 1 \tag{2.2}$$

and it can always be so written if $f(x)$ is allowed to be continuously zero outside the actual ranges of the variate.

It is convenient to consider a special form of the integral of the p.d.f. where:

$$\int_{-\infty}^{x} f(x)\,dx = p(x) \tag{2.3}$$

AN APPROACH TO RELIABILITY ASSESSMENT

This function, which carries the symbol $p(x)$, is called the cumulative distribution function. It is readily seen that:

$$p(x) \text{ is the probability that } x \leqslant x$$

or:

$$p(x) = \text{prob}[x \leqslant x]$$

and:

$$0 \leqslant p(x) \leqslant 1$$

Generally, then:

$$dp(x) = f(x)\,dx \qquad (2.4)$$

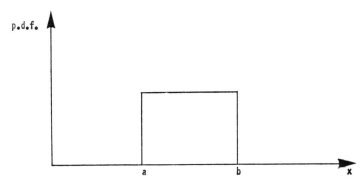

Figure 2.5 Simple rectangular distribution

Similar ideas and relationships hold for discrete variates. There is no limiting curve in this case, but a limiting set of ordinates at the proper values of the variate may be imagined in which case the probability function may be expressed as:

$$f(x_j) = \text{prob}[x = x_j] \quad \text{for all } j = 1, ..., n$$

and the corresponding integral becomes a summation:

$$\sum_{j=1}^{n} f(x_j) = 1 \qquad (2.5)$$

and, similarly, the cumulative function may be expressed as:

$$\sum_{j=1}^{r} f(x_j) = p(x_r) \qquad (2.6)$$

A simple example of a continuous distribution is the rectangular or uniform distribution. Here x is a variate which is equally likely to take any value between two limits a and b. The p.d.f. for this distribution is shown in Figure 2.5. It has important practical implications since it may be quite a fair representation of many technological devices which are specified as having their performance lying between an upper and a lower tolerance limit. The probability density function may be written as:

$$f(x) = k \quad \text{(a constant)}$$

and since:

$$\int_a^b f(x)\,dx = 1$$

then:

$$k = \frac{1}{b-a}$$

therefore:

$$f(x) = \frac{1}{b-a} \tag{2.7}$$

There are many other forms of statistical distributions and other properties that they possess, but these will be discussed further in Chapter 7.

Example 2.2

A type of electric motor which is produced in large quantities is found to have a power output which is randomly distributed according to a continuous rectangular distribution. The distribution has a lower limit of 200 W and an upper limit of 250 W. What is the probability that a particular motor has a power output of (a) less than 220 W and (b) greater than 240 W?

From Equation (2.7) the p.d.f. for this example is:

$$f(x) = \tfrac{1}{50}$$

Hence, the probability that the power output is less than 220 W from Equation (2.3) is:

$$p(x) = \int_{200}^{220} \frac{dx}{50}$$
$$= 0.4$$

Similarly it will be seen that the probability of the power being less than 240 W is given by:

$$p(x) = \int_{200}^{240} \frac{dx}{50}$$
$$= 0.8$$

and, hence, the complementary probability of the power being greater than 240 W is 0.2.

System Reliability

At the beginning of this chapter the overall reliability concept was seen to be a matter of correlation between the required and achieved performance of a system. The probabilistic aspects of this concept were explained and an introduction has now been given to the way in which both discrete and continuous performance characteristics or variates can be expressed in probabilistic terms. It is perhaps useful at this stage to illustrate the application of these ideas by considering the simple example system shown in Figure 2.6. This system has been depicted in electrical terminology but it

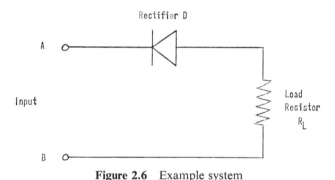

Figure 2.6 Example system

could equally be an electrical analogue of a mechanical, hydraulic or pneumatic system. It consists of two input terminals, A and B, connected by wires to a rectifier, D, in series with a resistor, R_L.

The example system, as yet, does not constitute a reliability problem because of the unknowns that exist in relation to the following questions:

(a) What is the system supposed to do?
(b) Under what input conditions is it supposed to do it?
(c) Under what environmental conditions is it supposed to do it?
(d) Where and when is it supposed to do it?
(e) What are the characteristics of the system in terms of the specifications of each component part under all the relevant conditions?

Questions (a) to (d) inclusive form the basis of the required performance and question (e) the basis for the achieved performance as discussed previously in connexion with Figure 2.1. In order to proceed, therefore, with an illustration of the correlation exercise of Figure 2.1 it is necessary to specify some further characteristics of the example system of Figure 2.6.

Suppose, therefore, that normally there is no input to the system but whenever and under whatever conditions an input occurs there is a requirement for a current to flow through the resistor R_L. At this time and under these conditions the current is required to be greater than or equal to 0.96

amperes d.c. and less than or equal to 1.04 amperes d.c. This required performance may be represented by a performance or d.c. current axis on which are erected requirement lines corresponding to the lower and upper limits of required performance as shown in Figure 2.7. Suppose also that

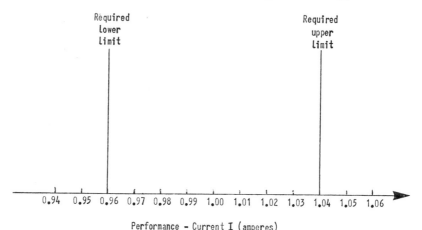

Figure 2.7 Requirement lines

the characteristics of input are:
a voltage V applied across the terminals A and B where:
$V = 0$ under normal conditions
$V = 11$ V d.c. when the input occurs and this value is always exactly 11 V under all conditions of input occurrence;
and an internal resistance R_i where:
$R_i = 0$ under all conditions
and the characteristics of the system are:
$R_L = 10\ \Omega$ exactly under all conditions
$D = 1\ \Omega$ exactly in the forward direction under all conditions
$D = \infty\ \Omega$ exactly in the reverse direction under all conditions
$R_c = 0\ \Omega$ under all conditions where R_c represents the resistance of all wires and connexions.

Then an assessment of the achieved performance may be made by simply applying Ohm's law in the following form:

$$I = \frac{V}{R_L + D}$$
$$= \frac{11}{10+1}$$
$$= 1.00\ \text{A}$$

This may now be illustrated pictorially, with reference to the requirement, as shown in Figure 2.8.

It appears from the foregoing that the achieved performance always meets the required performance and the logical conclusion from the correlation exercise is that the reliability is perfect. If this perfection is called 100% or, in probabilistic terms, simply 1.0 then:

$$\text{Reliability} \quad R = 1$$

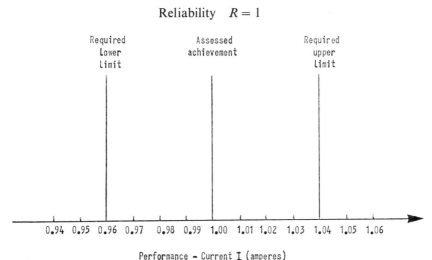

Figure 2.8 Requirement and achievement lines

This answer, of course, is dependent upon the validity of the assumptions made in deriving it. Some of these assumptions are as follows:

(i) That the system is logically capable of functioning in the manner intended.

(ii) That the basic laws upon which the functioning depends are both valid in themselves and valid in this particular application (i.e. the mathematical model of the system is correct).

(iii) That the input and system parameters used in the mathematical model are correct.

(iv) That there is no variability in the logic, or the laws, or the various parameter values under all the conditions under consideration.

These assumptions will only all be correct in a perfect world. In any practical situation, R, although it may approach unity, will always be less than unity. The assumptions are related to both the required performance and the achieved performance. The following sections deal with the aspects that arise under both these main headings. The achieved performance, however, is further subdivided to consider both capability of performance and variability of performance. The meaning of the terms 'capability' and 'variability' in this context will be explained as the discussion proceeds.

Capability of Performance

Consider, first of all, assumption (i) of the previous section. As applied to the present problem, this assumes that if any electrical voltage of some finite value is applied across some electrical network of finite resistance then some finite current will flow. In fact, this assumption may break down in the system under consideration if the d.c. input voltage is so polarized (positive to terminal A) that no current can flow through the rectifier D. The actual

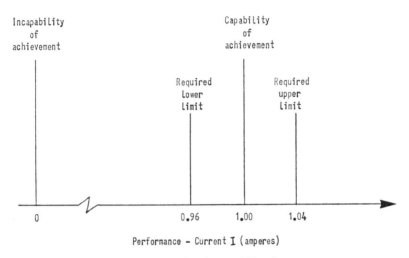

Figure 2.9 Simple capability lines

polarization of the input voltage was not specified in the original problem, so, in fact, two reliability answers are possible due to this particular logical functioning of the system:

$R = 1$ if negative input to terminal A

$R = 0$ if positive input to terminal A

The lack of capability may arise not only from an error in the assumed input conditions but also from any inherent design error in the system itself. Taking this form of capability into account, the performance picture now takes the form of Figure 2.9 where the achievement line can take up one of two positions representing capability and incapability respectively.

Secondly, consider assumption (ii). Here also are factors affecting capability. The basic law of functioning of the system depends upon there being a known and fixed relationship between the electrical current flowing through a resistive network and the voltage developed across it. Ohm's law, used in the calculation, is a particular example of such a basic relationship. It may be that the current–voltage relationship has not been understood and that Ohm's law is incorrect. In the particular case under consideration, one may say that all this is extremely unlikely. However, some system concepts which are at the van of technological advancement may depend for their functioning upon natural laws which are not completely understood and here caution may well need to be applied in assessing capability. More relevant, in most cases, is whether such fundamental or natural laws that do exist are applied correctly to the case in hand. This is now a question of the mathematical model for the complete system being correct. In the modelling process previously illustrated, the calculation was completely dependent upon Ohm's law. Strictly, this law only applies to a linear resistive system under steady-state conditions. There could be non-linearities in the system, for instance in the rectifier, D, which would make the application of Ohm's law inappropriate. In addition, if there were a dynamic requirement for the current through resistor, R_L, such that it should reach its stipulated value in, say, 10^{-12} s after the occurrence of the input voltage then a simple model using Ohm's law might be completely incorrect. Such considerations may affect the understanding of the system's capability and reduce the effective reliability to zero.

It is apparent, therefore, that even if the example system were perfect in terms of actual hardware there are capability facets associated with the basic logic of the system, the basic laws of functioning of the system and the application of such logic and laws which could completely nullify the reliability of the system to perform in the manner desired.

Next, under assumption (iii), comes the question of parameter values. The parameter values used in the modelling process may be wrong for two main reasons. First, the knowledge of the characteristics of the input or the system properties may be so lacking that the assumed parameter values are grossly in error. Secondly, the parameter values may be known and understood but a wrong value is inadvertently inserted into the mathematical model. The fact that a system is designed in the first place generally means that the knowledge of the system parameter values is not likely to be grossly in error. However, this cannot always be said about the input parameters as in many cases the actual values may be difficult to determine or calculate. The input voltage, for instance, to the example system might be derived from the measurement of a process variable under fault conditions in a large process plant. The direct measurement of such conditions may be impossible and even very difficult to simulate or model. The assumed input values, therefore, may be completely wrong. The second facet of a mistaken value applies to all parameter values fed into the mathematical model. Finally a mistake may arise in the actual calculation process be it automatic or manual.

Once again, a gross error or mistake in the foregoing aspects could affect the actual or assumed capability of the system and reduce the effective reliability to zero.

So far, the discussion has centred round those gross errors or mistakes which could change the hoped for achievement of the system from the nominal value of 1.00 A d.c. to a value of 0. Capability errors of this type can occur in practice, but, in addition, there are likely to be a whole range of errors which can change the hoped for achievement but do not lead to what might be called complete incapability. These errors come under the heading of variability as mentioned in assumption (iv).

Variability of Performance

The capability of performance discussed in the preceding section may be looked upon as deterministic in nature. An analysis of a system in this respect determines whether the system either can or cannot fundamentally perform its intended function, given that the actual hardware of the system is perfect. The corresponding assessed reliability has only two possible values which are 1 or 0 respectively. Although such a deterministic evaluation of capability is an important first step in any reliability assessment, it assumes a perfect situation in which all characteristics are invariable and the reliability can have a value of unity.

The existence of variability in any practical system means that the reliability, R, can never be unity but has some value which lies between 0 and 1. It is the crux of the reliability assessment problem to evaluate this particular value of R. Variations in system performance may take place in what may be termed both the 'software' and the 'hardware' of the system. The variations may be predictable, random or a combination of the two. In addition, they may cover small changes in performance or complete and catastrophic failures in the elements of the system.

Some consideration will first be given to variations in the software of a system. The software covers those aspects of system performance which are distinct from the characteristics of the physical elements of the system. Hence the basic laws by which the system functions, the logic of the system and the input conditions are particular examples of the system software characteristics. These factors have already been discussed in connexion with capability where they were considered as either being correct or completely incorrect. The state of correctness is now taken to be subject to some kind of imperfection or variation. So, for instance, the mathematical model may not be completely wrong but may be just an approximation to the true state of affairs. The application of Ohm's law may not be completely invalid but

AN APPROACH TO RELIABILITY ASSESSMENT

rather valid over a certain range of conditions. The parameter values may not be completely in error but rather subject to variation within known limits of accuracy. In general, all these variations will combine to provide some overall distribution of possible achieved values.

By way of illustration, a particular aspect may be examined where the value of the input voltage is known to be subject to a $\pm 5\%$ variation. This could be interpreted as meaning that the input voltage is equally likely to take on any value from 11 V minus 5% to 11 V plus 5%, i.e. 10.45–11.55 V. This may be represented by a rectangular probability density function of the type discussed in connexion with Figure 2.5 and now illustrated, for this case, in Figure 2.10 where $f(V)$ is the p.d.f. and has a constant value a

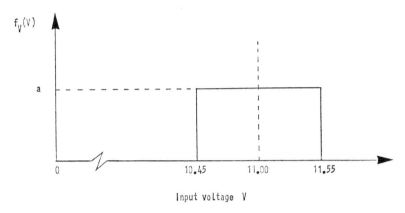

Figure 2.10 p.d.f. for input voltage variation

between the voltage limits 10.45 V and 11.55 V. If all voltage values lie between 10.45 V and 11.55 V, the total area under the curve must be unity. Hence:

$$a = \frac{1}{1.1} \text{ in voltage units}$$

For instance, the cumulative probability of the input voltage being less than or equal to 11.00 V is given by the area under the $f(V)$ curve between 10.45 V and 11.00 V, that is:

$$p(V) = (11.00 - 10.45) \times \frac{1}{1.1}$$

$$= \tfrac{1}{2}$$

as would be expected.

If the variation in input voltage represents the only uncertainty in the example system and Ohm's law applies, then the current I can be shown to be distributed in the following way:

$$f_I(I) = R_T f_V(R_T V) \tag{2.8}$$

where:

$$R_T = R_L + D \tag{2.9}$$

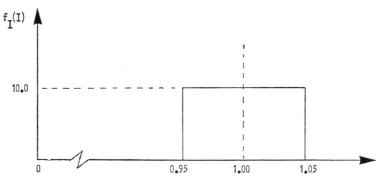

Figure 2.11   p.d.f. for current variation

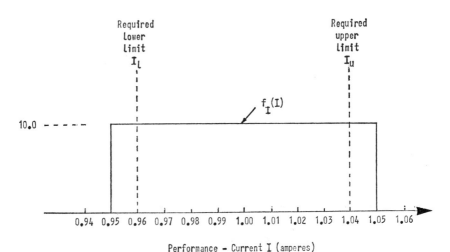

Figure 2.12   Requirement lines and current p.d.f.

The derivation of the type of relationship given in Equation (2.8) is discussed in Chapter 8 in connexion with Equation (8.38). Equation (2.8) can be represented graphically as shown in Figure 2.11 and the complete performance picture will take the form of Figure 2.12.

AN APPROACH TO RELIABILITY ASSESSMENT

It is now apparent that the probability of success (or reliability, R) of the system is given by the area of the $f(I)$ curve between the lower limit of 0.96 A and the upper limit of 1.04 A. Hence:

$$R = (1.04 - 0.96) \times 10.0$$
$$= 0.8 \quad \text{or} \quad 80\%$$

Mathematically, as shown earlier, the value of R can be written in the more general form:

$$R = \int_{I_1}^{I_u} f(I) \, dI \tag{2.10}$$

and this applies irrespective of the form of the $f(I)$ distribution.

Generally for any system which has some performance characteristic x under the appropriate conditions and where there are fixed values of x representing the upper and lower limits of required performance, the reliability due to the variations in the software characteristics becomes:

$$R = \int_{x_1}^{x_u} f_0(x) \, dx \tag{2.11}$$

where $f_0(x)$ is the p.d.f. representing the distribution of the appropriate software variability. Referring to the example system, $f_0(x)$ would take into account any variations associated with the logic of the system, the basic laws of operation, the mathematical model and its application and the input function.

Example 2.3

A simple resistive element of fixed value 10 ohms is known to obey Ohm's law. The current flowing through this element is randomly distributed according to a rectangular distribution which has lower and upper limits of 4 A and 6 A respectively. What is the probability that the voltage developed across the element meets the requirement of being at least 45 V?
Here:

$$f(I) = \tfrac{1}{2} \quad \text{for } 4 \leqslant I \leqslant 6$$

From the work done in connexion with Equation (8.37) in Chapter 8, it can be shown that the corresponding distribution of voltage becomes:

$$f(V) = \tfrac{1}{20} \quad \text{for } 40 \leqslant V \leqslant 60$$

The probability that the voltage is less than 45 V is then given by:

$$p(V) = \int_{40}^{45} \frac{dV}{20}$$
$$= 0.25$$

Hence, the probability of the requirement being met or the reliability of the situation is:

$$R = 1 - p(V)$$
$$= 0.75$$

There is now the question of the hardware of the system. This resolves itself into the chance of the actual physical system being available to perform the required performance function under the conditions and at the time when it is required so to perform.

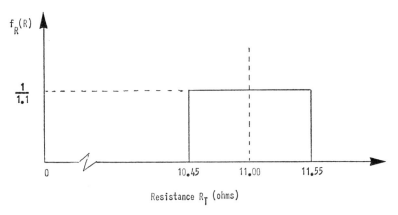

Figure 2.13 p.d.f. for resistance variation

The components representing the hardware of the system may have variations in their parameter values which on occasions take them outside their specification limits. For instance, the value of the resistor R_L may drift randomly in time, the rectifier D may change its forward characteristic due to changing environmental conditions and the connexions may be subject to corrosion or ageing.

Suppose, therefore, that the system capability is perfect and that the only imperfection is the variability, which may occur randomly in time, of the resistor R_L and D. If, for simplicity of example, this variation is assumed to cause a value of the total resistance, R_T, which is equally likely to occur anywhere between a lower value of 10.45 Ω and an upper value of 11.55 Ω, then the performance picture of R_T will follow the rectangular p.d.f. shown in Figure 2.13. The corresponding variation in current due to this variation in R_T can be shown, from the work done in Chapter 8, to be given by:

$$f_I(I) = \frac{V}{I^2} f_R\left(\frac{V}{I}\right) \qquad (2.12)$$

and since:
$$f_R(R) = \frac{1}{1.1} \quad \text{for } 10.45 \leqslant R \leqslant 11.55 \tag{2.13}$$
and:
$$V = 11.0 \tag{2.14}$$
then:
$$f_I(I) = \frac{10}{I^2} \quad \text{for } \frac{1}{1.05} \leqslant I \leqslant \frac{1}{0.95} \tag{2.15}$$

However, the variation of I is relatively small compared with the mean value of I and so Equation (2.15) may be approximated to a rectangular distribution where:
$$f_I(I) = 10.0 \quad \text{for } 0.95 \leqslant I \leqslant 1.05 \tag{2.16}$$
The reliability of the example situation is then given by:
$$R = \int_{I_1}^{I_u} f_I(I)\,dI \tag{2.17}$$
and taking $I_1 = 0.96$ A and $I_u = 1.04$ A as before, then:
$$R = \int_{0.96}^{1.04} 10\,dI$$
$$= 0.8$$

So, with everything else perfect, the assumed variability in R_L and D leads to a performance success or reliability of 80%.

Generally, if $f_a(x)$ is a p.d.f. representing the variability in hardware performance of the performance characteristics x, then:
$$R = \int_{x_1}^{x_u} f_a(x)\,dx \tag{2.18}$$
all other things being perfect.

There is now the question of combining the variability in software performance with the variability in hardware performance. It is apparent that the reliability function due to these two effects is of the following form:
$$R = \int_{x_1}^{x_u} f[f_0(x), f_a(x)]\,dx \tag{2.19}$$
where $f[f_0(x), f_a(x)]$ represents some function of the software variation p.d.f., $f_0(x)$, and the hardware variation p.d.f., $f_a(x)$.

In the particular system example chosen, the basic functional equation is:
$$I = \frac{V}{R} \tag{2.20}$$

It is shown in Chapter 8 that the p.d.f. for a resulting distribution of variations in I, $f(I)$, which is dependent upon the quotient of two other parameter distributions such as $f(V)$ and $f(R)$ is given approximately by:

$$f(I) = L^{-1}[\![\exp[-(V_1/R_1)s]\{L[f(V)]\}_{s=s/R_1}\{L[f(R)]\}_{s=(-V_1/R_1^2)s}]\!] \quad (2.21)$$

where:

L = Laplace transformation

s = Laplace operator

V_1 = lower limit of distribution of V

R_1 = lower limit of distribution of R

and the approximation is valid for small variations of V and R. Practically, this amounts to variations less than about 5%.

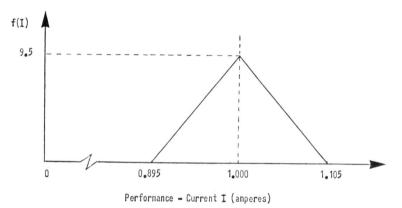

Performance – Current I (amperes)

Figure 2.14 p.d.f. for current variation due to variations in both voltage and resistance

The solution of the Laplace transformation equation for the two distributions of $f(V)$ and $f(R)$ previously assumed in Figures 2.10 and 2.13 respectively comes out to be of the following approximate form:

$$f(I) = 90.5(I - 0.895) \quad \text{for } 0.895 < I < 1.000$$

$$f(I) = 90.5(1.105 - I) \quad \text{for } 1.000 < I < 1.105 \quad (2.22)$$

and:

$$\int_{0.895}^{1.105} f(I)\, dI = 1 \quad (2.23)$$

The pictorial form of the probability density function, $f(I)$, for this case is shown in Figure 2.14.

AN APPROACH TO RELIABILITY ASSESSMENT

This, then, represents the overall performance picture due to variability in the software and hardware aspects of the example system. The reliability in the light of these combined variations may be found by inserting the appropriate requirement limits in the way illustrated in Figure 2.15.

Example 2.4

For an overall pattern of current variation as derived in Equation (2.22) and Figure 2.15, find the reliability of the situation for a current requirement

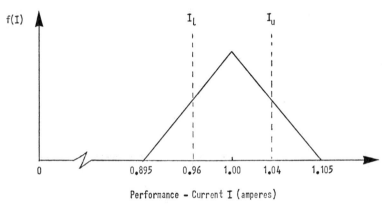

Figure 2.15 Overall current p.d.f. and requirement lines

given by (a) lower and upper limits of 0.96 A and 1.04 A respectively and (b) lower and upper limits of 0.0 A and 1.04 A respectively.
The reliability, R, is given by:

$$R = \int_{I_l}^{I_u} f(I)\,dI$$

Substituting for the form of $f(I)$ from Equation (2.22) and the example requirement limits from part (a) yields:

$$R = \int_{0.96}^{1.00} 90.5(I - 0.895)\,dI + \int_{1.00}^{1.04} 90.5(1.105 - I)\,dI$$

$$= 0.3077 + 0.3077$$

$$\simeq 0.615$$

and for the second part of the example:

$$R = 0.5 + 0.3077$$

$$\simeq 0.808$$

For a multiple element or multiple parameter system, Equation (2.19) can, of course, be extended to include variations due to all aspects of software performance and all aspects of performance relating to the various hardware elements. In this case both $f_0(x)$ and $f_a(x)$ will be complex distributions made up of combinations of individual distributions relating to the various variational aspects of interest. Methods of dealing with this type of situation, in a more general way, are discussed in the later chapters and, in particular, in Chapter 8.

A multivariate p.d.f. such as $f_a(x)$ may be used to describe all types of variation in the hardware elements of any system. However, it is often convenient, as shown in Chapter 5 and later, to restrict the application of $f_a(x)$ to small variations about the mean or nominal performance values. Any larger variations where the performance of a hardware element moves right outside its specification limits may then be treated as a complete and catastrophic failure. This may be considered as an extreme type of variation but it has certain special characteristics which warrant a particular method of treatment. The concept will be examined in more detail in Chapters 10, 11 and 12 but for the moment it may be simply illustrated with reference to the example system of Figure 2.6.

In the example system, catastrophic failures of the component parts may be typically of the following types:

(a) open circuit of any of the electrical connexions,
(b) open circuit or breakage of any of the connecting leads,
(c) open circuit or short circuit of the resistor R_L,
(d) open circuit of the rectifier D.

Any of these fault types and others similar to them will lead to a zero current through resistor R_L at the time of an input signal. That is, they will lead to a complete loss of performance or unavailability of the system.

If the chance of the system being free from these types of faults under all the required conditions is denoted by p, then, from this hardware point of view only, the reliability of the system may be written simply as:

$$R = p \qquad (2.24)$$

As an example of the way in which the probability, p, is made up, the chance of the various circuit components being in the completely failed state may be considered as follows.

Let the probability of being in the failed state at the time and under the conditions of the input voltage occurring be:

(i) for open circuit of any connexion $= 1 \times 10^{-3}$,
(ii) for open circuit or short circuit of resistor $R_L = 2 \times 10^{-3}$,
(iii) for open circuit of rectifier $D = 1 \times 10^{-2}$.

AN APPROACH TO RELIABILITY ASSESSMENT

For the circuit to be working, none of these effects must contribute to a failed state—that is everything must be working. Therefore, from the laws for the combinations of probabilities discussed in Chapter 1:

$$p = (1 - 10^{-3}) \times (1 - 2 \times 10^{-3}) \times (1 - 10^{-2})$$
$$\simeq 1 - 1.3 \times 10^{-2}$$
$$= 0.987 \qquad (2.25)$$

If the conditions considered represent all possible modes of failure, then this figure represents the probability of the system being successful when a demand arises for its operation. Such a probability may be termed the availability of the system where, in this case, the availability denotes the freedom from catastrophic faults in the hardware of the system.

It is a simple matter now to combine the total variation function of Equation (2.19) with the effects of catastrophic loss of availability. The general overall relationship for reliability then becomes:

$$R = p \int_{x_1}^{x_u} f[f_0(x), f_a(x)] \, dx \qquad (2.26)$$

For the example system, using the assumed values, the various aspects of this function have been calculated in Example 2.4 and Equation (2.25) as follows:

$$\int_{x_1}^{x_u} f[f_0(x)], f_a(x)] \, dx = 0.615$$

and:

$$p = 0.987$$

Therefore:

$$R = 0.987 \times 0.615$$
$$= 0.607$$

and the complete performance reliability diagram now takes the form shown in Figure 2.16 where the overall reliability corresponds to the shaded area.

If the p.d.f. representing all types of small variations in a particular performance parameter x for both the software and hardware aspects of the system can be expressed in terms of a single achievement function $f_h(x)$ where:

$$f[f_0(x), f_a(x)] = f_h(x) \qquad (2.27)$$

then, the general reliability function can be written as:

$$R = p \int_{x_1}^{x_u} f_h(x) \, dx \qquad (2.28)$$

Example 2.5

An electronic amplifier, when normally functioning, is found to have random variations in power output from all causes which follow a rectangular distribution between the limits of 45 mW and 55 mW. In addition, the amplifier has a probability at any time of 10^{-2} of being in the catastrophic or completely unavailable state where the power output is effectively zero. What is the reliability of the amplifier in meeting a requirement for the power output to be greater than 47 mW?

For the random variations alone:

$$R = \int_{47}^{55} \frac{d(mW)}{10}$$

$$= 0.8$$

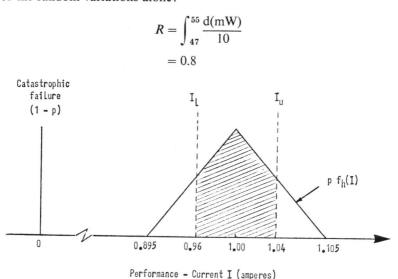

Figure 2.16 Overall reliability diagram with fixed requirement lines

For the catastrophic failures alone:

$$R = p$$
$$= 1 - 10^{-2}$$
$$= 0.99$$

Hence, applying Equation (2.28), the total reliability is:

$$R = 0.99 \times 0.8$$
$$= 0.792$$

Required Performance

Another aspect of variability now arises in connexion with the requirement. So far, the required performance has been specified by a fixed lower and upper limit of performance and these limits have therefore defined the integral range of the p.d.f. for the complete performance variability function, $f_h(x)$. The requirements themselves, however, may be subject to variability.

For instance, in the example system, the current flowing through resistor R_L may be required to operate some other device such as an indicating system or alarm. Suppose that the resistor R_L represents an indicating meter on which lower and upper currents of 0.96 A and 1.04 A respectively are marked by red lines. If an operator has now to interpret the reading of the meter, these lower and upper performance values will be subject to error or

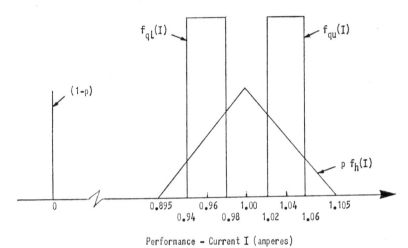

Figure 2.17 Achieved and required current probability density functions

variation due to such things as the meter pointer thickness, the red line thickness, the parallax and the relative positions of operator and meter. It may be said, therefore, that the requirement for system performance as imposed by the operator is subject to variation. By way of example the lower imposed requirement limit may be taken as lying anywhere between 0.94 A and 0.98 A with equal likelihood and the upper imposed requirement as lying anywhere between 1.02 A and 1.06 A with equal likelihood. Both the lower and upper requirement limits may now be represented by rectangular probability density functions. This is illustrated in Figure 2.17 where $f_{ql}(I)$ and $f_{qu}(I)$ are the p.d.f.'s for the lower and upper limits of performance requirement respectively and $f_h(I)$ is the p.d.f. for the performance achievement previously derived for the example system in Figure 2.16.

Examination of Figure 2.17 shows that in the range of I from 0.895 A to 0.94 A, success is not possible; in the range 0.94–0.98 A success is conditional on the relative values of achieved current against required current; in the range 0.98–1.02 A success is always possible; in the range 1.02–1.06 A success is again conditional; and in the final range 1.06–1.105 A success is not possible.

The reliability for the central range can be dealt with along the lines previously discussed because this is effectively a reliability calculation between the fixed limits of 0.98 A and 1.02 A. If this proportion of the reliability is designated R_2, then:

$$R_2 = p \int_{0.98}^{1.02} f_h(I)\,dI \tag{2.29}$$

and calculating this out with the appropriate example values gives a reliability value of:

$$R_2 = 0.987 \times 0.344$$

$$\simeq 0.339$$

In the lower range of possible success from $I = 0.94$ A to 0.98 A, the reliability is given by the fact that for any particular value of achieved current the required current must be less than or equal to this achieved value. Expressed mathematically, this reliability for the lower range, R_1, is:

$$R_1 = p \int_{0.94}^{0.98} \left[\int_{0.94}^{I} f_{ql}(I)\,dI \right] f_h(I)\,dI \tag{2.30}$$

The derivation of this type of integral relationship is discussed at the beginning of Chapter 10.

Now for the range 0.94–0.98 A:

$$f_{ql}(I) = \frac{1}{0.98 - 0.94} = 25.0$$

and for this same range as was seen in Equation (2.22):

$$f_h(I) = 90.5(I - 0.895)$$

Therefore:

$$R_1 = p \int_{0.94}^{0.98} \left(\int_{0.94}^{I} 25.0\,dI \right) 90.5(I - 0.895)\,dI$$

and this works out to be:

$$R_1 = p \times 0.130$$

$$= 0.987 \times 0.130$$

$$= 0.128$$

AN APPROACH TO RELIABILITY ASSESSMENT 75

In the upper range of possible success from $I = 1.02$ A to 1.06 A, the reliability is given by the fact that for any particular value of achieved current the required current must be greater than or equal to this achieved value. Hence, in the same way as before, this reliability, R_3, is given by:

$$R_3 = p \int_{1.02}^{1.06} \left[\int_{I}^{1.06} f_{qu}(I) \, dI \right] f_h(I) \, dI \qquad (2.31)$$

$$= p \int_{1.02}^{1.06} \left(\int_{I}^{1.06} 25.0 \, dI \right) 90.5(1.105 - I) \, dI$$

$$= 0.128$$

Therefore, the total reliability, R, is:

$$R = R_1 + R_2 + R_3 \qquad (2.32)$$

$$= 0.128 + 0.339 + 0.128$$

$$= 0.595$$

which is slightly worse than the value of 0.607 calculated previously for the case where there was no variation associated with the required limits.

In the more general case where the total variability function $f_h(x)$, the lower requirement function, $f_{ql}(x)$, and the upper requirement function, $f_{qu}(x)$ are all complete and continuous functions over the whole range of interest from say x_1 to x_2, then:

$$R = p \int_{x_2}^{x_1} \left[\int_{x_1}^{x} f_{ql}(x) \, dx \right] \left[\int_{x}^{x_2} f_{qu}(x) \, dx \right] f_h(x) \, dx \qquad (2.33)$$

which may be written as:

$$R = p \int_{x_1}^{x_2} p_{ql}(x) [1 - p_{qu}(x)] f_h(x) \, dx \qquad (2.34)$$

or:

$$R = p \int_{x_1}^{x_2} p_{ql}(x) [1 - p_{qu}(x)] \, dp_h(x) \qquad (2.35)$$

but the derivation and interpretation of these relationships are discussed more fully in Chapter 10 in connexion with Equations (10.16), (10.17), (10.18) and (10.62) and the general picture corresponding to these relationships is shown in Figure 10.3.

The Overall Reliability Function

Just as the variability of the system in its nominal working mode may be denoted by an overall variability function, $f_h(x)$, so the catastrophic unavailability instead of being single valued at, say, zero may also be represented by a variability function, $f_{hc}(x)$. This is saying that the complete or catastrophic failure of the system does not always result in exactly zero performance but may give rise to a distribution of values of performance which are close to zero. The relationship between these two functions follows the same lines as previously described. That is, the chance that, at any one time and under any one set of conditions, the performance lies in the active variability régime is p and is distributed according to $f_h(x)$. The corresponding chance that the performance lies in the inactive variability régime (or catastrophic régime) is $1-p$ and is distributed according to $f_{hc}(x)$.

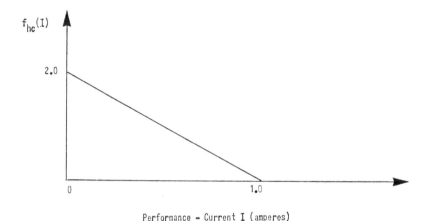

Figure 2.18 Current p.d.f. for catastrophic failure régime

If all functions are complete and continuous throughout the full range from 0 to ∞, then the overall reliability expression becomes:

$$R = \int_0^\infty p_{ql}(x)\,[1-p_{qu}(x)]\,[(1-p)f_{hc}(x)+pf_h(x)]\,dx \qquad (2.36)$$

since $1-p$ and p are, of course, mutually exclusive.

Consider, for the example system, the case where the catastrophic loss of performance does not result in zero current but in some distribution of current values from zero upwards. This could arise in practice, for instance, if the short-circuit condition of resistor R_L resulted in low resistance values rather than zero values. For simplicity, consider a distribution, $f_{hc}(I)$, which from a combination of all catastrophic unavailability causes, has the form shown in Figure 2.18. The corresponding equation for the p.d.f. is:

$$f_{hc}(I) = 2(1-I)$$

AN APPROACH TO RELIABILITY ASSESSMENT

The overall performance picture for the example system using the distribution characteristics and parameter values so far assumed is now as shown in Figure 2.19. For this situation the reliability equation of Equation (2.36) may be written in two parts:

$$R = \int_0^\infty p_{ql}(I)[1-p_{qu}(I)](1-p)f_{hc}(I)\,dI + \int_0^\infty p_{ql}(I)[1-p_{qu}(I)]pf_h(I)\,dI \tag{2.37}$$

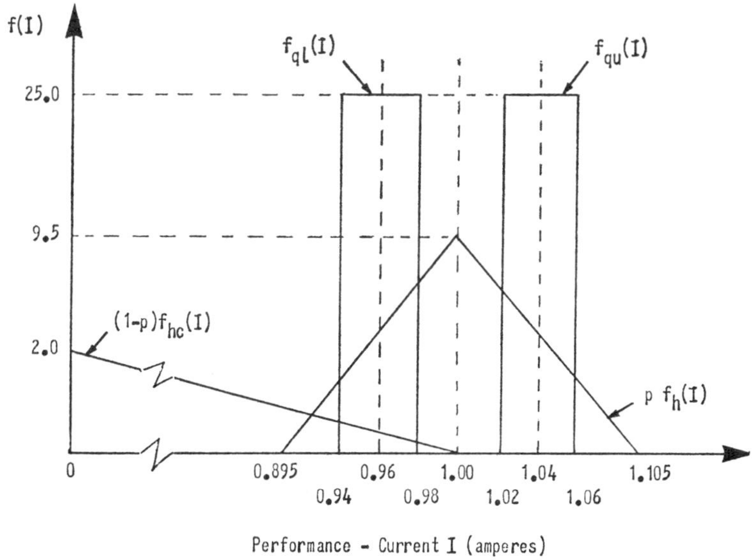

Performance - Current I (amperes)

Figure 2.19 Complete current p.d.f.'s for achievement and requirement

Example 2.6

Find the numerical value for the reliability of the example system with the characteristics depicted in Figure 2.19 and compare it with the reliability obtained when there were no variations in the catastrophic régime. The reliability is obtained from the expression given in Equation (2.37). The second part of this expression represents the reliability with no variations in the catastrophic régime and was found in Equation (2.32) to have a value of 0.595. The first part of the expression, therefore, represents the increase in reliability due to the present set of postulated conditions. In the first part, $p_{qu}(I) = 0$ for all values of $f_{hc}(I)$, $p_{ql}(I) = 0$ for all values of $f_{hc}(I)$ below 0.94 and $p_{ql}(I) = 1$ for all values of $f_{hc}(I)$ between 0.98 and 1.00. Hence, due to first part only:

$$R = (1-p)\int_{0.94}^{0.98} p_{ql}(I)f_{hc}(I)\,dI + (1-p)\int_{0.98}^{1.00} f_{hc}(I)\,dI$$

Substituting for $p_{ql}(I) = 25.0 (I-0.94)$ and $f_{hc}(I) = 2.0(1-I)$ gives:

$$R = (1-0.987)\,50.0 \int_{0.94}^{0.98} (I-0.94)(1-I)\,\mathrm{d}I + (1-0.987)\,2.0 \int_{0.98}^{1.00} (1-I)\,\mathrm{d}I$$

$$= 0.013\,[50 \times 0.0000267 + 2 \times 0.0002]$$

$$= 0.013 \times 0.0017$$

$$= 0.00002$$

So, in fact, the assumed variation in catastrophic performance has made very little difference to the reliability of the system.

The result of the preceding example is likely to be true in the majority of practical cases and hence the more general reliability formula of Equation (2.36) can often be simplified to:

$$R = p \int_0^\infty p_{ql}(x)\,[1 - p_{qu}(x)]f_h(x)\,\mathrm{d}x \qquad (2.38)$$

which in the case of fixed requirement limits reduces to:

$$R = p \int_{x_{ql}}^{x_{qu}} f_h(x)\,\mathrm{d}x \qquad (2.39)$$

The whole of the discussion so far has centred round finding the reliability of a system at a particular time and under a particular set of conditions that may prevail at that time. This means that the values of all the functions used such as $p, f_h(x), f_{hc}(x), f_{ql}(x)$ and $f_{qu}(x)$ are those which are evaluated under the conditions stipulated. Different conditions may lead to other values for all the individual functions and hence to a different value for R. If any of the conditions vary during the life of the system then R may become a time-dependent function. In cases of this sort it may be of interest to know the instantaneous value of R or, perhaps, some mean value of R over a stipulated interval of time. This gives rise to the concepts of 'mean availability' or its complement 'mean fractional dead time' which are discussed in Chapter 10.

Another restriction imposed on the discussion in this present chapter has been to confine equations such as Equations (2.38) and (2.39) to a single performance parameter, x. In the example system of Figure 2.6 this single performance parameter has been that of the current, I, flowing through the load resistance, R_L, under certain sets of input conditions. It is obvious, in a complex system, that there may be many performance parameters of interest. The corresponding reliability equations then become multi-dimensional with possible inter-relationships between the various dimensions. These aspects are reviewed in Chapters 3, 4 and 5 where a transfer matrix method of

AN APPROACH TO RELIABILITY ASSESSMENT

representation is introduced. Various properties of this type of transfer matrix are discussed in more detail in Chapter 6.

It is seen, then, that in order to evaulate the reliability of a system, the following processes are necessary:

(1) A logical process for establishing the precise required performance of the system under all the relevant conditions. This normally requires the determination of two requirement functions, $f_{ql}(x)$ and $f_{qu}(x)$, representing the distributions of required performance values at the lower and upper limits respectively. More generally, each of these functions may be space and time dependent and cover a range of individual performance parameters.

(2) A logical process for analysing a system in order to determine its inherent capability. This means examining the system with regard to its basic laws of operation and its basic method of construction in order to determine whether or not it is nominally capable of correct operation.

(3) A variation analysis of all software aspects of the system in order to arrive at a combined or overall variability function $f_0(x)$ which may be space and time dependent and cover a range of individual performance parameters. This involves methods for combining distributions of variation.

(4) A variation analysis of all hardware aspects of the system in order to arrive at a combined or overall variability function for the system hardware $f_a(x)$ which may also be space and time dependent and cover a range of individual performance parameters. This, again, involves methods for combining distributions.

(5) A logical process of fault mode analysis to determine the system availability in the context of freedom from catastrophic modes of failure in the hardware of the system. This gives a value of p which may be of a multi-dimensional nature.

(6) A process of combining $f_0(x)$ and $f_a(x)$ to give the complete performance variability function $f_h(x)$ which may then have to be combined with the corresponding p function.

(7) A similar process to (3), (4) and (6) in order to arrive at $f_{hc}(x)$.

(8) A process of then obtaining a particular reliability value or a reliability value averaged over the appropriate time and space domains.

Item (1) represents the evaluation of the Required Performance, items (2)–(7) inclusive the Achieved Performance and item (8) the Correlation which leads to the complete Reliability Assessment (see Figure 2.1).

Very often reliability is only evaluated and interpreted according to the very simple relationship:

$$R = p$$

As can be seen from the foregoing, this is only a small part of the whole process of reliability assessment. It is the purpose of this book to examine the complete process of reliability evaluation and the chapters which follow discuss in turn the appropriate processes outlined in (1) to (8) above.

Questions

1. An electrical generating station has a two-fold requirement to meet a normal demand of 350 MW and a peak demand of 750 MW. The probability of a normal demand arising at any time is 0.75 and the probability of a peak demand arising at any time is 0.25. The station has two independent generating sets running in parallel and each set has an output of 400 MW. The probability of a generating set being 'off line' at any time is 0.1. What is the reliability of the generating station in meeting its requirement?
2. Two aircraft, A and B, have probabilities of 0.9 and 0.8 respectively of being available for flying from airport W at a particular time each day. Aircraft A has a range of 1000 miles and the range of aircraft B is 2000 miles. The probability, at the particular time, that one of the aircraft is required to fly to airport X, which is 800 miles from airport W, is 0.4. The alternative probability at this particular time, that one aircraft is required to fly to airport X and the other to airport Y, which is 1500 miles from airport W, is 0.6. No other requirement exists. What is the reliability of the aircraft schedules being met?
3. A discrete variate, r, representing the number of failure events that take place in a system in a given length of time has a probability function of the following form:

$$f(r) = \frac{\alpha^r e^{-\alpha}}{r!}$$

   where $r = 1, 2, \ldots, \infty$ and α is a constant of value 2.0. What is the probability that (a) exactly two failure events occur and (b) two or less failure events occur in the prescribed length of time?
4. A continuous variate, x, representing the times to failure of a system has a probability density function of the following form:

$$f(x) = \frac{e^{-x/\lambda}}{\lambda}$$

   where λ is a constant of time value 1000 h. What is the probability that a particular system will have a time to failure of (a) less than 100 h and (b) less than 1000 h?
5. The values of nominal 100-Ω resistors coming off a production line are found to be randomly variable between a lower limit of 95 Ω and an upper limit of 105 Ω. The distribution of this variation is rectangular.

What is (a) the equation for the probability density function describing the resistance variation and (b) the probability that a particular resistor has a value lying between 96 Ω and 102 Ω?

6. Describe what you understand by 'capability of performance' in connexion with a technological system. Taking any proposed system design of your choice, indicate those areas which may need to be assessed for capability and how this may effect any subsequent reliability analysis.

7. A resistance element has variations in resistance value which may be described by a p.d.f. of the following form:

$$f_R(R) = 0 \quad \text{for } 0 \leqslant R \leqslant 7.5$$
$$f_R(R) = 0.2 \quad \text{for } 7.5 \leqslant R \leqslant 12.5$$
$$f_R(R) = 0 \quad \text{for } 12.5 \leqslant R \leqslant \infty$$

A constant and invariable current of value 2.0 A flows through this resistance element. What is (a) the form of the p.d.f. for the distribution of power dissipated in the resistor and (b) the reliability of meeting a requirement that the power dissipated should be greater than 34 W but less than 48 W?

8. The natural frequency of a strut is found to be inversely proportional to its length where the constant of proportionality has a value of 100 units. A range of such struts have their lengths randomly distributed according to a rectangular distribution with lower and upper limits of 10 and 20 units respectively. The requirement for a particular strut selected from the range is that its natural frequency should be less than 8 units. What is the reliability in meeting this requirement?

9. A control system has a power output measured in watts, W, which, as a result of variations in the elements within the system, is randomly distributed with respect to time according to the p.d.f. $f_W(W)$ where:

$$f_W(W) = 0 \quad \text{for } 0 \leqslant W \leqslant 42.5$$
$$f_W(W) = 0.032\,W - 1.36 \quad \text{for } 42.5 \leqslant W \leqslant 45$$
$$f_W(W) = 0.08 \quad \text{for } 45 \leqslant W \leqslant 55$$
$$f_W(W) = 1.84 - 0.032\,W \quad \text{for } 55 \leqslant W \leqslant 57.5$$
$$f_W(W) = 0 \quad \text{for } 57.5 \leqslant W \leqslant \infty$$

Draw the shape of this p.d.f. and calculate the reliability of the control system if the requirement for the power output at a particular time is (a) that it should be between 45 W and 57 W, (b) that it should be between 43 W and 57 W and (c) that it should be less than 55 W.

10. The same control system as that specified in Question 9 has requirements for power output which are not constant but which are randomly

distributed with respect to time. The requirements are in the form of a lower and upper limit where the p.d.f. for the lower requirement limit, $f_{ql}(W)$, is:

$$f_{ql}(W) = 0 \quad \text{for } 0 \leqslant W \leqslant 43$$
$$f_{ql}(W) = 1.0 \quad \text{for } 43 \leqslant W \leqslant 44$$
$$f_{ql}(W) = 0 \quad \text{for } 44 \leqslant W \leqslant \infty$$

and that for the upper requirement limit, $f_{qu}(W)$ is.

$$f_{qu}(W) = 0 \quad \text{for } 0 \leqslant W \leqslant 55$$
$$f_{qu}(W) = 0.5 \quad \text{for } 55 \leqslant W \leqslant 57$$
$$f_{qu}(W) = 0 \quad \text{for } 57 \leqslant W \leqslant \infty$$

What is the reliability of the control system in meeting an overall requirement of this form?

11. A fluid control system consists of two main elements. These are an electrically operated control valve and a control amplifier which feeds signals to the control valve. The control valve has a catastrophic mode of failure whereby it may move to the closed position. The probability of the control valve being in this state at any time is 0.05. The amplifier has a similar but independent catastrophic mode of failure which also closes the control valve and the probability of the amplifier being in this state at any time is 0.1. Whilst the control system is in the nominal working régime, with the control valve open, variations with time take place in the controlled flow of fluid. These variations follow a rectangular distribution with lower and upper limits of 5 m³/s and 7 m³/s respectively. What is the reliability of the control system in meeting a flow requirement of (a) greater than 4.8 m³/s, (b) between 5 m³/s and 6.5 m³/s and (c) greater than 6 m³/s?

12. An electronic amplifier, when functioning normally, is found to have random variations in voltage output with time which are distributed according to the p.d.f., $f_V(V)$, where:

$$f_V(V) = 0 \quad \text{for } 0 \leqslant V \leqslant 38$$
$$f_V(V) = 0.25 \text{ V}{-}9.5 \quad \text{for } 38 \leqslant V \leqslant 40$$
$$f_V(V) = 10.5{-}0.25 \text{ V} \quad \text{for } 40 \leqslant V \leqslant 42$$
$$f_V(V) = 0 \quad \text{for } 42 \leqslant V \leqslant \infty$$

In addition, the amplifier has a probability at any time of 0.02 of being in the catastrophic failed state where the output voltage is zero. The requirement is that the voltage output should lie between 39 V and 41.5 V. What is the reliability of the amplifier in meeting this requirement?

CHAPTER 3
The Performance Requirement

Requirement Concepts

The previous chapter concluded with a list of evaluation processes which were deemed to be necessary in order to carry out a complete reliability assessment. The first of these processes involved a need for a method to establish the precise required performance of a system under all the relevant conditions. The definition of reliability given in Chapter 1 implied that the reliability of an item cannot be assessed unless its required function in the desired manner and under all the relevant conditions is known. In other words it is essential to know what is wanted before a reliable meaning can be attached to what is achieved. This was also illustrated by the 'Required Performance' block of Figure 2.1.

Some discussion on the ways in which the required performance may be evaluated and represented has already been given in Chapter 2, but it is the purpose of this chapter to take these basic ideas and examine them in more detail.

As shown schematically in Figure 2.1 and mathematically in Equations (2.36)–(2.39) inclusive, reliability is always a function of required performance and achieved performance. This fact may be written symbolically in the following form:

$$R = f(Q, H) \qquad (3.1)$$

where:

Q = the complete required performance function

H = the complete achieved performance function

$f(Q, H)$ = the function representing the probability of the achieved performance meeting the required performance, i.e. the reliability.

Obviously, the assessment of the reliability, R, depends upon the determination of both the required performance function, Q, and the achieved performance function, H. The distinction needs to be emphasized between the overall reliability function R and parameters of reliability such as failure-rate, repair time, availability, fractional dead time and probability of failure or success. Such parameters may often be calculated and expressed without any detailed knowledge of the required performance function Q. An introduction to some of the methods of calculating reliability parameters was given in Chapter 1. For instance, Example 1.1 calculated that the probability of a customer receiving a reject fuse was 10^{-5}. This figure is just the estimated value of a reliability parameter since nothing was specified in the example as to the customer's requirements. If the customer's requirement happened to be that he could not accept a single reject fuse, then the unreliability of the situation is 10^{-5} for every fuse received. In this case the overall reliability, R, is the complement of this value, namely, 0.99999, that each fuse is successful. If, however, the customer was willing to accept, say, one reject fuse in every 10,000 received then the overall reliability of the situation is no longer on the basis of 0.99999 per fuse. In a total batch of 15×10^7 fuses, the average number of expected rejects was 1579. The customer's requirement would now allow this figure to rise to 15,000 before the situation became unreliable. So, the reliability question, under these conditions, relates to the chance of the customer receiving more than 15,000 rejects in, say, a batch of 15×10^7 fuses when the average number of expected rejects is 1579. This involves an entirely new calculation and a knowledge of the distribution of rejects in the total population of fuses. Methods of dealing with this type of analysis will be discussed in later chapters, but the important aspect at the moment is to recognize that the requirement needs to be specified in order to evaluate R. Changes in the requirement will obviously alter the value of R even though the values of the reliability parameters remain the same.

It is seen that similar ideas apply to all the examples given in Chapter 1. For instance, in Example 1.6 the reliability parameter for the performance of a fleet of taxis is calculated at five breakdowns per year. The corresponding probability of failure of 0.714 for a 3-monthly period is based on the implicit assumption that no breakdowns are required. If, on the other hand, some breakdowns are acceptable and this number of acceptable breakdowns is used as the requirement, then the final overall reliability answer will be different. In those cases, however, where the probability of a hazard or accident condition is calculated, for instance in Examples 1.9 and 1.11, then the implicit requirement for no accident may be the only legitimate requirement to consider. However, particular reliability parameters, like that of fractional dead time in Example 1.9, may have no absolute meaning unless they are compared with a requirement.

These aspects of requirements are not, of course, particular to reliability technology. Requirements are generally needed in order to interpret any performance parameter for a system. For instance, it may be stated that a particular model of a motor car is capable of achieving a speed of, say, 95 km/h. This speed is a performance parameter and in itself does not lead to any performance conclusion. For the driver who never wants to travel at a speed greater than 80 km/h, the performance of such a car is perfectly adequate. But for the driver who wants to cruise regularly at 110 km/h the car's performance is totally inadequate.

Overall reliability, then, is always a function of a requirement which may be either implicitly or explicitly expressed. The relationship given in Equation (3.1) is quite fundamental to this overall aspect of reliability evaluation. The meaning of and the estimate of the value of Q must be determined before it is possible to evaluate R. Sometimes, however, in technological system analysis, no explicit information is given as to the characteristic form or value of Q. In cases of this sort, the reliability analyst may have to assume or estimate some appropriate value of the Q function in order to proceed with a determination of R. An absence of a Q value means that Equation (3.1) is insoluble, a situation which in terms of general problem solution is sometimes termed an 'open problem'. An assumed value of Q, on the other hand, enables Equation (3.1) to be solved and the problem of evaluating R becomes 'artificially closed'. Most of the examples given in Chapter 1 were of the open problem type in that no specific requirements were laid down. However, as has already been indicated in this present chapter, the assumed requirement of no failures or no accidents turns them into artificially closed problems capable of some absolute solution with regard to the overall reliability.

Many practical problems in reliability assessment have to be turned into artificially closed problems at some stage in their evaluation. The assessment must, of course, make the meaning of its conclusions abundantly clear when an artificially closed system of evaluation is used. If the adoption of some assumed value of Q leads to the conclusion that the reliability, R, is adequate or acceptable, then the final acceptance of this adequacy depends upon substantiating that the originally assumed requirement value, Q, was correct. Sometimes, however, it may be feasible or expedient to assume a value of Q which represents the most stringent requirement that is likely to occur. If, with such an upper 'boundary' value of Q it is possible to show that the eventual reliability R is adequate, then there is no need to evaluate the actual value of Q which may pertain. This boundary approach is often of value in high risk situations or in those cases where the detailed specification of Q is difficult to obtain. On the other hand, it is always preferable, if possible, to specify a precise and unambiguous value of Q as a basis for the reliability evaluation.

Since the final evaluation of the reliability R is sensitive to the form and value of the required performance Q, it is important that great care should be taken in the detailed specification of Q for any particular application. This applies irrespective of whether the value of Q is proved or assumed. Ideas need to be clarified and the setting up of the requirement approached in a logical fashion. It is convenient in this process to consider the requirement as initially falling into two main parts. The first part is concerned with the basic definition or concept of the required performance. This is synonymous with the 'required function' aspect of the reliability definition already given in Chapter 1. The second part of the performance requirement is concerned with those terms or conditions under which the basic concept of performance is to be fulfilled. These conditions or terms are equivalent to the restrictions implied by the phrase 'in the desired manner under all the relevant conditions...' which also formed part of the definition of reliability originally

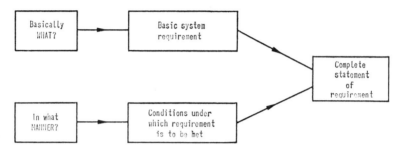

Figure 3.1 Basic requirement and conditions

given. In other words, there is a two part question to be answered. 'What is basically required?' and 'Under what conditions or in what manner is it required?' This break down of ideas is illustrated schematically in Figure 3.1. The conditions may be further divided into more specific groups and associated questions as shown in Figure 3.2, but first it is prudent to examine in more detail the meaning of the basic requirement.

Basic Requirement

The basic requirement of any technological system is determined by the answer to the simple question, 'What, fundamentally, is the system supposed to do?' The answer should match the question in its simplicity and be a basic statement without any provisos or conditions attached. For instance, a requirement might be expressed in the following form:

The designer of an automatic control system for an application in controlling a particular process at a certain location with certain characteristics requires an electronic amplifier with a stipulated gain, frequency response and transfer function with a view to it being used for certain lengths of time on specified occasions during the expected life of the process plant for an outlay of capital and annual expenditure which is less than some prescribed maximum.

THE PERFORMANCE REQUIREMENT

Such a statement of a requirement is woolly and confused. Some of the various terms and conditions are intermingled and in total they tend to hide the basic objective. Stripping the statement of all its conditions both specified and implied, leaves a more basic requirement or concept of simply:

An electronic amplifier.

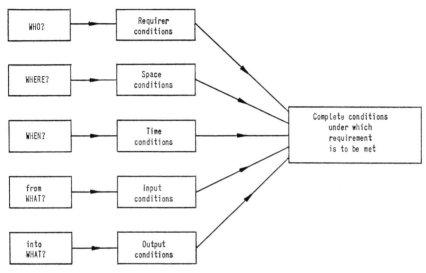

Figure 3.2 Breakdown of conditions

Although this is the basic element of the original statement, it is still to some extent conditional. It may now be questioned whether an electronic amplifier is really what is required. As a concept it is simply expressed, but is the concept itself correct? If the requirement is fundamentally examined, it may be found that an 'electronic' amplifier is not necessarily required but that other types of amplifier (pneumatic, hydraulic, mechanical, etc.) may be equally suitable. If this is true, then the requirement may be basically simplified further to:

An amplifier.

It also depends, of course, on what is being looked for in the final reliability evaluation. If it is the reliability of an electronic amplifier in this application which is really being sought then, obviously the reliability needs to be evaluated against a basic requirement for an 'electronic amplifier'. This means, though, that if the actual system uses a 'pneumatic amplifier' then the reliability would be evaluated at $R = 0$ since a pneumatic amplifier is not able to meet a positive and basic requirement for an electronic amplifier. On the other hand, if the requirement is more truly for just 'an amplifier', then

88   RELIABILITY TECHNOLOGY

any type of amplifier, including an electronic amplifier, is capable of being evaluated with a finite value of R.

There is always a tendency for a person to express a requirement in terms of his own discipline or, more generally perhaps, in terms of what he thinks can be most easily achieved. For instance, a man may say 'I require a motor car' when what he basically means is that he requires a facility for getting from A to B. An aircraft owner may feel that he wants a triplex automatic landing control in his aircraft whereas basically he requires to always land his aircraft with a certain degree of security. Such fundamental ideas need to be sorted out at this stage since they will markedly affect the meaning which is attached to the final reliability adjudication.

It is nearly always possible to shift any expression of a 'basic' requirement into the more particular or the more general. A more particular case of the present example would be:

An electronic transistorized amplifier.

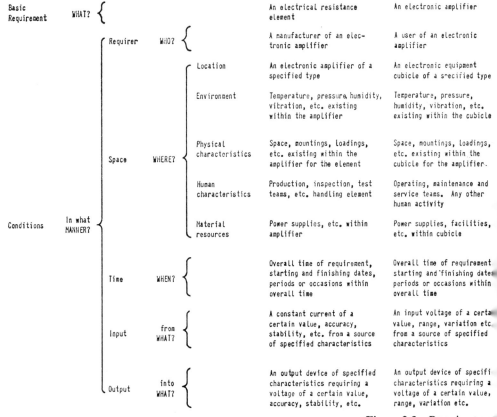

Figure 3.3 Requirement

THE PERFORMANCE REQUIREMENT

and a more general case would be:
   A signal processing device.

In some cases it may be necessary to proceed even further into generalizations. It is conceivable that the designer who specified a need for an electronic amplifier did so on the basis of certain preconceived ideas as to how he was going to design his automatic control system. An alternative concept of control might not necessarily require any type of amplifier at all. This is, in fact, saying that the requirement has not so far been examined at the right functional level. Perhaps it should, therefore, be moved into a more general hierarchy such as the expression of a requirement for:
   An automatic control system.

It is obvious that the concept of the basic requirement can go on being examined in this way. The top line of Figure 3.3—in answer to the basic requirement WHAT?—gives an illustration of a few of the various functional levels into which the requirement may fall. Many other levels and branches

signal processor	A control system	A process plant
user of a signal processor	A manufacturer of a process plant	An owner of a process plant
equipment mounting cility of a specified type	Process plant X in place A	Place A
mperature, pressure, midity, vibration, etc. sociated with the mounting cility	Temperature, pressure, humidity, vibration, etc. associated with the process plant	Temperature, pressure, humidity, vibration, etc. associated with place A
ace, fixings, loadings, c. associated with the unting facility	Space, loadings, etc. associated with the process plant	Area, topography, terrain, buildings, other structures etc. associated with place A
erating, maintenance, and rvice teams. Any other man activity	Production, inspection, test teams, etc. Other human activity	Managerial, operating, maintenance, test teams, etc. Other relevant human activity
wer supplies, facilities, c. available	Power supplies, facilities, etc. available	Power supplies, services, communications etc. available at A
erall time of requirement, tarting and finishing dates, riods or occasions within erall time	Overall time of requirement, starting and finishing dates, periods or occasions within overall time	Overall time of requirement, starting and finishing dates, periods or occasions within overall time
input signal of a certain lue, range, type, variation c. from a source of ecified characteristics	Certain specified characteristics in a certain process variable of the plant	An input supply of certain materials, of certain characteristics, of a certain amount at certain times
output device of specified aracteristics requiring a gnal of a certain value, nge, type, variation etc.	Maintaining a certain process variable of the plant within specified limits	Delivering to a customer certain output materials of specified characteristics at certain times

erarchies

within the levels can obviously be considered. In the ultimate, only the particular application and the interpretation to be applied to the reliability evaluation will determine whether or not the basic requirement is expressed at the correct functional level. The important thing for each application is to determine which is the correct level or hierarchy.

A final point for consideration is whether there are any aspects or conditions which may affect or change the basic requirement. This means conditions which may alter the basic requirement itself rather than those conditions under which the basic requirement is to be met. It might be considered, for instance, that the requirement might change with time, place or environment. For a man who normally requires to drive his car at 110 km/h, such a requirement may be considerably reduced during poor visibility conditions. The poor visibility, therefore, could be argued as being a factor which affected the basic requirement. On the other hand, the man may require a normal achievement of 110 km/h on different types of road surfaces. The road surfaces would then seem to represent conditions under which the requirement must be fulfilled. A logical examination of these statements, however, will reveal that the requirement to travel at 110 km/h is not the true basic requirement. The basic requirement must be associated with the need to travel from one place to another or, perhaps more abstractly, with the man's level of pleasure in driving. With the basic requirement at this sort of level, both the poor visibility and the road surfaces are conditions under which the basic requirement must be met. Neither of them affects nor alters the true basic requirement. The basic requirement, therefore, should be generally invariable. If it is not, then it should be examined to see if it is truly basic from this point of view.

Example 3.1

A motor car manufacturing company is planning to introduce a new model of motor car in 5 years' time. The general design of the body has been completed and a general set of overall performance requirements, relating to speed, acceleration, handling, size and selling price, have been laid down. It is intended that the car should sell at home and abroad and have an average life expectancy of at least 10 years or the equivalent of 160,000 km. The main purpose of the car is to appeal to the average family man. The chief designer of the car has still to finalize his choice of a suitable engine for the car and is about to approach the engine division of the company with his requirements. What basic requirements could arise out of this situation?

A number of possible requirements are related to this overall situation. The company will obviously require the car to be a success or, in other words, they want to make money out of it. The company and the average family man require a car or, perhaps more generally, a road vehicle. The chief

THE PERFORMANCE REQUIREMENT

designer is looking for a suitable engine or, may be, any suitable prime mover. So it is possible, by ignoring all the conditional aspects, to formulate at least three basic requirements at different hierarchic levels. These are:

>A money-making system
>
>A road vehicle
>
>A prime mover

It is not possible to decide from the example as stated which of these or any other hierarchic level is the correct one. So further questions would need to be asked to clarify this point. In fact, of course, a requirement may exist at all three levels which would then entail three separate problems to be solved.

It may now be said that a complete examination of the basic requirement involves obtaining affirmative answers to the following type of questions:
Is the statement of the basic requirement free from all conditions under which the requirement has to be met?
Is the basic requirement stated in its simplest form?
Is the statement free from any unnecessary descriptive or practically derived limitations?
Is the basic requirement the correct one?
Is the basic requirement expressed at the right functional level or hierarchy for the purpose in hand?
Is the basic requirement invariable?
Only when the basic requirement has been examined along these lines can it be effectively used in setting up the complete requirement and hence used in leading to a meaningful interpretation of the final reliability assessment. The complete requirement, however, is made up of the basic requirement and the terms or conditions which bound it. It is now necessary to examine these terms and conditions.

Requirement Conditions

The schematic diagram of Figure 3.1 illustrates the complementary nature of the conditions under which the basic requirement has to be met compared with the basic requirement itself. However, the conditions are also dependent to some extent upon the form of the basic requirement and hence the need for having considered the basic requirement first.

The basic requirement, although fundamental to the process of reliability evaluation, is not capable of leading to a solution without the conditions which surround it. In the example statement of a requirement given at the start of the last section a number of typical conditions were illustrated. These were the requirer conditions such as the designer of an automatic control system, the characteristic conditions of the amplifier such as frequency response, the space conditions such as the process plant and the time conditions such as the occasions on which the amplifier was to be used. These example conditions were neither exhaustive nor adequately defined. Some, like the frequency response of the amplifier, were not fundamental conditions but derived conditions. Figure 3.2 shows a break down of conditions into five main categories which normally cover the range of conditions likely to be met in any technological system. An examination of these five categories or the detailed answers to the related questions WHO?, WHERE?, WHEN?, from WHAT? and into WHAT?, should enable the conditions to be covered logically and exhaustively. Under each heading the appropriate conditions need to be basically defined and then specified in detail. Also it is necessary to ensure for each condition, as with the basic requirement, that it is the correct one for the application in mind. Some of these aspects may be further illustrated by considering each main condition in turn.

Requirer conditions

The aspect of whose requirement it is that is being specified is sometimes neglected in the consideration of the overall conditions. This, however, is a very important aspect. It may also be considered as a fundamental condition since it tends to dictate the form and the course of the remaining conditions. Suppose, for instance, that a basic requirement is for a television set. This requirement could be set up by man A who, because of his characteristics, is quite content with a black-and-white picture and average sound reproduction. Another man B may have the same basic requirement but insist on a colour picture and 'hi-fi' sound. Obviously the set that meets the requirer conditions of man A will not meet those of man B and, possibly, vice versa. Still taking the same basic requirement, the requirer conditions may be set up by neither man A nor man B. They may be implemented by the retailer, distributor, manufacturer or designer of television sets. Each time the person or corporate body of the requirer is changed, so are the conditions imposed upon the basic requirement even though the basic requirement itself may remain the same.

In the requirement example given in the previous section, it was the designer of an automatic control system who required an electronic amplifier. For the same basic requirement, it could, of course, have been a user of electronic amplifiers or a manufacturer of electronic amplifiers who had imposed the requirer conditions. Obviously then, in defining the requirer conditions it is not only necessary to know who the requirer is but also to determine whether he is the right one for the purpose in hand. The second line of Figure 3.3 shows a few typical requirers and it can be seen how the

THE PERFORMANCE REQUIREMENT   93

overall requirement could change, even with the same basic objective, if the requirer were to change.

As a further example, consider the basic requirement of an aircraft. The requirers could be, for instance, the manufacturer, the owner, the fleet operator, the crew or the passengers. Each requirer would instigate an entirely different set of conditions under which the basic requirement should be met. It is the same technological system but in each case it is viewed in a different light. That it is important to determine this viewpoint is seen when it is considered that what may ultimately be an acceptable level of reliability to a fleet operator will not necessarily be acceptable to a passenger or vice versa.

Example 3.2

In Example 3.1, who is the requirer or who are the possible requirers that may exist?

It is apparent in this example that the range of possible requirers who may exist, within the terms of the situation as stated, are as follows:

The motor car manufacturing company.
Some official or officials of the company.
The chief designer of the car or his staff.
The engine division of the company represented by certain of its staff.
The average family man.
A particular family man.

As with Example 3.1, it is obvious that more details of the situation would be needed in order to decide which is the correct requirer or whether a separate problem needs to be solved for each possible requirer.

General questions that need to be affirmatively answered then, are:
Is the requirer precisely defined?
Is he the right one for the application being considered?
Are his characteristics, which are relevant to the remaining conditions, adequately known?

This last question leads into the next set of conditions which are concerned with the spatial aspects of the item defined by the basic requirement.

Space conditions

The requirer by his own nature, characteristics or location will have largely influenced the location or general spatial conditions in which the item defined by the basic requirement has to behave. As with all conditions, there is a need to define the condition, specify its detailed characteristics and, at the same time, determine that each characteristic is the correct one. The basic requirement for a computer coupled with the condition that it must operate in an air-conditioned room will lead to a different overall requirement from that for the same computer installed in an ocean-going liner. The requirement for an aircraft to operate within one country will be different from the requirement for the same aircraft to ply a world-wide service. If the reliability of a pressure gauge were assessed against the condition of its use on a domestic boiler it would give a completely erroneous value of reliability if the actual intended location of the gauge were a submarine engine room. The cry is often heard that technological devices have 'failed' because they have been misapplied. Very often, this misapplication is due to a misunderstanding or ignorance of the detailed spatial conditions under which the basic requirement has to be met.

A typical series of spatial condition characteristics are shown in the third series of lines in Figure 3.3. Obviously, under any one column, an almost infinite variety of locations and detailed characteristics of locations is possible. The need for precision and correctness in the specification is paramount since any error will lead to the wrong overall requirement and hence to an incorrect reliability evaluation. The spatial conditions are conveniently divided into a number of sub-groups as shown in Figure 3.3 which are designed to help in the detailed specification of the various aspects which may apply.

The 'location' represents a simple definition of the immediate surroundings in which the required item has to perform. In some cases it may be sufficient, as with the resistance element in the first column of Figure 3.3, to define the location as an electronic amplifier of a certain type. In other instances it may be necessary to be more specific because, for example, such an electronic amplifier may not in itself be isolated from its own surroundings. The definition would then require to be expanded, for instance, into the form of an electronic amplifier of a certain type, in a certain building, in a certain place, etc. Some technological systems may be associated with several locations instead of just one. Others, like transportation systems, may be mobile and move from place to place. In these cases, each of the different locations and the relevant routes between the locations will need to be defined. Since movement will generally imply some time dependence, this aspect will also have to be carefully specified.

The 'environment' represents the ambient characteristics of the surrounding media at the specified location or locations. As shown in Figure 3.3, these include such aspects as temperature, pressure, humidity, vibration, shock and radiation. These properties may always be at fixed levels or they may vary with time. This means that, for every relevant property, a complete and correct dynamic behaviour pattern may need to be obtained. It is important also to remember that abnormal variations in the environmental characteristics may have to be considered as well as the normal variations. Cooling air, for instance, may be available in the location for the required item. This would lead to a certain level of temperature being normally maintained but, if the cooling fans were to stop, the temperature might rise to some abnormally higher value. This sort of situation, in formulating the requirement for some systems, could be a crucial aspect to take into account.

The 'physical characteristics', which are the third element of the space conditions shown in Figure 3.3, represent those aspects covered by the material description of the location. These include the shape and dimensions of the space available for the required item, the loadings and weight that such a space structure can tolerate and, for the larger locations, the geography, topography, terrain and man-made structures or objects that are contained within the defined location or locations. Once again, a correctness and preciseness in the definition and specification of these characteristics is important. A basic requirement may be for a diesel generator set and the required location a basement room of a hospital building. Obviously, the shape and size of the room, the acoustic properties of the area, the loadings that can be taken by the floor are all going to influence the overall requirement. An automatic landing system for an aircraft is conditional upon the physical properties of the airports at which the aircraft is intended to land. In this case, some of the relevant physical properties would be the dimensions, strength and direction of the runway, the terrain surrounding the runway, the disposition of hills and trees, the nature and form of nearby buildings and structures, and the population and distribution of other flying objects such as birds and other aircraft. It is apparent that different airports will have different physical characteristics in these respects. So a basic requirement which is conditional upon one airport will lead to a different reliability value from one which is conditional upon another airport. This is just emphasizing that the correct detailed characteristics or series of characteristics need to be determined in order to form a meaningful overall requirement. Sometimes the physical characteristics may be time dependent as is almost certainly the case with any moveable object within the defined location. Time may also enter in where there are deteriorations or other types of changes in the physical properties. It is necessary to examine, therefore, any time dependencies which may exist.

Moving to the fourth sub-division of space conditions shown in Figure 3.3, the 'human characteristics' represent the type and form of human activity which is contained within the location. Human activity or human influence factors of this nature may fall into two categories. First, there are those factors or human resources which are available in the location to directly contribute towards the success of the required item or system. Secondly, there are those human influence factors which are necessarily present in the location but which may bound or limit the eventual success. If the basic requirement is for a journey by a road vehicle, an example of the first type would be the maintenance and service teams who are available to look after the upkeep of the vehicle, the roadways and all the road facilities. An example of the second type would be pedestrians who stray across the road or the farmer moving his cows. Neither type will necessarily wholly expand or wholly limit the requirement, since the characteristics of the maintenance or service personnel may be that they are prone to error and the characteristic of the pedestrian may be that he is considerate to the motorist. All types of human beings contribute towards the space conditions and generally, for any technological system, these types will consist of production, inspection, testing, operating, maintenance and service teams, administrative and managerial bodies and, to some extent, elements of the general population. For all such teams or bodies which are present in the location, the resources, skills and other relevant characteristics need to be determined. As with other space conditions, many of the human influence factors will be time dependent and this aspect will require careful survey and understanding.

Apart from the fact that the basic requirement is conditional upon human resources, it will also be conditional upon 'material resources'. This leads to the last of the space condition sub-divisions shown in Figure 3.3. For most technological systems, the relevant material resources come under the more specific heading of power supplies. The power supplies that may be available in the location are typically electrical, mechanical in the form of prime movers, hydraulic in the form of water, air or other fluid heads, and fuel such as coal, gas and oil. For larger systems or enterprises the material resources may cover communications, transportation facilities and all kinds of available wealth. Both normal and abnormal variations are likely to occur in electrical supplies, other power supplies, availability of fuel and general facilities. As before, each material resource needs to be defined, specified and investigated for its time dependency so that it can be used in a scientific manner to supplement the basic requirement.

Example 3.3

What are the appropriate space conditions relevant to Example 3.1 if it is assumed that the basic requirement is 'A road vehicle' and the requirer is 'A particular family man'?

It should be noted that the general question 'What are the space conditions associated with Example 3.1?' could not be answered unless the basic requirement and the requirer had been previously defined in detail. With the definition of basic requirement and requirer assumed, the general space conditions would be related to:

Location: The places and routes to and by which the particular family man requires to be transported by the road vehicle.

Environment: The atmospheric conditions associated with such places and routes.
Human: The human characteristics, resources and activity associated with the places and routes.
Material: The geography, topography, facilities, services and relevant material resources associated with the places and routes.

These space conditions will, of course, be entirely different if either the basic requirement or the requirer is changed.

As a typical summary of the space conditions that may require to be considered, it should be ascertained that the following questions can be answered in the affirmative:

Is the location, in which the item or system defined by the basic requirement is required to operate, precisely defined and is it the correct location?

Are all the normal and abnormal environmental conditions which are relevant to the location correctly known and completely specified at all relevant times?

Are all the normal and abnormal physical characteristics of the location correctly known and completely specified at all relevant times?

Are all aspects of human activity and human resources which are relevant to the location correctly known and completely specified for both normal and abnormal conditions at all relevant times?

Are all the relevant material resources in the location correctly known and their values and time variations completely specified under both normal and abnormal conditions?

Time conditions

Just as the requirer influences the space conditions so he tends to govern the conditions of time which apply. The time conditions of interest are those times during which or at which the item or system as defined by the basic requirement is required to perform. The dependence of the final reliability of a system upon time was emphasized in the original definition of reliability given in Chapter 1. Time is also fundamental in the whole concept of the requirement conditions. Even though it is complementary to the other conditions of interest it tends to be an independent variable upon which many of the other conditions are based. This aspect has already been brought out in studying the space conditions.

It is obviously important that the relevant time conditions should be accurately defined. First, there is the general time period over which the basic system requirement has to be met. This generally corresponds to the expected useful life period of the system which is designed to meet the basic requirement. So, if a particular man has a basic requirement for a motor car this may cover a general period of say 2 years or 5 years or perhaps 10 years. If the requirer is the prospective owner or operator of a process plant, typical overall periods of time may be 20–30 years. Radical changes in these required lengths of time will obviously have a marked effect upon the overall performance requirement.

Secondly, it is important to define the starting date on which the overall period of required time is to commence. The requirer of the motor car, for instance, may be thinking in terms of an immediate purchase of such a vehicle or he may be considering an acquisition or replacement in 1 or 2 years' time. The process plant operator could be buying an existing plant in order to continue with the operation of the process or he could be planning for a new plant 5 or even 10 years ahead. The possible influences of any change in the starting date on the overall requirement are immediately apparent. What is required now is nearly always different from what may be required in the future.

Thirdly, it is necessary to define the frequency and pattern of required use within the overall period of time. The prospective motor car owner may want to use his car every day or he may just want to take it out at weekends or holiday periods. He might even only want to keep it in his garage as a show piece! Each time he takes it out, he may require it for short journeys, for long journeys or for varying lengths of journeys. The times that he requires his car may occur at regular and systematic intervals or they may approach complete randomness. There is obviously a vast difference between the requirement of the man who on any day of the week at any time during the day wants to perform journeys up to several hundred miles and of the man who only wants to travel 30 miles every other weekend to see his relations. Similar ranges of time patterns can occur with many technological systems. The requirement for large stationary plants, however, is likely to be more on a continuous basis. The continuum will not be complete, though, as all plants will be required to come 'off line' for maintenance, testing and modifications at some periods during their life cycle. In general, the frequency and pattern of required usage will be a complex function but this is all the more reason why it is important to define the pattern in all its relevant details.

Example 3.4

What are the appropriate time conditions relevant to Example 3.1 if it assumed that the basic requirement is 'A road vehicle' and the requirer is 'The motor car manufacturing company'?

As with the space conditions in Example 3.3, the time conditions are dependent upon the basic requirement and the requirer being adequately defined. Under the terms of Example 3.4 it can be deduced from Example 3.1

THE PERFORMANCE REQUIREMENT

that the general time conditions are as follows:
A starting date 5 years from now.
An overall time period in excess of 10 years.
Probably a continuous commitment to produce the road vehicle for the complete overall time.

The exact overall time that the motor car company requires the particular type of road vehicle and the continuity of this requirement would depend upon factors not disclosed in the original Example 3.1. These aspects would need, therefore, to be ascertained.

A general summary of the time conditions that require investigating is given in the fourth main line of Figure 3.3. The corresponding questions that need to be answered affirmatively are as follows:
Is the overall period of time, for which the basic requirement is to be met, correctly and precisely known?
Is the exact starting time and date for the required system correctly defined?
Is the frequency and pattern of time usage of the required system correctly and exactly specified?

Time has also occurred in the discussion on the space conditions in the previous sub-section. In fact, the answers to the space condition questions cannot be completely verified until the answers have been obtained to the time condition questions as well. The two sets of questions are, therefore, to some extent complementary.

It may be argued that a comprehensive space and time definition of conditions is complete and exhaustive in itself in that all other characteristics spring fundamentally from these two dimensions. However, for technological systems, it is useful to consider two other sets of dependent conditions which are related to the functional input and output conditions of the required system. These could be implied from the space and time considerations but it is generally necessary to examine their detailed derivation.

Input conditions

The requirement for a technological system normally arises from the fact that something needs to be converted into something else. A control system takes some kind of input signal from the process to be controlled and converts it into an output signal capable of actuating a controller of the process. A process plant takes in a range of 'raw' materials and converts them into different or 'processed' output materials. An engine requires an input of fuel in order to change it into some form of useful power output. An aircraft uses the skills of the pilot and crew and converts them into movement from one place to another. In the ultimate, any enterprise, of which technological systems may be a part, converts some form of 'wealth' into another form of 'wealth'. The whole system is generally only viable when the equivalent wealth of the output equals or exceeds that of the input. From these considerations, a technological system may be defined in the following way:

> A technological system is that engineered complex which converts a given quantity of a certain range of facets such as wealth, materials, resources, facilities and information with specified characteristics into a given quantity of a certain range of information, power, products, services or wealth with specified characteristics under a certain set of space and time conditions.

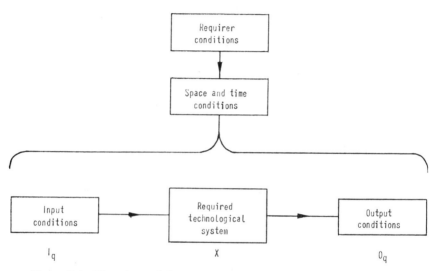

Figure 3.4 The place of the required system in the required conditions

Such a system is shown schematically at the centre of Figure 3.4. It is seen that, starting from the requirer conditions, the whole complex is influenced by the space and time conditions. Within these overall required conditions exist the specific required inputs and outputs of the technological system.

The input conditions, then, are concerned with the wealth, materials, resources, facilities and information that require to be converted into something else. It could be said that the basic inputs have already been covered when dealing with the human characteristics and material resources which were discussed under the space conditions. In the widest sense this is true, but on the more specific technological plane, it is necessary to consider the form of the particular functional inputs. So, with an electronic amplifier

THE PERFORMANCE REQUIREMENT

for instance, a somewhat artificial but practically useful distinction may be drawn between the power supplies for the amplifier and the operational input signal. The available power supplies are an instance of the more general space condition resources and the operational input signal an example of the more specific functional input requirement. This distinction may be drawn with most technological systems. With a basic requirement for a motor car, the functional input conditions are the skills of the driver since these are necessary for its direct and commanded method of working. The resources of fuel supplies, maintenance facilities, road facilities, etc., although also necessary for working, are deemed to be part of the general space environment in which the working is to take place. An automatic control system has a functional input in the form of some operating signal—that is, a command to function. The power supplies and fuels which enable it to function are not in the form of a command but are a characteristic of the space in which it functions.

The input conditions can be usually isolated, then, by considering what conditions exist which may be used as a command for the required system to operate. It is important to remember that if possible commands exist in the form of human skills then the human being or groups of beings who possess these skills are part of the required input conditions. They are not necessarily the same as the being or beings represented by the requirer or other conditions. An airline operator may have a requirement for an aircraft and such an aircraft may need a pilot to give it technological commands in order to fly it and a group of maintenance men to maintain it in an airworthy state. The airline operator is the requirer in this case, the pilot's skills the input conditions and the maintenance men part of the human characteristics of the space conditions.

Having isolated the meaning of input conditions it is now necessary to consider their correct specification. The basic requirement will have laid down the precise point at which the required technological system 'starts'. In other words, at what exact 'level' the requirement exists. This aspect has already been fully discussed but now its importance in relation to the input conditions becomes evident. The electronic amplifier in the second column of Figure 3.3 will have input conditions associated with voltage or current signals. On the other hand, the signal processor in the third column of Figure 3.3 may have input conditions associated with various process characteristics such as temperature, pressure, flow and chemical composition. A speedometer may have input conditions concerned with the rotation of a shaft or wheel but the input conditions for a car, of which the speedometer is part, could be the driver of the car. It is very important, therefore, to refer back to the basic requirement before defining the relevant input conditions. The input conditions should be the commands available at the point where the required technological system begins. There should be no 'gap' or 'overlap' between the two.

In general, a whole range of possible command facilities may require specifying. For each one it will be necessary to define the type of input and its characteristics with reference not only to steady-state values but also to any variations which may take place in the space and time domains. Time tends to be the predominant independent variable in these applications but any space factors that exist should be accounted for. If a requirement exists for an electronic voltage amplifier, then the input conditions require to be defined as a voltage with specified variations due to time, source characteristics or other relevant factors. Some examples of different types of input conditions, as they are related to various basic requirements, are shown in the fifth main line of Figure 3.3.

At the input level of a technological system requirement, it is very often possible to express the overall required input conditions, I_q, in the form of a mathematical characteristic function or of a number of such functions. A typical function could be written in the form:

$$I_q = f_i(x_j | t, z) \tag{3.2}$$

where x_j represents a range of input variables and t and z represent the general time and space conditions on which they may be all dependent.

Input variables could include not only the characteristics of the various input signals but also the characteristics of the immediate sources from which the input functions are derived. For instance, if an input voltage function were derived from an electrical battery, then the voltage under load would be dependent, amongst other things, on the internal impedance of the battery. Any practical system actuated by such an input voltage would load the battery to some extent and hence take some current and power from the battery. The required input conditions need to specify, therefore, not only the voltage of the battery but also its voltage/current relationship.

Generally this means that some requirement exists for the 'matching' of the input conditions to the required system. There will nearly always be some restrictions on the way in which the required system may be allowed to reflect back on the input conditions. In the majority of cases the input conditions should not be appreciably changed by the system. With the example of the battery input, this would mean that the required input impedance of the system should be much greater than that of the battery. If the input conditions are the temperature of a fluid, then the required system would be connected to the input conditions at this point by some sort of temperature-sensing device. This device might now distort the temperature of the fluid unless its thermal capacity was low compared with that of the

THE PERFORMANCE REQUIREMENT

fluid medium. Hence, in this case the degree of permitted temperature distortion would require specifying.

An input function of the type given in Equation (3.2) will normally be allowed only small variations as a result of the connexion of the input items to the required system. This means that the system may be allowed to absorb only a small fraction of the available input power or, alternatively, that there is a requirement for the system input impedance to be much greater than that of the input items. These sort of considerations lead not only to the specification of the available functional characteristics of the input conditions but also to certain required input characteristics for the proposed system. Such input characteristics are typically related to the required input matching, input power absorption or input impedance of the system.

If, for instance, an input function consists of a voltage, V, and a power, W, then Equation (3.2) can be written in the form:

$$I_q = f_i(V, W \mid t, z) \qquad (3.3)$$

or, if the voltage and power characteristics are separable, as two complementary functions:

and:
$$\left. \begin{array}{l} I_q = f_i(V \mid t, z) \\ \\ I_q = f_i(W \mid t, z) \end{array} \right\} \qquad (3.4)$$

In the example electrical circuit of Figure 2.6, which was used to illustrate the general principles of system reliability evaluation in Chapter 2, the main input condition was a voltage of a specified form. The voltage was dependent upon time and no other relevant input variable existed. Hence, the input function could have been expressed in the form:

$$\left. \begin{array}{l} f_i(V \mid t) = 0 \quad \text{for } 0 \leqslant t \leqslant t_1 \\ f_i(V \mid t) = 11 \text{ V for } t_1 \leqslant t \leqslant t_2, \quad \text{etc.} \end{array} \right\} \qquad (3.5)$$

where times t_1–t_2, etc. were the times of input occurrence.

An electronic voltage amplifier may have an available input signal in the form of an alternating voltage. In this case, the input function could be:

$$f_i(V \mid t) = V \sin \omega t \qquad (3.6)$$

where V is the maximum amplitude of the voltage and ω the frequency of alternation. In practice, however, due to lack of stability, accuracy or other factors, both V and ω may also be time dependent. In this case:

$$f_i(V, \omega \mid t) = V(t) \sin \omega(t) t \qquad (3.7)$$

The ramifications of Equation (3.7) should also be borne in mind when assessing input conditions. The characteristics, such as V and ω, should never be assumed to be constant unconditionally. Some time or space variations are always likely to occur in any parameter of interest. Care also needs to be taken, as with the space conditions, in examining any abnormal variations in the input conditions as well as in examining all the normal variations.

Example 3.5

If the basic requirement of Example 3.1 is 'A prime mover' and the requirer is 'The designer of the motor car', what are the sort of input conditions that may apply and how could these be functionally expressed?

The requirement and requirer in this example would first of all define the appropriate space and time conditions that apply in the same way as typically illustrated in Examples 3.3 and 3.4. Under these terms it is necessary to determine what 'commands' the prime mover in order to arrive at the input conditions. The direct commands to the prime mover would generally be:

An electrical signal or rotational torque of defined characteristics to start the prime mover (function x_1).

A supply intercept signal of defined characteristics to stop the prime mover (function x_2).

An electrical signal or mechanical movement signal of defined characteristics representing demanded speed, torque or power (function x_3).

All these three functions would generally be space and time dependent in the sense of variation with place of operation, time of operation, frequency of operation, rates-of-change of operation, etc. Hence a general functional expression would be:

$$I_q = f_i(x_1, x_2, x_3 | t, z)$$

The questions which have been raised so far in the discussion of input conditions and which require an affirmative answer are:

Are all the relevant input conditions precisely defined?

Is such a definition of input conditions consistent with the 'level' or hierarchy of the basic requirement?

Do the defined input conditions represent an input command to the required technological system as opposed to general space resources?

Is the time and space pattern of input condition variation, for each input function, adequately specified for any normal or abnormal conditions that may exist?

Are the permissible variations in the input conditions, as a result of the connexion of the required system to the input items, fully and completely specified?

THE PERFORMANCE REQUIREMENT

Output conditions

The general pertinence of the output conditions was outlined at the beginning of the previous sub-section. A technological system is required to carry out a conversion process from one thing into something else. The 'from WHAT?' has been dealt with under the input conditions, the 'into WHAT?' is concerned with the output conditions. This is the last main requirement condition of Figure 3.2 and represents the end-product of the situation in Figure 3.4.

The required output may be in the form of information, power, products, services or wealth as discussed in the definition of a technological system already given. The general nature of the output conditions, like the input conditions, may have been covered under such facets as the physical characteristics, human characteristics and material resources of the space conditions. For instance, the man who has a basic requirement for a motor car may require to use it for transporting himself from A to B. The characteristics of A and B and of the route between them are part of the space conditions but some aspects of the man himself are the output conditions. In other words, the movement of the man is the basic functional output. An output condition is distinguished by determining what directly is being acted upon or, in the widest sense of the word, what is being commanded.

An electronic voltage amplifier may be required for actuating a voltmeter. The voltmeter and its associated characteristics represent the output conditions. It is this meter which is to be commanded. The environment in which the meter is situated, the man who maintains it and the man who reads it are part of the space conditions and not part of the output conditions except in so far as they may affect the characteristics of the meter itself. The required functional output of an automatic control system may be to operate a flow control valve. This valve and its characteristics in the space and time domains now represent the output conditions. Or, again, the output conditions of a chemical process plant may well be concerned with the quantity, quality and composition of a certain chemical if it is these facets which are being 'commanded'.

There must, as with the input conditions, be a defined point at which the required system finishes and the output conditions start. It is necessary, therefore, to refer back again to the basic requirement in order to determine this point of demarcation. If the basic requirement was for an electronic voltage amplifier, it could well be that the output conditions relate to a voltmeter. But if the basic requirement was for an amplified voltage indication, then the voltmeter disappears from the required conditions and interest focuses on the man or men who are going to assimilate the indication. Similarly, the characteristics of a flow control valve are the right output conditions to consider for a control system which is basically required to operate such a valve. On the other hand, if the basic requirement is really to control flow then the valve is no longer relevant. It is now the characteristics of the medium which is flowing and, possibly, its containment which govern the output conditions. Hence, once again, the importance of the basic requirement being at the right 'level'.

Some aspects of typical output conditions and their change with requirement level are illustrated in the sixth main line of Figure 3.3. Not only, of course, is the right level important but also that there should be no discontinuity or overlap between the end of the system as defined by the basic requirement and the start of the item or aspect as defined by the required output conditions.

The output of the required system may need to be received by a whole range of devices or persons. For each one it will be necessary to define the type of output required and the characteristics of the devices themselves with reference to any variations that may occur in the space and time domains. Just as with the input conditions, it may be possible to express the required output conditions, O_q, in the form of a mathematical function or number of functions similar to the function given in Equation (3.2). This could be represented in the following symbolic way:

$$O_q = f_o(y_i | t, z) \qquad (3.8)$$

where y_i represents a range of relevant output variables in the space and time domains of z and t. The function in Equation (3.8) should be interpreted as implying a certain independence of as well as dependence upon the appropriate variables. For instance, the output voltmeter for the electronic voltage amplifier may require a certain voltage level and power level with certain time-dependent characteristics. On the other hand, it may require in addition that this form of output be maintained independent of a range of variations in the voltmeter's electrical impedance and other allied characteristics.

This means, once again, that there may be the question of 'matching' the required output of the system to the available output device. For an automatic control system which is required to operate a flow control valve, a function of the type expressed in Equation (3.8) will generally be needed when the required system is connected to the flow control valve. Any characteristics of the valve, such as its varying power requirement, which might reflect back into the required system must not alter the required output function outside certain limits of variation. If a basic requirement is for a vehicle traction motor, then the conditions of the output load may require that an output function of the motor, such as its speed, is maintained within a certain range irrespective of such varying loads as traction weights or gradients. Generally, these sort of considerations lead not only to the formation of a required functional output for the system but also for the

THE PERFORMANCE REQUIREMENT

required system to possess certain output characteristics. These output characteristics are typically related to the required output matching, output power reserve or output impedance of the system.

Where only one relevant output function exists, such as a current I, and where time is the only or principal variable of interest, then the functional output conditions may be represented simply by:

$$O_q = f_o(I|t) \tag{3.9}$$

In the example electrical circuit of Figure 2.6, the simplest required output condition was a current lying in the range 0.96–1.04 A coupled with the occurrence of a certain input condition. Hence, in this case:

$$\left. \begin{array}{l} f_o(I|t) = 0 \quad \text{for } 0 \leqslant t \leqslant t_1 \\ f_o(I|t) = 0.96\text{–}1.04 \text{ A} \quad \text{for } t_1 \leqslant t \leqslant t_2, \quad \text{etc.} \end{array} \right\} \tag{3.10}$$

where t_1–t_2, etc. were the times of input occurrence.

Obviously, in the more general case, $f_o(I|t)$ could be a much more complex or series of complex functions. The effects of space as well as time variations could be important as also could be the effects of abnormal conditions compared with normal conditions.

Example 3.6

What are the equivalent output conditions for the case of Example 3.5?

Typical equivalent output conditions, i.e. items that need to be 'commanded', would be:

A transmission shaft of known characteristics which requires to be turned with specified speed and torque characteristics under:
  A defined set of starting conditions (function y_1);
  A defined set of stopping conditions (function y_2);
  A defined set of running conditions (function y_3)
and the time and space dependencies of these three output functions could be represented in the notational form:

$$O_q = f_o(y_1, y_2, y_3 | t, z)$$

Affirmative answers, then, are required at this stage to the following questions related to the output conditions:

Are all the relevant output conditions precisely defined?
Is such a definition of output conditions consistent with the 'level' or hierarchy of the basic requirement?
Do the defined output functional conditions represent a command from the required technological system to the items that need to be commanded?
Is the time and space pattern of the output condition variation, for each

output function, adequately specified for both normal and abnormal conditions?

Are the permissible variations in the output conditions, as a result of the connexion of the required system to the output items, fully and completely specified?

Overall Requirement

The basic requirement together with all the conditions under which the requirement is to be met form the complete statement of the overall requirement as shown in Figure 3.1. If all the questions raised in the preceding discussions on the basic requirement and the requirement conditions have been adequately answered, then a basis now exists for a complete overall requirement specification. There is, in fact, no absolute need to take the matter further since all the basic constituent parts of such a requirement specification have been covered. The main aim, however, in the method of coverage has been to proceed from point to point in a logical and consistent fashion so that no relevant stone should be left unturned. This has meant, figuratively, going round the outside of the required system to determine all the external requirements, influences and conditions which may exist. Such a method is fundamental and desirable for the many reasons already given, but it leaves the final requirement in an indirect form.

A simple example may help to illustrate the point. A man of known characteristics is standing on a floor of known characteristics and, subsequently, at a known time requires to sit down in the same place with a certain required posture, degree of comfort and stability. Some 'system' is obviously required for him to do this. The overall requirement for this system has been completely specified in an indirect form by virtue of all the fundamental but external characteristics which exist. However, a more direct requirement for the system can be deduced or derived on the basis of the fundamental and external requirements. In the case of the present example, this would be a requirement for a device which would stand on the known floor and support the known man in the required way. A direct requirement of this nature may, in some instances, be more succinct and useful than a series of indirect though more fundamental requirements. It is important to remember, however, that this so called direct requirement is deduced or derived from the basic facts and hence may be subject to errors of deduction. Also, the temptation needs to be avoided of specifying for the system something which is known to be already practically and readily available. For instance, the direct requirement in the present example is not for a 'chair' although, of course, it may be eventually found that a 'chair' meets the requirement.

THE PERFORMANCE REQUIREMENT

When discussing the input and output conditions, the simple electrical system of Figure 2.6 was used as an example. One set of the functional external requirements for this example was that an input of 11 V should produce a mean output of 1.0 A. The corresponding deduced or direct requirement for the system is then that it should have a mean functional conversion characteristic of $\frac{1}{11}$ A/V. It would appear, of course, that a resistor of value 11 Ω would have such a characteristic. However, it is not a resistor but a conversion characteristic which is basically required for the specified input and output conditions. Of course, if the input and output conditions had other limitations associated with them then so would the required conversion characteristic. For instance, the 11-V input signal might have a source impedance of 10 Ω and there may be a requirement that any system should not alter the 11-V input to the system by more than, say, 1%. The output current of 1.0 A may be required to flow through a resistor of 10 Ω and be relatively unaffected by, say, a 10% change in the value of this resistor. The first thing to note is that there is still a required functional conversion characteristic for the system of $\frac{1}{11}$ A/V. However, there is now an additional required system input characteristic that it should not absorb more than about 120 mW with an input of about 11 V and an additional required output characteristic that the system should supply between 9 W and 11 W at a current of 1.0 A into the output device. This means that there now exists another functional conversion characteristic which may be expressed as a power gain requirement of approximately 83. Obviously a simple resistor would no longer fulfil this requirement.

For the example being pursued, an output current/time function is required to be related to an input voltage/time function. This could be expressed notationally as:

$$f_o(I|t) = \phi_1 f_i(V|t) \tag{3.11}$$

where ϕ_1 represents the required conversion characteristic for this particular aspect. More particularly, at the appropriate times,

$$I_o = \tfrac{1}{11} V_i \tag{3.12}$$

In addition, it is also required that the output power/time condition be related in some way to the input power/time condition. An appropriate conversion characteristic, ϕ_2, could be defined for this relationship where:

$$f_o(W|t) = \phi_2 f_i(W|t) \tag{3.13}$$

or, again, more particularly:

$$W_o = 83 W_i \tag{3.14}$$

Since, Equations (3.12) and (3.14) represent a group of simultaneous equations they can be combined in a matrix notation of the following form:

$$\begin{bmatrix} I_o \\ W_o \end{bmatrix} = \begin{bmatrix} \frac{1}{11} & 0 \\ 0 & 83 \end{bmatrix} \begin{bmatrix} V_i \\ W_i \end{bmatrix} \qquad (3.15)$$

Extending this concept to a number of input and output variables, x_j and y_i respectively, leads to the more general case of:

$$\begin{bmatrix} y_1 \\ y_2 \\ \vdots \\ y_m \end{bmatrix} = \begin{bmatrix} \phi_{11} & \phi_{12} & \cdots & \phi_{1n} \\ \phi_{21} & \phi_{22} & \cdots & \phi_{2n} \\ \cdot & \cdot & \cdots & \cdot \\ \phi_{m1} & \phi_{m2} & \cdots & \phi_{mn} \end{bmatrix} \begin{bmatrix} x_1 \\ x_2 \\ \vdots \\ x_n \end{bmatrix} \qquad (3.16)$$

where the elements ϕ_{ij} represent the appropriate conversion characteristics. These elements, like the individual input and output functions x_j and y_i may well be time and space dependent. Equation (3.16) can therefore be represented as:

$$\{f_o(y_i|t,z)\} = [\phi_{ij}|t,z]_q \{f_i(x_j|t,z)\} \qquad (3.17)$$

or:

$$O_q = [\phi_{ij}|t,z]_q I_q$$

where the matrix:

$$[\phi_{ij}|t,z]_q = \frac{O_q}{I_q} \qquad (3.18)$$

represents the complete required conversion or transfer characteristic between all the relevant output functions and all the relevant input functions. This means that the overall required performance of the system may now be capable of being expressed as a single complex matrix function of the type represented by Equation (3.18). The use of this type of matrix expression and the ways that it can be manipulated will be discussed again in the later chapters of the book. For the present, it may just be noted that where such a transfer characteristic can be formulated it represents a succinct method of bringing together all the various functional requirements for any proposed system.

So far it has been understood that the matrix function in Equation (3.18) is comprised of elements ϕ_{ij} which, although they may be time and space dependent, are in the form of real coefficients. This need not necessarily apply. For instance, it may well be that some output function is required to be related to some input function by a differential equation. A case in point would be where an output shaft position is required to be derived from a rate of change of input shaft position. The required equation would then be

THE PERFORMANCE REQUIREMENT

of the form:

$$y_1 = k_1 \frac{dx_1}{dt} \tag{3.19}$$

Differential equations may be reduced to algebraic forms by applying some appropriate transformation. A useful transformation, which will be discussed in more detail in Chapter 6, is the Laplace transformation. This transformation, under certain conditions, would reduce Equation (3.19) to the form:

$$L[y_1] = k_1 s L[x_1] \tag{3.20}$$

where $L[y]$ represents the Laplace transformation of function y and s is the complex Laplace operator.

If another functional relationship between input and output is required where:

$$y_2 = \frac{dx_2}{dt} + k_2 x_2 \tag{3.21}$$

then a Laplace transformation of this equation, under the appropriate conditions, would yield:

$$L[y_2] = (s + k_2) L[x_2] \tag{3.22}$$

Both Equations (3.20) and (3.22) are now in an algebraic form and, therefore, the corresponding matrix notation linking y_1 and y_2 to x_1 and x_2 is:

$$\begin{bmatrix} L[y_1] \\ L[y_2] \end{bmatrix} = \begin{bmatrix} k_1 s & 0 \\ 0 & (s+k_2) \end{bmatrix} \begin{bmatrix} L[x_1] \\ L[x_2] \end{bmatrix} \tag{3.23}$$

The terms in the transfer characteristic matrix such as $k_1 s$ and $(s+k_2)$ are already known in the Laplace notation as 'transfer functions'. The appropriateness of this existing designation to the present application is immediately apparent. In general, therefore, the matrix which represents the required transfer characteristic for any technological system is made up of elements, $\Phi_{ij}(s)$, which are each in the form of a transfer function. Applying the concept of Equation (3.23) to the more general relationship of Equation (3.17), it is seen that such a matrix must satisfy the following condition:

$$\{L[f_o(y_i | t, z)]\} = [\Phi_{ij}(s) | t, z]_q \{L[f_i(x_j \ y, z)]\} \tag{3.24}$$

or:

$$L[O_q] = [\Phi_{ij}(s) | t, z]_q L[I_q]$$

where the general matrix:

$$[\Phi_{ij}(s) | t, z]_q = \frac{L[O_q]}{L[I_q]} \tag{3.25}$$

contains elements $\Phi_{ij}(s)$ which are usually polynomials or ratios of polynomials in the complex Laplacian operator s. In common with general matrix notation, this complex matrix and its use will be discussed further in the later chapters.

Example 3.7

If the input and output functions of Examples 3.5 and 3.6 are independent of and not conditional upon each other, what is the form of the required overall transfer characteristic matrix?

The general form of the appropriate matrix would be as in Equation (3.16). If, however, each output function is required to bear a simple linear relationship to each input function, then it would be of the form:

$$\begin{bmatrix} \phi_{11} & 0 & 0 \\ 0 & \phi_{22} & 0 \\ 0 & 0 & \phi_{33} \end{bmatrix}$$

where each ϕ_{ij} is time and space dependent, but has instantaneous values given by:

$$\phi_{11} = \frac{y_1}{x_1}, \quad \phi_{22} = \frac{y_2}{x_2}, \quad \phi_{33} = \frac{y_3}{x_3}$$

Very often, then, the complete performance requirement for a technological system may be expressed in the form of the operational function of Equation (3.25) together with the appropriate definitions of the relevant time and space conditions as described earlier. There are, however, some specific types of variations in the function of Equation (3.25) which it is useful to consider in a little more detail at this stage. The first type of variation is concerned with the number of 'limits' that may be implicit in any overall performance requirement.

Requirement Limits

When a performance requirement is fully expressed, it could conceivably mean that the required system must exactly meet the requirement in every detail. In other words, that every element of the required matrix function of Equation (3.25) should be identically repeated in the characteristic matrix of the proposed system with no permissible variation in any way. As a simple illustration, this is equivalent to saying that if a requirement exists for a 1.5-V battery then unless the proposed battery has a value of exactly 1.5 V it does not meet the requirement. This is patently unrealistic and in practice any requirement must contain, either explicitly or implicitly, some acceptable tolerance or limit on all the required values.

THE PERFORMANCE REQUIREMENT

With a 'single-valued' requirement, the implicit assumption may well be that the required system should be equal to or better than the requirement in all respects. So the battery requirement might be interpreted as a need for a battery with a voltage of greater than or equal to 1.5 V. In some cases, of course, the requirement value might represent an upper limit rather than a lower limit. For instance, the need could be for a battery which has a voltage no greater than 1.5 V. The concept of a single upper limit or a single lower limit in a requirement has already been previously introduced in Chapter 2. Figure 2.2 and Example 2.1 were illustrations of this. However, such cases are slightly misleading since the expression of only a lower requirement limit generally implies the existence of some practical or desirable upper limit and vice versa. In Example 2.1 the lower limit was some specified level of smoothness in landing an aircraft, the implied upper limit was the practical interpretation of 'perfect' smoothness. If an upper requirement limit is given for the size of a motor car, a lower limit implicitly exists due to the practical nature of a vehicle of this type and of the people it has to carry. Generally, then, a requirement for any performance value will have both an upper limit and a lower limit. The battery requirement should therefore be expressed, say, as 1.5 ± 0.2 V or as a required lower limit of 1.3 V and a required upper limit of 1.7 V.

Simple lower and upper limits were also introduced in Chapter 2 in connexion with the example system of Figure 2.6 and were illustrated in Figure 2.7. With this example, all the elements of the required transfer characteristic matrix as represented by Equation (3.18) are zero except ϕ_{11}. The mean value of ϕ_{11} has already been seen to be:

$$\phi_{11} = \tfrac{1}{11} \tag{3.26}$$

However, reference to Figure 2.7 shows that ϕ_{11} could be expressed as having both a lower limit and an upper limit where:

$$(\phi_{11})_l = \frac{0.96}{11} \tag{3.27}$$

and:

$$(\phi_{11})_u = \frac{1.04}{11} \tag{3.28}$$

This means, for this case, that the complete system performance requirement can be described with reference to two matrices of the form:

$$[\phi_{11} | t, z]_{ql} \tag{3.29}$$

and:

$$[\phi_{11} | t, z]_{qu} \tag{3.30}$$

where the matrix function of Expression (3.29) represents a lower set of conditions and the matrix function of Expression (3.30) an upper set of conditions. The proposed system is now required to have an equivalent matrix function which lies between the function of Expression (3.29) and the function of Expression (3.30). Frequently, however, it is sufficient to define the general form of the requirement matrix together with the required upper limits and lower limits for each element in the matrix such as Equations (3.27) and (3.28).

An additional factor, though, arises in multi-element matrices. For instance, consider again the example matrix of Equation (3.15). This arose from a requirement for a mean output current to input voltage relationship as defined by Equation (3.12) and a mean output power to input power relationship as defined by Equation (3.14). The corresponding mean values of the two non-zero elements of the matrix were:

$$\phi_{11} = \tfrac{1}{11} \tag{3.31}$$

and:

$$\phi_{22} = 83 \tag{3.32}$$

If it is now taken, as implied previously, that ϕ_{11} has a lower required limit of $(\phi_{11}-4\%)$ together with an upper limit of $(\phi_{11}+4\%)$, and ϕ_{22} has a lower required limit of $(\phi_{22}-10\%)$ together with an upper limit of $(\phi_{22}+10\%)$, then:

$$(\phi_{11})_l = \frac{0.96}{11} \tag{3.33}$$

$$(\phi_{11})_u = \frac{1.04}{11} \tag{3.34}$$

$$(\phi_{22})_l = 74.7 \tag{3.35}$$

$$(\phi_{22})_u = 91.3 \tag{3.36}$$

So, coupled with the defined form of the required transfer characteristic matrix, there are four limit values which are required to be met in connexion with the two matrix elements. No difficulty arises here if the required values for the two elements are completely independent of each other. However, this may not necessarily be the case. It is possible, for instance, that the requirer may not tolerate both elements being at their lower limit together. The requirement may be conditional so that $(\phi_{11})_l$ can be as low as $0.96/11$ provided that ϕ_{22} is not less than, say, 83. Or conversely, that $(\phi_{22})_l$ can be as low as 74.7 provided that ϕ_{11} is greater than $\tfrac{1}{11}$. There now exists some dependent or conditional relationship between the limits of the element values in the required matrix.

Different types of conditional relationships of this sort are quite likely to arise in practice. The prospective buyer of a motor car may initially lay down a requirement for a minimum top speed and a minimum acceptable acceleration. In the event, however, he may be willing to accept a lower top speed in exchange for a higher acceleration or vice versa. His requirement has created a conditional relationship between the various performance parameters. An airline operator may lay down a tentative requirement for an aircraft with a certain minimum range, minimum seating capacity and minimum speed but he may additionally require that he does not want a minimum value associated with every aspect. On the other hand, he might accept a reduction in his minimum speed requirement for, say, an increased payload or an increased range. These type of requirement limits, and many others like them, are reflected back into a conditional relationship between the various elements of the transfer characteristic matrix. They represent a trade-off of one requirement limit against another.

Obviously, where any such conditional relationships exist, they need to be defined and precisely specified in conjunction with the methods of requirement limit specification already discussed. Sometimes the conditional specification will consist of a series of statements which define the interrelationship between the limits of the matrix elements. On other occasions, where a simple or continuous relationship exists, it may be possible to specify this by means of a series of equations which relate the matrix element limits to each other.

One particular trade-off which finds its way, at some stage, into all technological problems is that associated with cost. So, a set of performance requirement limits is almost certain to be coupled with some maximum cost requirement. The trade-off occurs, though, when a requirer is willing to accept some degraded performance for a lower cost or perhaps some exceptional performance for an increased cost. The next section reviews the effects of such financial requirements on the general technological system requirements.

Financial Requirements

The estimation of reliability in connexion with financial problems is not part of the main theme of this book where the principal emphasis is on technological problems. However, there tends to be some overlap between the two types of problems and the reliability analysis of any technological system always has some financial overtones.

There is, in fact, no basic difference between the methods of reliability evaluation as applied to technological systems and those as applied to financial systems. In addition, a definition of a financial system could well be made along the same lines as the definition of a technological system given earlier in this chapter at the beginning of the section on 'Input conditions'. Any distinction between financial and technological systems that does exist is really connected with the hierarchic level of the system. The five columns in Figure 3.3 are an illustration of five hierarchic system levels starting at the technological component part level and proceeding up to a technological control system in the fourth column. The final column, representing a process plant, is still a technological system but it also has other facilities and money-making processes associated with it. If a step is now taken upwards into a higher level, this could be, for example, 'a chemical industry', then perhaps 'a business or commercial enterprise' and ultimately 'a money-making system'. This pattern of hierarchic levels is common to most technological systems and, in practice, nearly all such systems lead up to a money-making system. Therefore, the requirements for a money-making system or the financial requirements for any system follow the same form as typically illustrated in the rows of Figure 3.3 and already discussed.

In particular, the input conditions of a financial system are associated with the money available to be spent or invested and the output conditions are associated with the money that is required to be earned or accrued. The required transfer characteristic will then be simply expressed in terms, for instance, of so many pounds per pound although, of course, this characteristic may be complex and also time and space dependent.

There can be a large number of intermediate levels between what may be defined as a 'pure' technological system, on the one hand, and a 'pure' financial system, on the other hand. A typical intermediate level is where the requirer has input conditions in the form of finance but requires output conditions in the form of goods, services, convenience, entertainment, pleasure, etc. The prospective private owners of domestic appliances, television sets and motor cars or the direct users of public services and transportation can be cases of this sort. The input is money and the output is some kind of value. Hence, the required system transfer characteristic is in the form of a required level of value for money.

When any question of costs arise in a system requirement these costs are almost certainly related to some financial system requirement which lies in a higher level in the hierarchy than a purely technological requirement. A particular technological performance aspect, on the other hand, is almost certainly connected with a lower hierarchic level which is governed by the basic technological requirement of which it is a part. Hence, if there is to be some trade-off between cost, and say, technical performance, this trade-off takes place between one hierarchic level and another. Figure 3.5 helps to illustrate how these technical and financial levels may be typically related. The financial system in Figure 3.5 follows exactly the general pattern of the overall reliability concept as expressed in Figure 2.1. The 'financial system achievement', however, is itself dependent upon the technological system and its reliability. At this lower level, the technological system reliability again follows the pattern of Figure 2.1. This transition from one level to another

THE PERFORMANCE REQUIREMENT

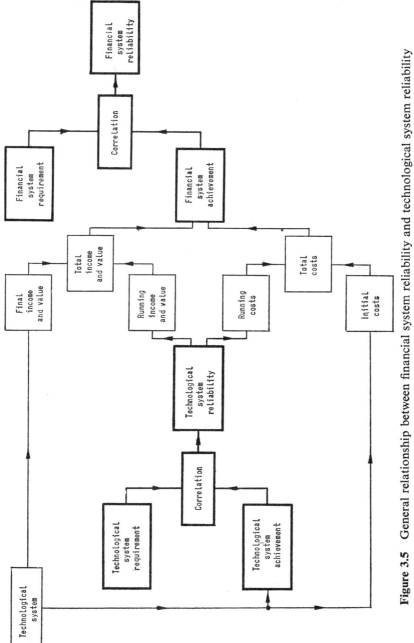

Figure 3.5 General relationship between financial system reliability and technological system reliability

can, of course, be carried on either downwards to the smallest conceivable item or upwards to the largest conceivable system. The main point being that the eventually assessed reliability at one level affects the achievable reliability at higher levels, or the required reliability at one level influences the required reliability at lower levels.

It is apparent from Figure 3.5 that a cheap technological system with a low technological reliability could lead to the same financial system reliability as a costly technological system with a very high technological reliability. This is the sort of consideration or possible trade-off which may have to be taken into account when dealing with reliability assessments of technological systems. However, it is important when financial requirements enter into the argument to treat them with the same rigour and logical approach as has been applied earlier to technological requirements. Misconceptions about basic requirements, hierarchic levels and condition specifications could lead to erroneous reliability assessments and misleading conclusions which might bring another major cost factor into Figure 3.5.

Requirement Variations

The procedure so far has been to lay down a basic requirement which, provided it is the correct requirement for the problem in hand, should be invariable. This basic requirement then needs to be met under a certain range of terms or conditions as previously illustrated in Figure 3.1. These conditions are quite likely to be subject to variations. The principal independent variable is very often that of time although some variations which take place in the space domain may also be considered as independent. The method, therefore, following the schematic of Figure 3.2 has been to define and specify, first of all, the relevant requirements for the time and space conditions that apply. Following these conditions and dependent upon them are the appropriate input and output conditions. Both these latter conditions define a range of functional parameters, such as x_j and y_i respectively, each of which may be related to the independent time variable t and the space variable z. The complete functional expression for the input conditions and output conditions can therefore be described in the notational form of Equations (3.2) and (3.8) respectively or their equivalent Laplace transformations. Finally, the required input and output functions can be related by a required transfer characteristic matrix of the form given in Equation (3.18) and as shown in Equation (3.17). For further completeness this matrix may take on the complex operational form of the function in Equation (3.25). Each element of the complex matrix, $\Phi_{ij}(s)$, is likely to be dependent upon t and z. Also, where requirement limits are specified, there may be a conditional relationship between the individual elements themselves.

THE PERFORMANCE REQUIREMENT 119

The concept of variation, therefore, has already been built into the requirement model so far derived. In general, variations may be either systematic or random. Systematic variations may be defined as those which lead to precise and predictable values of the variable at any point in the time and space domains. Random variations, on the other hand, are those which have no precise and predictable value in time or space or both but the overall pattern or distribution of random variations may be known and predictable. The mathematical concepts associated with systematic and random variations are enlarged upon in later chapters. The reader has already been introduced to the idea of distributions of random variables in the brief discussion on probability density functions in Chapter 2.

If a battery voltage represents an input condition to a technological system, then any steady drop in the voltage with time as the battery discharges could be considered as a systematic variation. The mean characteristic of such an effect is likely to be known and predictable. If the input voltage were to come from a source of alternating current, then the instantaneous voltage would be systematic because its time behaviour would be fairly precisely known. Equation (3.6) represents a systematic variation of this type. Equation (3.7) could also define a systematic variation with time if the precise behaviour of the amplitude and frequency variations were known. Similar systematic effects can take place in the space domain. If one battery was situated in the middle of the Sahara Desert and another, nominally identical battery, was situated at the South Pole at the same time, their output voltages would almost certainly be different due to the different environmental conditions. But this difference or variation in the space domain could well be predictable and therefore systematic.

Although the general or mean trend of all the preceding effects may be fairly exactly known, it is unlikely that the precise instantaneous value is known. All things tend to be subject to unpredictable differences, errors or random variations. So the drop of battery voltage with time may fluctuate about some mean value in a way which cannot be precisely predicted. The alternating voltage may vary randomly due to slight changing characteristics in the source machine, cables, interference or load conditions. The two batteries at different points in the space domain might be nominally identical but subject to small individual and random differences as a result of such things as variations in materials and production techniques.

Any systematic variation in a requirement parameter means that the actual value of the parameter or its required limit may be represented as a single-valued quantity at any point in time and space. So, the requirement lines discussed in Chapter 2 and shown in Figure 2.7 are instances of a requirement with a systematic variation. These requirement lines might, of course, have had a different value at some different point in space and time but, provided such a change was predictable, they would still have been systematic.

Any random variation in a requirement parameter means that its actual value or required limit is unpredictable but it may be represented by a probability density function or a distribution of values which describes the 'chance' of it being between some particular range of values. This concept was also introduced in Chapter 2 and Figure 2.17 shows two 'rectangular' distributions for the random variations in the lower and upper limit of the required parameter value. In fact, Figure 2.17 combines the systematic or mean value of the required parameter together with the distribution of random variations about this mean.

The time and space variations, therefore, related to any element in the transfer characteristic matrix need to be in a two-part form. One part describing, by use of appropriate equations, the systematic variation of the element with time and space. The other part describing, by use of probability density functions, the random variations of the element in the space and time domains. For a transfer characteristic matrix with a single non-zero element which has no systematic variations and which has a mean lower and upper required limit with random variations about these means related to, say, the time domain only, the requirement curves can be represented in a one-dimensional form of the type shown in Figure 10.3. Any increase in the dimensions of time or space, or in the number of combinations of systematic or random variations, or in the numbers of relevant matrix elements will lead to a situation which it is almost impossible to display pictorially. In fact, the overall required matrix of n elements can be looked upon as two n-dimensional surfaces, representing the lower and upper required limits, where each surface is changing systematically and randomly in space and time. Such a 'picture' or its mathematical representation is the Q of Equation (3.1) which this chapter set out to discuss and determine in a general way. More detailed aspects of the determination of Q will be dealt with in later chapters, but the next step is to review the meaning of H in a similar general way.

Questions
1. Explain, with illustrations, why any device of your choice whose characteristics are completely known cannot be assessed for reliability unless its complete purpose is also known. What steps or methods are open to the reliability assessor if an exact knowledge of this purpose remains unobtainable?
2. Given a statement of a basic requirement in the form 'A hydraulically operated shut-off valve', describe the reasoning you would adopt and the

THE PERFORMANCE REQUIREMENT 121

questions you would ask in order to determine whether this statement was the true and correct basic requirement.

3. Taking the concept of a complete performance requirement being made up of a basic requirement and a set of requirement conditions, explain by means of illustrations how you would determine whether particular facets of the requirement were basic or part of the conditions under which some basic requirement should be met.
4. If a basic requirement is for a means of transport, show by assessing at least three different requirers how the complete performance requirement can be affected by the nature of the requirer.
5. Discuss and illustrate the type of space conditions that may require defining in connexion with a basic requirement for (a) a process plant, (b) a ship, (c) an electricity generating station.
6. Discuss the relevance of time and its implication in connexion with any particular basic requirement that has to be met. Enumerate with examples the particular facets of time that would need to be specified and defined.
7. Describe, in relation to a control system for an aircraft, how you would isolate and specify the correct functional input and output conditions for such a control system.
8. Given a basic requirement for a process plant with defined functional input conditions of two chemical compounds and a required functional output of a third chemical compound, discuss what characteristics of the input and output conditions would require investigating. Show how the required transfer characteristic of the plant could be represented in matrix form and detail the meaning that would be attached to the elements of such a matrix.
9. Explain why an overall functional requirement may need to be expressed in the form of two sets of limits. Taking any required system with two required input parameters and two required output parameters, show how the upper and lower limits of the elements in the transfer characteristic matrix may be conditionally dependent upon each other.
10. Discuss the place of finance in the general requirements for technological systems and illustrate how technical and financial requirements may be inter-related.
11. Explain, with illustrations, the meanings of systematic variations and random variations as they may affect the complete performance requirement for a technological system.
12. Take any type of control system of your choice and write down a typical basic requirement together with a logically derived and complete set of typical conditions under which such a basic requirement should be met.

CHAPTER 4

The Performance Achievement

Achievement Concepts

Introduction

As described in the previous chapter, the reliability R of any system is dependent upon both the definition of the requirement function, Q, and the achievement function, H. This dependence was symbolized in the relationship of Equation (3.1). The discussion in Chapter 3 was principally concerned with the evaluation of Q and it is now necessary to consider the corresponding determination of H in a similar way.

For any required system, the requirement, Q, may be specified in terms of the required input, I_q, the required output, O_q, and the required overall conditions as shown in Figure 3.4. However, with reference to an actual system, X, which is purported to meet the requirement, the achieved output of the system will be some function O_h for an input of I_q. The reliability consideration is then centred around the assessment of how nearly the achieved output, O_h, matches the required output, O_q. Another way of viewing the problem is to take the required transfer characteristic, which was defined in Chapter 3 as the output to input ratio, O_q/I_q, and to compare it with the achieved transfer characteristic, O_h/I_q. Generally, however, the overall achievement function, H, may be described by a relationship such as:

$$H = f(Q \text{ conditions}, I_q, O_h) \qquad (4.1)$$

The interface problem

In practice, the requirement, Q, may not define I_q in terms of the actual input to the system, X. Furthermore, the output requirement may not be expressed in the same terms as the actual system output. This is a similar problem to that which arose in Chapter 3 in connexion with assuring that no 'gap' or 'overlap' existed in the method of describing the input and output conditions of Q. Hence, arising from the manner of defining Q, interfaces may exist at the input and output of the actual system, X, as shown in Figure 4.1. This involves matching the system at these interfaces to the requirements as set by Q.

THE PERFORMANCE ACHIEVEMENT

As an illustration of the problems that may arise at the input interface the case may be considered of a certain chemical process plant which is prone, under certain circumstances, to an explosion hazard. Because of this characteristic the plant may require an automatic protective system which has a basic requirement to shut down the plant if explosive conditions arise. The basic input function for such a protective system, as set by the requirement, is obviously the measurement of the onset of explosive conditions. In practice, it may be very difficult, if not impossible, to make such a measurement.

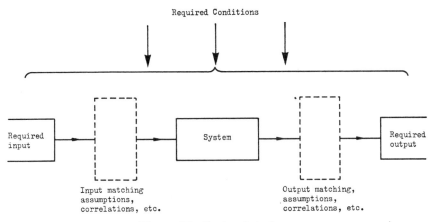

Figure 4.1 System interfaces

Consequently, the actual inputs to the protective system may be derived from quantities which are related in some way to the conditions leading up to an explosion.

One such quantity that might be derived in this way, for instance, is the measurement of temperature at some selected point in the process plant. This now involves an assumption that the measurement at one point is representative of all points throughout the relevant area of the plant. Also, of course, is the more basic assumption that a measurement of temperature itself is an adequate indication of the onset of an explosion under all conditions for an explosion that may pertain. This latter could be a critical assumption if there is some mechanism for leading to an explosion which does not produce the expected temperature conditions which are being sensed by the automatic protective system. It is very necessary, at the input interface, to consider the confidence that can be placed in the validity of these types of assumptions.

In general, the more the actual input functions become remote in character, or degraded, from the required input functions the more the assumptions increase. Such degradation is often forced upon the system designer because the techniques of measuring the required quantity are not available or, perhaps, because the available techniques would not give adequate performance. Other factors may lead to degradation, for example there may be a greater practical reliability in a particular indirect measurement than in a direct one. Often these factors lead to actual system inputs based upon anticipatory methods of sensing the cause of a plant condition rather than on direct methods of measuring the effect.

Whatever the reasons for degradation the reliability analyst is interested in the interface relationships between the actual input quantities and the required input quantities under all relevant conditions. He is also vitally concerned with the inherent assumptions which are made in the process of degradation as these may well be a limiting factor in the final reliability achieved by the overall technological system.

Similar problems may arise at the output interface where, again, the actual system output function is not quite of the same form or type as that laid down in the original requirement. One aspect of this is where the system itself generates outputs and conditions which are over and above those specified in the requirement function. Such generated outputs may not only be functional in the direct sense but may include components which add to or modify the characteristics of the environment as originally defined.

As an example, the required functional output for a system may be that of a mechanical torque. However, the actual system may also generate vibration which is, in effect, an unwanted or spurious output. Such spurious outputs may be defined as those which were not described or specified in the requirements as originally set. No acceptable level of vibration, for instance, may have been defined in the requirement and this could be taken to imply that any form or level of actual vibration is acceptable. In this case, no further evaluation of vibration or of the output interface relationships in this respect would be necessary. On the other hand it may be necessary to re-examine the original requirement in order to take any spurious output, such as vibration, into account. This is the question of ensuring that the description of the functional behaviour of the actual system conforms in form and type to the description of required behaviour and will be discussed further in this chapter and in Chapter 5.

In general, the definition of output interface relationships and any assumptions that such relationships involve is as important as the understanding of the input interface problems already discussed. More particularly, spurious outputs may be considered as arising from some transformation of the input and other general conditions as given by Q. In this respect, spurious outputs may be included in the output function, O_h, and related to their corresponding inputs in the overall achievement function of Equation (4.1).

This latter point leads to a general method of dealing with both input and output interface relationships. The method is to expand the definition of the actual system, X, so that it embraces all the interfaces. The characteristics of the interfaces then become, for analytical purposes, part of the characteristics of the system. At the same time the new system 'boundaries' set up by this method should conform exactly with the boundaries set by the requirement. This enables the correlation process to proceed with the necessary condition fulfilled that no 'gap' or 'overlap' should exist between the requirement and the achievement.

The next problem, therefore, is the definition and specification of the system, X. It now being understood that such a definition may include any relevant interfaces that may exist or may have to be taken into account.

System definition

Obviously, before the complete performance achievement, H, of a system and thus its reliability can be determined the system itself must be adequately defined and specified. If the system is already in existence, then a definition of its configuration may be obtained from a practical examination and a definition of its performance from the results of past operational experience. For any other system in the stages from an initial proposal to the completion of construction, the definition would normally be obtained from engineering drawings, performance specifications, reports and test results. Even for an existing system this latter form of 'documentary' definition may be the most expedient process to adopt provided that it is backed up and confirmed by the practical evidence.

Two aspects of system definition are important. One is an accurate description of the system hardware and its configuration. The other is a detailed knowledge of how the system performs under the various conditions as laid down by the Q requirements. As mentioned in the previous section, these aspects relate not only to the elements within the system's physical boundaries but also to any interfaces that may exist.

The substantiation of a system's performance characteristics may be carried out by either direct or indirect methods of proof. The direct method involves practical measurements of the performance parameters and can obviously only be used when a system is actually in existence or has been in existence over an adequate period of time. For instance, if it is required to know the power output of an existing engine then the engine can be run and the power output actually measured. If, in addition, the variation of this output with time is needed then the measurements could be continued over the appropriate time scale.

The indirect method of proof involves some deductive or theoretical method of performance estimation and generally comes into play where direct methods of proof are not possible. Cases for the indirect approach obviously arise when a system is not yet in being or is still in the tentative design stage. Less obviously, there may be facets of performance, which even for an existing system, cannot be directly measured. Examples of this could be the performance of devices designed to sense hazardous or emergency conditions such as fire, explosion or other extreme system states. Putting the system into such extreme conditions in order to test the performance of such sensors may not be possible or economical. Other difficulties may arise with direct measurements and, in particular, the performance variation of a system or its elements over long periods of time may not be amenable to a direct approach. The time and effort involved may be too costly and a knowledge of the expected performance may be required long before the relevant time has elapsed. The substantiation of performance variations, therefore, is often based either wholly of partly on some form of indirect estimation.

Whether the relevant performance aspects of a system are derived by a direct or an indirect method of proof, the evidence on which the proofs are based needs to be carefully and logically analysed. Methods of logical analysis will be discussed later in the chapter, but there are some general points of practice associated with the examination of the evidence which are worth reviewing at this stage.

The general points related to the input and output interface problems have already been covered in the previous section of this chapter and also, under the appropriate headings, in Chapter 3. The important factor to remember is that any relevant interface relationship should be part of either the required performance or the achieved performance. In other words, that there should be perfect matching between the boundaries of the required conditions and those of the defined system.

The starting point of investigation for the system itself is to obtain a complete performance specification covering all aspects of system behaviour. Typical items and performance values for each system parameter which should be included in any such specification are shown in Table 4.1. The idea of such a specification is to define for the reliability analyst the claimed or intended performance of the system in all its aspects. The next step is for the analyst to assure himself that the system will meet this intended performance in its actual installed position, under all conditions and at all relevant times. In seeking this assurance of the system's performance, studies will normally be made of various points related to:

(a) design,
(b) manufacture and production testing,
(c) type tests,
(d) previous experience,
(e) present installation,
(f) conditions of operation.

Table 4.1 General specification requirements

Description
    Make
    Type No.
    General assembly drawing No.
    Circuit or electrical diagram No.
    Test specification No.

Static characteristics under specified conditions
    Size
    Weight
    Power consumption
    Sensitivity over normal range, at set points, alarm points, etc.
    Accuracy over normal range, at set points, alarm points, etc.
    Linearity over normal range, at set points, alarm points, etc.
    Hysteresis over normal range, at set points, alarm points, etc.
    Zero error—dead band
    Range—threshold and limits—range stops
    Transfer function
    Signal/noise ratio—discrimination
    Warm-up time

Effect of changing conditions, such as:

Supplies:	magnitude
	frequency
	waveform
Environments:	temperature
	pressure
	humidity
	contamination
	radiation
	acceleration, vibration, shock
	electric and magnetic fields
Loads:	input loads, earthing, superimposed signals
	output loads, earthing, superimposed signals
Time:	life, wear-out characteristic
	soak period, initial characteristics
	period and frequency of usage

It is essential to see that the design is capable of producing a system and its equipment which will generally fulfil their intended functions. Further it is important to determine whether the design permits an adequate uniformity of performance to be achieved between nominally identical items of equipment. Various other aspects of the design may require investigation such as the provision of sufficient accessibility for maintenance, adjustments and proof testing.

The investigation of the system will in part be concerned with establishing that good manufacturing practice is being employed in the production of the various elements of the system. For example, the control of uniformity by the use of production test specifications to monitor each part of the manufacturing process and assembly and the adequacy of the inspection and quality control procedures.

Where appropriate, the information from type tests should be backed up by the results of previous experience. Care needs to be taken here, though, in ensuring that the results of past experience are relevant to the application in mind. Changes in model types, the environmental conditions and conditions of use may not make all previous performance behaviour directly applicable to the case in hand. The use and analysis of information from type tests and previous experience will be discussed more fully in Chapters 9 and 13.

This leads to the suitability of the design for the intended application and aspects related to whether the elements of the system were specifically designed for the system in the proposed installation. An element or device which is designed for one specific purpose may well suffer performance degradation when it is used in a slightly different application. It is also important that the elements of the system as designed are capable of meeting the appropriate specification limits after any transporting or handling procedures which lead up to the ultimate installation. The final installation procedures and corresponding commissioning tests should be thoroughly understood in order to ensure that the installed system and all its elements will, in fact, function as intended.

Once the system is installed, the conditions of operation may affect the continuous achievement of its required performance. In Chapter 3 the appropriate conditions of operation which apply under the general requirement have already been discussed. It remains to determine whether or not the system will maintain its required performance under these conditions. At this stage, additional environmental tests may have to be carried out in order that the case may be substantiated. When it has been proved that the system will perform adequately in its particular environment, then further assurance must be sought that this performance will be maintained throughout the intended life of the system. Here it is necessary to examine the maintenance routines, the test procedures and the feedback of actual operating experience. This operating experience should be recorded on a routine basis so as to provide a continual surveillance of the performance being achieved by the system and all its constituent elements.

From the foregoing, it is seen that the assessment of system performance can continue right through the design, commissioning and operation of the system and only really concludes when the system has reached the end of its useful life.

THE PERFORMANCE REQUIREMENT   129

Capability and variability

Although different systems may have the appearance of great differences because of their materialistic forms, they may be reduced to common types of flow diagrams. Any system, for instance, will have some set of inputs and outputs dependent upon the appropriate time and space conditions that pertain. Generally, a system will have a set of n inputs upon which some functional operations are performed to produce a set of m outputs. Such an input to output relationship may be expressed in a similar manner to that already discussed in connexion with Equation (3.16). This relationship may define outputs which are entirely dependent upon the present inputs only. However, in other cases, the outputs may be a function of both present and past inputs. This means that the system output may be dependent upon its internal state which, in turn, is determined by the past inputs. Most dynamic systems with finite states can be described by these ideas.

Whatever the relationship between outputs and inputs may be, the overall system achievement may be expressed in a similar form to that given in either Equation (3.18) or (3.25) for the overall system requirement. This results in a transfer characteristic matrix for the achievement, H, of the following form:

$$[\phi_{ij}|t,z]_h = \frac{O_h}{I_q} \quad (4.2)$$

which, in operational form, becomes:

$$[\Phi_{ij}(s)|t,z]_h = \frac{L[O_h]}{L[I_q]} \quad (4.3)$$

where the elements $\Phi_{ij}(s)$ are set up by the definition of the system configuration and achieved performance under the required conditions as discussed in the previous section. The correlation between Equation (4.3) for the system achievement and Equation (3.25) for the system requirement can then lead to the reliability function of interest. It is important, though, that both the expressions for requirement and achievement should be in the same form and in the same set of dimensions. This aspect will be dealt with further in the next chapter.

In general, however, an n-dimensional matrix representation of H can be looked upon as an n-dimensional surface which is changing systematically and randomly in time and space. This is the same concept as outlined for the representation of Q at the end of Chapter 3. Although H is a multi-dimensional variable, it is obviously possible to evaluate it at some single point in the time and space domains. This point may be chosen so that all the elements of the matrix, H, are evaluated at some nominal or singularly representative value of system performance within the appropriate dimensions of the system. Such an evaluation represents the deterministic approach originally outlined in Chapter 2 where the system is looked upon as operating without variability or failure. Comparing such a deterministic value of H with the requirement, Q, expressed in the same way leads to an assessment of the system's functional capability. The reader will remember that the meaning of this particular aspect of capability was also previously introduced in Chapter 2. In terms of the transfer characteristic of Equation (4.3), the representative values of time and space which are chosen for the deterministic approach may be designated by T and Z respectively. These values, in their turn, particularize the values of $\Phi_{ij}(s)$ according to the system being studied. Hence, the singularly representative value of H may be described by the following overall transfer characteristic:

$$[\Phi_{ij}(s)|\ T,Z]_h \qquad (4.4)$$

The variability that exists in any practical system will, of course, require the ultimate use of the general transfer characteristic in Equation (4.3) rather than the particular characteristic of Expression (4.4) in order to evaluate the appropriate reliability. Some aspects of the more general variability in system performance were discussed in Chapter 2 and they will be taken up again in Chapter 5. The intention of this present chapter is to examine only the aspects of capability in performance achievement. As has been seen, this process is determininstic rather than probabilistic in nature.

It is generally expedient to examine the functional capability of a system before carrying out any detailed variational analysis. If it is found, for instance, that the system is incapable of meeting the requirement when operating at some nominally representative values then the reliability is effectively zero and the variability analysis is unlikely to be required. Examination of capability also tends to be less complex than an examination based upon variability. However, there are various techniques for dealing with a capability analysis and these form the basis for the discussion in the remainder of this chapter.

Functional Capability

General concepts

An assessment of a system's functional capability will, because of its deterministic nature, lead to an answer which is in the affirmative or the negative. As mentioned earlier when discussing the definition of general system performance, it may be possible to obtain this answer by a direct method of proof. Tests which can be carried out on certain performance aspects of an existing system should obviously be capable of showing whether the system can or cannot perform in the manner desired. Provided such tests are relevant to the representative point in space and time this leads to a direct and practical determination of capability.

THE PERFORMANCE REQUIREMENT

As has already been seen, however, direct methods of proof may not be possible for a variety of reasons. There is a need, therefore, for a method of examining the evidence that may exist in order to determine the logical functioning of the system with a view to assessing capability. This deterministic approach to system examination can be generally described under the heading of 'Logistics'.

Logistics

The basic definition of a system's configuration and performance will obviously be dependent upon the disciplines on which the system is based. Such disciplines could require a knowledge of stressing, electronics, mechanical engineering, physics, etc., and these are already all dealt with in the world's literature. However, basically the processing of the various laws and the methods of deriving a solution within the constraints of the requirement normally follow a logical approach from which emerges some overall basic methods. The methods can be referred to as 'Logistics' which is defined as the science of the general principles of deduction. Use is made in the method of ideographic symbols developed in such a way as to show the interconnexion between the various concepts involved. Not only are logistics useful in the assessment of capability but the techniques have a more general application in the overall process of reliability assessment as will be shown in the later chapters of the book.

In considering the performance capability of any technological system, propositions can be answered as being true or false without any difficulty if they have been sufficiently simplified for this purpose. Hence a sound method of approach is to 'model' the system and arrange the facts so that only elementary propositions are offered for consideration. For example, if the requirement, Q, for a new motor car has been laid down in the form of a comprehensive specification then perhaps the task of assessing the performance achievement could appear to be formidable. However, by considering the individual elements or constraints of the requirement the task may be made easier. If one constraint in the requirement is for a speed of 140 km per hour and the car's performance gives only 110 km per hour then clearly without any further evaluation the car cannot operate in the manner desired, i.e. it has not the capability. Perhaps in the field of general technological systems it may, at first sight, appear not to be so simple to evaluate performance. This may be seen in large control systems where many active units require to operate in a particular sequence and to follow automatically selected checking routines. It is for these reasons that a need arises to break down complex systems into basic elements so that the overall performance may be synthesized.

In demonstrating the functional capability of a system it is necessary to consider various meanings and to arrange the considerations with reference to the requirement. Obviously, it is necessary to make use of any knowledge previously known on the subject. Furthermore the development of the performance achievement requires organizing so that overall unity is preserved.

In every day affairs one tends to use only intuitive thinking which is reasoning having the appearance of plausibility. In deciding upon some method for evaluating the performance this type of approach can play an important part in using one's experience coupled with a specialized training in a particular discipline. This thinking is analogous to the circumstantial evidence in a court of law which is subject to chance and to controversy and which can only be considered as temporarily established.

In strict or logical reasoning the aim is to distinguish proof from plausibility, and this, in the science of logic, will be beyond doubt. It is this deterministic attitude with which this chapter is concerned. Many techniques are available for the determination of the various characteristics of performance. In general, it has been found that techniques based on a concise language which is readily understood and which involves the minimum number of rules are often the most successful. The language of mathematics is an illustration of the use of a shorthand language which enables rigid standards to be maintained. Not only does it allow one to classify but also to quantify.

Hence the common techniques for which one searches on examining functional capability will be expected to be based on a practice which will enable the classification of things and the arrangement of ideas in orderly fashion. This arrangement of ideas will then permit the recognition of inconsistencies and uncertainties and permit justification in a logical way.

Logic

Aristotle may be considered as the founder of the science of logic which is concerned with language. It will be well known that this logic deals with sentences and their relationships. The aim being to determine truth from falsehood and the conclusions follow with certainty from the premises (words represent phonograms and in themselves are not so precise as ideograms). An example of this type of reasoning would be:
(1) All turbines are large machines.
(2) All large machines are mechanical assemblies.
(3) All mechanical assemblies are reliable.
  Hence the conclusion: All turbines are reliable.
This is known as a syllogism and could be laid out as a basic scheme commencing with premises (1), (2) and (3):
(1) All A are B.
(2) All B are C.
(3) All C are D.
  Therefore: All A are D which is the conclusion.

THE PERFORMANCE ACHIEVEMENT

It will be noted that common terms appearing in the premises disappear from the conclusion.

In this approach to language it will be found certain words keep appearing such as AND, OR, NOT, IF — THEN, etc., also words such as ALL, SOME, NONE which indicate quantity. Hence basic elements can be identified with conventional signs such as *A*, *B* or *C*, etc. and these are related one to the other by logical operations such as AND, OR, etc. Generally these operations go together with conditional expressions such as 'if not *A*, then not *B*', thereby permitting a logical flow of thought to be represented. The words which are phonograms can be given ideogrammatic representation as in the use of a flow chart which is a well-known method for gaining a ready appreciation of the structure of a situation.

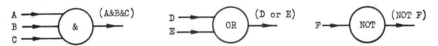

Figure 4.2   Logic gates

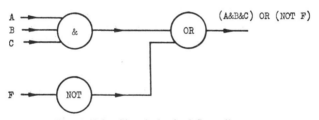

Figure 4.3   Simple logical flow diagram

Flow diagrams of the types shown in Figures 1.6 and 1.13 of Chapter 1 give an appreciation which represents a good starting point for understanding the performance capability but in such schematic diagrams the 'logic' of the situation may not be readily apparent. Hence the value of preparing logical flow diagrams. Elements *A*, *B*, *C* combined as AND can be written as (*A* and *B* and *C*) then represented by a block or gate as in Figure 4.2. There is only an output from the gate when the elements *A*, *B*, *C* are applied simultaneously. Similarly for an OR gate there is an output when *D* or *E* is applied to the gate. If it is required not to include a particular element during the process, for example element *F*, then this is denoted by NOT *F*, as shown by the NOT gate.

It is now possible to combine elements in a logical manner such that the flow is maintained towards an objective. The three gates shown in Figure 4.2 may be combined to represent some logical statement such as (*A* and *B* and *C*) OR (NOT *F*), as shown in Figure 4.3.

Complete logical flow diagrams may be built up for the particular system under examination thereby permitting conditions to be logically analysed. Time delay may also be introduced by having a time delay gate ΔT.

In Chapter 1, the use of a 2-out-of-3 redundancy arrangement was mentioned. This could be represented as shown in Figure 4.4(a). Here, if

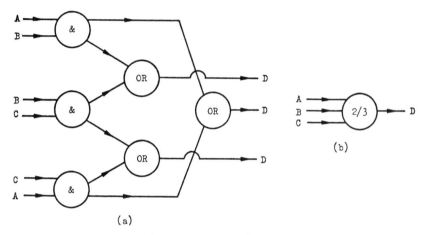

Figure 4.4 2-out-of-3 logic

any two of the three inputs A, B, C exist together or all three exist together then there is an output D. However, if only one input exists then there is no output. More simply this can be represented as shown in Figure 4.4(b) where the logic is implied by convention.

Quantity can be also assigned to the symbolism and in two-state logic it can be stated that an input or output does or does not exist. This can be represented by a 1 for existence and a 0 for non-existence as shown in Figure 4.5 and is a binary representation.

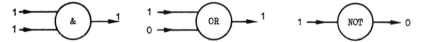

Figure 4.5 Binary representation

Other representations of binary or two states will be well known such as putting a tick or cross against an item which indicates acceptance or rejection.

From the logic flow diagram cause and effect can be studied and any event can be influential or consequential, that is an output from one logical operation can be the input of another one.

It will be readily appreciated that if in a pipe circuit along which water flows there are two stop-valves A and B in series then the closing of either

THE PERFORMANCE ACHIEVEMENT

valve A or B will lead to no flow of water. Here there is a logical OR operation being carried out by the stop-valves. Similarly, if the two valves were in parallel then to stop the flow of water both A and B would be required to be closed together, hence a logical AND operation is involved.

Clearly in the case of electrical circuits where there is a flow of current similar cases could be quoted. Figure 4.6 shows the case for some arrangements of electromagnetic relay circuits. Each relay coil is indicated by the symbol (A) and its associated electrical contact by the symbol A.

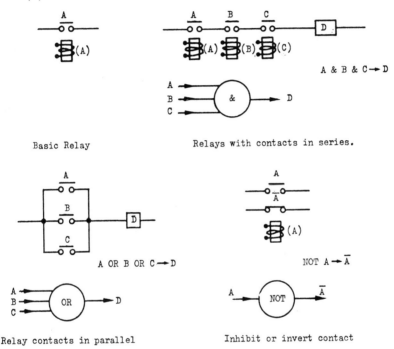

Figure 4.6 Logic operations with relay contacts

It will be seen that by arranging the relay contacts in series an AND operation takes place, in parallel gives an OR operation and by inverting the action of a contact a NOT operation is performed.

Example 4.1

A motor circuit is shown in Figure 4.7 and assuming the supply is energized what is the logical function given by pressing the push buttons? The logic of this circuit is:

if [(*PB*1 & *PB*2) & (*A* OR *B*)] then *M*

or, if required, the logical flow diagram could be drawn as in Figure 4.8.

In analysing the performance capability of a system and limiting it to two-state operation it is clear that logical analysis can be undertaken. Symbolic algebra can play a useful part and the Boolean type of algebra appears in this book in various forms. Hence an outline of this type of algebra is given at this stage.

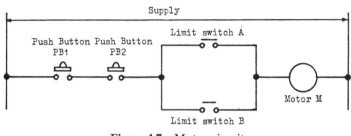

Figure 4.7 Motor circuit

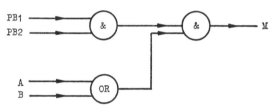

Figure 4.8 Simple logic flow diagram

Boolean Algebra

In a number of types of system the examination of capability may be carried out on a two-state basis. Often this may only approximate to the true states. Boolean algebra can be used to analyse the logic of two-state operation. This type of algebra deals with classes where a class is a 'set of elements' and each element has a value limited to 1 or 0, i.e. it is binary.

The general theorems and definitions enable a mathematical model to be used in applications where it is deemed to fit. In order to have a framework for discussion some electrical circuits are used as examples but it should be appreciated that generally these are networks which could be analogous to many other applications.

A closed circuit contact could be given the value 1 and an open circuit contact the value 0, no other value being permitted. Each contact is represented by a single letter e.g., a, b, c, etc. For contacts in parallel such as a, b, this condition is represented by the algebraic expression $a+b$ and the addition sign $(+)$ indicates a parallel connexion. For a series connexion the multiplication sign $(.)$ is used and this gives $a.b$ although the '.' is usually omitted to give ab as shown in Figure 4.9.

THE PERFORMANCE ACHIEVEMENT 137

From the functions representing these circuits the value 1, 0 can be assigned to a, b as shown in the following examples.

PARALLEL
$\begin{cases} 1+1 = 1 & \text{Closed circuit contact in parallel with closed circuit contact} \\ 0+1 = 1 & \text{Open circuit contact in parallel with closed circuit contact} \\ 0+0 = 0 & \text{Open circuit contact in parallel with open circuit contact} \end{cases}$

SERIES
$\begin{cases} 0\ 0 = 0 & \text{Open circuit contact in series with open circuit contact} \\ 0\ 1 = 0 & \text{Open circuit contact in series with closed circuit contact} \\ 1\ 1 = 1 & \text{Closed circuit contact in series with closed circuit contact} \end{cases}$

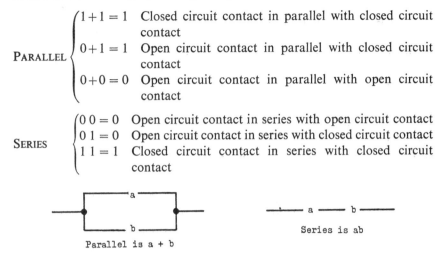

Figure 4.9 Parallel and series representation of logic

Figure 4.10 Representation of complementary contacts

It may be noted that apart from $1+1 = 1$ the other examples follow the ordinary rules of arithmetic. If there are two contacts which open or close together they are designated by the same letter or symbol. Where the two contacts are such that when one is open the other is closed and vice versa then one contact is the negative or complement of the other. In this case, if one contact is designated a then the other is designated $\bar{a}$ as shown in Figure 4.10. A is the relay coil which operates both contacts. It is now possible to consider combinations of two contacts of the same relay as follows:

$aa = a$ since this represents two contacts of the same relay in series and can logically be reduced to one contact

$a+a = a$ this is similar to the previous case since the two contacts in parallel can be logically replaced by one contact

$a+\bar{a} = 1$ since a closed circuit always exists

$a\bar{a} = 0$ since an open circuit always exists

A transfer contact may be used to replace the last two combinations as shown in Figure 4.11.

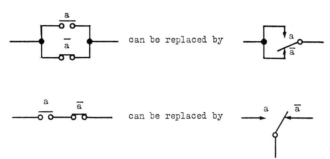

Figure 4.11 Replacement by a transfer contact

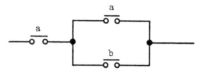

Figure 4.12 Series–parallel arrangement of contacts

Figure 4.12 now shows a series–parallel arrangement of contacts and the appropriate Boolean functions.

$$\text{Boolean function} = a(a+b)$$
$$= a^2 + ab$$
$$= a + ab \quad \text{since } a^2 = a$$
$$= a(1+b)$$
$$= a \quad \text{since } 1+b = 1$$

Hence:
$$a(a+b) = a$$

Another circuit arrangement as shown in Figure 4.13 may now be considered.

$$\text{Boolean function} = a + ab$$
$$= a(1+b)$$
$$= a \quad \text{since } 1+b = 1$$

Hence the two circuits have the same functional capability.

Negatives for a circuit can be realized as shown in Figure 4.14 or it may be stated in Boolean notation as:

$$\overline{[a+b]} = \bar{a}\bar{b} \tag{4.5}$$

$$\overline{[ab]} = \bar{a}+\bar{b} \tag{4.6}$$

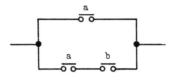

Figure 4.13 Alternative series–parallel arrangement

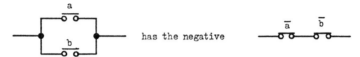

Figure 4.14 Negative or complementary arrangement

Example 4.2

Form the negative of the circuit shown in Figure 4.15.

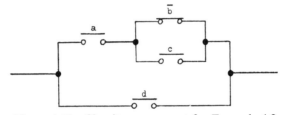

Figure 4.15 Circuit arrangement for Example 4.2

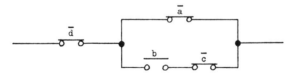

Figure 4.16 Solution of Example 4.2

The circuit realizes the function $a(\bar{b}+c)+d$ and its negative is:

$\overline{[a(\bar{b}+c)+d]}$

$= [\overline{a(\bar{b}+c)}][\bar{d}]$ from the previous relationship for negatives, Equation (4.5)

$= [\bar{a}+\overline{(\bar{b}+c)}]\bar{d}$ applying the relationship of Equation (4.6)

$= [\bar{a}+b\bar{c}]\bar{d}$

Hence the negative is as shown in Figure 4.16.

The foregoing examples have used various rules and theorems which will be observed in many cases to be similar to those of ordinary algebra. The analogy of electrical circuits has enabled a visual picture to emerge of the manipulations undertaken. However, as previously mentioned, Boolean algebra is an algebra of sets and the various laws and theorems can be summarized for easy reference in the following manner. It will be observed that a number of the properties have been already applied.

Commutative laws
$$AB = BA \tag{4.7}$$
$$A+B = B+A \tag{4.8}$$

Associative laws
$$A(BC) = (BC)A \tag{4.9}$$
$$A+(B+C) = (A+B)+C \tag{4.10}$$

Distributive laws
$$A(B+C) = AB+AC \tag{4.11}$$
$$A+BC = (A+B)(A+C) \tag{4.12}$$

Other laws are
$$AA = A \tag{4.13}$$
$$A+A = A \tag{4.14}$$
$$A(A+B) = A \tag{4.15}$$
$$A+AB = A \tag{4.16}$$
$$A\bar{A} = 0 \tag{4.17}$$
$$A+\bar{A} = 1 \tag{4.18}$$
$$\overline{(\bar{A})} = A \tag{4.19}$$

Laws of De Morgan
$$\overline{(AB)} = \bar{A}+\bar{B} \tag{4.20}$$
$$\overline{(A+B)} = \bar{A}\bar{B} \tag{4.21}$$

Operations involving 1 and 0
$$0A = 0 \tag{4.22}$$
$$1+A = 1 \tag{4.23}$$
$$1A = A \tag{4.24}$$
$$0+A = A \tag{4.25}$$
$$\bar{0} = 1 \tag{4.26}$$
$$\bar{1} = 0 \tag{4.27}$$

By appealing to the appropriate circuits an interpretation of the laws may be made e.g. the commutative law $AB = BA$ implies that two contacts are in series and it does not matter which comes first. Similarly $A+B = B+A$

implies that two contacts are in parallel and it does not matter which contact is in which parallel branch. The differences between ordinary algebra and this algebra will be seen for example by the fact that A^2 and $2A$ do not appear in Boolean algebra. It may be noted that from the principle of duality as given by the laws of De Morgan, Equations (4.20) and (4.21), that an identity will be unchanged if in the relationship the 1 and 0 and $(+)$ and $(.)$ are interchanged. Hence if $A\bar{A} = 0$ then $\bar{A} + A = 1$. Some writers use this other symbolic notation which is identical in outcome if the same interpretation is consistently maintained.

The various laws which can be viewed as analogous electrical circuits can also be illustrated by the use of a Venn diagram. This type of diagram is useful in carrying out intuitive reasoning and represents sets as areas. Appropriate areas may be shaded and all combinations of sets represented graphically.

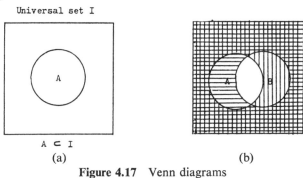

Figure 4.17 Venn diagrams

In the Venn diagram, Figure 4.17(a), the set of points inside the rectangle is taken as the universal set so that any other set which may be arbitrarily chosen is a set of points within the enclosed area shown as A. In other words the set A has been chosen from some larger collection or universal set (I). The shape of the area is irrelevant but a circle is convenient to illustrate. Hence Figure 4.17(a) illustrates that the set A does not contain all the members of the universal set I and this may be symbolically written $A \subset I$.

In Figure 4.17(b) is shown an interconnexion of sets A and B. The unshaded area represents AB. The complement of A, i.e. $\bar{A}$, is shaded vertically and the complement of B, i.e. $\bar{B}$, is shaded horizontally. Hence AB must be the complement or negative of $\bar{A} + \bar{B}$. Therefore, $\overline{(AB)} = \bar{A} + \bar{B}$ which is the law of De Morgan, Equation (4.20). Similarly the law of Equation (4.21) $\overline{(A+B)} = \bar{A}\bar{B}$ may be shown.

The foregoing laws may be applied as is usual in an algebra of this type to the various operations of expanding products and factoring. The aim is usually to obtain an appropriate form of expression or to simplify. One of

Example 4.3

Obtain by reduction a more simplified form of the function $A(\bar{A}+B) + B(B+C)$:

$$
\begin{aligned}
A(\bar{A}+B)+B(B+C) &= A\bar{A}+AB+B(B+C) & &\text{by Equation (4.11)} \\
&= 0+AB+B(B+C) & &\text{by Equation (4.17)} \\
&= AB+B(B+C) & &\text{by Equation (4.25)} \\
&= AB+B & &\text{by Equation (4.15)} \\
&= B+BA & &\text{by Equations (4.7) and (4.8)} \\
&= B & &\text{by Equation (4.16)}
\end{aligned}
$$

Example 4.4

Expand the function $AB+C$ into a polynomial so that each product term involves the three variables of the function. Multiply logically each term of the function by the variables which are absent expressed in the form $(A+\bar{A})$.

$$
\begin{aligned}
AB+C &= AB(C+\bar{C})+C(A+\bar{A})(B+\bar{B}) & &\text{by Equation (4.18)} \\
&= ABC+AB\bar{C}+ABC+A\bar{B}C+\bar{A}BC+\bar{A}\bar{B}C & &\text{by Equation (4.11)} \\
&= ABC+AB\bar{C}+A\bar{B}C+\bar{A}BC+\bar{A}\bar{B}C & &\text{by Equation (4.14)}
\end{aligned}
$$

This type of manipulation enables functions to be expressed in a basic form and can be useful for comparison term by term.

Example 4.5

Describe the properties of the circuit of Figure 4.18 which is shown in the unoperated condition.

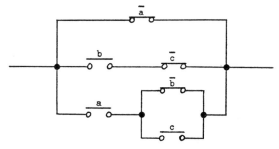

Figure 4.18 Circuit for Example 4.5

THE PERFORMANCE ACHIEVEMENT

The circuit realizes the function

$$\bar{a}+b\bar{c}+a(\bar{b}+c) = \bar{a}+b\bar{c}+a\bar{b}+ac \quad \text{by Equation (4.11)}$$

$$= 1+0+0+0 \quad \text{by inserting 1 or 0 for closed or open for the variables and by Equations (4.22) and (4.24)}$$

$$= 1 \quad \text{by Equations (4.23) and (4.25)}$$

Similarly, for the operated state it may be shown that the circuit realizes a function equal to 1. Hence the property of this circuit is a permanently closed circuit i.e. a short circuit and could be equivalent to a length of wire.

Truth tables

In any evaluation it is important to have a full understanding of the various logical considerations. This requires a determination of the truth of the various statements involved and a consideration of the different logical outcomes. Something which is acceptable or true can be defined as 1 and something which is unacceptable or false can be defined as 0. Hence in the case of an electrical circuit the acceptable or operational state may be defined as 1.

Truth tables may be constructed step by step to show the different logical outcomes or possibilities and then particular properties may be selected. Assume, for example, that two propositions A and B could independently be assigned to be 1 or 0 then the value of some function of A and B can be determined such as $A+B$ or AB and shown as in Table 4.2.

Table 4.2 Truth table for a function

Row	A	B	$A+B$	AB
1	1	1	1	1
2	1	0	1	0
3	0	1	1	0
4	0	0	0	0

Example 4.6

Derive the exact circuit configuration for controlling a single lamp from three independent switches. Multi-contact switches may be used.

Let each switch be A, B, C and assume that when a switch is in the operated state this be represented by 1 and when unoperated by 0. The function is to control the lighting of a lamp.

The problem is to find a circuit which has given properties and logically involves finding a proposition which has a given truth table. Hence a truth table is constructed showing the various states as in Table 4.3.

Table 4.3 Truth table for three switches and lamp

Row	Switch A	Switch B	Switch C	Lamp	Function for a row
1	0	0	0	Unlit	ABC
2	0	0	1	Lit	$AB\bar{C}$
3	0	1	0	Lit	$A\bar{B}C$
4	1	0	0	Lit	$\bar{A}BC$
5	0	1	1	Unlit	$A\bar{B}\bar{C}$
6	1	0	1	Unlit	$\bar{A}B\bar{C}$
7	1	1	0	Unlit	$\bar{A}\bar{B}C$
8	1	1	1	Lit	$\bar{A}\bar{B}\bar{C}$

In the truth table, row 1 is all switches unoperated and the state of the lamp is shown unlit although this is arbitrary. Having defined row 1 then row 2 shows a single change of state of one switch and this requires the corresponding state of the lamp to change i.e. to be lit. Similarly each row may be derived on the basis of considering one change of state. The corresponding functions for each row are shown. It has now to be decided that the conditions to be met are those for putting the lamp into the unlit state and these are given by rows 1, 5, 6 and 7. Hence the complete function f required is given by combining these rows and is given by:

$$f = ABC + A\bar{B}\bar{C} + \bar{A}B\bar{C} + \bar{A}\bar{B}C$$
$$= A(BC + \bar{B}\bar{C}) + \bar{A}(B\bar{C} + \bar{B}C)$$

The corresponding circuit is shown in Figure 4.19.

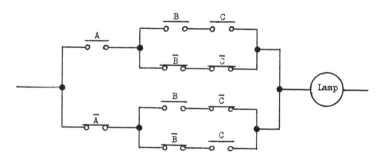

Figure 4.19 Solution of Example 4.6

Wherever there are contact pairs $(A, \bar{A})$, $(B, \bar{B})$, $(C, \bar{C})$ these may be replaced by transfer contacts if required. Such a rearrangement of the circuit of Figure 4.19 is shown in Figure 4.20.

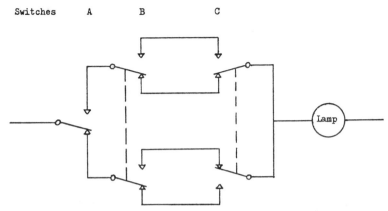

Figure 4.20 Circuit using transfer contacts for Example 4.6

Logic Diagrams

Mention has already been made of the use of flow diagrams showing the flow of information and how it is combined logically. The various logical operations can be represented in the Boolean algebra form or by using logical symbols. Some basic symbols are shown in Table 4.4. Logical flow diagrams can be prepared using such symbolic representations and this allows analysis techniques to be applied and the results shown at various levels in the system.

Figure 4.21 shows a simple example of an automatic protective system for shutting down a nuclear reactor under certain fault conditions. Shut-down is required to be completed within a certain time of either the measured temperatures or the measured neutron flux levels reaching particular specified limits.

Three thermocouples t_a, t_b and t_c are used to sense the appropriate temperatures. The thermocouples feed the temperature trip amplifiers whose output relays are triplicated and used to actuate contacts in guard lines A, B and C in the manner shown.

The flux level is measured by three ion chambers s_a, s_b and s_c in conjunction with their appropriate shut-down amplifiers. Each shut-down amplifier has an output relay whose contact is connected into the appropriate guard line. The guard lines have two sets of end relays. One set has its contacts arranged in a '2-out-of-3' configuration to actuate the D contactor and the other set, in a similar way, actuates the E contactor.

Contactor D is arranged to break the supply to all control rod release mechanisms by interrupting the main supply bus-bar. Contactor E has five contacts which interrupt the supply to each release mechanism separately. Adequate shut-down of the reactor is achieved if any four or more of the release mechanisms function correctly.

Table 4.4 Logical symbols

Rectangle
This identifies a particular element

AND

$x = ab$

Gives an output when all the inputs are simultaneously present

OR

$x = a+b$

Gives an output when at least one input is present

NOT

$x = \bar{a}$

Gives an output when the input is absent

NOR (NOT–OR)

$x = \overline{(a+b)} = \bar{a}\bar{b}$

Gives an output when all inputs are absent

NAND (NOT–AND)

$x = \overline{(ab)} = \bar{a}+\bar{b}$

Gives an output when at least one input is absent

Special Operations

2-out-of-3

Gives an output if any two or three inputs are simultaneously present. This is a particular case of an *m*-out-of-*n* operation which may be represented as shown.

Time Delay

x is given as an output of A delayed by time ΔT

The system shown in Figure 4.21 may be redrawn to show the logic as in Figure 4.22. In studying the logical functioning of the system the logic diagram is a most useful tool for analysis and forms a stepping stone in deriving the overall reliability of the system.

Alternatively, a logical diagram may be constructed based upon a truth table approach. For instance, various events at a particular level in a system

THE PERFORMANCE ACHIEVEMENT

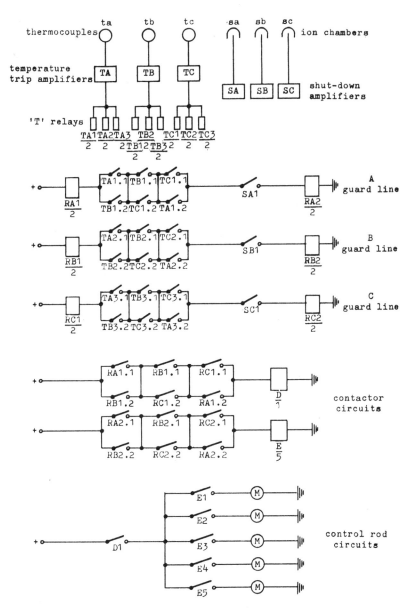

Figure 4.21 Automatic protective system

148 RELIABILITY TECHNOLOGY

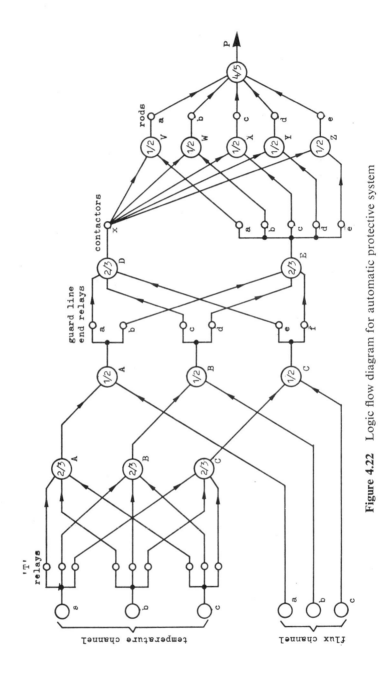

Figure 4.22 Logic flow diagram for automatic protective system

THE PERFORMANCE ACHIEVEMENT

may be shown to lead to a particular event at the system level. In arranging the events and the logical operations there is a similarlity to a tree, hence this type of representation is known as a tree diagram. The path of an event may be traced from one level to another. This allows critical paths to be identified for further study. Techniques such as Boolean algebra may be used for reduction and numerical analysis permits quantification to be given to a particular event. The tree diagram may be related to a logic flow diagram as seen in the simple illustration of Figure 4.23 the circles R to X represent particular logical operations and the rectangles A to I are events.

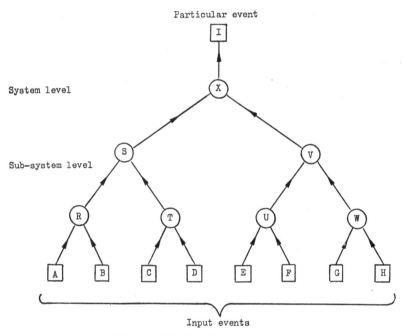

Figure 4.23 Simple tree diagram

Matrix Representation of a Network

In analysing a system, a network may be derived showing the relationship between two-state variables. Boolean matrices may be used to describe such a network and to derive a 'transmission function' from the input to the output of the network in a systematic manner.

The matrix notation can be compact and it enables the algebraic transmission of the network being studied, together with its structure, to be uniquely specified. A primitive connexion matrix of the network, which gives the identification and location of each element, may be given as shown in the following example.

A typical network is shown in Figure 4.24 which could be a configuration of electrical contacts or similar devices. Each element here is in Boolean form and is considered to be 'bilateral' so that the flow through each element

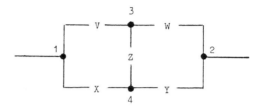

Figure 4.24 Typical network

can be in either direction. The primitive connexion matrix is written as follows:

$$
\begin{array}{c}
 & \text{To} \\
\text{From} \quad
\begin{array}{c} \text{Node} \\ 1 \\ 2 \\ 3 \\ 4 \end{array}
\begin{array}{c} \begin{array}{cccc} 1 & 2 & 3 & 4 \end{array} \\
\begin{bmatrix}
1 & 0 & V & X \\
0 & 1 & W & Y \\
V & W & 1 & Z \\
X & Y & Z & 1
\end{bmatrix}
\end{array}
\end{array}
$$

In order to specify the identity and location of each variable, i.e. element, the nodes are numbered 1–4 as shown. A 1 is entered for a connexion between a node and itself giving the principal diagonal of the primitive connexion matrix and a 0 for no direct connexion. Each element of the matrix specifies the element of the network but not the transmission between nodes. It will be noted that the matrix is symmetrical about its principal diagonal and this is generally true for all networks containing bilateral elements only. However, if the elements are not bilateral and have only one direction of flow then this direction must be specified from node to node uniquely and the matrix will not be symmetrical about the principal diagonal.

The primitive connexion matrix may now be reduced by a process of 'node removal'. To remove a node x from a matrix of n columns and m rows then each element J_{nm} requires to be replaced by the form:

$$K_{nm} = J_{nm} + (J_{nx} J_{xm})$$

where the elements K_{nm} form a new reduced matrix having one less column and row.

THE PERFORMANCE ACHIEVEMENT

Hence if node 3 is removed the new entries for column 1 of the reduced matrix are as follows:

$$K_{11} = J_{11} + (J_{13} J_{31}) = 1 + VV = 1$$

$$K_{12} = J_{12} + (J_{13} J_{32}) = 0 + VW = VW$$

$$K_{14} = J_{14} + (J_{13} J_{34}) = X + VZ$$

The other entries are modified in a similar manner and the reduced matrix is:

$$\begin{array}{c} \\ 1 \\ 2 \\ 4 \end{array} \begin{array}{c} 1 \quad\quad 2 \quad\quad 4 \\ \begin{bmatrix} 1 & VW & X+VZ \\ VW & 1 & Y+WZ \\ X+VZ & Y+WZ & 1 \end{bmatrix} \end{array}$$

Interpreting the information given by the reduced matrix, e.g. row 2, column 1, this implies that when node 3 is removed then it is equivalent to considering V and W in series as a connexion between nodes 1 and 2. This process can now be repeated to remove say node 4 and the new entries can be designated by L's as follows:

$$L_{11} = K_{11} + (K_{14} K_{41}) = 1 + (X+VZ)(X+VZ) = 1$$

$$L_{12} = K_{12} + (K_{14} K_{42}) = VW + (X+VZ)(Y+WZ)$$

$$= VW + XY + XWZ + VYZ + VWZZ$$

$$= VW + XY + VYZ + WXZ$$

The reduced matrix is now finally:

$$\begin{bmatrix} 1 & VW+XY+VYZ+WXZ \\ VW+XY+VYZ+WXZ & 1 \end{bmatrix}$$

The expression $VW+XY+VYZ+WXZ$ is the transmission between nodes 1 and 2, thus the output of this network may be written as:

$$\begin{bmatrix} 1 & T \\ T & 1 \end{bmatrix}$$

Where T represents the transmission function of the complete network.

The use of matrices in the analysis of the capability of a system can in some circumstances be advantageous and this will be seen in various sections of this book. A typical example is given in Chapter 12.

Functional Diagrams

The appreciation of the capability of a system is largely dependent upon the method of presenting the facts. Involved in the system will be some form of energy such as mechanical, electrical, hydraulic or combinations of these and the usual practice is to prepare a schematic diagram. Quite often in complex schematic diagrams it is difficult to understand the function. Not only is there no obvious starting point but there is no obvious objective. It is useful, therefore, to have some proforma method of describing the intended function of the system. Consider the very simple example of a schematic diagram for an electric motor with a circuit for starting and stopping as shown in Figure 4.25.

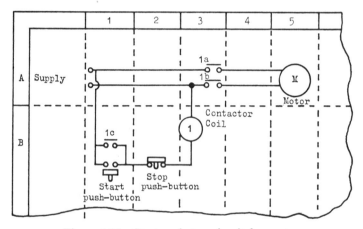

Figure 4.25 Start and stop circuit for motor

The functioning of this circuit can be easily described in words as follows. With the supply energized the start push-button is pressed which energizes the contactor coil 1. This then closes contacts 1a and 1b to energize the motor thereby starting the motor. At the same time contact 1c closes and the start push-button can be released and the motor continues to run. To stop the motor the stop push-button is pressed which opens the contactor coil circuit thereby de-energizing the coil. Contacts 1a and 1b open and de-energize the motor so that it stops. At the same time contact 1c opens and the contactor coil circuit now remains de-energized so that the stop push-button can be released. Imagine for a complex system how involved a description in words would be and hence the need for a functional diagram.

This simple system can be considered as a two-state sequential type involving a very simple memory. Hence a binary-type representation has been used in the functional diagram shown in Figure 4.26. Column I shows the item or circuit being considered and in Column II its drawing grid

THE PERFORMANCE ACHIEVEMENT

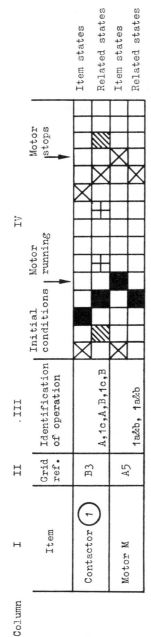

Figure 4.26 Functional diagram

Information List

A Start push-button
B Stop push-button

reference as given in Figure 4.25. Sequential item states are shown in Column IV. A closed or energized state is given the symbol ■. Each item state has related operations associated with it. A related operation may be either automatic or manual and for an automatic operation the same symbolism is used as for the item state. However, for a manual closing or energizing operation the symbol ▧ is used and for a manual opening or de-energizing operation the symbol ⊞ is used. The identification of a related operation is given in Column III which is read from left to right corresponding with the related operational symbols in Column IV. The identifications using capital letters, e.g. A, lead to further information given in the information list. The other identifications refer directly to the circuit shown in Figure 4.25. The steps and operations for starting and stopping the motor may now be read directly from the functional diagram.

Table 4.5 Symbols for function diagram

Symbol	Description
■	Closed or energised automatically
▧	Closed or energized manually
⊠	Opened or de-energized automatically
⊞	Opened or de-energized manually
◢	Closes after a time delay
⟨	Opens after a time delay
⊿⊿	Increase in voltage, current, speed, flow or volume of liquid etc.
⊾⊾	Decrease of voltage, current, speed, flow or volume of liquid etc.
◸◸	Voltage etc., building up in opposite sense to the above
⊿⊿	Change in rate of growth of a quantity
⊿⊿	Indicates related state and immediate item state. The latter being displaced to avoid confusion

Obviously this motor circuit is so simple that it can be easily explained but this does not necessarily apply to large complex systems. In addition, in generalized systems, analogue quantities which can vary may be involved and not just two-state conditions. The general symbols permitting the representation of systems are shown in Table 4.5 and may be used for mechanical, electrical, hydraulic, etc. types of function.

As a more detailed example Figure 4.27 shows a functional diagram for a sequence in the operation of a large haulage motor. The particular sequence is the movement of the main controller to the full forward speed position. A manual operation of the main controller commences the sequence of events. The various states may be seen such as the build up of current and voltage. The control scheme incorporates limits for armature current and acceleration. It is left for the reader to follow this functional diagram in more detail, but it is obvious that with reference to direction and amplitude these can be observed at a glance. The description of this sequence in words would be more difficult and a clear picture would not be easily obtained.

In the analysis of capability some ideogrammatic representations of the type described are essential when dealing with complex systems.

Generalized Systems

In two-state-type systems involving combinational logic with, for example, n inputs and m outputs, some method of deriving truth tables and obtaining the equations for each output, in terms of the input variable, can be applied. This type of system would give an output dependent entirely on the inputs present. However, in more general types of system the output may be a function of both the present and past inputs. These systems may be illustrated simply by a combination lock in which the correct sequence or code must be set up before the lock can be opened. The simple motor circuit of Figure 4.25 was another example.

A general example of a two-state-type system which is sequential and includes a memory or storage of past inputs is shown in Figure 4.28.

The output of such a system depends upon its internal states, which may have been determined by past inputs, and its present inputs. In general terms this is included in the overall class of dynamic systems with finite states. The reliability analyst must choose according to the particular system and its discipline the analytical method of examining capability. Clearly Boolean matrices can be very useful. However, the system may not have finite states and may involve general dynamic solutions and other problems. Generally in systems' engineering sets of equations are derived and often conveniently expressed in matrix form. In the analysis of complex systems it is the ability to set up computer programs and to use computerized techniques which dictates the method of analysis. Matrix manipulations can often, when properly applied, be of use in computerized techniques.

As mentioned earlier in this chapter, the matrix representation of the transfer characteristic H for the overall system achievement leads to a deterministic evaluation of the elements ϕ_{ij} in Equation (4.2). This singular determination is a measure of the deterministic capability of achievement and in the argument of this chapter is invariant. However, this is not the real-world situation due to variability.

Often the reliability analyst has to find means of interconnecting partial solutions of large systems. This requires synthesis from the sub-system level

Item	Grid Ref.	Identificati of operation		
Main controller	B7	A		
Master contactor (MC)	G9			
Forward contactor (FC)	F2	MC/1	SD/1	
Motor field interposing contactor (IC)	E11	FC/1		
Motor field weakening contactor (WC)	D6	IC/1		
Main motor field				
Speed reference circuit output		B		
Generator field current				
Generator armature voltage				
Motor armature current				
Haulage drum speed				
Brakes				
Speed detector (SD)	C4			
Information List				
A	Main controller to position "full forward"			
B	Regulator moves to full speed forward			

Figure 4.27 Functional diagram for forw

THE PERFORMANCE ACHIEVEMENT 157

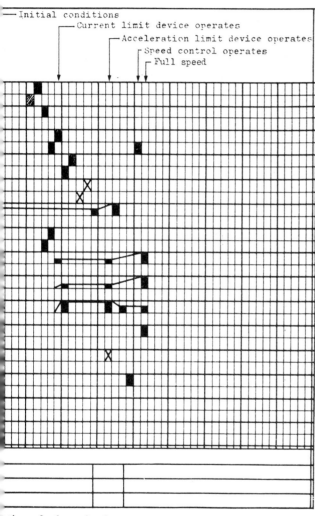

ation of a large haulage motor

to be undertaken. Ideally the methods should be chosen so that the overall system may be 'torn up' at any section and if any changes are made in one section of the system then they may be easily programmed into the process of solution.

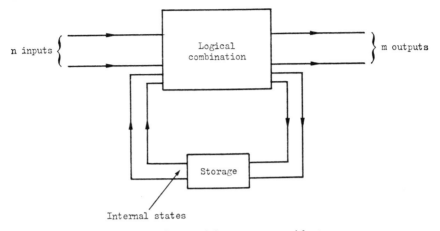

Figure 4.28 Sequential type system with storage

From a topological point of view it is a generic problem of dealing, for example, with nodes, meshes, branches and planes. For instance, electrical circuits subject to the constraints of Kirchhoff's laws and electromagnetic field theory require more structural topology than just simple graph theory. Hence it may be considered that the networks are surrounded by fields which are electromagnetic, thermal, etc., the whole system being subject to the constraints of various fundamental equations.

However, matrices and graph theory with, for example, Boolean functions as already illustrated may adequately deal with certain types of systems involving, say, transport, signal and information flow, switching and automation and logic.

Therefore, along these lines of thinking the reliability analyst is modelling in n-dimensions and is attempting to define the capability of the overall system in co-ordinates in generalized spaces. Unfortunately the vectors so derived are not fixed but are subject to variability. This involves multi-dimensional statistics and probability if it is necessary to answer the question as to whether the system will have the capability to function everytime it is called upon to do so. This is further discussed in the next chapter.

Questions

1. Given for any general system a requirement defined by Q, state what would be required to be defined for the transfer characteristic in analysing

the performance achievement. What effect could input and output interfaces have in the analysis?
2. In considering the functional capability of an equipment or system, describe the various points which may be studied.
3. What is meant by the logistics of a system? State the general logical operations which are encountered and construct a logic flow diagram showing (A or B or C) AND (D or E) OR (not F and G) THEN H.
4. Represent by means of electrical circuits the following logical functions:
   (a) $ab + \bar{a}\bar{b}$,
   (b) $(x+y+z)(\bar{x}+y+\bar{z})$.
5. Show by electrical contact representation the meaning of the following theorems:
   (a) $AA = A$,
   (b) $A + A = A$,
   (c) $A + AB = A$,
   (d) $AB + AC = A(B+C)$.
6. Simplify the Boolean expression $A(\bar{A}+C)+B(B+C)+D$.
7. Define the properties of the circuits shown in Figures 4.29(a) and 4.29(b).

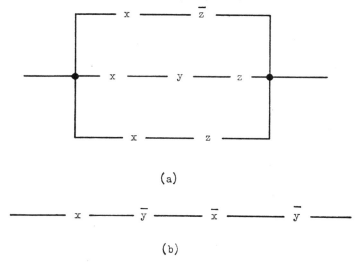

Figure 4.29 Example circuits for Question 7

8. An expansion formula is given for a Boolean function of (X, Y) as:
$$f(X, Y) = \bar{X}\bar{Y}f(0,0) + \bar{X}Yf(0,1) + X\bar{Y}f(1,0) + XYf(1,1)$$
Expand by this formula the function $XY + Y\bar{Z}[\bar{X}YZ + XY\bar{Z} + XYZ]$ also use the alternative method shown in Example 4.4.

9. Construct a truth table for the functions:
   (a) $A + \bar{B}$,
   (b) $A\bar{B}$.
10. Derive the exact circuit configuration for controlling a single lamp L from two independent switches A and B such that the control by any one switch is independent of the state of the other. Multi-contact switches may be used. Give a truth table and derive the function which describes the circuit.

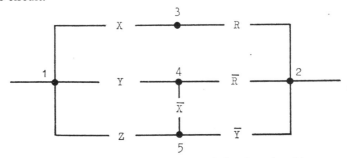

Figure 4.30 Example network for Question 11

11. Describe how a Boolean matrix may be used to represent a two-state network in terms of its nodes. Derive the transmission function for the contact network shown in Figure 4.30 by setting up a primitive connexion matrix and applying a node-removal technique commencing with node 4 then nodes 5 and 3. All the elements may be considered to be of the bilateral type.
12. Describe what is meant by:
    (a) logic flow diagram,
    (b) tree diagram,
    (c) functional diagram.
    Give illustrations of each type of diagram.
13. What are the features which distinguish a sequential system from other types? Give an example. Prepare an appropriate diagram to represent the logical sequencing of a system of your own choice.

CHAPTER 5
Variations in the Performance Achievement

General Concepts

The reader has already been introduced, in the previous chapter, to the need for a complete evaluation of the performance achievement. Reference to Equation (3.1) and Figure 2.1 will serve as a reminder that a knowledge of H, the achieved performance, is just as vital to a reliability analysis as a knowledge of Q, the required performance. Moreover, for the reliability function, R, to be capable of solution both Q and H must be determined in the same form and in the same set of dimensions.

At the beginning of Chapter 4 a useful distinction was made between a deterministic or a singularly representative value of H and the variations that may take place about this singular value. The methods of determining the singular value of H and the application of this value in describing the inherent capability of the achieved performance in meeting the required performance has been adequately covered in the aforementioned chapter. The purpose of this present chapter is to consider the variational aspects of H, the methods of describing such variations and their use in determining the overall reliability of any proposed system.

With a representative and singular value of H and with a value of Q in the same form, Equation (3.1) leads to an expression of the chance of H meeting Q at one representative location in the relevant space and time domains. This leads to a point or deterministic value of R, equivalent to the system's capability of meeting the requirement, which is either 0 or 1 as previously discussed. When variations are taken into account, the corresponding value of R can take on any value in the range 0 to 1 and this represents the overall value of reliability which is eventually required to be evaluated. This value of R is equivalent to the chance of H meeting Q at any location or range of locations in the relevant space and time domains.

One form in which the performance achievement has already been expressed in the previous chapter is:

$$[\Phi_{ij}(s)|T,Z]_h \tag{5.1}$$

where T and Z represent the appropriate singular location.

The requirement now is to determine a more general expression of performance achievement which, as given in Equation (4.3), could be of the form:

$$[\Phi_{ij}(s)|t,z]_h = \frac{L[O_h]}{L[I_q]} \qquad (5.2)$$

where t and z now represent the range of all relevant locations.

Since, ultimately, H and Q must be expressed in the same form, the function in Equation (5.2) will need to be compared directly with a requirement in the form given in Equation (3.25). Variations in connexion with this requirement expression have already been discussed, in a general way, in Chapter 3. These general aspects of variation are now to be enlarged upon with particular reference to the function H. However, the methods of describing the variations in H apply equally well to Q due to the inherent and necessary similarity between Equations (5.2) and (3.25).

Achievement Variations

The need for a complete and comprehensive definition of the system has already been emphasized in Chapter 4. Any variational aspects of H, just as the singular value of H, are fundamentally based upon this definition. The specification of any component part of the system, for instance, may either inherently define its characteristics of variation or there may be a need to carry out further work to deduce such characteristics on the basis of the specification. An inherent knowledge of variation generally arises when the component part is a well-tried and frequently used item on which a large amount of past experience has been accumulated. The known and established record of this past experience then defines the pattern and form of the variations that are likely to exist.

With relatively new or untried component parts, there may be no past experience on which to base any variational assessment. So, although the part may be defined and specified, this specification may not be absolutely complete due to a lack of operational experience. It is necessary, in cases of this sort, to determine the characteristics of variation either by carrying out tests or by resorting to some form of theoretical analysis. Any tests that are carried out are likely to be type tests or limited sample tests. They will provide, therefore, only some 'estimation' of the appropriate variational characteristics due to the limited results or limited sample size. Some aspects of this type of estimation and the confidence that can be placed in it will be discussed in later chapters, and, in particular, in Chapter 9.

Any theoretical analysis of variation is generally based on being able to synthesize the behaviour of an item from a knowledge of the behaviour of its constituent parts. Although many complete assemblies in a system may be relatively new, they are often composed of parts which have accumulated an adequate past history of operation. For instance, an electronic assembly may be the first of its type but also be made up of parts, such as resistors or capacitors, whose variational characteristics are known. A new engine may be produced, but there may be an extensive knowledge of the variations that are likely to take place in the bearings, shafts, pins and castings of which it is comprised. The general techniques of this type of theoretical analysis or prediction are also discussed in Chapters 8, 11 and 12.

Whether directly specified, deduced from tests or theoretically predicted, the complete characteristic of variation needs to be determined for the whole system. All variations are normally dependent upon two basic variables. These are the variables of time and space as already indicated in Chapter 3. Any derivation of the characteristics of variation, therefore, must recognize that change may be related to time, situation or a combination of the two. This, of course, has been illustrated in the achieved transfer characteristic matrix of the function in Equation (5.2).

Although all variations in the system are likely to be functions of time and space, it is useful to consider the principal factors that may cause these space and time variations in the system. The first thing, of course, is to relate the system and each item in the system to the required characteristics of the space and time variables themselves. This means referring back to the space and time conditions as laid down in the performance requirement of Chapter 3. For instance, if a system is required to run continuously for 10 years, then it is the system variational characteristic over this specific 10-year period which is of interest. Or again, if the system is required for operation in place A, then it is the variation of the system due to the characteristics of place A which must be considered.

A second cause of variation could be related to the required input conditions. So, for instance, changes in the source impedance or the source signal level could cause changes in some of the input elements of the system. The gain or the degree of distortion in an amplifier may be affected by changes in the input signal level. The characteristics of an engine throttle control may vary with the force available to move it. All the characteristics of the input should have been assessed as laid down in Chapter 3, and it is a question of seeing how these characteristics may reflect into the system itself and cause variations. Also of importance is any 'gap' or 'interface' that may exist between the required input and the actual input of the system. The significance of any such interface problems was discussed in connexion with Figure 4.1 in Chapter 4 as it was related to a single and representative value of achievement. But now the singular value of the relationship between required input and actual input must be considered as being subject to variation. For example, a required input might consist of the average temperature of a substance in a large vessel and the actual input be derived from the temperature sensed at one particular point by a particular thermocouple. The relationship between average temperature of the substance and the thermocouple e.m.f. could vary with changes in, say, the pressure, constitution and flow of the substance within the vessel. The characteristics of this type of 'link' between the required input and the actual input of the system can, in many cases, form the principal reason for variations in the overall system achievement.

Thirdly, and in a similar way to the effects of input conditions just considered, there may be variations due either to the output conditions or to the 'matching' of the actual output conditions with those that are required. The way in which an output may load a system is a typical case to be considered. For instance, the varying pressure conditions around the output control surfaces of an aircraft are likely to cause variations in the characteristics of the relevant control system. Variations in the characteristics of any 'gap' or 'interface' between the actual output and the true required output may also be important. A required output, for example, might be the achievement of a certain fluid flow in a pipe whereas the actual output of the system may be to produce a pressure drop along the pipe. Changes in the fluid itself or the characteristics of the pipe will obviously cause variations in the overall achieved performance. The characteristics of the required output will have been directly specified as shown in Chapter 3 and the corresponding facets of the interface will have been deduced as discussed in Chapter 4. The need, then, is to examine the effects of these aspects on the changes that they may produce in the actual system as defined.

The fourth and last cause of principal variation arises within the system itself. If all the known and foreseeable conditions surrounding the system apart from, say, time are constant there are still variations that can take place in the component parts of the system. These are ageing effects, drifts in performance values and failures of components which appear to be just inherent characteristics of the component parts themselves. A battery may slowly discharge, a bearing gradually wear out, a resistance value fluctuate, a structure vibrate in position or a lamp suddenly burn out. Also with complex devices or systems there may be a certain lack of detailed knowledge or understanding of their complete method of operation which causes apparent variations about the expected behaviour pattern.

To summarize, then, variations in the achieved performance of a system with respect to time or space may be expected to arise due to the characteristics of:

> the required space or time conditions,
> the required input conditions,
> the relationship between the required input and the actual input of the system (input interface),

the required output conditions,

the relationship between the actual output of the system and the required output (output interface),

the component parts of the system,

the knowledge of the laws which govern the ways in which the system functions.

All this implies that the system and its constituent parts need to be examined under all the relevant conditions as discussed in Chapters 3 and 4 in order to determine the type, shape and form of all the variations that may exist. The variations then require to be expressed in such a way as is amenable to a final evaluation of the system's reliability. One general method of leading to this is to describe the system's achieved performance in the matrix form of Equation (5.2) and ascribe the appropriate variational characteristics to each element, $\Phi_{ij}(s)$, of the matrix. If then the performance requirement is expressed in a similar form, then the way is open for a technique of reliability evaluation.

The Transfer Function Approach

As mentioned in Chapters 3 and 4, each element of the transfer characteristic matrix may be expressed in the operational form of a general transfer function, $\Phi_{ij}(s)$. Such a transfer function represents the complete operational relationship between the appropriate output and input functions. In many cases the transfer function can be expressed in the form of a ratio of polynomials in the operator s, for example:

$$\Phi_{ij}(s) = \frac{a_m s^m + a_{m-1} s^{m-1} + \ldots + a_0}{b_n s^n + b_{n-1} s^{n-1} + \ldots + b_0} \qquad (5.3)$$

where, for a real linear system, the coefficients a and b are real constants and m and n are real positive integers.

As a very simple example, consider the case where:

$$a_0 = b_0 = 1$$
$$b_1 = \tau$$

all other coefficients $= 0$

then:

$$\Phi_{ij}(s) = \frac{1}{1 + \tau s} \qquad (5.4)$$

As shown in the general discussion on transfer functions in Chapter 6 this represents a 'simple lag' between the input function and the output function. The effect is to give a delay in the output of time constant τ under transient input conditions and a direct equivalence between output and input under steady-state conditions. The output and input could be of any form such as pressure, temperature, power, voltage or torque values. A simple electrical analogue of Equation (5.4), relating output voltage to input voltage, is shown in Figure 5.1. This electrical analogue consists of a resistor, R, and a capacitor,

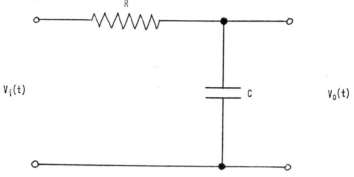

Figure 5.1 Electrical analogue of a 'simple lag'

C, connected as shown. Hence, for this circuit:

$$\Phi(s) = \frac{V_o(s)}{V_i(s)} = \frac{1}{1+\tau s} \tag{5.5}$$

where, as shown in Chapter 6,

$$\tau = RC \tag{5.6}$$

In a singular value approach, as discussed in Chapter 4, the coefficient τ would be taken as a constant. Where variations are present, however, the values of R and C will be subject to change. Hence, in the general case, τ will become a function of time and space and may be written in the form:

$$\tau(t, z) \tag{5.7}$$

This implies that any variational characteristics in the system are associated with the coefficients in the general transfer function of Equation (5.3) and not with the form or operational characteristics of the transfer function. It might be thought, however, that the actual form of the transfer function could change, say, with time. Over one period of time, for instance, the system might be represented by the analogue of Figure 5.1 and over some other period of time by the analogue of Figure 5.2. That is, the capacitance value develops some parallel inductance, L, or its equivalent. Reference to the techniques described in Chapter 6 and Example 6.9 show that the transfer function of the circuit of Figure 5.2 is given by:

$$\Phi(s) = \frac{V_o(s)}{V_i(s)} = \frac{\tau_2 s}{\tau_1 \tau_2 s^2 + \tau_2 s + 1} \tag{5.8}$$

where:
$$\tau_1 = RC \qquad (5.9)$$
$$\tau_2 = L/R \qquad (5.10)$$

or, in terms of the coefficients in Equation (5.3):

$$a_0 = 0$$
$$a_1 = \tau_2$$
$$b_0 = 1$$
$$b_1 = \tau_2$$
$$b_2 = \tau_1 \tau_2$$

all other coefficients $= 0$

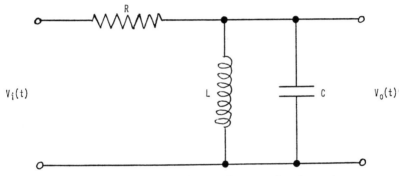

Figure 5.2 Electrical analogue of a second-order system

Part of the aspect of variation considered, therefore, could be expressed by saying that the transfer function takes on the form of Equation (5.5) with its defined coefficients for part of the time and that it takes on the form of Equation (5.8) with its defined coefficients for another part of the time. Such a complex form of statement, however, is superfluous and unnecessary. All that is necessary is, first, to define the transfer function in its most complete form such as Equation (5.8) in this case. Secondly, to define the coefficient change which reduces the transfer function to its simpler form. For the example being considered, this means allowing the value of τ_2 to approach infinity which then converts Equation (5.8) into Equation (5.5).

The general method, therefore, is to examine all aspects of variation in the proposed system irrespective of whether any aspects of variation are coincident in space or time or not. Then, from the consideration of all possible variations, to write down the transfer function which takes them all into account. The form of this transfer function is now invariate and all possible changes can be described by simply changing the values of the transfer function coefficients.

A similar factor, which may also be mentioned at this stage, arises when comparing the achieved transfer characteristic matrix with the required transfer characteristic matrix in order to evaluate the overall reliability. As stressed earlier, these matrices must be of the same form. This implies that each element and thus each corresponding transfer function must also be of the same form. Suppose, for instance, that the system of Figure 5.1 was submitted as a solution to the requirement that the output voltage must always equal the input voltage. The required transfer function would then be:

$$\frac{V_o(s)}{V_i(s)} = 1 \qquad (5.11)$$

It is obvious that the achieved transfer function, as represented by Equation (5.5), can only meet the required transfer function of Equation (5.11) under steady-state conditions. No comparison is possible under transient or dynamic conditions because the two transfer functions are of a different form. However, the requirement as expressed by Equation (5.11) could be written in the form:

$$\frac{V_o(s)}{V_i(s)} = \frac{1}{1 + \tau_3 s} \qquad (5.12)$$

where, in the problem as originally expressed, τ_3 would have the particular value of zero. The achieved coefficient value of τ_1 can now be compared with the required coefficient value of τ_3 under all conditions even if τ_3 tends to zero. Most practical requirements, of course, would have stipulated an equality between output and input with a certain permissible tolerance or accuracy. In such an instance, the required transfer function could be written in the form of Equation (5.12) with some possibly small but finite value of τ_3.

It is generally necessary, therefore, to increase the order of the corresponding transfer function for either the requirement or the achievement until it is in a form which, by simply changing coefficients, can express all variations in the achievement or the equivalent requirement. With this type and order of expression, all variational characteristics and the comparisons between different variational characteristics can be carried out by concentrating only on the pattern of coefficient changes. It is now pertinent to proceed to an examination of the different patterns of variation that are likely to occur in these coefficients.

It is convenient first of all to divide variations into the two main types which were mentioned in the discussion on requirement variations in Chapter 3. These are the systematic type of variation and the random type of variation.

Systematic Variations

Systematic variations were defined in Chapter 3 and they represent those variations which follow some known or ascertainable law. In other words they have predictable values at any point in the time and space domains. In order to evaluate a systematic variation, it is necessary to compare the variation with some reference value. This reference value may be a 'true' value, if such a value is known, an agreed standard value or a representative singular value as discussed in Chapter 4. For the purpose of this discussion, the singular value of Chapter 4 will be taken as the reference value.

The systematic variations about the reference value may be constant, discrete or continuous. A constant variation may seem, at first sight, to be a contradiction in terms. However, what is meant is a constant shift or error away from the reference value. This could occur, for instance, when the average value of a component's characteristic, taken over the measurements made on a large number of the components, deviates from the reference value continually irrespective of the number of measurements made. So, the reference value taken for a resistor might be 10 Ω whereas the average value of the resistor derived from readings on a large number of such resistors might be 9.9 Ω. This would represent a constant systematic variation of 0.1 Ω. It could, of course, be looked upon as an error in the choice of the reference value. On the other hand, it might be possible in many cases to reduce such a constant error by calibration or selection of components. The main point, however, is to recognize that such errors can exist. If they are not 'designed out' of the system, then they need to be taken into account either by adjusting the reference value used or by treating them as a constant shift in the systematic range of variations.

The combination of constant systematic errors or variations is relatively simple involving the normal algebraic functions of addition, subtraction, multiplication and division. If the reference value of an item is a and the measured value is a_0 with a constant error Δa_0, then:

$$a = a_0 + \Delta a_0 \tag{5.13}$$

Similarly, if another item has a reference value b with a measured value b_0 and constant error Δb_0, then:

$$b = b_0 + \Delta b_0 \tag{5.14}$$

The relationships defined by Equations (5.13) and (5.14) may also be expressed in the form of a fractional error, ε, where:

$$\varepsilon_a = \frac{\Delta a_0}{a_0} \tag{5.15}$$

and:
$$\varepsilon_b = \frac{\Delta b_0}{b_0} \tag{5.16}$$

Hence:
$$a = a_0(1+\varepsilon_a) \tag{5.17}$$

and:
$$b = b_0(1+\varepsilon_b) \tag{5.18}$$

If now the items represented by the values a and b are combined so that their values are additive, subtractive, etc., then the following algebraic relationships will hold:

Addition:
$$a+b = (a_0+b_0)\left[1 + \frac{a_0\varepsilon_a + b_0\varepsilon_b}{a_0+b_0}\right] \tag{5.19}$$

Subtraction:
$$a-b = (a_0-b_0)\left[1 + \frac{a_0\varepsilon_a - b_0\varepsilon_b}{a_0-b_0}\right] \tag{5.20}$$

Multiplication:
$$ab = a_0 b_0 [1 + (\varepsilon_a + \varepsilon_b) + \varepsilon_a \varepsilon_b]$$

or:
$$ab = a_0 b_0 [1 + (\varepsilon_a + \varepsilon_b)] \quad \text{if } \varepsilon_a \varepsilon_b \ll 1 \tag{5.21}$$

Division:
$$\frac{a}{b} = \frac{a_0}{b_0}[1 + (\varepsilon_a - \varepsilon_b)] \quad \text{if } \varepsilon_a \varepsilon_b \ll 1 \tag{5.22}$$

So, a combination of any constant systematic errors within a system may be found by applying the relationships of Equations (5.19) to (5.22) inclusive.

Example 5.1

In the example system of Figure 2.6 the rectifier D has a reference value of 1 Ω and a constant error of $+1\%$, the resistor R_L has a reference value of 10 Ω and a constant error of $+2\%$ and the input voltage has a reference value of 11 V with a constant error of $+2\%$. What is the reference value of the current flowing through resistor R_L and the percentage constant error on this reference value?

From Ohm's law, the reference value of the current, I, is:

$$I = \frac{V}{R_L + D} = \frac{11}{10+1}$$

$$= 1.0 \text{ A}$$

From Equation (5.19), the fractional error in total resistance, ε_{rt}, is:

$$\varepsilon_{rt} = \frac{D_0 \varepsilon_d + R_0 \varepsilon_r}{D_0 + R_0}$$

$$= \frac{1.01 \times 0.01 + 10.2 \times 0.02}{1.01 + 10.2}$$

$$= 0.019099$$

$$\simeq 1.91\%$$

and, from Equation (5.22), the fractional error in current, ε_i, is:

$$\varepsilon_i = \varepsilon_v - \varepsilon_{rt}$$

$$= 0.02 - 0.0191$$

$$= 0.0009$$

$$\simeq 0.09\%$$

The next type of systematic variation to be considered is that which is described as discrete. This, in fact, is an extension of the constant type of variation where the value of the constant error changes in discrete steps for different known locations in space and time. An example would be where an item has a certain fixed fractional error in its value over one defined period of time and has another different but also fixed fractional error over another defined time period. Or, again, the item might possess one constant error value in one place of operation and another constant error value when it is moved to an alternative place of operation. In cases of this sort, the combinations of errors are carried out in the same manner as previously indicated for constant errors but a separate calculation may need to be carried out for each different 'state'.

Example 5.2

The system of Example 5.1 is to operate for a total period of 1 year. Over this time, the errors in the resistance values are invariate but the error in the input voltage value remains at $+2\%$ for the first 6 months and then changes to -3% for the second 6 months. What is the pattern and magnitude of the fractional error in current?

The pattern of current error obviously follows that of the voltage error. Over the first 6 months the fractional current error is the same as that previously calculated in Example 5.1, namely:

$$\varepsilon_i = +0.0009 \quad \text{or} \quad +0.09\%$$

Over the second 6-monthly period, the voltage error, ε_v, is:

$$\varepsilon_v = -0.03$$

hence:

$$\varepsilon_i = -0.03 - 0.0191$$
$$= -0.0491$$
$$\simeq -5\%$$

Obviously, any number of changes in error state can be accommodated in this way provided that each change and the value resulting at the change can be precisely defined. If the discrete changes from one state to another follow some regular, periodic or cyclic form then the combination of errors may be expressed in a single functional form which represents the addition or subtraction of the individual functional forms.

This leads to the consideration of the third type of systematic variation where the change in error value follows some well-defined continuous function. An example of this would be where a characteristic value of a component suffers some small cyclic error in value in step with some cyclic changes, say, in environmental conditions. The change in diameter of an engine cylinder during the complete combustion cycle, the change in an electrical supply network frequency with the time of the day or the change in bearing load produced by the slight out-of-balance of rotating machinery would be particular examples of continuous systematic variations.

Such variations can normally be expressed in terms of some continuous time or space function or, in many cases, adequately approximated to some simple function of this type. Typical functions representing the patterns of change could be, say, linear, exponential or sinusoidal.

Example 5.3

Assume in Example 5.1 that the total resistance error, ε_{rt}, has a maximum value as previously calculated of 1.91% but that this varies sinusoidally with an angular frequency of ω_1. Assume also that the error on the voltage, ε_v, also varies sinusoidally with an angular frequency ω_2, then what is the form of the resulting current error?

Here,

$$\varepsilon_{rt} = 0.0191 \sin \omega_1 t$$

VARIATIONS IN THE PERFORMANCE ACHIEVEMENT

and,
$$\varepsilon_v = 0.02 \sin \omega_2 t$$

Hence:
$$\varepsilon_i = \varepsilon_v - \varepsilon_{rt}$$
$$= 0.02 \sin \omega_2 t - 0.0191 \sin \omega_1 t$$

or, if $\omega_1 = \omega_2 = \omega$, then:
$$\varepsilon_i = 0.0009 \sin \omega t$$

or, as another consideration, if:
$$\varepsilon_{rt} = 0.02 \sin \omega_1 t$$

then:
$$\varepsilon_i = 0.04 \left[\sin \left(\frac{\omega_2 - \omega_1}{2} \right) t \right] \left[\cos \left(\frac{\omega_2 + \omega_1}{2} \right) t \right]$$

It is apparent, therefore, that whatever the functional form of the systematic variations, they can be analysed and combined according to the same general methods. It is sometimes convenient to examine the variations in the constant, discrete or continuous classifications but the general relationships of Equations (5.13)–(5.22) apply to all systematic variations.

The overall effect of systematic variations in the assessment of achieved performance generally means applying a correction factor to the singular value of achievement as derived in Chapter 4. This correction factor may be a constant or may, of itself, be time and space dependent in a defined way. Where only systematic variations are considered, the evaluation of reliability depends upon comparing the systematically corrected value of achievement with the corresponding value or values of the requirement. These latter requirement values may also, of course, be subject to systematic variation.

Consider again the achieved transfer function of Equation (5.5) which was related to the system of Figure 5.1. Suppose that the required transfer function was of the same form with the coefficient τ_q having a lower required value of $\tau_{ql} = 95$ ms and an upper required value of $\tau_{qu} = 105$ ms. The singular value of the achieved τ_h derived for this system might well have been deduced from a nominal value of $R = 0.1$ MΩ and of $C = 1$ μF, that is:

$$\tau_h = RC$$
$$= 0.1 \times 1.0$$
$$= 0.1 \text{ s}$$
$$= 100 \text{ ms}$$

Obviously, the singular value of achievement, τ_h, lies midway between the two requirement limits and the reliability of the system is, therefore, 100% against this requirement. However, suppose it was known that R varied systematically in that its value varied alternately between 0.1 MΩ and 0.11 MΩ dwelling for 1 s at each value and that C was still invariate. Two discrete values of τ_h would now occur, namely:

$$\tau_h = 100 \text{ ms}$$

and:

$$\tau_h = 110 \text{ ms}$$

Whereas one value still gives a reliability of 100%, the other value leads to a reliability of zero. Since equal times are spent at each value, the average reliability over an extended time period now becomes 50%. So the systematic variation has reduced the reliability from 100% to 50%.

Example 5.4

Taking the general system configuration and requirements just discussed, what is the average reliability over a long period of time if the systematic variation in τ_h is known to follow the law:

$$\tau_h = 100[1 + 0.1 \sin \omega t]$$

(assume that the time period defined by ω is very much shorter than the overall time period being considered).

The form of the systematic function defined in the example implies that τ_h has a mean value of 100 ms with a superimposed sinusoidal variation of peak amplitude 10 ms. The requirement of $\tau_{qu} = 105$ ms and $\tau_{ql} = 95$ ms will only be met, however, when the sinusoidal variation of τ_h is 5 ms or less. That is, less than half-amplitude. The value of ωt for half-amplitude, from sine tables, is 30 degrees or one-third of a quarter cycle. A quarter cycle is being considered since the half-amplitude point is reached every quarter cycle. Since each quarter cycle of variation is identical from the point of view of the proportion of time that the variation remains below half-amplitude, this means that the overall average reliability is:

$$\text{Reliability} = \tfrac{1}{3} \quad \text{or} \quad 33.3\%$$

Other effects of systematic variations may be readily deduced by the reader. Attention is now turned to the other main type of variation that can exist, namely, random variations.

Random Variations

A definition of random variations was given in Chapter 3 and they represent those variations which do not of themselves follow a known or ascertainable

VARIATIONS IN THE PERFORMANCE ACHIEVEMENT

law. They have, therefore, the implication of chance. As such they may be understood in the mass and their properties derived by using the theories of probability and statistics.

Random variations may be looked upon as being 'distributed' or 'grouped' in some pattern about the systematic values or representative values which have already been discussed. The first step in dealing with random variations is to understand the forms of distribution that may typically arise. In a somewhat similar manner to systematic variations, these distributions fall basically into two types which may be described as 'continuous' and 'discrete'. A continuous distribution of random variations is one where the magnitude of the variation can take on any value between certain limits. A discrete distribution implies that the variation can only take on one of a series of discrete values. It is pertinent now to consider these two basic types of distribution in a little more detail.

Continuous distributions

The reader has already been introduced to the concepts of a continuous distribution of random variates in the discussion in Chapter 2 under the heading of 'Some further probability concepts'. As previously shown, variates may be classified in histogram form or, in the limit, they may be represented by a probability density function. A probability density function may take on any form and a rectangular type of such a function was illustrated in Figure 2.5. It is now relevant to see what other forms may typically exist or may be of use in the evaluation of achieved performance.

It may be intuitively felt that random variations, where they exist, are likely to be grouped around some 'mean' value in such a way that small variations are more likely than large variations. This was the type of circumstance which was illustrated in Figure 2.4. The limiting case of this intuitive picture is the 'bell-shaped' probability density function which formed the last diagram in Figure 2.4. The intuitive concepts associated with this sort of curve may be used to deduce an appropriate mathematical expression for such a 'bell-shaped' distribution of variates.

Suppose that n measurements $x_1, x_2, ..., x_n$ are made of the same quantity under the same conditions and the deviations of these measured values from some mean value, μ, are denoted by $z_1, z_2, ..., z_n$. Then:

$$z_1 = x_1 - \mu$$
$$z_2 = x_2 - \mu$$
$$\cdots\cdots$$
$$z_n = x_n - \mu \qquad (5.23)$$

Suppose also that the overall pattern of random variations results from a large number of small variations distributed randomly and that positive and negative variations from the mean value are equally probable. This latter supposition implies that the mean of the deviations $z_1, z_2, ..., z_n$ is zero, that is:

$$\sum_{j=1}^{n} z_j = 0 \qquad (5.24)$$

If, in the interval z to $z+dz$ the probability of a variation lying in this interval is denoted by $f(z)\,dz$, then the probability of n variations as a combined event $z_1, z_2, ..., z_n$ is:

$$P = f(z_1)f(z_2)...f(z_n)(dz)^n \qquad (5.25)$$

from the law of combination of probabilities given in Equation (1.15).

Assume that the function $f(z)$ is to be such that P meets the conditions of the maximum probability of the event for the mean value μ of the quantity measured.

Now from Equation (5.25):

$$\log P = \log f(z_1) + \log f(z_2) + ... + \log f(z_n) + n\log(dz) \qquad (5.26)$$

Differentiating Equation (5.26):

$$\frac{1}{P}\frac{dP}{d\mu} = \frac{f'(z_1)}{f(z_1)}\frac{dz_1}{d\mu} + \frac{f'(z_2)}{f(z_2)}\frac{dz_2}{d\mu} + ... + \frac{f'(z_n)}{f(z_n)}\frac{dz_n}{d\mu} \qquad (5.27)$$

Since the last term in Equation (5.26) is a constant, its derivative is zero. Equating Equation (5.27) to zero as a necessary condition for a maximum, either:

$$\frac{1}{P} = 0$$

or

$$\frac{dP}{d\mu} = 0$$

but, $1/P$ is not zero or otherwise, P would tend to infinity. Therefore, the condition for a maximum is:

$$\frac{dP}{d\mu} = 0 \qquad (5.28)$$

Now from the group of Equations (5.23):

$$\frac{dz_1}{d\mu} = \frac{dz_2}{d\mu} = ... = \frac{dz_n}{d\mu} = -1 \qquad (5.29)$$

VARIATIONS IN THE PERFORMANCE ACHIEVEMENT

So, from Equations (5.27), (5.28) and (5.29):

$$\frac{f'(z_1)}{f(z_1)} + \frac{f'(z_2)}{f(z_2)} + \ldots + \frac{f'(z_n)}{f(z_n)} = 0$$

or, more succinctly:

$$\sum_{j=1}^{n} \frac{f'(z_j)}{f(z_j)} = 0 \tag{5.30}$$

A solution is required for Equations (5.24) and (5.30) so that they may be true at the same time for each large set of n measurements of variation.
Suppose:

$$\frac{f'(z_j)}{f(z_j)} = k z_j \tag{5.31}$$

and that Equation (5.31) holds for all j's where k is a constant. Then:

$$\sum_{j=1}^{n} \frac{f'(z_j)}{f(z_j)} = k \sum_{j=1}^{n} z_j \tag{5.32}$$

Hence, from Equations (5.24) and (5.32), it follows that:

$$\sum_{j=1}^{n} \frac{f'(z_j)}{f(z_j)} = 0$$

which was the required condition for a maximum established in Equation (5.30). It follows, therefore, that Equation (5.31) represents a solution of the conditions implied by Equations (5.24) and (5.30). In particular, then:

$$\frac{f'(z)}{f(z)} = kz \tag{5.33}$$

which, on integrating, leads to:

$$\log f(z) = \tfrac{1}{2} k z^2 + \log A$$

where A is a constant of integration, therefore:

$$f(z) = A \exp(\tfrac{1}{2} k z^2) \tag{5.34}$$

The total probability of all such events as defined in Equation (5.25) can be written as:

$$\int_{-\infty}^{\infty} f(z) \, dz = 1 \tag{5.35}$$

which will be recognized as one of the basic characteristics of a density function, $f(z)$, as described in Chapter 2.

The constant A can be defined in terms of k from the combined solution of Equations (5.34) and (5.35) and it is found that the integral converges for negative k. The solution can be shown to lead to the expression:

$$f(z) = \frac{1}{\sigma\sqrt{(2\pi)}} \exp\left(-\frac{z^2}{2\sigma^2}\right) \qquad (5.36)$$

where σ is defined as the standard deviation of the distribution and is discussed further in Chapter 7.

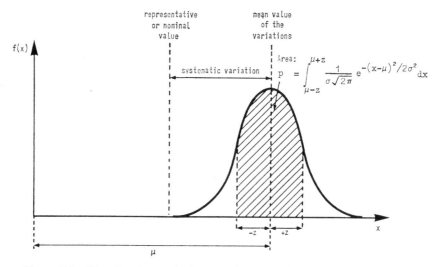

Figure 5.3 Plot showing typical types of systematic and random variations

The density function deduced in the form of Equation (5.36) is known as the 'Gaussian' or 'normal' distribution of variates. It is not a proven law in the strictest sense, but is essentially a deduction based on the ideas of zero mean deviation, maximum probability values and large numbers of small independent variations.

Equation (5.36) may be written in terms of the variate x instead of z by employing the defined relationship between these quantities given in Equation (5.23):

$$f(x) = \frac{1}{\sigma\sqrt{(2\pi)}} \exp\left[-\frac{(x-\mu)^2}{2\sigma^2}\right] \qquad (5.37)$$

Figure 5.3 shows the curve plotted from Equation (5.37) and it is seen to be bell shaped as previously discussed. In common with all density functions, the total area under the curve is unity as mentioned before in connexion with Equation (2.2). In the same way, the cumulative distribution, $p(x)$, as defined

VARIATIONS IN THE PERFORMANCE ACHIEVEMENT

in Equation (2.3), is given in this case by:

$$p(x) = \int_{-\infty}^{x} \frac{1}{\sigma\sqrt{(2\pi)}} \exp\left[-\frac{(x-\mu)^2}{2\sigma^2}\right] dx \qquad (5.38)$$

which represents the probability of the random variations lying in the range from the lower limit of $-\infty$ to some specific value x. The probabilities of the variate lying in other ranges can be expressed in a similar way by considering the area under the density function curve between the values required. For instance, the probability of a variate occurring between, say, plus and minus z is given by:

$$p(\mu-z, \mu+z) = \int_{\mu-z}^{\mu+z} \frac{1}{\sigma\sqrt{(2\pi)}} \exp\left[-\frac{(x-\mu)^2}{2\sigma^2}\right] dx \qquad (5.39)$$

which is illustrated by the shaded area in Figure 5.3.

It may be also seen from Figure 5.3 how in some cases the representative value of the quantity under consideration can be related to the mean of the random variation by the value of the appropriate systematic variation.

The probability integrals in Equations (5.38) and (5.39) cannot be evaluated in terms of elementary functions and a set of tables is usually used. For this purpose a standardized form of the integral is employed. If a standard substitution of the form:

$$t = \frac{x-\mu}{\sigma} \qquad (5.40)$$

is made in Equation (5.37), this gives:

$$f(t) = \frac{1}{\sqrt{(2\pi)}} \exp\left(-\frac{t^2}{2}\right) \qquad (5.41)$$

This distribution may be described as normal (0, 1) that is, the mean, μ, is zero and the standard deviation, σ, is unity. The corresponding form of the probability integral becomes:

$$p(-t, t) = \frac{1}{\sqrt{(2\pi)}} \int_{-t}^{t} \exp\left(-\frac{t^2}{2}\right) dt \qquad (5.42)$$

It is useful at this stage to consider another property of the normal density function, namely that it is symmetrical about its mean value. Hence:

$$\frac{1}{\sqrt{(2\pi)}} \int_{-t}^{0} \exp\left(-\frac{t^2}{2}\right) dt = \frac{1}{\sqrt{(2\pi)}} \int_{0}^{t} \exp\left(-\frac{t^2}{2}\right) dt \qquad (5.43)$$

and in particular, of course:

$$\frac{1}{\sqrt{(2\pi)}} \int_{-\infty}^{0} \exp\left(-\frac{t^2}{2}\right) dt = \frac{1}{\sqrt{(2\pi)}} \int_{0}^{\infty} \exp\left(-\frac{t^2}{2}\right) dt = \tfrac{1}{2} \qquad (5.44)$$

So that Equation (5.42) can be written in the form:

$$p(-t, t) = \frac{2}{\sqrt{(2\pi)}} \int_0^t \exp\left(-\frac{t^2}{2}\right) dt \qquad (5.45)$$

and $p(-\infty, t)$ in the form:

$$p(-\infty, t) = \tfrac{1}{2} + \frac{1}{\sqrt{(2\pi)}} \int_0^t \exp\left(-\frac{t^2}{2}\right) dt \qquad (5.46)$$

All ranges of the normal probability integral can be derived, therefore, from the particular integral:

$$p(0, t) = \frac{1}{\sqrt{(2\pi)}} \int_0^t \exp\left(-\frac{t^2}{2}\right) dt \qquad (5.47)$$

and values of this latter integral are tabulated in Table A.1.

Example 5.5

In the system of Figure 5.1, the value of the resistor, R, is 0.1 MΩ and it is invariate under the required conditions but the value of C is subject to random variations about a mean value of 1 μF. These random variations are known to be normally distributed with a standard deviation of 0.05 μF. If the values of the upper limit and the lower limit of the coefficient τ_q in the required transfer function are 105 ms and 95 ms respectively, what is the reliability of the system under the required conditions?

The mean value of the achieved τ_h, from Equation (5.6), is:

$$\mu_{\tau_h} = RC$$
$$= 0.1 \times 1.0$$
$$= 100 \text{ ms}$$

The standard deviation of the achieved τ_h is:

$$\sigma_{\tau_h} = RC_\sigma$$
$$= 0.1 \times 0.05$$
$$= 5 \text{ ms}$$

The chance of the achieved value of τ_h lying between the limits of 95 ms and 105 ms from the relationship of Equation (5.39) is:

$$p = \int_{95}^{105} \frac{1}{\sigma_{\tau_h} \sqrt{(2\pi)}} \exp\left[-\frac{(x - \mu_{\tau_h})^2}{2\sigma_{\tau_h}^2}\right] dx$$

which, on making the standard substitution:

$$t = \frac{x - 100}{5}$$

leads to:

$$p = \int_{-1}^{+1} \frac{1}{\sqrt{(2\pi)}} \exp\left(-\frac{t^2}{2}\right) dt$$

$$= \frac{2}{\sqrt{(2\pi)}} \int_{0}^{1} \exp\left(-\frac{t^2}{2}\right) dt$$

$$= 2 \times 0.3413 \quad \text{from Table A.1}$$

$$= 0.6826$$

Hence, the reliability under the required conditions is 68.26%. It may be noticed, in passing, that this figure represents the probability of lying in the range of ± 1 standard deviation of a normal distribution.

The discussions, so far, on random variations have illustrated how the reliability with reference to one particular parameter of performance may be calculated when the variations follow a normal distribution. The techniques, however, are the same irrespective of the form of the continuous distribution which represents the random variations. For instance, the manipulation of some other simple distributions such as rectangular and triangular were discussed in Chapter 2. There are, in addition, a number of other distributions such as lognormal, exponential and Weibull which are useful in describing variations in the different types of technological system. A more comprehensive review of all the typical distributional forms is given in Chapter 7.

Where a range of performance parameters are required to be evaluated or where variations in different component parts combine to form an overall variation in one particular parameter, it is necessary to understand the ways in which distributions of variations may be combined. The process of combining random variations of the continuous type is not as straightforward as the combinational processes discussed in connexion with systematic variations. For this reason, the consideration of the combination of random variates of the continuous form is deferred to Chapter 8.

For the present, attention is now directed towards the other main type of random variation where the variate of interest may take on a series of discrete, rather than continuous, values.

Discrete distributions

The concept of random and discrete changes in achieved performance was outlined at the beginning of Chapter 2. The simplest example of this concept is that of a two-state system. For instance, an achieved performance parameter may take on just one of two discrete and distinct values and the change from one value to the other follows a random process in either space or time.

Such a process may be described, on average, by allocating one probability to the performance parameter being at one value and another probability to it taking on the alternative value. Since the performance parameter cannot generally assume both values at the same time, the two probabilities are mutually exclusive. Probabilities of this type were discussed in Chapter 1, hence if p_1 is the probability of state 1 and p_2 is the probability of state 2 then:

$$p_1 + p_2 = 1 \qquad (5.48)$$

or, alternatively, in line with the concept and notation of Equation (1.10):

$$p_1 + \bar{p}_1 = 1 \qquad (5.49)$$

The two-state model, as has already been seen in Chapters 1 and 2, can be a fair representation of a technological system where the main criteria of either complete success or complete failure are the only ones of interest. The changes between two such states can, of course, be systematic and they would then be dealt with in the manner described earlier in this chapter. In practice, however, the particular transition from a state of success to a state of failure is more likely to follow a random process.

In an electronic amplifier, a resistor may suddenly go open-circuit. A pipe in a hydraulic system may suddenly burst, a support in a structure may suddenly collapse or a bearing surface may suddenly seize. Because such transitions from one state to another represent large and rapid changes in the variate of performance, they are often described as 'catastrophic failures'.

Example 5.6

Suppose that the value of the capacitor, C, in the system of Figure 5.1 is invariate and of value 1 μF but that the value of the resistor, R, can randomly take on one of two states. These two states for R are the 'normal' state where the value is 0.1 MΩ and a 'catastrophic failure' state where the value is infinite (i.e. open-circuit). The probability of the resistor being in the normal or successful state, under all the required conditions, is 0.99. Assuming that the required limits on the transfer function coefficient, τ_q, are 0 ms to 105 ms, what is the reliability of the system?

It is immediately obvious that in the successful state of the resistor:

$$\tau_h = 100 \text{ ms}$$

and in the failed state:

$$\tau_h = \infty$$

So the reliability of the situation is governed purely by the state of the resistor. Hence, the value of reliability is the same as the resistor's probability of success, that is 0.99. The corresponding chance of system failure or the unreliability of the situation is obviously, from Equation (5.49), 0.01.

VARIATIONS IN THE PERFORMANCE ACHIEVEMENT   183

Example 5.7

In addition to the set of conditions stipulated in Example 5.6, the value of the capacitor is now assumed to have a probability of 0.98 of maintaining its nominal value of 1 μF. Apart from this nominal state the capacitor also has another state where its value is infinite (i.e. short-circuit). How does this affect the reliability of the system as calculated in Example 5.6?

The first thing to note is that, even though each component in the system now has two states, the system as a whole has four sub-states. The probabilities of each of these four sub-states may be calculated using the probability calculus developed in Chapter 1.

System sub-state number 1 is:

> Resistor has nominal value of 0.1 MΩ
> Capacitor has nominal value of 1 μF
>
> $\therefore \quad \tau_{h_1} = 100$ ms

and

$$p_1 = 0.99 \times 0.98$$
$$= 0.9702$$

System sub-state number 2 is:

> Resistor has nominal value of 0.1 MΩ
> Capacitor has 'catastrophic' value of ∞
>
> $\therefore \quad \tau_{h_2} = \infty$

and

$$p_2 = 0.99 \times 0.02$$
$$= 0.0198$$

System sub-state number 3 is:

> Resistor has 'catastrophic' value of ∞
> Capacitor has nominal value of 1 μF
>
> $\therefore \quad \tau_{h_3} = \infty$

and

$$p_3 = 0.01 \times 0.98$$
$$= 0.0098$$

System sub-state number 4 is:

> Resistor has 'catastrophic' value of ∞
> Capacitor has 'catastrophic' value of ∞
>
> $\therefore \tau_{h_4} = \infty$
>
> and
>
> $$p_4 = 0.01 \times 0.02$$
> $$= 0.0002$$

All four sub-states are mutually exclusive and exhaustive. Therefore, as would be expected:

$$p_1 + p_2 + p_3 + p_4 = 1$$

The system as a whole can still be considered as having two main states. One which meets the requirement (success) and one which does not meet it (failure). If these two main states are denoted by p_s' and p_t respectively, then it is apparent that:

$$p_s = \quad p_1 \quad = 0.9702$$
$$p_t = p_2 + p_3 + p_4 = 0.0298$$

Hence the reliability of the system is now 0.9702.

The last example is a particular illustration of a more general case. If a system is made up of n component parts where each component has two possible states, then the total number of sub-states for the system is 2^n. If all these sub-states lead to one of two main states for the overall system then a proportion m, say, of the sub-states may lead to success and a proportion $1-m$ to failure.

More generally still, each of the n component parts of the system may have N states. The total number of system sub-states is then N^n. At the overall system level there may then be not only two but a number of consequential main states which are of interest. This will involve grouping certain proportions of N^n states into the appropriate main-state categories.

In calculating the probabilities for any sub-state the component part state probabilities are combined according to the laws for independent probabilities as discussed in Chapter 1. There may be cases, of course, where the component part probabilities are not independent. In which case the calculus needs to be modified to accommodate this fact as already indicated in Chapter 1. When it comes to combining a proportion of the sub-states to arrive at the probability of a system main state, the probabilities are normally combined on the basis of exclusive probabilities with the corresponding calculus which this implies.

VARIATIONS IN THE PERFORMANCE ACHIEVEMENT

In any system subject to random state changes of the type being discussed, the various states and their associated probabilities may be enumerated and written in the manner used in Example 5.7. Alternatively, the states may be ranked in tabular form after the manner used in Example 1.11 in Chapter 1 and illustrated in Table 1.6. This latter form of expression, as already indicated in Chapter 4, is sometimes known as a 'truth table' and it can be a very valuable approach in checking that all possible states have been covered. However, when systems become large and complex the value of N^n (or the total number of possible states) may be too large to be manageable by any listing or tabular process. Techniques are therefore required for grouping or combining states, and their associated probabilities, in a more succinct form. These techniques, as applied to different types of technological systems, are described in Chapters 11 and 12.

One particular combinational method, however, can be usefully dealt with at this stage. Suppose a room is lit by a total of n nominally identical but independent electric light bulbs. Each bulb is subject to a random and discrete variation between one state of normal light emission and another state of no emission or failure. The probability of a light bulb being in the normal state is p and, hence, the probability of it being in the failed state is $(1-p)$. Suppose now the requirement is such that an illumination in the room produced by m or more bulbs out of n ($m \leqslant n$) is satisfactory but that produced by $m-1$ or less is unsatisfactory. It is then required to find the reliability of the satisfactory situation. The particular characteristic of this problem is that all the component parts of the system (the light bulbs), their possible states and the probabilities of being in any one individual state are of the same value.

Consider, first of all, one sub-state where a particular m bulbs are functioning and $(n-m)$ bulbs are failed. The probability of this one state which may be the jth state, say, out of the total of 2^n states is obviously given by:

$$p_j = p^m (1-p)^{n-m} \qquad (5.50)$$

There are, however, a number of states, all with m bulbs functioning and $(n-m)$ failed, which are numerically equal to the value given by Equation (5.50). The total number of such states is the number of ways that m items can be selected from n. This is the combinational number:

$$^nC_m \quad \text{or} \quad \binom{n}{m} \qquad (5.51)$$

where:

$$\binom{n}{m} = \frac{n!}{m!(n-m)!} \qquad (5.52)$$

Therefore, the total probability that the number of functioning bulbs (represented, say, by the variate x) is exactly equal to m is:

$$p(x = m) = \binom{n}{m} p^m (1-p)^{n-m} \qquad (5.53)$$

If $(1-p) = q$, then Equation (5.53) will be recognized as a term in the binomial expansion of $(p+q)^n$. For this reason Equation (5.53) is normally known as the 'binomial distribution' for discrete random variates. The function $p(x = m)$, or $f(m)$, represents a particular p.d.f. for discrete variates of the type already briefly mentioned in Chapter 2.

In the problem being considered at the moment, the requirement would be met not only if exactly m bulbs were working but if m or more bulbs out of n were working. The probability of m, $(m+1)$, $(m+2)$, ... or n bulbs functioning is obviously derived from the sums of the probabilities for each of these sub-states since each sub-state is mutually exclusive. Hence:

$$p(x \geqslant m) = \sum_{j=m}^{n} \binom{n}{j} p^j (1-p)^{n-j} \qquad (5.54)$$

Since, however, the sum of all sub-states must be unity or, put another way in this case, that $(p+q)^n = 1$, then:

$$\sum_{j=0}^{n} \binom{n}{j} p^j (1-p)^{n-j} = 1 \qquad (5.55)$$

and:

$$\sum_{j=0}^{m-1} \binom{n}{j} p^j (1-p)^{n-j} + \sum_{j=m}^{n} \binom{n}{j} p^j (1-p)^{n-j} = 1 \qquad (5.56)$$

or:

$$p(x \geqslant m) = 1 - \sum_{j=0}^{m-1} \binom{n}{j} p^j (1-p)^{n-j} \qquad (5.57)$$

or again:

$$p(x \geqslant m) = 1 - p[x \leqslant (m-1)] \qquad (5.58)$$

A function of the type $p[x \leqslant (m-1)]$ is a cumulative probability distribution of the form originally defined in Equation (2.6). In the present example, $(m-1)$ bulbs or less functioning is the same condition as r or more bulbs failed where:

$$r = n - m + 1 \qquad (5.59)$$

So Equation (5.57) could be written in the form:

$$p(x \geqslant m) = 1 - \sum_{j=r}^{n} \binom{n}{j} (1-p)^j p^{n-j} \qquad (5.60)$$

VARIATIONS IN THE PERFORMANCE ACHIEVEMENT

or:
$$p(x \geq m) = \sum_{j=0}^{r-1} \binom{n}{j}(1-p)^j p^{n-j} \quad (5.61)$$

or:
$$p(x \geq m) = \sum_{j=0}^{r-1} \binom{n}{j} q^j (1-q)^{n-j} \quad (5.62)$$

Equations (5.54) to (5.62), therefore, illustrate some of the various properties of the binomial cumulative distribution function. Values of this function and the corresponding p.d.f. are available in a number of sets of reference tables. In particular, where high accuracy is not required, the functional values can be obtained from the graphs plotted in Figures A.1 to A.27 inclusive.

Example 5.8

A particular type of aircraft engine fitted to an aircraft which operates a regular service over a regular route has been found to have a probability of failure per flight of 0.005. If the requirement for a successful flight is that not more than two engines out of a total of four should fail, what is the reliability of a successful flight?

The probability of a system failure, p_f, is:

$$p_f = \sum_{r=3}^{4} \binom{4}{j} q^j (1-q)^{4-j}$$

where $q = 0.005$

$$\therefore \quad p_f = \binom{4}{3}(0.005)^3(0.995) + \binom{4}{4}(0.005)^4$$

$$p_f = 4(125 \times 10^{-9})(0.995) + (625 \times 10^{-12})$$

$$\simeq 5 \times 10^{-7}$$

Hence, probability of system success p_s, or reliability, is:

$$p_s = 1 - 5 \times 10^{-7}$$

$$\simeq 99.99995\%$$

Other distributions of discrete random variates, apart from the binomial, can be of interest in the assessment of performance achievement and reliability but these are dealt with in Chapter 7. Also, in the same chapter, will be discussed other general properties associated with discrete variates.

Combinations of continuous and discrete distributions

In general, the random variations that take place in any system are likely to result from a combination of individual variations of both the continuous and discrete type. General methods by which such combinations may be analysed and used are discussed later, but at this stage it is useful to mention one particular and typical type of combination.

It is conceivable that a performance parameter of a system, item of equipment or component part could randomly take on any value in the range zero to infinity. In the ideal, it would be possible to express this continuum of variation by an appropriate p.d.f. extending over the same range of possible values. There may, however, be practical difficulties in determining a distributional characteristic of this type. Apart from this, many items or devices in technological systems can be approximated to a simpler form of behaviour.

This more simplified behaviour pattern has already been implicitly expressed in the discussions in this chapter and in Chapter 4. It was also illustrated in connexion with the 'catastrophic' failures of the example system in Chapter 2. For instance, the achieved performance in relation to any parameter may be considered as being represented by some singular or nominal value. Around this nominal or 'mean' value, small random variations are likely to take place which can be represented by a continuous distribution. Very often, in practice, this continuous distribution of variation is bounded on either side by the so-called specification limits associated with the particular parameter. Variations outside these limits are not normally expected but when they do occur they are quite likely to be of a 'catastrophic' nature. This means that the value of the performance parameter moves to some extreme value which may approximate to the 'zero' or 'infinite' condition.

In standard engineering parlance, all this means that the item is either 'working' within its specification limits or it is 'not working' at all. In the statistical sense, the item is approximating to a two-state model. One discrete random state is that of the item 'not working' or 'failed' and the other discrete random state is that of the item 'working' or 'successful' Any other variations which may take place in the failed state are generally of little interest, but the variations which may occur in the successful state can be important. So, the failed state can normally be defined by a single extreme value and the successful state by a continuous random distribution about the nominal working value. This 'picture' of behaviour as it is related to random variations is illustrated in Figure 5.4. It is seen that such a picture represents a particular combination of the ideas of discrete and continuous random variates.

Obviously, the idea shown in Figure 5.4 can be extended to more than two discrete states and to different types of continuous distributions of variates within each discrete state.

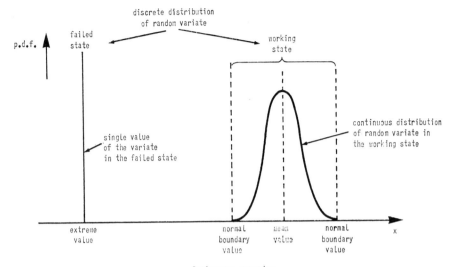

Figure 5.4 Illustration of a typical pattern of discrete and continuous random variations in a system performance parameter

Example 5.9

In the system of Figure 5.1, the value of the resistor, R, can randomly take on one of two states. One state where the value of R is 0.1 MΩ and the other where the value is infinite (open-circuit). The probability of the resistor having the infinite value is 0.01. The value of the capacitor C is subject to continuous random variations about a mean value of 1 μF. These random variations are known to be normally distributed with a standard deviation of 0.05 μF. No other variations exist. If the values of the upper and lower limits of the coefficient τ_q in the required transfer function are 105 ms and 95 ms respectively, what is the reliability of the system under the required conditions?

The value of the coefficient τ_h in the achieved transfer function is, first of all, subject to two discrete states due to the variability behaviour of the resistor. The probability of τ_h being in the successful of these two states is 0.99 as shown in Example 5.6.

In the successful state the variations are due to the capacitor only. The probability of these variations meeting the requirement is 0.6826 as shown in Example 5.5.

Hence:

$$\text{Overall reliability} = 0.99 \times 0.6826$$

$$= 0.6758$$

Other extensions of this type of example can now be readily appreciated.

Achieved Performance

This chapter has shown that the complete performance achievement as represented by H in Equation (3.1) is made up of an overall nominal value around which various types of variation can take place. The complete value of H may be described, although obviously this is not essential, by a characteristic transfer matrix whose elements are made up of transfer functions. Each transfer function represents the performance achievement between a set of input and output parameters. Very often the form of the transfer function can be represented by a ratio of polynomials in the Laplacian operator s. In the reliability analysis of any system where the achievement is being compared with the requirement, the form of each transfer function can be made invariate and a direct correspondence in form produced between the appropriate required and achieved functions. The determination of any singular value of performance and all variations that may exist about this value can then be described by determining the characteristics of variation in the transfer function coefficients.

Such variations in coefficients have been shown to be of either a systematic or random type. The functions which describe each of these types may be either discrete or continuous in nature. An introduction has been given to typical functions and some of the ways in which they may be manipulated has been described. The reader should now be aware of how a complete performance characteristic for H can be built up. In addition, the chapter has shown, in certain cases, how H may be compared with the corresponding Q in order to arrive at an overall reliability evaluation.

The comparisons carried out in this way have been of a one-dimensional nature. In general, however, the representation of H may cover a large number of achieved performance parameters each of which may be varying randomly and systematically in space and time. Such an overall achievement function may be looked upon as an n-dimensional surface in a similar way to that described in connexion with the requirement, Q, at the end of Chapter 3. The overall reliability, R, is then given by the probability of the H surface remaining within the limits defined by the corresponding Q surfaces when all these surfaces are subject to random and systematic changes.

The review of the general techniques of reliability evaluation has now been completed. Problems still arise in dealing with more complex systems and systems with certain specific types of characteristics. These aspects will be enlarged upon, with examples, in the later chapters. There are, however, a number of mathematical, probabilistic and statistical theories and methods which the reliability analyst needs in order to deal with these problems. The immediate succeeding chapters, therefore, are devoted to a review of these theories in relation to reliability technology. The reader who feels that

VARIATIONS IN THE PERFORMANCE ACHIEVEMENT   191

he is already familiar with these concepts may like, therefore, to omit Chapters 6 to 9 and proceed with his reading to Chapter 10.

Questions
1. Describe, with illustrations, the significance of any variations that may take place in the achieved performance of a system. Show what needs to be known about the variations in order to proceed to a reliability evaluation of a system.
2. Discuss the general methods by which variations in performance achievement may be deduced. In attempting to define the variational characteristics of a system, what sort of causes of variation would you examine?
3. Describe what is meant by the 'gap' or 'interface' problem between a performance requirement and a performance achievement of a system. Illustrate, with examples, the significance of any variations that may take place at such interfaces.
4. Assuming that the performance achievement of a system may be described by a transfer function, show in what way the form of such a function may be dictated by the variational characteristics of a system itself and of its corresponding requirement. What is the significance of the coefficients of a transfer function in any variational analysis?
5. A system for displaying a reading of steady-state temperature consists of a thermocouple, a voltage amplifier and an indicating meter which have simple transfer function characteristics of k_1, k_2 and k_3 respectively. If the constant systematic percentage error in k_1 is -1%, in k_2 is $+2\%$ and in k_3 is -0.5%, what is the overall systematic error of the system? If k_2 now varies systematically and discretely in time between two values of $k_2+2\%$ and $k_2-2\%$ with an equal duration at each value, what is the new value and form of the system's systematic error? What is the reliability of the system in each case if the requirement for the system's overall transfer function, k, is $k \pm 3\%$? (Assume $k = k_1 k_2 k_3$.)
6. The achieved shearing stress on a structural member is equal to the force on the member, f, divided by the cross-sectional area, a. The force f is known to have a nominal value of 100 kN with a superimposed sinusoidal systematic variation of peak amplitude equivalent to 5 kN. The cross-sectional area, a, has a nominal value of 1000 mm² with a superimposed sinusoidal systematic change of peak amplitude of 100 mm² which varies at the same frequency and in phase with the force f. If the required shearing stress is 100 MN/m² $\pm 4\frac{1}{2}\%$, what is the average reliability of the situation over a long period of time?
7. The flow of a liquid in a pipe is found to follow a normal distribution of random variation with time. The mean value of the flow is 50 litres/s and the standard deviation 20 litres/s. Calculate the chance at any time

of finding the flow (a) less than 25 litres/s, (b) greater than 80 litres/s, (c) less than 5 litres/s, (d) greater than 95 litres/s, (e) between 40 and 60 litres/s, (f) between 30 and 65 litres/s.

8. A flywheel attached to a motor is required to have a time constant between 90 s and 105 s. The actual flywheel has a nominal value of inertia, I, of 10 kg m^2 and a viscous damping coefficient, M, of nominal value 0.1 N m s. No other restraints exist on the system. If the value of M is invariate but the value of I is subject to normal random variation about its mean value with a standard deviation of 0.5 kg m^2, what is the reliability of the system against its requirement? (Take the time constant, τ, as $\tau = I/M$.)

·9. A system consists of two resistors, R_1 and R_2, connected in series. The nominal value of R_1 is 100 Ω and that of R_2 is 250 Ω. Each resistor has three possible states which it may randomly take on in time. The states correspond to the nominal value, the zero value (short-circuit) and the infinite value (open-circuit). The probabilities of R_1 and R_2 being in the nominal state at any time are 0.99 and 0.96 respectively. The probability of R_1 being in the short-circuit state is 0.001 and of R_2 being in the open-circuit state is 0.02. If the system requirement at any time is for the total resistance to lie between 50 Ω and 360 Ω, what is the reliability of the system? Tabulate the probabilities for each possible state of the system and show that the sum of these probabilities is unity.

10. If the same two resistors as those described in Question 9 are connected in parallel, what is the reliability of this new system if the requirement is for a system resistance of (a) between 90 Ω and 110 Ω, (b) between 70 Ω and 110 Ω and (c) between 70 Ω and 260 Ω?

11. A standby generating system consists of three nominally identical but independent diesel alternator sets. The emergency power requirement can be met if any two of the diesel sets starts successfully on demand. If the probability of failure to start for each set is 0.04, what is (a) the reliability of the system? How is the reliability affected if the failure to start per set is (b) 0.1 and (c) 0.6?

12. An automatic control system for a plant is capable of entering one of three possible states. The probability of being in the normal control state is 0.985. The probability of being in the failed passive state, where the control justs stops working, is 0.01. The probability of being in the failed active state, where the plant runs away out of control, is 0.005. In addition, the performance of the system in the normal state varies randomly according to a normal distribution of variation about the mean control function. Normal control is still maintained when these random variations are contained within ± 2 standard deviations. What is the reliability of the system if (a) only normal control is acceptable, (b) only the plant run-away condition is unacceptable?

CHAPTER 6

The Transfer Characteristic

Transfer Characteristic Concepts

It was pointed out in Chapter 3 that the overall performance requirement for any system could be expressed, in the functional sense, by a characteristic which described the relation between the required functional outputs and inputs. This is in the nature of a derived characteristic since the requirement only basically specifies the available inputs and the required outputs. However, such a characteristic, which may be termed a transfer characteristic, has certain advantages from the point of view of mathematical notation and manipulation. This notational representation was first introduced in Equations (3.17), (3.18) and (3.24). Subsequent chapters have shown the use of the transfer characteristic in describing the functional achieved performance of a system both in its deterministic and variational modes of behaviour. The reader is referred to the discussion in Chapter 4 in connexion with Equation (4.3) and in Chapter 5 in connexion with Equation (5.2).

Generally, then, whether the output relationship is that which is required or achieved, a general notation of the following form may be used:

$$\{L[f_o(y_i|t,z)]\} = [\Phi_{ij}(s)|t,z]\{L[f_i(x_j|t,z)]\} \quad (6.1)$$

where the function, f_i, represents the complete array of functional inputs in the time and space domains, function, f_o, represents the corresponding array of functional outputs and the matrix, $[\Phi_{ij}(s)|t,z]$, is a complex operator which acts upon f_i to produce or to convert it into f_o.

It is the purpose of this present chapter to examine the general properties of the transfer matrix and some of the particular properties of its elements when they are expressed in operational form. The discussion in this context is limited to what may be termed linear systems. Cases do arise in practice, of course, where the conversion from input to output functions contains non-linearities or discontinuities. In many of these instances, however, the conversion process can be represented by a model based on linear analysis provided that the reliability assessor is critically aware of the assumptions involved. Where an approach is adopted based on boundary values of performance, the linear analysis may always be an adequate representation of the practical case.

As has been shown in the previous chapters, the transfer characteristic matrix is composed of elements which represent 'transfer functions'. A transfer function may be generally described as an operator which when acting upon a defined input condition produces a defined output condition. With many technological systems, particular outputs are related to particular inputs by differential equations. The transfer function is, therefore, intimately related to the methods of representing and solving such equations. Differential equations may be solved and manipulated by classical methods and by use of various forms of operational calculus. Of these calculi, the Laplace transformation yields an approach which is both convenient in this context and has a general use in the statistical theory of reliability evaluation. So although the calculus arising from the Laplace transform is by no means the only approach, it is the one adopted and enlarged upon in this book. The following sections of this chapter now discuss the various characteristics of the Laplace transform as they are applicable to reliability analysis.

The Laplace Transform

If a function of a real variable, x, is $f(x)$ then it may be represented by a function $F(s)$ of a complex variable s where:

$$s = \alpha + i\omega \tag{6.2}$$

The direct transformation from $f(x)$ to $F(s)$ may represented in notational form by:

$$L[f(x)] = F(s) \tag{6.3}$$

where L stands for the Laplace transformation. Similarly, the inverse transformation may be expressed as:

$$L^{-1}[F(s)] = f(x) \tag{6.4}$$

The actual form of the Laplace transformation may be defined by:

$$F(s) = \int_{-\infty}^{\infty} f(x) e^{-sx} dx \tag{6.5}$$

This defined relationship is valid if, for some choice of α, the integral is absolutely convergent from $-\infty$ to ∞. In this case, the function $F(s)$ is known as the 'two-sided' Laplace transform of $f(x)$.

For functions which are very small on one half of the x-axis or have other particular characteristics, it may be possible to choose α so that $f(x) e^{-\alpha x}$ remains small over the whole range from $-\infty$ to ∞ and the integral is absolutely convergent. A particular function of this type, which will be discussed in more detail later, is the normal probability density function of Equation (5.37).

THE TRANSFER CHARACTERISTIC

Other and larger ranges of functions, however, do not have this property and it is often found that a value of α which makes the integral converge in the range 0 to ∞ makes it diverge in the range $-\infty$ to 0 and vice versa. There are, however, many useful functions which do not exist for negative x. If, for these functions $f(x)$ increases rapidly as x approaches ∞ then multiplication by $e^{-\alpha x}$, where α is positive, will generally lead to Equation (6.5) being absolutely integrable from 0 to ∞. An alternative transform may therefore be defined by:

$$F(s) = \int_0^\infty f(x) e^{-sx} dx \tag{6.6}$$

where this integral is generally valid for some positive α and where $f(x) = 0$ for $x < 0$. The defined relationship of Equation (6.6) is sometimes known as the 'one-sided' Laplace transformation or more frequently, as simply the Laplace transform of $f(x)$. Further discussion on the limits and conditions under which the Laplace transformation exists will be found in the standard works on the subject.

The transformation of Equation (6.6) generally allows a function $f(x)$ to be transformed from a function of a real variable x to a function of a complex variable s. The transform can then often be manipulated in purely algebraic terms and be retransformed back into the real variable x at the end of the manipulation. This has obvious advantages which will be illustrated in the succeeding sections.

Transforms of Particular Functions

A number of particular functions are now examined by way of illustration of the Laplace transformation.

Step function

Let $f(x) \equiv 1$ for $x \geqslant 0$ and $f(x) \equiv 0$ for $x < 0$. This function, which is zero for negative x, changes to a value of unity at the origin and remains at unity for all positive x, is called a 'unit step function' and is often represented by $f(x) = U(x)$.

$$L[U(x)] = \int_0^\infty e^{-sx} dx$$

$$= \left[\frac{e^{-sx}}{-s}\right]_0^\infty$$

This converges for $\alpha > 0$, therefore:

$$L[U(x)] = \frac{1}{s} \tag{6.7}$$

Such a unit step function may be displaced along the x-axis such that $f(x) \equiv 1$ for $x \geqslant a$ and $f(x) \equiv 0$ for $x < a$. This is represented by $f(x) = U(x-a)$.

$$L[U(x-a)] = \int_a^\infty e^{-sx} \, dx$$

$$= \left[\frac{e^{-sx}}{-s} \right]_a^\infty$$

which also converges for $\alpha > 0$, hence

$$L[U(x-a)] = \frac{e^{-as}}{s} \tag{6.8}$$

The unit step function may generally be multiplied by a constant and this will lead to the equivalent transform multiplied by the same constant, for instance:

$$L[kU(x)] = \frac{k}{s} \tag{6.9}$$

Example 6.1

A system has an input function for a parameter x given by $f_i(x) \equiv k_1$ for $x \geqslant 0$ and $f_i(x) = 0$ for $x < 0$. The corresponding output function is $f_o(y) = k_2$ for $y \geqslant 0$ and $f_o(y) = 0$ for $y < 0$. What are the Laplace transforms of the input and output functions and what is the form of the transfer function for the system in Laplace notation?

$$F_i(s) = L[f_i(x)] = \frac{k_1}{s}$$

$$F_o(s) = L[f_o(y)] = \frac{k_2}{s}$$

Expressing the relationship between input and output in the form of Equation (6.1) gives:

$$L[f_o(y)] = [\Phi(s)] L[f_i(x)]$$

In this case, the matrix, $[\Phi(s)]$, is a simple one-element matrix and the element represents the transfer function between the particular output and the particular input. Hence:

$$L[f_o(y)] = \Phi(s) L[f_i(x)]$$

or:

$$\Phi(s) = \frac{L[f_o(y)]}{L[f_i(x)]} = \frac{F_o(s)}{F_i(s)} \tag{6.10}$$

THE TRANSFER CHARACTERISTIC

Substituting the expressions for $F_i(s)$ and $F_o(s)$ gives:

$$\Phi(s) = \frac{k_2}{k_1}$$

So the transfer function for this system is a simple constant. It can be seen, of course, that this must be a general result where any $F_o(s)$ only differs from the corresponding $F_i(s)$ by a multiplying constant.

Impulse function

As an extension of the concept of a step function, it may be noted that the rectangular distribution function discussed in Chapter 2 in connexion with Figure 2.5 may be expressed by the difference of two step functions. If the rectangular function lies between a lower limit of $x = a$ and an upper limit of $x = b$ and has a constant magnitude k between these limits, that is:

$$f(x) \equiv 0 \quad \text{for} \quad x < a$$
$$f(x) \equiv k \quad \text{for} \quad a \leqslant x \leqslant b$$
$$f(x) \equiv 0 \quad \text{for} \quad x > b$$

then:

$$f(x) = k[U(x-a) - U(x-b)]$$

whence, from Equation (6.8):

$$L[f(x)] = \frac{k}{s}(e^{-as} - e^{-bs}) \tag{6.11}$$

If the function is a probability density function, then as shown in Equation (2.7):

$$k = \frac{1}{(b-a)}$$

and therefore:

$$L[f(x)] = \frac{1}{s(b-a)}(e^{-as} - e^{-bs}) \tag{6.12}$$

and for the particular case where $a = 0$, that is the function starts at the origin of $x = 0$, then:

$$L[f(x)] = \frac{1}{sb}(1 - e^{-bs}) \tag{6.13}$$

The limiting case of Equation (6.13) as $b \to 0$ is of interest in a number of applications. In the field of probability density functions it represents the approach to a single-valued 'density' function or to a discrete probability function. In general mathematics, the limit is described as the Dirac delta function or unit impluse function and is denoted by the symbol $\delta(x)$. Its 'physical' characteristics correspond to a pulse approaching zero width, infinite height and unit area. Mathematically this may be defined as:

$$\delta(x) = 0 \quad \text{for} \quad x \neq 0 \tag{6.14}$$

and:

$$\int_{-\infty}^{\infty} \delta(x)\, dx = 1 \tag{6.15}$$

from which the particular property arises that:

$$\int_{-\infty}^{\infty} \delta(x) f(x)\, dx = f(0) \tag{6.16}$$

Expanding the exponential of Equation (6.13) and allowing b to approach zero gives:

$$L[\delta(x)] = \lim_{b \to 0} \frac{1}{sb}\left[1 - \left(1 - sb + \frac{s^2 b^2}{2!} - \ldots\right)\right]$$

and therefore:

$$L[\delta(x)] = 1 \tag{6.17}$$

The function $\delta(x)$ is a unit impulse function located at the origin. In a similar way, $\delta(x-a)$ is a unit impulse function located at the point $x = a$. The transform of $\delta(x-a)$ may be obtained from Equation (6.12) by allowing $(b-a)$ to approach zero. From which it is seen that:

$$L[\delta(x-a)] = e^{-as} \tag{6.18}$$

Ramp function

A ramp function of a parameter x may be considered as representing a constant rate of change of the parameter value. The function can be such that $f(x) \equiv kx$ for $x \geqslant 0$ and $f(x) \equiv 0$ for $x < 0$. That is, $f(x) = kxU(x)$, hence:

$$L[kxU(x)] = \int_{0}^{\infty} kx\, e^{-sx}\, dx$$

and integrating by parts:

$$L[kxU(x)] = \left[\frac{-kx e^{-sx}}{s}\right]_0^\infty - k\int_0^\infty -\frac{e^{-sx}}{s}\, dx$$

$$= k\left[\frac{-e^{-sx}}{s^2}\right]_0^\infty$$

THE TRANSFER CHARACTERISTIC

therefore:
$$L[kxU(x)] = \frac{k}{s^2} \qquad (6.19)$$

Example 6.2

The available functional input for a system consists of a signal in the form of a step function such that $f_i(x) = kU(x)$. The output signal from the system is required to be a ramp function of the form $f_o(x) = kxU(x)$ for the same variable x. What is the form of the required transfer function for the system?

If $\Phi(s)$ is the system's required transfer function, then, from Equation (6.10):
$$\Phi(s) = \frac{F_o(s)}{F_i(s)}$$

In this case, from Equations (6.9) and (6.19):
$$F_i(s) = \frac{k}{s} \quad \text{and} \quad F_o(s) = \frac{k}{s^2}$$

Therefore:
$$\Phi(s) = \frac{1}{s}$$

For the system as defined, the physical meaning of the output is that it is the direct integral of the input signal. It appears, therefore, that a system which performs a simple integration has a transfer function of the form $1/s$ under certain conditions. The general application of this concept and the conditions under which it applies will be discussed later.

A general extension of the ramp function which starts at the origin is the function $f(x) = kx^n U(x)$ where n is a positive integer. Here:
$$L[kx^n U(x)] = \int_0^\infty kx^n e^{-sx} dx$$

and integrating by parts:
$$L[kx^n U(x)] = \left[\frac{-kx^n e^{-sx}}{s}\right]_0^\infty + \frac{n}{s}\int_0^\infty kx^{n-1} e^{-sx} dx$$
$$= \frac{n}{s} L[kx^{n-1} U(x)]$$

Hence, by induction:
$$L[kx^n U(x)] = \frac{kn!}{s^{n+1}} \qquad (6.20)$$

If n is not an integer, then:

$$L[kx^n\, U(x)] = \frac{k}{s^{n+1}} \int_0^\infty y^n e^{-y}\, dy$$

where:

$$y = sx$$

But:

$$\int_0^\infty y^n e^{-y}\, dy = \Gamma(n+1) \tag{6.21}$$

from the definition of the gamma function.
Therefore:

$$L[kx^n\, U(x)] = \frac{k\Gamma(n+1)}{s^{n+1}} \tag{6.22}$$

for $n > -1$.

Exponential functions

Consider, first of all, the simple exponential function where $f(x) \equiv k e^{-ax}$ for $x \geqslant 0$. That is $f(x) = k e^{-ax}\, U(x)$. Then:

$$L[k e^{-ax}\, U(x)] = k \int_0^\infty e^{-(a+s)x}\, dx$$

$$= \frac{k}{s+a} \tag{6.23}$$

Trigonometrical functions can be expressed in terms of exponential functions, so that for instance:

$$k \sin \omega x = \frac{k}{2i}(e^{i\omega x} - e^{-i\omega x})$$

Whence, from Equation (6.23):

$$L[k \sin \omega x\, U(x)] = \frac{k}{2i}\left[\frac{1}{s-i\omega} - \frac{1}{s+i\omega}\right]$$

$$= \frac{k\omega}{s^2 + \omega^2} \tag{6.24}$$

Example 6.3

A system has a sinusoidal input signal of the form $f_i(x) = \sin \omega x\, U(x)$ and the system's transfer function is of the form $\Phi(s) = s$. What is the form of the output signal?

THE TRANSFER CHARACTERISTIC

Now, from Equation (6.10):

$$F_o(s) = \Phi(s) F_i(s)$$

$F_i(s)$ may be obtained from Equation (6.24) and since $\Phi(s) = s$, then:

$$F_o(s) = \frac{\omega s}{s^2 + \omega^2}$$

which may be written as:

$$F_o(s) = \frac{\omega}{2}\left[\frac{1}{s-i\omega} + \frac{1}{s+i\omega}\right]$$

Therefore, from Equation (6.23):

$$f_o(x) = \frac{\omega}{2}[e^{i\omega x} + e^{-i\omega x}]U(x)$$

$$= \omega \cos \omega x \, U(x)$$

Since this output signal, $f_o(x) = \omega \cos \omega x$, is the simple differential of the input signal, $f_i(x) = \sin \omega x$, it would appear that a system with a transfer function of $\Phi(s) = s$ performs a simple differentiation on the input signal. This concept will be examined further in the next section.

The types of particular functions that could be considered is almost inexhaustible. However, the transforms of many additional functions can be derived from the examples already given. In addition, Table A.2 gives a list of Laplace transforms for a fairly wide range of the more common functions which are met in practice. Table A.2 does, of course, work both ways and can be used either to find the Laplace transform of a particular function or to find the inverse transform of a particular transformation. Where a function is not specifically given in Table A.2, it may often be derived by re-expressing it in terms of standard functions or by splitting it up into partial fractions. Illustrations of this technique have already been given in the discussion on exponential functions and in Example 6.3.

Transforms of Particular Operations

Differentiation and integration

It was seen in Example 6.3 that multiplying a transform by s produced a result equivalent to differentiating the corresponding function of this transform. Consider more generally now a function $f(x)$ where $f(x) \equiv 0$ for $x < 0$ and where $f(x)$ is continuous for $x \geq 0$. Consider also the function $f'(x)$ where:

$$f'(x) = \frac{d}{dx}f(x)$$

and let $f'(x) \equiv 0$ for $x < 0$. The Laplace transforms of both $f(x)$ and $f'(x)$ are taken to exist and be valid for some range of values of α.

Now:
$$L[f'(x)] = \int_0^\infty f'(x) e^{-sx} \, dx$$

also:
$$L[f(x)] = \int_0^\infty f(x) e^{-sx} \, dx$$

which, on integrating by parts, gives:
$$L[f(x)] = \left[\frac{-f(x) e^{-sx}}{s} \right]_0^\infty + \frac{1}{s} \int_0^\infty f'(x) e^{-sx} \, dx$$

As $x \to \infty$ the function $f(x) e^{-sx}$ must have a zero limit for the transform integral of this function to converge as initially assumed. Therefore:
$$L[f(x)] = +\frac{1}{s} f(0+) + \frac{1}{s} \int_0^\infty f'(x) e^{-sx} \, dx$$

and hence:
$$L[f'(x)] = sL[f(x)] - f(0+) \tag{6.25}$$

where $f(0+)$ is the value of $f(x)$ as $x \to 0$ from the positive side.

By induction, it may be shown generally that:
$$L\left[\frac{d^n f(x)}{dx^n} \right] = s^n L[f(x)] - s^{n-1} f(0+) - s^{n-2} f'(0+) - \ldots \tag{6.26}$$

It also follows from Equation (6.25) that:
$$L[f^{-1}(x)] = \frac{1}{s} L[f(x)] + \frac{1}{s} f^{-1}(0+) \tag{6.27}$$

where:
$$f^{-1}(x) = \int f(x) \, dx$$

or, more generally, that:
$$L[f^{-n}(x)] = \frac{1}{s^n} L[f(x)] + \frac{1}{s^n} f^{-1}(0+) + \frac{1}{s^{n-1}} f^{-2}(0+) + \ldots \tag{6.28}$$

If all values of $f^n(0+)$ are zero, then Equations (6.26) and (6.28) reduce to:
$$L\left[\frac{d^n f(x)}{dx^n} \right] = s^n L[f(x)] \tag{6.29}$$

THE TRANSFER CHARACTERISTIC

and:

$$L[f^{-n}(x)] = \frac{1}{s^n} L[f(x)] \tag{6.30}$$

Example 6.4

The input function to a system can be expressed by:

$$f_1(x) = 2\sqrt{\frac{x}{\pi}}$$

Given that the Laplace transformation of $f_1(x)$ is:

$$F_1(s) = \frac{1}{s^{3/2}}$$

what would be the Laplace transformation of the input function if it were changed to:

$$f_2(x) = \frac{1}{\sqrt{(\pi x)}} ?$$

It is easily recognized that

$$f_2(x) = \frac{d}{dx} f_1(x)$$

Therefore, from Equation (6.25):

$$F_2(s) - sF_1(s)$$
$$= \frac{1}{\sqrt{s}}$$

Another aspect of differentiation and integration may now be considered in connexion with the Laplace transforms themselves. The transform of a function is defined as:

$$F(s) = \int_0^\infty f(x) e^{-sx} dx$$

Under certain conditions this may be differentiated with respect to the complex variable s and will give:

$$\frac{dF(s)}{ds} = \int_0^\infty -xf(x) e^{-sx} dx$$
$$= -L[xf(x)] \tag{6.31}$$

and repeating the process will yield:

$$\frac{d^n F(s)}{ds^n} = (-1)^n L[x^n f(x)] \tag{6.32}$$

It also follows in particular that:

$$\int_s^\infty F(s)\,ds = L\left[\frac{f(x)}{x}\right] \tag{6.33}$$

and generally that:

$$[F^{-n}(s)]_s^\infty = L\left[\frac{f(x)}{x^n}\right] \tag{6.34}$$

Example 6.5

Given that the Laplace transform of $\sin \omega x$ is:

$$\frac{\omega}{s^2+\omega^2}$$

what are the transforms of $x \sin \omega x$ and $(1/x) \sin \omega x$?

From Equation (6.31):

$$L[x \sin \omega x] = -\frac{d}{ds}\frac{\omega}{s^2+\omega^2}$$

$$= \frac{2\omega s}{(s^2+\omega^2)^2}$$

and, from Equation (6.33):

$$L\left[\frac{\sin \omega x}{x}\right] = \int_s^\infty \frac{\omega\,ds}{s^2+\omega^2}$$

$$= \left[\tan^{-1}\frac{s}{\omega}\right]_s^\infty$$

$$= \frac{\pi}{2} - \tan^{-1}\frac{s}{\omega}$$

$$= \tan^{-1}\frac{\omega}{s}$$

Multiplication by an exponential

Consider the function $f(x)$ multiplied by e^{-ax}, then:

$$L[e^{-ax}f(x)] = \int_0^\infty [e^{-ax}f(x)]e^{-sx}\,dx$$

$$= \int_0^\infty e^{-(s+a)x}f(x)\,dx$$

$$= F(s+a) \tag{6.35}$$

THE TRANSFER CHARACTERISTIC

Example 6.6

Knowing the Laplace transform of sin ωx as given in Example 6.5, what is the transform of $e^{-ax}\sin\omega x$?

$$L[\sin\omega x] = \frac{\omega}{s^2+\omega^2}$$

Therefore, from Equation (6.35):

$$L[e^{-ax}\sin\omega x] = \frac{\omega}{(s+a)^2+\omega^2}$$

Change of variable

If the variable, x, in the function, $f(x)$, is multiplied by a constant equal to, say, $1/a$, then:

$$L\left[f\left(\frac{x}{a}\right)\right] = \int_0^\infty e^{-sx} f\left(\frac{x}{a}\right) dx$$

$$= \int_0^\infty e^{-as(x/a)} f\left(\frac{x}{a}\right) \frac{d(ax)}{a}$$

$$= a \int_0^\infty e^{-as(x/a)} f\left(\frac{x}{a}\right) d\left(\frac{x}{a}\right)$$

$$= aF(as) \tag{6.36}$$

On the other hand, if the variable x is shifted by a constant amount a in order to create a new variable y such that:

$$y = x-a \quad \text{for} \quad a \leqslant x \leqslant \infty$$

then:

$$L[f(x-a)] = \int_0^\infty e^{-sx} f(x-a) dx$$

$$= \int_0^a e^{-sx} f(x-a) dx + \int_a^\infty e^{-sx} f(x-a) dx$$

and since $f(x-a) \equiv 0$ for $x < a$, then:

$$L[f(x-a)] = \int_a^\infty e^{-sx} f(x-a) dx$$

$$= e^{-as} \int_0^\infty e^{-sy} f(y) dy$$

$$= e^{-as} F(s) \tag{6.37}$$

Example 6.7

From a knowledge of the Laplace transform of $e^{-ax}U(x)$ and of the relationship given in Equation (6.37), find the Laplace transform of $e^{-ax}U(x-b)$.

From Equation (6.23):

$$L[e^{-ax}U(x)] = \frac{1}{s+a}$$

Therefore:

$$L[e^{-a(x-b)}U(x-b)] = \frac{e^{-bs}}{s+a}$$

and hence:

$$L[e^{-ax}U(x-b)] = \frac{e^{-bs}e^{-ab}}{s+a}$$

Convolution

Consider two functions of x, $f_1(x)$ and $f_2(x)$, where $f_1(x) \equiv 0$ and $f_2(x) \equiv 0$ for $x < 0$, then the convolution of $f_1(x)$ and $f_2(x)$, denoted usually by $f_1 * f_2$, is defined by:

$$f_1 * f_2 = \int_0^x f_1(u) f_2(x-u) \, du \tag{6.38}$$

from which the following relationships may be deduced:

$$f_1 * f_2 = f_2 * f_1 \tag{6.39}$$

$$f_1 * (af_2) = (af_1) * f_2 = a(f_1 * f_2) \tag{6.40}$$

$$f_1 * (f_2 + f_3) = f_1 * f_2 + f_1 * f_3 \tag{6.41}$$

$$e^{ax}(f_1 * f_2) = e^{ax}f_1 * e^{ax}f_2 \tag{6.42}$$

$$f_1 * (f_2 * f_3) = (f_1 * f_2) * f_3 \tag{6.43}$$

These convolutions are useful in describing some of the combinations of distributions of random variates which are met with in technological systems and it is convenient, at this stage, to consider the corresponding Laplace transformations.

$$L[f_1 * f_2] = \int_0^\infty (f_1 * f_2) e^{-sx} \, dx$$

$$= \int_0^\infty \int_0^x f_1(u) f_2(x-u) e^{-sx} \, du \, dx$$

Now, the integration with respect to u from 0 to x followed by the integration

THE TRANSFER CHARACTERISTIC

of x from 0 to ∞ is the same in this case as integrating x first from u to ∞ followed by integrating u from 0 to ∞, hence:

$$L[f_1 * f_2] = \int_0^\infty \int_u^\infty f_1(u) f_2(x-u) e^{-sx} \, dx \, du$$

$$= \int_0^\infty f_1(u) e^{-su} \left[\int_u^\infty f_2(x-u) e^{-s(x-u)} \, dx \right] du$$

which, on making the substitution $y = x - u$, gives:

$$L[f_1 * f_2] = \int_0^\infty f_1(u) e^{-su} \left[\int_0^\infty f_2(y) e^{-sy} \, dy \right] du$$

$$= L[f_1(x)] L[f_2(x)] \tag{6.44}$$

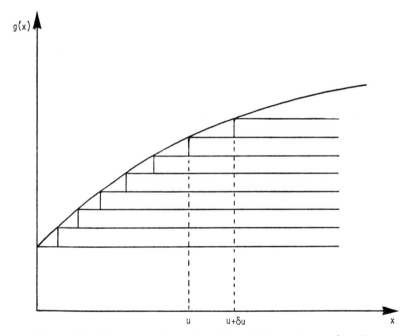

Figure 6.1 Representation of function $g(x)$ by unit step functions

A particular example of convolution arises in the case of Duhamel's integral. This is an integral which enables the functional output of a system to any input stimulus to be evaluated knowing, say, the functional output produced by a unit step function input. The concept is to consider that any shape of stimulus may be comprised of a succession of unit step function forms as illustrated in Figure 6.1. Applying the Principle of Superposition, it follows that the functional output to any such input stimulus is the sum of the functional outputs for each corresponding unit step function input. This is conditional, of course, on the assumption that the systems act in a linear fashion.

Suppose that the input stimulus is defined by a function $g(x)$ where $g(x) \equiv 0$ for $x < 0$. Suppose also that for the system to which the functional input $g(x)$ is to be applied, the functional output to a unit step function input is $f(x)$. From Figure 6.1, the functional output produced by the initial step is:

$$g(0) f(x)$$

At any other value of the parameter x where, say, $x = u$, the height of the additional small step is:

$$g'(u) \delta u$$

and the corresponding functional output produced by this small step is:

$$g'(u) \delta u f(x-u)$$

Hence, the functional output $h(x)$ to the complete series of steps representing $g(x)$ is given by:

$$h(x) = g(0) f(x) + \sum_{u=0}^{u=x} g'(u) f(x-u) \delta u$$

and proceeding to the limit as $\delta u \to 0$ gives:

$$h(x) = g(0) f(x) + \int_0^x g'(u) f(x-u) \, du \qquad (6.45)$$

which, in terms of the convolutions just discussed, may be written as:

$$h(x) = g(0) f(x) + (g' * f) \qquad (6.46)$$

and the convolution $(g' * f)$ is Duhamel's integral.

From the relationship of Equation (6.44), the Laplace transform of Equation (6.46) is:

$$L[h(x)] = g(0) L[f(x)] + L[g'(x)] L[f(x)] \qquad (6.47)$$

But from Equation (6.25), $L[g'(x)] = sL[g(x)]$ provided that $g(0+) = 0$. Therefore, under these conditions:

$$L[h(x)] = sL[g(x)] L[f(x)] \qquad (6.48)$$

Example 6.8

Suppose that it is known that a particular system's response in time to a unit step input is given by:

$$f(t) = (1 - e^{-t/\tau})$$

THE TRANSFER CHARACTERISTIC 209

where τ is the time constant of the system. What would be the system's functional output if a ramp function of the form $g(t) = kt$ were applied to the same system?

Here, both $f(t) \equiv 0$ and $g(t) \equiv 0$ at $t = 0$. From Equations (6.19) and (6.23) the transforms of $g(t)$ and $f(t)$ are:

$$L[g(t)] = \frac{k}{s^2}$$

$$L[f(t)] = \frac{1}{s} - \frac{1}{s + 1/\tau}$$

Hence, from Equation (6.48):

$$L[h(t)] = \frac{k/\tau}{s^2(s + 1/\tau)}$$

which, on splitting into partial fractions, becomes:

$$L[h(t)] = k\left[\frac{1}{s^2} - \frac{\tau}{s} + \frac{\tau}{(s + 1/\tau)}\right]$$

Hence:

$$h(t) = k[t - \tau(1 - e^{-t/\tau})]$$
$$= g(t) - k\tau f(t)$$

It is not necessary to discuss any further transforms of particular operations at this stage. Some additional transform theorems will be dealt with as they arise in the subsequent discussion. However, for ready reference, a table of the more important transforms of operations is given in Table A.3.

It is now possible to proceed to some further discussion of the transfer function which relates the operational output of any system to its corresponding input.

Transfer Function

It has already been shown simply in Chapter 3 that the functional output of a system, $f_o(x)$, may be related to the corresponding input, $f_i(x)$, by a differential equation. If such a differential equation has constant coefficients, this concept may be expressed more generally by:

$$a_n \frac{d^n f_o(x)}{dx^n} + a_{n-1} \frac{d^{n-1} f_o(x)}{dx^{n-1}} + \ldots + a_1 \frac{df_o(x)}{dx} + a_0 = f_i(x) \quad (6.49)$$

If the initial conditions are assumed to be zero and a Laplace transformation is carried out on Equation (6.49), then:

$$(a_n s^n + a_{n-1} s^{n-1} + \ldots + a_1 s + a_0) F_o(s) = F_i(s)$$

therefore:

$$\frac{F_o(s)}{F_i(s)} = \frac{1}{a_n s^n + a_{n-1} s^{n-1} + \ldots + a_1 s + a_0}$$
$$= \Phi(s) \tag{6.50}$$

where $\Phi(s)$ is called the transfer function of the system. The concept can be extended to integrodifferential equations, whence $\Phi(s)$ becomes of the more general form:

$$\Phi(s) = \frac{b_m s^m + \ldots + b_0}{a_n s^n + \ldots + a_0} \tag{6.51}$$

as previously mentioned in connexion with Equation (5.3). Generally, for real linear systems, the coefficients a and b are real constants and m and n are real positive integers.

From Equation (6.50):

$$F_o(s) = \Phi(s) F_i(s) \tag{6.52}$$

or:

$$f_o = (\phi * f_i) \tag{6.53}$$

where:

$$\phi = L^{-1}[\Phi(s)] \tag{6.54}$$

If, in Equation (6.49), the initial conditions are not zero, then:

$$F_o(s) = \Phi(s) F_i(s) + \Phi(s) Q(s) \tag{6.55}$$

where $Q(s)$ is a polynomial in s of degree not more than $(n-1)$ and depends on the initial values $f_o(0), f_o'(0), f_o''(0)$ etc...

Consider now, by way of a simple example, the system analogue of Figure 5.1. The voltage equation for this circuit is:

$$V_i(t) = V_R(t) + V_C(t) \tag{6.56}$$

where $V_R(t)$ is the instantaneous voltage drop across the resistor R with respect to time and $V_C(t)$ is the corresponding voltage drop across the capacitor C. If the current flowing round the circuit is $I(t)$, then:

$$V_R(t) = RI(t) \tag{6.57}$$

and:

$$V_C(t) = \frac{1}{C} \int_0^t I(t) \, dt \tag{6.58}$$

or

$$C \frac{dV_C(t)}{dt} = I(t) \tag{6.59}$$

THE TRANSFER CHARACTERISTIC

Also, of course, in this case:

$$V_C(t) = V_o(t) \tag{6.60}$$

Hence, from Equations (6.56) to (6.60), the differential equation relating voltage output to voltage input is:

$$\tau \frac{dV_o(t)}{dt} + V_o(t) = V_i(t) \tag{6.61}$$

where:

$$\tau = RC \tag{6.62}$$

Therefore, taking the Laplace transforms of Equation (6.61) and assuming $V_o(0) = 0$:

$$(1 + \tau s) V_o(s) = V_i(s) \tag{6.63}$$

and hence:

$$\Phi(s) = \frac{V_o(s)}{V_i(s)} = \frac{1}{1 + \tau s} \tag{6.64}$$

Equation (6.64) now represents the transfer function of the simple electrical analogue illustrated in Figure 5.1 which was originally quoted in Equation (5.5). Reference to the differential equation, Equation (6.61), from which this transfer function was derived will show that $1/(1 + \tau s)$ represents the transfer function of any system, electrical or other wise, which has 'simple lag' characteristics with a time constant τ.

It is interesting to note that the transforms of Equations (6.57) and (6.58) yield:

$$\frac{V_R(s)}{I(s)} = R \tag{6.65}$$

and:

$$\frac{V_C(s)}{I(s)} = \frac{1}{sC} \tag{6.66}$$

and, in the case of an inductor, since:

$$V_L(t) = L \frac{dI(t)}{dt}$$

then:

$$\frac{V_L(s)}{I(s)} = sL \tag{6.67}$$

So that Equations (6.65)–(6.67) represent the transfer functions relating the voltage developed across an electrical component (or the component of which it is the electrical analogue) to the current flowing through it. It will

8

be noticed that the form of these transfer functions is the same as that for the complex impedance of the respective electrical components. For instance, the complex impedance of an inductor is $i\omega L$ and its transfer function is sL. It can be shown generally that the substitution of s for $i\omega$ will convert any complex impedance into the corresponding transfer function. This is often a convenient method of finding the overall transfer function of a system which can be represented by an electrical analogue.

Example 6.9

Find the overall transfer function for the system whose electrical analogue is given by Figure 5.2.

In terms of a.c. steady-state impedances, the ratio of output voltage, $V_o(t)$, to input voltage, $V_i(t)$, is given by:

$$\frac{V_o(t)}{V_i(t)} = \frac{\frac{L}{C} / \left(i\omega L + \frac{1}{i\omega C}\right)}{R + \frac{L}{C} / \left(i\omega L + \frac{1}{i\omega C}\right)}$$

Making the substitution, $i\omega = s$:

$$\frac{V_o(s)}{V_i(s)} = \frac{\frac{L}{C} / \left(sL + \frac{1}{sC}\right)}{R + \frac{L}{C} / \left(sL + \frac{1}{sC}\right)}$$

$$= \frac{sL/R}{s^2 LC + sL/R + 1}$$

$$= \frac{\tau_2 s}{\tau_1 \tau_2 s^2 + \tau_2 s + 1} \qquad (6.68)$$

where:

$$\tau_1 = RC \quad \text{and} \quad \tau_2 = L/R$$

The reader might like to verify for himself that the same result would have been achieved by transforming the appropriate differential equation for this circuit.

The system whose transfer function has been calculated in Example 6.9 represents a 'second-order' system. Some further properties of the transfer function for this type of system will now be discussed. Consider the mechanical system illustrated in Figure 6.2. Here an output shaft having inertia and viscous damping is connected to an input shaft by means of a torsion spring. The inertia, represented by the symbol J, has basic units of (mass) (length)^2,

THE TRANSFER CHARACTERISTIC

the viscous damping, N, has units of torque per unit of angular velocity or basically (mass) (length)^2 (time)$^{-1}$ and the basic units of the spring restraint K in terms of torque per unit angular displacement are (mass) (length)^2 (time)$^{-2}$. The torque balance equation for the system of Figure 6.2 is therefore:

$$K[\theta_i(t) - \theta_o(t)] = N\frac{d\theta_o(t)}{dt} + J\frac{d^2\theta_o(t)}{dt^2} \tag{6.69}$$

or:

$$\theta_i(t) = \theta_o(t) + \frac{N}{K}\frac{d\theta_o(t)}{dt} + \frac{J}{K}\frac{d^2\theta_o(t)}{dt^2} \tag{6.70}$$

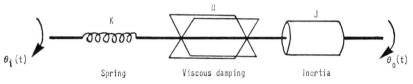

Figure 6.2 Second-order mechanical system

Assuming that the initial conditions are zero, then the Laplace transform of Equation (6.70) leads to a transfer function of the form:

$$\Phi(s) = \frac{\theta_o(s)}{\theta_i(s)} = \frac{1}{(J/K)s^2 + (N/K)s + 1} \tag{6.71}$$

which may be written as:

$$\Phi(s) = \frac{1}{\tau_1\tau_2 s^2 + \tau_1 s + 1} \tag{6.72}$$

where:

$$\tau_1 = N/K \tag{6.73}$$

$$\tau_2 = J/N \tag{6.74}$$

and the similarity may be noticed between Equation (6.72) and Equation (6.68).

Alternatively, Equation (6.71) may be written in the more standard form for a 'second-order' transfer function where:

$$\Phi(s) = \frac{\omega_0^2}{s^2 + 2\zeta\omega_0 s + \omega_0^2} \tag{6.75}$$

and where:

$$\omega_0 = \sqrt{\frac{K}{J}} \tag{6.76}$$

$$\zeta = \frac{N}{2}\sqrt{\frac{1}{JK}} \tag{6.77}$$

The constant, ω_0, is known as the natural frequency of the system and the constant, ζ, as the damping factor. Amongst others, three particular cases are of interest. First, where the damping factor is zero due to the damping coefficient, N, being zero. Here:

$$\Phi(s) = \frac{\omega_0^2}{s^2 + \omega_0^2} \tag{6.78}$$

Secondly, where the damping factor ζ is unity. Here:

$$\Phi(s) = \frac{\omega_0^2}{(s+\omega_0)^2} \tag{6.79}$$

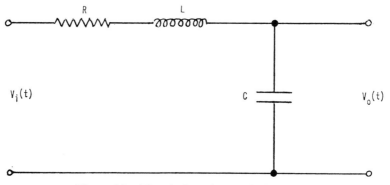

Figure 6.3 Electrical analogue of Figure 6.2

Thirdly, where the inertia, J, is zero. Here:

$$\Phi(s) = \frac{1}{1+\tau_1 s} \tag{6.80}$$

and it is seen, in this case, that the system's transfer function has reverted to the form of a 'simple lag'.

The system of Figure 6.3 represents an electrical analogue of the mechanical system of Figure 6.2. This may be seen as follows. The differential voltage equation for the system of Figure 6.3 is:

$$V_i(t) = V_o(t) + RC\frac{dV_o(t)}{dt} + LC\frac{d^2 V_o(t)}{dt^2} \tag{6.81}$$

Whence:

$$\Phi(s) = \frac{V_o(s)}{V_i(s)} = \frac{1}{LCs^2 + RCs + 1} \tag{6.82}$$

or, in terms of time constants:

$$\Phi(s) = \frac{1}{\tau_1 \tau_2 s^2 + \tau_1 s + 1} \tag{6.83}$$

THE TRANSFER CHARACTERISTIC

where:

$$\tau_1 = RC \tag{6.84}$$

$$\tau_2 = L/R \tag{6.85}$$

Comparing the mechanical Equations (6.72)–(6.74) with the electrical Equations (6.83)–(6.85), it is now apparent that the following mechanical/electrical equivalents apply:

K (spring restraint) $= 1/C$ (1/capacitance) (6.86)

N (damping coefficient) $= R$ (resistance) (6.87)

J (inertia) $= L$ (inductance) (6.88)

and this may be used in determining the electrical analogue of many mechanical systems.

The electrical transfer function of Equation (6.82) may also be written in terms of a natural frequency, ω_o, and a damping factor, ζ. Here:

$$\Phi(s) = \frac{\omega_o^2}{s^2 + 2\zeta\omega_o s + \omega_o^2} \tag{6.89}$$

where, in this case:

$$\omega_o = \sqrt{\frac{1}{LC}} \tag{6.90}$$

$$\zeta = \frac{R}{2}\sqrt{\frac{C}{L}} \tag{6.91}$$

Letting the inductance value, L, be zero, reduces the electrical transfer function to:

$$\Phi(s) = \frac{1}{1 + \tau_1 s}$$

which, as would be expected, is the transfer function already obtained for the electrical analogue of Figure 5.1.

Example 6.10

Consider a mechanical system such as shown in Figure 6.2 where the damping coefficient, N, is zero. Derive the transfer function relating output shaft angle to input shaft angle using the a.c. impedance method with an electrical analogue.

The electrical analogue is the same as shown in Figure 6.3 with a resistance value, R, of zero. Hence, the a.c. impedance equation is:

$$\frac{V_o(t)}{V_i(t)} = \frac{(1/i\omega C)}{i\omega L + (1/i\omega C)}$$

Letting $i\omega = s$ gives the equivalent transfer function:

$$\Phi(s) = \frac{V_o(s)}{V_i(s)} = \frac{\theta_o(s)}{\theta_i(s)} = \frac{\omega_o}{s^2 + \omega_o^2}$$

where:

$$\omega_o = \sqrt{\frac{1}{LC}} = \sqrt{\frac{K}{J}}$$

Effects of Particular Transfer Functions

Many technological systems or their elements have characteristics which approximate to 'first-order' (simple lag) or 'second-order' transfer functions. It is useful, therefore, to examine the sort of input to output conversions that are produced by these typical functions.

First-order system

The transfer function for this type of system has been shown to be:

$$\Phi(s) = \frac{F_o(s)}{F_i(s)} = \frac{1}{1 + \tau s} \tag{6.92}$$

or:

$$F_o(s) = \frac{F_i(s)}{1 + \tau s}$$

If the input is in the form of a unit step function, that is $f_i(x) = U(x)$ or $f_i(x) \equiv 1$ for $x \geq 0$ and $f_i(x) \equiv 0$ for $x < 0$, then:

$$F_i(s) = \frac{1}{s} \tag{6.93}$$

and:

$$F_o(s) = \frac{1}{s(1 + \tau s)} \tag{6.94}$$

and the inverse transform of Equation (6.94) from Table A.2 is:

$$f_o(x) = 1 - e^{-x/\tau} \tag{6.95}$$

So the output function is in the form of an exponential rise where $f_o(x) = 0$ for $x = 0$ and $f_o(x) = 1$ for $x = \infty$. The input and output functions for this situation are plotted in Figure 6.4. It is seen that the time constant, τ, represents the value of x (which could be time) for the output, $f_o(x)$, to reach $[1 - (1/e)]$ or 63.2% of its final value. It is also interesting to note that:

$$\frac{df_o(x)}{dx} = \frac{1}{\tau} e^{-x/\tau} \tag{6.96}$$

THE TRANSFER CHARACTERISTIC 217

and:

$$\frac{df_o(0)}{dx} = \frac{1}{\tau} \tag{6.97}$$

so that the initial slope of the output function is $1/\tau$ and hence, if this rate of rise was maintained, the value of τ would be the value of x for which the final output level was achieved.

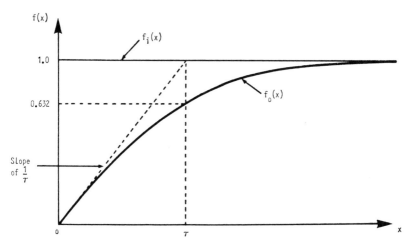

Figure 6.4 Response of a first-order system to a step function

As a second conversion characteristic of a first-order system consider the application of a ramp input where $f_i(x) = kxU(x)$, then:

$$F_i(s) = \frac{k}{s^2}$$

and hence:

$$F_o(s) = \frac{k}{s^2(1+\tau s)} \tag{6.98}$$

Taking the inverse transform of Equation (6.98) from Table A.2 gives:

$$f_o(x) = k[x - \tau(1 - e^{-x/\tau})] \tag{6.99}$$

which is the same result as was derived in Example 6.8 by a different method. In particular:

$$f_o(0) = 0$$

$$f_o(x) = k(x - \tau) \quad \text{as} \quad x \to \infty \tag{6.100}$$

The values of $f_i(x)$ amd $f_o(x)$ for the ramp input case are plotted in Figure 6.5. The output function is within 5% of following the slope of the input function when the value of x reaches the equivalent of about three time constants. Also for large x, as shown by Equation (6.100), the output function 'lags' behind the input function by a value equal to the time constant.

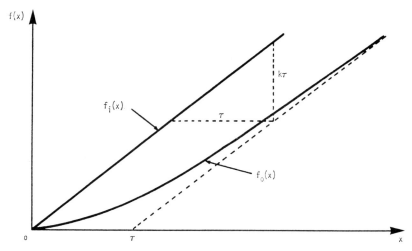

Figure 6.5 Response of a first-order system to a ramp function

A third and final case is now examined for the first-order system where the input function is of sinusoidal form, that is $f_i(x) = \sin \omega x$ for $x \geq 0$. Here:

$$F_i(s) = \frac{\omega}{s^2 + \omega^2}$$

and:

$$F_o(s) = \frac{\omega}{(s^2 + \omega^2)(1 + \tau s)} \tag{6.101}$$

from whence the reader can verify that:

$$f_o(x) = e^{-x/\tau} A^2 \tan \theta + A \sin(\omega x - \theta) \tag{6.102}$$

where:

$$A = \frac{1}{\sqrt{(1 + \omega^2 \tau^2)}} \tag{6.103}$$

and:

$$\theta = \tan^{-1} \omega \tau \tag{6.104}$$

THE TRANSFER CHARACTERISTIC

The exponential term in Equation (6.102) is a transient term and so, for large x, the steady-state solution is:

$$f_o(x) = A \sin(\omega x - \theta) \tag{6.105}$$

which is the same form as the input but with a reduced amplitude and with a phase lag. The output to input amplitude ratio and the phase lag depend upon the angular frequency ω. For zero frequency, $A = 1$ and $\theta = 0$ and as

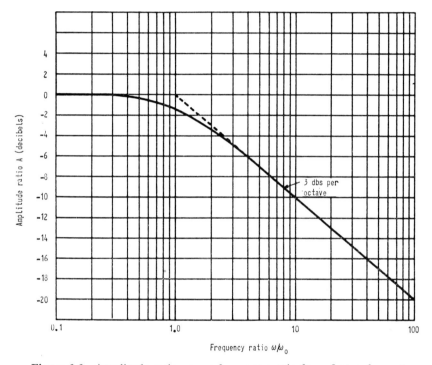

Figure 6.6 Amplitude ratio versus frequency ratio for a first-order system

the frequency becomes very large, $A \to 0$ and $\theta \to 90°$. In particular, when the frequency, ω, is numerically equal to $1/\tau$, the amplitude ratio is $1/\sqrt{2}$ and the phase lag is 45°. This frequency is sometimes known as the equivalent 'natural frequency', ω_0, of a first-order system where:

$$\omega_0 = \frac{1}{\tau} \tag{6.106}$$

In Figures 6.6 and 6.7, the values of the amplitude ratio, A, expressed as ten times the logarithm of the ratio or in what may be termed 'decibels' and θ expressed in degrees are plotted against the frequency ratio ω/ω_0 on a logarithmic scale. The presentation of the two characteristics A and θ may also be combined by using a polar plot as shown in Figure 6.8. In this latter case, it is seen that the frequency plot represents a semi-circle of radius 0.5 centred at the polar co-ordinate point (0.5, 0).

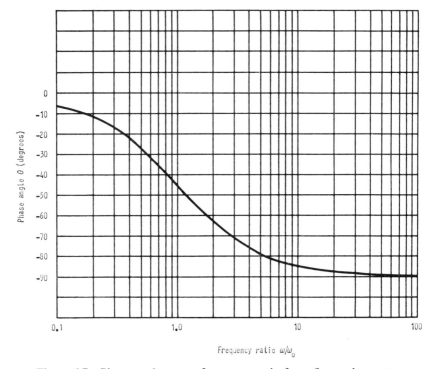

Figure 6.7 Phase angle versus frequency ratio for a first-order system

Second-order system

The transfer function of a second-order system has been seen to be of the general form:

$$\Phi(s) = \frac{F_o(s)}{F_i(s)} = \frac{\omega_o^2}{s^2 + 2\zeta\omega_o s + \omega_o^2} \quad (6.107)$$

For a unit step function input:

$$F_o(s) = \frac{\omega_o^2}{s(s^2 + 2\zeta\omega_o s + \omega_o^2)} \quad (6.108)$$

The inverse transform of Equation (6.108) may be obtained by splitting into partial fractions and making use of the standard forms in Table A.2. However, in this process, three cases need to be considered.

THE TRANSFER CHARACTERISTIC

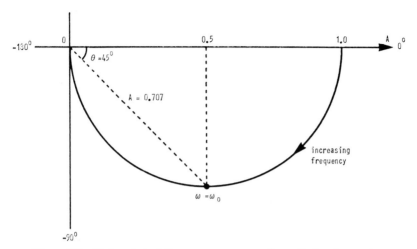

Figure 6.8 Polar plot of frequency response for a first-order system

First, where $\zeta > 1$. Here:

$$f_o(x) = 1 - \frac{1}{2\sqrt{[\zeta^2 - 1]}} [[\zeta + \sqrt{(\zeta^2 - 1)}] \exp\{\omega_o x [\sqrt{(\zeta^2 - 1)} - \zeta]\}$$
$$+ [\sqrt{(\zeta^2 - 1)} - \zeta] \exp\{-\omega_o x [\sqrt{(\zeta^2 - 1)} + \zeta]\}] \qquad (6.109)$$

and:

$$f_o(x) = 1 \quad \text{as} \quad x \to \infty$$

also, if ζ is very much greater than unity, then Equation (6.109) approximates to:

$$f_o(x) = 1 - e^{-\omega_o x/2\zeta} \qquad (6.110)$$

which is equivalent, as would be expected, to a first-order system of time constant τ where:

$$\tau = \frac{2\zeta}{\omega_o} \qquad (6.111)$$

Secondly, if $\zeta = 1$, then:

$$F_o(s) = \frac{\omega_o^2}{s(s + \omega_o)^2}$$

and:

$$f_o(x) = 1 - e^{-\omega_o x}(1 + \omega_o x) \qquad (6.112)$$

and again:

$$f_o(x) = 1 \quad \text{as} \quad x \to \infty$$

Thirdly, if $\zeta < 1$, then:
$$f_o(x) = 1 - e^{-\zeta \omega_o x} A \sin[\omega_o x \sqrt{(1-\zeta^2)} + \theta] \tag{6.113}$$
where:
$$A = \frac{1}{\sqrt{(1-\zeta^2)}} \tag{6.114}$$
$$\theta = \tan^{-1} \frac{\sqrt{(1-\zeta^2)}}{\zeta} \tag{6.115}$$
and:
$$f_o(x) = 1 \quad \text{as} \quad x \to \infty$$

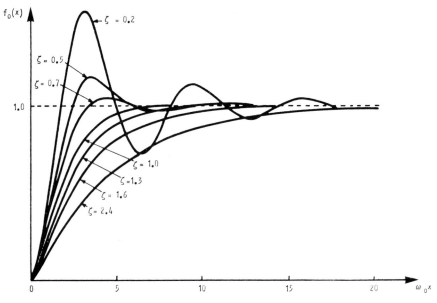

Figure 6.9 Response of a second-order system to a step function

Equation (6.113) is in the form of a damped oscillation about the final steady-state value. As ζ approaches zero, $f_o(x)$ approaches a sinusoidal oscillation of the form:
$$f_o(x) = 1 - \sin\left(\omega_o x + \frac{\pi}{2}\right) \tag{6.116}$$

The complete set of results for the response of a second-order system to a unit step function input is illustrated in Figure 6.9.

For a sinusoidal input:
$$F_o(s) = \frac{\omega \omega_o^2}{(s^2 + \omega^2)(s^2 + 2\zeta \omega_o s + \omega_o^2)} \tag{6.117}$$

THE TRANSFER CHARACTERISTIC

of which the steady-state solution is:

$$f_o(x) = A \sin(\omega x - \theta) \tag{6.118}$$

where, in this case:

$$A = \frac{1}{\sqrt{\{[1-(\omega/\omega_o)^2]^2 + 4\zeta^2(\omega/\omega_o)^2\}}} \tag{6.119}$$

and:

$$\theta = \tan^{-1}\frac{2\zeta(\omega/\omega_o)}{1-(\omega/\omega_o)^2} \tag{6.120}$$

Particular cases are, if $\zeta = 1$, then:

$$A = \frac{1}{1+(\omega/\omega_o)^2} \tag{6.121}$$

$$\theta = \tan^{-1}\frac{(\omega/\omega_o)}{1-(\omega/\omega_o)^2} \tag{6.122}$$

and if $(\omega/\omega_o) = 1$, then:

$$A = \frac{1}{2\zeta} \tag{6.123}$$

$$\theta = \frac{\pi}{2} \tag{6.124}$$

The general characteristics of Equation (6.118) for the output function $f_o(x)$ with a sinusoidal input of unit amplitude are illustrated in the polar plot of Figure 6.10.

Example 6.11

A mechanical system of the type illustrated in Figure 6.2 is found to have a steady-state output function of the form $\theta_o(t) = 1.5 \sin(100t - \pi/3)$ when given an input signal of the form $\theta_i(t) = \sin 100t$. What is the form of the system's transfer function and what are the values of the coefficients in this transfer function? If the spring restraint of the mechanical system is known to be 10 N m (newton metres per radian) what are the values of the damping coefficient, N, and the inertia, J?

The form of the transfer function is:

$$\Phi(s) = \frac{\omega_o^2}{s^2 + 2\zeta\omega_o s + \omega_o^2}$$

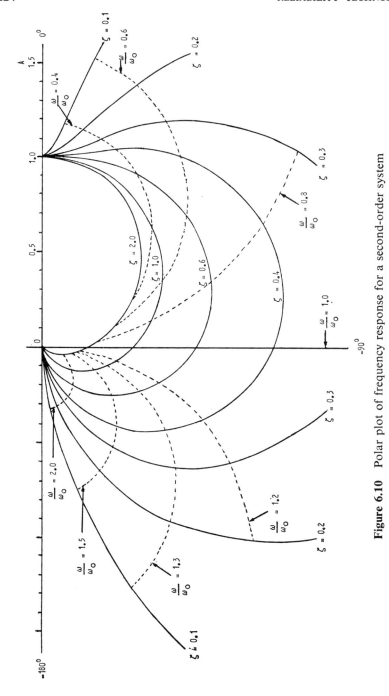

Figure 6.10 Polar plot of frequency response for a second-order system

THE TRANSFER CHARACTERISTIC

Eliminating ζ from Equations (6.119) and (6.120) gives:

$$\left(\frac{\omega}{\omega_o}\right)^2 = 1 - \frac{\cos\theta}{A} \qquad (6.125)$$

so:

$$\left(\frac{\omega}{\omega_o}\right)^2 = 1 - \frac{\cos(\pi/3)}{3/2}$$

$$\left(\frac{\omega}{\omega_o}\right)^2 = \tfrac{2}{3}$$

hence:

$$\omega_o^2 = \tfrac{3}{2} \times 100^2$$

or:

$$\omega_o = 122.5 \text{ radians per second}$$

Also:

$$\zeta = \frac{[1-(\omega/\omega_o)^2]\tan\theta}{2(\omega/\omega_o)}$$

$$= \frac{1}{2\sqrt{2}}$$

or:

$$\zeta = 0.3536$$

Now $K = 10$ N m, therefore from Equation (6.76):

$$J = \frac{K}{\omega_o^2}$$

$$= \frac{10}{(122.5)^2}$$

$$= 6.667 \times 10^{-4} \text{ kg m}^2$$

Also, from Equation (6.77):

$$N = 2\zeta\sqrt{(JK)}$$

$$= 5.773 \times 10^{-2} \text{ N m s}$$

Transfer Function Algebra

Although the final criterion of interest is often the form of the overall transfer function for a complete technological system, it is often useful to be able to build this up from a knowledge of the transfer functions of the various component parts. The simplest case is that of a series or cascade of elements as shown in Figure 6.11. Here it is apparent that:

$$\Phi(s) = \frac{F_o(s)}{F_i(s)} = \Phi_1(s)\,\Phi_2(s)\,\Phi_3(s) \tag{6.126}$$

provided that the connexions between blocks 1 and 2 and between blocks 2 and 3 do not alter the characteristics of any of the blocks themselves. This means that any number of transfer function elements connected in series will obey the multiplicative rule of Equation (6.126) if each of the elements is suitably 'buffered' from one another.

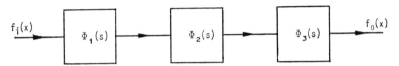

Figure 6.11 Series connexion of transfer function blocks

For instance, if the system of Figure 5.1 was connected in series with the system of Figure 6.3 by means of, say, a buffer element of infinite input impedance, zero output impedance and transfer function of unity, then the overall transfer function would be:

$$\Phi(s) = \frac{1}{1+\tau s} \times \frac{\omega_o^2}{s^2 + 2\zeta\omega_o s + \omega_o^2} \tag{6.127}$$

where:

$$\tau = R_1 C_1, \quad \omega_o^2 = \frac{1}{L_2 C_2}, \quad \zeta = \frac{R_2}{2}\sqrt{\frac{C_2}{L_2}}$$

that is, the simple multiplicative rule would apply.

However, if the two systems were directly coupled then the combined system would have to be treated as a whole yielding a transfer function of the form:

$$\Phi(s) = \frac{\omega_o^2}{(1+\tau s)(s^2+2\zeta\omega_o s+\omega_o^2)+s\tau'} \tag{6.128}$$

where:

$$\tau' = \frac{R_1}{L_2}$$

If transfer function blocks are connected in parallel as shown in Figure 6.12, and, if once again, there is no change in the characteristics of the elements produced by the method of connexion, then:

$$h_o(x) = f_o(x) + g_o(x)$$

or:
$$H_o(s) = F_o(s) + G_o(s)$$
whence:
$$H_o(s) = \Phi_1(s) F_i(s) + \Phi_2(s) G_i(s) \quad (6.129)$$

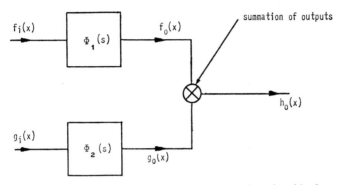

Figure 6.12 Parallel connexion of transfer function blocks

More particularly, if the parallel system is fed from a common input $f_i(x)$ so that $g_i(x) = f_i(x)$, then:

$$\Phi(s) = \frac{H_o(s)}{F_i(s)} = \Phi_1(s) + \Phi_2(s) \quad (6.130)$$

So, in a parallel system where each element is buffered from the other, where each is fed from a common input and where the outputs are summated, the additive rule of Equation (6.130) applies. The rule applies equally, of course, for subtractions as well as additions.

A specialized form of parallel connexion is the feed-back or closed-loop system shown in Figure 6.13. Here the following relationships apply:

$$\frac{H_o(s)}{H_i(s)} = \Phi_1(s)$$

$$\frac{G_o(s)}{H_o(s)} = \Phi_2(s)$$

and:
$$H_i(s) = F_i(s) - G_o(s)$$

whence:
$$\Phi(s) = \frac{H_o(s)}{F_i(s)} = \frac{\Phi_1(s)}{1 + \Phi_1(s) \Phi_2(s)} \quad (6.131)$$

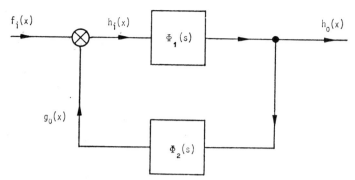

Figure 6.13 Closed-loop connexion of transfer function blocks

or, in the more particular case where $\Phi_2(s) = 1$,

$$\Phi(s) = \frac{\Phi_1(s)}{1+\Phi_1(s)} \tag{6.132}$$

Example 6.12

In a closed-loop control system of the type illustrated in Figure 6.13, the transfer function $\Phi_1(s)$ is made up of two elements with transfer functions $\Phi_a(s)$ and $\Phi_b(s)$ connected in series where:

$$\Phi_a(s) = k_1$$

$$\Phi_b(s) = \frac{k_2}{s(1+\tau s)}$$

Also:

$$\Phi_2(s) = 1$$

What is the transfer function of the overall system?

First:

$$\Phi_1(s) = \Phi_a(s)\Phi_b(s)$$

$$= \frac{k_1 k_2}{s(1+\tau s)}$$

then:

$$\Phi(s) = \frac{\Phi_1(s)}{1+\Phi_1(s)}$$

$$= \frac{\omega_0^2}{s^2 + 2\zeta\omega_0 s + \omega_0^2}$$

THE TRANSFER CHARACTERISTIC

where:

$$\omega_0^2 = \frac{k_1 k_2}{\tau}$$

and:

$$\zeta = \frac{1}{2}\sqrt{\left(\frac{1}{k_1 k_2 \tau}\right)}$$

So, the overall transfer function of this particular arrangement is equivalent to a second-order system.

The Transfer Matrix

So far, attention has been directed towards systems having a single functional input from which the corresponding functional output is derived. The transform relationship between the two has been considered in the form:

$$\frac{F_o(s)}{F_i(s)} = \Phi(s) \qquad (6.133)$$

where $\Phi(s)$ represents the transfer function of the system. The reciprocal relationship of Equation (6.133) may be defined as:

$$\frac{F_i(s)}{F_o(s)} = \Psi(s) \qquad (6.134)$$

where:

$$\Psi(s) = \frac{1}{\Phi(s)} \qquad (6.135)$$

The function $\Psi(s)$, which in this case is the reciprocal of the transfer function, is sometimes known as the 'characteristic function'.

If a system consists of a number of output functions $g_1(x), g_2(x), \ldots, g_n(x)$ which are derived independently from corresponding input functions $f_1(x), f_2(x), \ldots, f_n(x)$, then a separate and independent transfer function exists for each relationship. For instance the transfer function relating the ith output function to the ith input function is:

$$\Phi_i(s) = \frac{G_i(s)}{F_i(s)} \qquad (6.136)$$

Very often, however, a system may produce a range of output functions which are interrelated to each input function. It has been seen that a system with a single output to input relationship can be described, in many instances, by a linear integrodifferential equation with constant coefficients. Hence, the multiple output to input relationship is equivalent to a set of simultaneous linear integrodifferential equations with constant coefficients. It is pertinent, therefore, to consider a set of equations of the following form:

$$\psi_{11}(x)g_1(x)+\psi_{12}(x)g_2(x)+\ldots+\psi_{1n}(x)g_n(x) = f_1(x)$$
$$\psi_{21}(x)g_1(x)+\psi_{22}(x)g_2(x)+\ldots+\psi_{2n}(x)g_n(x) = f_2(x)$$
$$\cdot \quad \cdot \quad \cdot \quad \cdot \quad \cdot \quad \cdot \quad \cdot \quad \cdot$$
$$\psi_{n1}(x)g_1(x)+\psi_{n2}(x)g_2(x)+\ldots+\psi_{nn}(x)g_n(x) = f_n(x) \qquad (6.137)$$

where each $\psi_{ij}(x)$ represents an integrodifferential operator. If a Laplace transformation is carried out on the set of Equations (6.137), then in matrix notation:

$$\begin{bmatrix} \Psi_{11}(s) & \Psi_{12}(s) & \ldots & \Psi_{1n}(s) \\ \Psi_{21}(s) & \Psi_{22}(s) & \ldots & \Psi_{2n}(s) \\ \cdot & \cdot & & \cdot \\ \Psi_{n1}(s) & \Psi_{n2}(s) & \ldots & \Psi_{nn}(s) \end{bmatrix} \begin{bmatrix} G_1(s) \\ G_2(s) \\ \vdots \\ G_n(s) \end{bmatrix} = \begin{bmatrix} F_1(s) \\ F_2(s) \\ \vdots \\ F_n(s) \end{bmatrix}$$

or:

$$[\Psi_{ij}(s)]\{G_1(s) \ \ldots \ G_n(s)\} = \{F_1(s) \ \ldots \ F_n(s)\} \qquad (6.138)$$

where $[\Psi_{ij}(s)]$ represents the complete matrix of elements $\Psi_{ij}(s)$ and $\{G_1(s) \ \ldots \ G_n(s)\}$ represents a column matrix of the functions $G_1(s)$, $G_2(s)$, etc.

The matrix $[\Psi_{ij}(s)]$ is known as the 'characteristic matrix' and the characteristic equation for the set of differential equations is obtained by allowing the corresponding determinant of the characteristic matrix to be zero, that is:

$$|\Psi_{ij}(s)| = 0 \qquad (6.139)$$

and the roots of Equation (6.139) are the characteristic roots.

For further illustration it is convenient to consider a set of just two simultaneous differential equations where:

$$\psi_{11}(x)g_1(x)+\psi_{12}(x)g_2(x) = f_1(x)$$
$$\psi_{21}(x)g_1(x)+\psi_{22}(x)g_2(x) = f_2(x) \qquad (6.140)$$

and the corresponding transformation equations are:

$$\Psi_{11}(s)G_1(s)+\Psi_{12}(s)G_2(s) = F_1(s)$$
$$\Psi_{21}(s)G_1(s)+\Psi_{22}(s)G_2(s) = F_2(s) \qquad (6.141)$$

or:

$$[\Psi_{ij}(s)]\{G_1(s), G_2(s)\} = \{F_1(s), F_2(s)\} \qquad (6.142)$$

THE TRANSFER CHARACTERISTIC

Whence, solving for $G_1(s)$ in determinant form gives:

$$G_1(s) = \frac{\begin{vmatrix} F_1(s) & \Psi_{12}(s) \\ F_2(s) & \Psi_{22}(s) \end{vmatrix}}{\begin{vmatrix} \Psi_{11}(s) & \Psi_{12}(s) \\ \Psi_{21}(s) & \Psi_{22}(s) \end{vmatrix}}$$

or:

$$G_1(s) = \frac{\Psi_{22}(s)}{|\Psi_{ij}(s)|} F_1(s) - \frac{\Psi_{12}(s)}{|\Psi_{ij}(s)|} F_2(s) \tag{6.143}$$

which can be written in the form:

$$G_1(s) = \Phi_{11}(s) F_1(s) + \Phi_{12}(s) F_2(s) \tag{6.144}$$

where:

$$\Phi_{11}(s) = \frac{\Psi_{22}(s)}{|\Psi_{ij}(s)|}, \quad \Phi_{12}(s) = \frac{-\Psi_{12}(s)}{|\Psi_{ij}(s)|} \tag{6.145}$$

similarly:

$$G_2(s) = \Phi_{21}(s) F_1(s) + \Phi_{22}(s) F_2(s) \tag{6.146}$$

where:

$$\Phi_{21}(s) = \frac{-\Psi_{21}(s)}{|\Psi_{ij}(s)|}, \quad \Phi_{22}(s) = \frac{\Psi_{11}(s)}{|\Psi_{ij}(s)|} \tag{6.147}$$

whence, in matrix notation:

$$\begin{bmatrix} G_1(s) \\ G_2(s) \end{bmatrix} = \begin{bmatrix} \Phi_{11}(s) & \Phi_{12}(s) \\ \Phi_{21}(s) & \Phi_{22}(s) \end{bmatrix} \begin{bmatrix} F_1(s) \\ F_2(s) \end{bmatrix}$$

or:

$$\{G_1(s), G_2(s)\} = [\Phi_{ij}(s)]\{F_1(s), F_2(s)\} \tag{6.148}$$

The matrix $[\Phi_{ij}(s)]$ is then the 'transfer matrix', referred to in previous chapters, which converts the set of operational input functions into the corresponding set of operational output functions.

Each element, $\Phi_{ij}(s)$, of the transfer matrix, $[\Phi_{ij}(s)]$, is the transfer function of the ith output function $g_i(x)$ relative to the jth input function $f_j(x)$.

In the particular case where each output is related to the corresponding input and no other, then Equation (6.148) becomes:

$$\begin{bmatrix} G_1(s) \\ G_2(s) \end{bmatrix} = \begin{bmatrix} \Phi_{11}(s) & 0 \\ 0 & \Phi_{22}(s) \end{bmatrix} \begin{bmatrix} F_1(s) \\ F_2(s) \end{bmatrix} \tag{6.149}$$

which was the form originally discussed in connexion with Equation (3.23) in Chapter 3. Generally, for this situation $[\Phi_{ij}(s)]$ is known as a 'diagonal matrix' where each element $\Phi_{ij} = 0$ for $i \neq j$.

Example 6.13

A system is operated by two input functions $f_1(x)$ and $f_2(x)$. Two output functions $g_1(x)$ and $g_2(x)$ are required from the system where these functions are related to the inputs by the following differential equations:

$$\frac{g_1(x)}{k} - \frac{1}{k}\frac{dg_2(x)}{dx} = f_1(x)$$

$$2\frac{dg_2(x)}{dx} + 2\tau\frac{d^2 g_2(x)}{dx^2} - g_1(x) - \tau\frac{dg_1(x)}{dx} = f_2(x)$$

what is the form of the required transfer matrix of output relative to input and how can this be expressed using the block diagram representation of transfer function algebra?

The equivalent Laplace transforms of the two differential equations, assuming all initial conditions are zero, are:

$$\frac{1}{k}G_1(s) - \frac{s}{k}G_2(s) = F_1(s)$$

$$-(1+\tau s)G_1(s) + 2s(1+\tau s)G_2(s) = F_2(s)$$

from whence the characteristic matrix is

$$[\Psi_{ij}(s)] = \begin{bmatrix} \frac{1}{k} & -\frac{s}{k} \\ -(1+\tau s) & 2s(1+\tau s) \end{bmatrix}$$

and the corresponding determinant is:

$$|\Psi_{ij}(s)| = \frac{2s(1+\tau s)}{k} - \frac{s(1+\tau s)}{k}$$

$$= \frac{s(1+\tau s)}{k}$$

Hence:

$$\Phi_{11}(s) = \frac{\Psi_{22}(s)}{|\Psi_{ij}(s)|} = 2k$$

Similarly:

$$\Phi_{12}(s) = \frac{1}{(1+\tau s)}, \quad \Phi_{21}(s) = \frac{k}{s}, \quad \Phi_{22}(s) = \frac{1}{s(1+\tau s)}$$

so that:

$$G_1(s) = 2kF_1(s) + \frac{1}{(1+\tau s)}F_2(s)$$

$$G_2(s) = \frac{k}{s}F_1(s) + \frac{1}{s(1+\tau s)}F_2(s)$$

THE TRANSFER CHARACTERISTIC

or the form of the required transfer matrix is:

$$[\Phi_{ij}(s)] = \begin{bmatrix} 2k & \dfrac{1}{(1+\tau s)} \\ \dfrac{k}{s} & \dfrac{1}{s(1+\tau s)} \end{bmatrix}$$

and the equivalent block diagram in transfer function algebra is as shown in Figure 6.14.

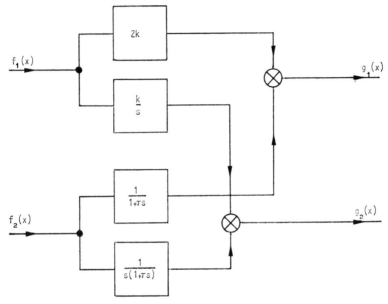

Figure 6.14 Transfer function block diagram for Example 6.13

The discussion in this chapter has now illustrated the general properties and use of the Laplace transformation. It has been shown that the relationship between a particular system output function and the corresponding input function can be expressed in the form of a transfer function. For a system in which multiple outputs are related to multiple inputs, the array of relevant transfer functions can be expressed in a matrix form whose characteristics have been briefly described.

The systems considered in this method of analysis have been restricted to those which may be described by sets of linear integrodifferential equations with constant coefficients. Many practical technological systems may be represented in this way or their overall performance approximated by such an approach. In addition, the input functions have generally been considered where they are zero for $x<0$ and where all initial values are zero at $x=0$. The appropriate Laplace transformations have been taken as being absolutely convergent for a choice of a sufficiently large value of α. On the basis of these assumptions, any system may be completely described in terms of Laplace transforms. If the Laplace transform of each input $f_j(x)$ is $F_j(s)$ and for each output $g_i(x)$ is $G_i(s)$, then:

$$\{G_i(s)\} = [\Phi_{ij}(s)]\{F_j(s)\} \qquad (6.150)$$

where $[\Phi_{ij}(s)]$ is the transfer matrix made up of elements $\Phi_{ij}(s)$. Each $\Phi_{ij}(s)$ then relates the ith output function to the jth input function.

As pointed out in Chapters 3, 4 and 5, the form of the required transfer matrix and the form of the achieved transfer matrix can be made the same and invariate. The manner in which the achieved system performance meets the required system performance (or the reliability of the situation) then depends solely on the relative values of the directly corresponding coefficients in each transfer function element of the required and achieved transfer matrices. Further methods of comparing these coefficients will be discussed in subsequent chapters.

Questions
1. Derive, from the definition of the Laplace transform given in Equation (6.6), the transforms of the following functions of x:
   (a) x^n,
   (b) $e^{-ax} - e^{-bx}$,
   (c) $e^{-ax}(1-ax)$,
   (d) $[1 - e^{-ax}(1+ax)]$,
   (e) $\sin^2 \omega x$,
   (f) $\sinh \omega x$.
   If in each case $f(x) = 0$ for $x<0$.
2. The input function to a system is $f_i(x) = kU(x)$. What is the transfer function of the system if it is found that the output function is:
   (a) $f_o(x) = k(1 - e^{-ax})/a$,
   (b) $f_o(x) = k(be^{-bx} - ae^{-ax})/(b-a)$,
   (c) $f_o(x) = k(\omega x - \sin \omega x)/\omega^3$?
3. Using the relationship of Equation (6.25) and the information that the Laplace transform of $x \sin \omega x$ is $2\omega s/(s^2 + \omega^2)^2$, what is the transform of:
   (a) $\sin \omega x + \omega x \cos \omega x$,
   (b) $2\omega \cos \omega x - \omega^2 x \sin \omega x$,
   (c) $3\omega^2 \sin \omega x + \omega^3 x \cos \omega x$?

THE TRANSFER CHARACTERISTIC

4. Using the relationships of Equations (6.32) and (6.34) and given that the Laplace transform of xe^{-ax} is $1/(s+a)^2$, what is the transform of:
   (a) $x^2 e^{-ax}$,
   (b) $x^3 e^{-ax}$,
   (c) e^{-ax}?

5. Using the relationships of Equations (6.35), (6.36) and (6.37) and given that the inverse transform of $a/s(s+a)$ is $(1-e^{-ax})$ find the inverse transforms of:
   (a) $a/(s+b)(s+a+b)$,
   (b) $a/s(s-a)$,
   (c) $a/s(bs+a)$,
   (d) $ae^{-as}/s(s+a)$,
   (e) $e^{-a^2s}/s(s+1)$.

6. By means of splitting up into partial fractions and by the use of standard forms, find the inverse transforms of:

   (a) $\dfrac{s+1}{s(s+a)}$,

   (b) $\dfrac{(a-2b)s^2 + ab^2}{s(s+a)(s+b)^2}$,

   (c) $\dfrac{ab}{s^2(s+a)(s+b)}$.

7. Derive the Laplace transforms of the following functions:

   (a) $\displaystyle\int_0^x e^{-au} \sin a(x-u)\, du$,

   (b) $\dfrac{1}{6}\displaystyle\int_0^x u^2 e^{-au}(x-u)^3\, du$,

   (c) $\displaystyle\int_0^x \dfrac{(x-u)^2 \sin au\, du}{u}$.

8. The output response of a particular system to a unit step function input is known to be of the form:
$$f_o(x) = 1 - e^{-ax}(1+ax)$$
what is the form of the output of the system if the input is replaced by a sinusoidal function of the form:
$$f_i(x) = \sin ax\, U(x)?$$

9. By reference to the a.c. steady-state impedances find the transfer function, relating output voltage to input voltage, for the following system electrical analogues:

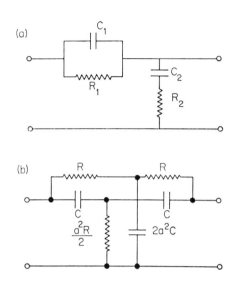

10. Given the following system of linear mechanical movement:

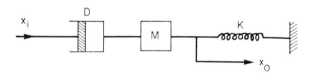

where x_i is the input linear displacement and x_o is the corresponding output linear displacement, find the transfer function relating output to input. The element D represents linear viscous damping in units of force per unit linear velocity, the element M represents mass and the element K is the linear spring restraint in terms of force per unit linear displacement.

What is the corresponding electrical analogue of this mechanical system?

11. If the mechanical system of Question 10 has zero mass, what is the form of the transfer function and what is the form of the output displacement with time if the input consists of a unit step function of displacement with time? What are the initial and final values of x_o and the time for x_o to reach $1/e$ of its initial value if $D = 1000$ N m^{-1} s and $K = 200$ N m^{-1}?

THE TRANSFER CHARACTERISTIC

12. The mechanical system of Question 10 has the following values associated with each element: $D = 50 \text{ N m}^{-1}\text{s}$, $M = 1 \text{ kg}$, $K = 2500 \text{ N m}^{-1}$ and the steady-state input function is $x_i(t) = \sin 25t$. What is:
    (a) the natural frequency of the system,
    (b) the damping factor of the system,
    (c) the steady-state phase lead of the output displacement relative to the input, and
    (d) the steady-state amplitude ratio of the output displacement relative to the input?

13. An overall control system is made up of two sub-systems A and B. Any connexions between A and B are such that system A does not interfere with the characteristics of system B and vice versa. The transfer functions for each system are:

    $$\Phi_A(s) = \frac{1}{1+\tau_1 s}$$

    $$\Phi_B(s) = \frac{\tau_2 s}{1+\tau_2 s}$$

    What is the overall transfer function of the system if:
    (a) Systems A and B are connected in series?
    (b) System B is turned into a closed-loop control system with a feed-back transfer function of -1 and this closed-loop system is then connected in series with system A?
    (c) Systems A and B are connected in series and the complete system converted into a closed-loop control system with a feed-back transfer function of $-\tau_1 s$?

14. Two output functions of a system, $g_1(x)$ and $g_2(x)$, are found to be related to two input functions, $f_1(x)$ and $f_2(x)$, by the following simultaneous differential equations:

    $$\int g_1(x)\,dx + \tau g_1(x) = f_1(x)$$

    $$\frac{1}{k}\frac{dg_2(x)}{dx} - \frac{dg_1(x)}{dx} - \tau\frac{d^2 g_1(x)}{dx^2} = f_2(x)$$

    If all initial conditions are assumed to be zero, what is the form of the system's transfer matrix?

CHAPTER 7

Properties of Distributions

The Distribution Function

The reliability evaluation of any system has been seen to be dependent upon the comparison between the system's required and achieved performance. Both these performance functions, in whatever form they are expressed, are subject to variation. The probabilistic nature of the system's reliability arises from the fact that a major aspect in all practical variations is that of randomness. These concepts have already been introduced in the previous chapters and, in particular, Chapter 5 discussed some of the general implications of variations in performance.

When a performance parameter, x, of a system is subject to random variation the actual value of x at any precise point in the space and time domains is not known. However, such a variate x may be more likely to take on some values rather than others so that the 'pattern' of its occurrence may be expressed in probabilistic terms. The function which describes this 'pattern' or 'distribution' of the values of x is known as a distribution function. When a variate is continuous the distribution function of basic interest is the probability density function (p.d.f.). The meaning and general characteristics of the p.d.f. were discussed in Chapter 2 and the particular case of the normal p.d.f. was examined in Chapter 5.

If $f(x)$ is a p.d.f. for a continuous variate x, then $f(x)\delta x$ is defined as the probability that values of the variate x fall between $(x-\frac{1}{2}\delta x)$ and $(x+\frac{1}{2}\delta x)$. In other words, $f(x)\delta x$ is the limiting ratio of the number of values of x in the range $(x-\frac{1}{2}\delta x)$ to $(x+\frac{1}{2}\delta x)$ to the total number of all possible values of x. The area under the p.d.f. curve, therefore, represents probability. Hence, the probability that x lies in some finite range from a to b is:

$$p(a \leqslant x \leqslant b) = \int_a^b f(x)\,dx \qquad (7.1)$$

and when a and b represent the extreme limits of all possible values of x, then it is obvious that:

$$\int_{-\infty}^{\infty} f(x)\,dx = 1 \qquad (7.2)$$

PROPERTIES OF DISTRIBUTIONS

The normal p.d.f. is an example of a function which has extreme limits of $-\infty$ and ∞. Many other p.d.f.'s, however, are completely described in the range 0 to ∞ or, more particularly, in the range a to b. If the integral of Equation (7.2) is taken to be identical to zero for all values of x outside the possible range of x, then:

$$\int_{-\infty}^{\infty} f(x)\,dx = \int_{a}^{b} f(x)\,dx = 1 \tag{7.3}$$

where, in Equation (7.3), a can be any value less than b but is generally either $-\infty$ or more often zero. The limit b can have any value greater than a but is generally $+\infty$.

The following are some typical probability density functions for continuous random variates:

Normal:

$$f(x) = \frac{1}{\sigma\sqrt{(2\pi)}} \exp\left[-\frac{(x-\mu)^2}{2\sigma^2}\right] \quad \text{for } -\infty < x < \infty \tag{7.4}$$

Lognormal:

$$f(x) = \frac{1}{x\sigma\sqrt{(2\pi)}} \exp\left[-\frac{(\log x - \mu)^2}{2\sigma^2}\right] \quad \text{for } 0 < x < \infty \tag{7.5}$$

Weibull:

$$f(x) = \frac{\beta}{\lambda^\beta} x^{\beta-1} \exp\left[-\left(\frac{x}{\lambda}\right)^\beta\right] \quad \text{for } x \geq 0,\ \beta > 0,\ \lambda > 0 \tag{7.6}$$

Gamma:

$$f(x) = \frac{1}{\lambda^\beta \Gamma(\beta)} x^{(\beta-1)} \exp\left(\frac{-x}{\lambda}\right) \quad \text{for } x \geq 0 \tag{7.7}$$

Exponential:

$$f(x) = \frac{1}{\lambda} \exp\left(\frac{-x}{\lambda}\right) \quad \text{for } x \geq 0 \tag{7.8}$$

Special Erlangian:

$$f(x) = \frac{x}{\lambda^2} \exp\left(\frac{-x}{\lambda}\right) \quad \text{for } x \geq 0 \tag{7.9}$$

It is seen that most of these distribution functions are related to each other. The lognormal is the normal function with the variate expressed logarithmically and the exponential and the special Erlangian functions are particular cases from either the Weibull or the gamma functions. Many other types of p.d.f. are possible and some additional cases will be discussed as and when the need arises to consider them.

What has been said about continuous random variates applies generally to discrete random variates. The reader has already been introduced to discrete random variates in Chapters 1 and 5. If $f(r)$ represents the distribution function for a discrete variate r, then $f(r_j)$ is defined as the probability of occurrence of the discrete value $r = r_j$. The variate r has only point values and typically $r = 0, 1, 2, \ldots, n$, where each r is some integer value and n may be ∞. In line with Equations (7.2) and (7.3), it is apparent that:

$$\sum_{r=0}^{n} f(r) = 1 \tag{7.10}$$

The two-state system of 'success' or 'failure' is a very particular example of a discrete distribution function where, say, $r = 0$ or $r = 1$. In general, any form of multi-state system may be represented by some form of discrete distribution function. A particular example of a multi-state function, known as the binomial distribution function, was described in Chapter 5, where:

Binomial,

$$f(r) = \binom{n}{r} p^r (1-p)^{n-r} \tag{7.11}$$

and the reader is referred back to Chapter 5 for the derivation and meaning of this function.

If, in Equation (7.11), n is allowed to become very large, p to become very small, whilst the product np remains finite, then:

$$\frac{n!}{(n-r)!} \to n^r$$

and:

$$(1-p)^{n-r} \to e^{-np}$$

whence:

$$f(r) = \frac{(np)^r}{r!} e^{-np}$$

or:

$$f(r) = \frac{\alpha^r}{r!} e^{-\alpha} \tag{7.12}$$

where:

$$\alpha = np \tag{7.13}$$

The relationship of Equation (7.12) is known as the Poisson probability function.

PROPERTIES OF DISTRIBUTIONS 241

As will be seen later, it is useful to consider the Laplace transformations of distribution functions. The two-sided transformation of a p.d.f. is:

$$L[f(x)] = \int_{-\infty}^{\infty} e^{-sx} f(x) \, dx \tag{7.14}$$

for the conditions of absolute convergence discussed in Chapter 6. If the function $f(x)$ only exists between the limits a and b, then the integral of Equation (7.14) is taken to be everywhere zero outside these limits. Hence, if $f(x) \equiv 0$ for $x < 0$, then:

$$L[f(x)] = \int_{0}^{\infty} e^{-sx} f(x) \, dx \tag{7.15}$$

If, now the complex variable s is set equal to zero, then:

$$\{L[f(x)]\}_{s=0} = \int_{-\infty}^{\infty} f(x) \, dx = 1 \tag{7.16}$$

and Equation (7.16) represents an inherent property of the Laplace transformations of all probability density functions.

A similar property exists for discrete distribution functions if the Laplace transformation is defined for these functions as:

$$L[f(r)] = \sum_{r=0}^{\infty} e^{-sr} f(r) \tag{7.17}$$

where the summation is taken to be zero for every value of r greater than n in those cases where r does not exist at values greater than n.
For $s = 0$:

$$\{L[f(r)]\}_{s=0} = \sum_{r=0}^{\infty} f(r) = 1 \tag{7.18}$$

Example 7.1
Find the Laplace transformations of the normal and the exponential p.d.f.'s and show that their values are unity for $s = 0$.
For the normal p.d.f.:

$$L[f(x)] = \frac{1}{\sigma \sqrt{(2\pi)}} \int_{-\infty}^{\infty} \exp(-sx) \exp\left[-\frac{(x-\mu)^2}{2\sigma^2}\right] dx$$

which, on rearranging the exponential terms, gives:

$$L[f(x)] = \exp[-(\mu s - \tfrac{1}{2}\sigma^2 s^2)] \left[\frac{1}{\sigma \sqrt{(2\pi)}} \int_{-\infty}^{\infty} \exp\left\{-\frac{[x-(\mu-\sigma^2 s)]^2}{2\sigma^2}\right\} dx\right]$$

The expression under the integral sign is a normal p.d.f. with a new mean value of $(\mu - \sigma^2 s)$ instead of μ, hence:

$$\frac{1}{\sigma \sqrt{(2\pi)}} \int_{-\infty}^{\infty} \exp\left\{-\frac{[x-(\mu-\sigma^2 s)]^2}{2\sigma^2}\right\} dx = 1$$

and consequently:

$$L[f(x)] = \exp[-(\mu s - \tfrac{1}{2}\sigma^2 s^2)] \tag{7.19}$$

Setting $s = 0$ obviously gives a value of unity to the Laplace transformation.

For the exponential p.d.f.:

$$L[f(x)] = \frac{1}{\lambda} \int_0^{\infty} e^{-sx} e^{-x/\lambda} dx$$

whence, from Table A.2:

$$L[f(x)] = \frac{1}{1+\lambda s} \tag{7.20}$$

and, once again:

$$L[f(x)] = 1 \quad \text{for } s = 0.$$

The Laplace transforms of these and other distributions are given in Table 7.1.

Moments of Distributions

The characteristics of any probability distribution may be described by its 'moments'. A single moment of a distribution may be defined as the 'weight' of every value of the variate in relationship to some fixed value. If this fixed value is x_0, the corresponding moment is given by:

$$M = \int_{-\infty}^{\infty} (x - x_0) f(x) dx \tag{7.21}$$

or:

$$M = \sum_0^{\infty} (r - r_0) f(r) \tag{7.22}$$

Considering the distribution functions as physical or geometrical shapes then Equations (7.21) and (7.22) can be seen to correspond to moments in the mechanical sense for similar shapes of unity density.

An mth moment is defined by:

$$M_m = \int_{-\infty}^{\infty} (x - x_0)^m f(x) dx \tag{7.23}$$

PROPERTIES OF DISTRIBUTIONS

with a similar relationship for discrete distributions. The particular value of the variate x_0 or r_0 can take on any value, but the zero value is generally the one of particular interest. Setting x_0 or r_0 to zero gives the mth moment about the origin where:

$$M_m = \int_{-\infty}^{\infty} x^m f(x)\,dx \qquad (7.24)$$

for continuous distributions, and:

$$M_m = \sum_{r=0}^{\infty} r^m f(r) \qquad (7.25)$$

for discrete distributions.

The first moment about the origin is:

$$M_1 = \int_{-\infty}^{\infty} x f(x)\,dx \qquad (7.26)$$

and this is obviously equal to the mean value, μ, of the distribution or, in the physical sense, to the position of the centre of gravity of the distribution shape. Hence:

$$\mu = M_1 = \int_{-\infty}^{\infty} x f(x)\,dx \qquad (7.27)$$

and:

$$\mu = M_1 = \sum_{r=0}^{\infty} r f(r) \qquad (7.28)$$

It is interesting now to consider the expansion of the exponential term in the Laplace transform of a distribution as defined in Equation (7.14). The expansion gives:

$$L[f(x)] = \int_{-\infty}^{\infty} \left(1 - sx + \frac{s^2 x^2}{2!} - \frac{s^3 x^3}{3!} + \ldots\right) f(x)\,dx \qquad (7.29)$$

which, from Equation (7.24), now becomes:

$$L[f(x)] = \left(1 - sM_1 + \frac{s^2}{2!}M_2 - \frac{s^3}{3!}M_3 + \ldots\right) \qquad (7.30)$$

or:

$$L[f(x)] = \sum_{m=0}^{\infty} (-1)^m \frac{s^m}{m!} M_m \qquad (7.31)$$

and this can be seen to apply for discrete distributions as well.

Since the Laplace transform of a distribution completely defines the distribution, then the distribution can also be completely defined in terms of all the moments about the origin.

Table 7.1 Probability distribution expressions

Type of distribution	Range of variate	Probability density function $f(x)$ or $f(r)$	Laplace transform of $f(x)$ $L[f(x)]$
General	$-\infty < x < \infty$	$(f)x$	$\int_{-\infty}^{\infty} \exp(-sx) f(x)\, dx$
Normal	$-\infty < x < \infty$	$\dfrac{1}{\sigma\sqrt{(2\pi)}} \exp[-(x-\mu)^2/2\sigma^2]$	$\exp[-(\mu s - \tfrac{1}{2}\sigma^2 s^2)]$
Lognormal	$0 < x < \infty$	$\dfrac{1}{x\sigma\sqrt{(2\pi)}} \exp[-(\log x - \mu)^2/2\sigma^2]$	$\sum_{j=0}^{\infty}(-1)^j \dfrac{s^j}{j!}\exp[(j\mu + \tfrac{1}{2}j^2\sigma^2)]$
Weibull	$x \geq 0$, $\beta > 0$, $\lambda > 0$	$\dfrac{\beta x^{\beta-1}}{\lambda^\beta}\exp[-(x/\lambda)^\beta]$	$\sum_{j=0}^{\infty}(-1)^j \dfrac{(\lambda s)^j}{j!} \Gamma\!\left(\dfrac{j+\beta}{\beta}\right)$
Gamma	$x \geq 0$, $\beta > 0$	$\dfrac{x^{\beta-1}}{\lambda^\beta \Gamma(\beta)}\exp(-x/\lambda)$	$\dfrac{1}{(1+s\lambda)^\beta}$
Special Erlangian	$x \geq 0$	$\dfrac{x}{\lambda^2}\exp(-x/\lambda)$	$\dfrac{1}{(1+s\lambda)^2}$
Exponential	$x \geq 0$	$\dfrac{1}{\lambda}\exp(-x/\lambda)$	$\dfrac{1}{1+s\lambda}$
Rectangular	$a \leq x \leq b$	$\dfrac{1}{b-a}$	$\dfrac{\exp(-as) - \exp(-bs)}{(b-a)s}$
Poisson	$0 \leq \alpha \leq \infty$, $r = 1, 2 \ldots n$	$\dfrac{\alpha^r}{r!}\exp(-\alpha)$	$\exp\{-\alpha[1-(\exp-s)]\}$
Binomial	$0 \leq p \leq 1$, $r = 1, 2 \ldots n$	$\binom{n}{r} p^r (1-p)^{n-r}$	$\sum_{j=0}^{\infty}(-1)^j \dfrac{s^j}{j!} M_j$

PROPERTIES OF DISTRIBUTIONS

	mth moment about the origin M_m	First moment about the origin (mean) M_1
:	$\int_{-\infty}^{\infty} x^m f(x)\,dx$ $(-1)^m \left\{\dfrac{d^m L[f(x)]}{ds^m}\right\}_{s=0}$	$\int_{-\infty}^{\infty} x f(x)\,dx$ or: $-\left\{\dfrac{dL[f(x)]}{ds}\right\}_{s=0}$
d:	$M_{2m} = \mu^{2m} \sum_{j=0}^{m} \dfrac{(2m)!}{(2m-2j)!\,j!} \left(\dfrac{\sigma^2}{2\mu^2}\right)^j$ $M_{2m-1} = \mu^{2m-1} \sum_{j=0}^{m-1} \dfrac{(2m-1)!}{(2m-1-2j)!\,j!} \left(\dfrac{\sigma^2}{2\mu^2}\right)^j$	μ
	$\exp[(m\mu + \tfrac{1}{2} m^2 \sigma^2)]$	$\exp[(\mu + \tfrac{1}{2}\sigma^2)]$
	$\lambda^m\, \Gamma\!\left(\dfrac{m+\beta}{\beta}\right)$	$\lambda \Gamma\!\left(\dfrac{1+\beta}{\beta}\right)$
ere:	$\lambda^m [\beta]^m$ $[\beta]^m = \beta(\beta+1)\dots(\beta+m-1)$	$\lambda\beta$
	$\lambda^m (m+1)!$	2λ
	$\lambda^m\, m!$	λ
	$\dfrac{b^{m+1} - a^{m+1}}{(m+1)(b-a)}$	$\dfrac{a+b}{2}$
ere: d:	$\sum_{j=1}^{m} a_j(m)\, \alpha^j$ $a_1(m) = 1$ $a_j(m) = \sum_{k=0}^{m-j} j^{(m-k-j)} a_{j-1}(j-1+k)$	α
	$\sum_{j=1}^{m} a_j(m) \dfrac{n!}{(n-j)!} p^j$ $a_j(m) =$ as Poisson	np

Table 7.1 (*cont.*)

Type of distribution	Second moment about the origin M_2	mth moment about the mean $\bar{M}_m$
General	$\int_{-\infty}^{\infty} x^2 f(x)\,\mathrm{d}x$ or: $\left\{\dfrac{\mathrm{d}^2 L[f(x)]}{\mathrm{d}s^2}\right\}_{s=0}$	$\int_{-\infty}^{\infty} (x-M_1)^m f(x)\,\mathrm{d}x$ or: $\sum_{j=0}^{m}(-1)^j \binom{m}{j} M_{m-j} M_1^j$
Normal	$\mu^2 + \sigma^2$	and: $\bar{M}_{2m} = \dfrac{(2m)!}{m!}\left(\dfrac{\sigma^2}{2}\right)^m$ $\bar{M}_{2m-1} = 0$
Lognormal	$\exp[2(\mu+\sigma^2)]$	$\exp(m\mu)\sum_{j=0}^{m}(-1)^j\binom{m}{j}\exp\{\tfrac{1}{2}\sigma^2[(m-j)^2 +$
Weibull	$\lambda^2\,\Gamma\!\left(\dfrac{2+\beta}{\beta}\right)$	$\lambda^m \sum_{j=0}^{m}(-1)^j\binom{m}{j}\Gamma\!\left(\dfrac{m-j+\beta}{\beta}\right)\Gamma^j\!\left(\dfrac{1+\beta}{\beta}\right)$
Gamma	$\lambda^2\,\beta(\beta+1)$	$\lambda^m([\beta]-\beta)^m$ where: $[\beta]^m = \beta(\beta+1)\ldots(\beta+m-1)$
Special Erlangian	$6\lambda^2$	$\lambda^m([2]-2)^m$ or: $\lambda^m\, m!\sum_{j=0}^{m}(-1)^j\dfrac{2^j}{j!}(m-j+1)$
Exponential	$2\lambda^2$	$\lambda^m([1]-1)^m$ or: $\lambda^m\, m!\sum_{j=0}^{m}(-1)^j\dfrac{1}{j!}$
Rectangular	$\dfrac{a^2+ab+b^2}{3}$	$\dfrac{[1+(-1)^m](b-a)^m}{2^{m+1}(m+1)}$
Poisson	$\alpha + \alpha^2$	$\sum_{j=1}^{m} b_j(m)\,\alpha^j$ where: $b_j(m) = \sum_{k=0}^{j-1}(-1)^k\binom{m}{k}a_{j-k}(m-k)$
Binomial	$np + n(n-1)p^2$	$\sum_{j=0}^{m}(-1)^j\binom{m}{j}M_{m-j}M_1^j$

PROPERTIES OF DISTRIBUTIONS

Second moment about the mean (variance) $\bar{M}_2 = \sigma_x^2$	Moment coefficient $K_m = \dfrac{\bar{M}_m}{(\bar{M}_2)^{m/2}}$	Coefficient of skewness K_3
$\displaystyle\int_{-\infty}^{\infty}(x - M_1)^2 f(x)\,dx$ $M_2 - M_1^2$	$\dfrac{\bar{M}_m}{(\bar{M}_2)^{m/2}}$	$\dfrac{\bar{M}_3}{(\bar{M}_2)^{3/2}}$
σ^2	$K_{2m} = \dfrac{(2m)!}{2^m m!}$	0
$p(2\mu + \sigma^2)[\exp(\sigma^2) - 1]$	$\dfrac{\bar{M}_m}{(\bar{M}_2)^{m/2}}$	$\dfrac{[\exp(3\sigma^2) - 3\exp(\sigma^2) + 2]}{[\exp(\sigma^2) - 1]^{\frac{3}{2}}}$
$\left[\Gamma\left(\dfrac{2+\beta}{\beta}\right) - \Gamma^2\left(\dfrac{1+\beta}{\beta}\right)\right]$	$\dfrac{\bar{M}_m}{(\bar{M}_2)^{m/2}}$	$\dfrac{\bar{M}_3}{(\bar{M}_2)^{\frac{3}{2}}}$
$\lambda^2 \beta$	$\dfrac{([\beta] - \beta)^m}{\beta^{m/2}}$	$\dfrac{2}{\sqrt{\beta}}$
$2\lambda^2$	$\dfrac{([2] - 2)^m}{2^{m/2}}$	$\sqrt{2}$
λ^2	$([1] - 1)^m$	2
$\dfrac{(b-a)^2}{12}$	$\dfrac{[1 + (-1)^m]\,12^{m/2}}{2^{m+1}(m+1)}$	0
α	$\alpha^{-m/2} \displaystyle\sum_{j=1}^{m} b_j(m)\,\alpha^j$	$\dfrac{1}{\sqrt{\alpha}}$
$np(1-p)$	$\dfrac{\bar{M}_m}{(\bar{M}_2)^{m/2}}$	$\dfrac{1 - 2p}{\sqrt{[np(1-p)]}}$

Table 7.1 (*cont.*)

Type of distribution	Coefficient of kurtosis K_4	K_5
General	$\dfrac{\bar{M}_4}{(\bar{M}_2)^2}$	$\dfrac{\bar{M}_5}{(\bar{M}_2)^{5/2}}$
Normal	3	0
Lognormal	$\dfrac{[\exp(6\sigma^2) - 4\exp(3\sigma^2) + 6\exp(\sigma^2) - 3]}{[\exp(\sigma^2) - 1]^2}$	$\dfrac{\bar{M}_5}{(\bar{M}_2)^{5/2}}$
Weibull	$\dfrac{\bar{M}_4}{(\bar{M}_2)^2}$	$\dfrac{\bar{M}_5}{(\bar{M}_2)^{5/2}}$
Gamma	$3 + \dfrac{6}{\beta}$	$\dfrac{20\beta + 24}{\beta\sqrt{\beta}}$
Special Erlangian	6	$16\sqrt{2}$
Exponential	9	44
Rectangular	$\dfrac{9}{5}$	0
Poisson	$3 + \dfrac{1}{\alpha}$	$\dfrac{10\alpha + 1}{\alpha\sqrt{\alpha}}$
Binomial	$3 - \dfrac{6}{n} + \dfrac{1}{np(1-p)}$	$\dfrac{\bar{M}_5}{(\bar{M}_2)^{5/2}}$

PROPERTIES OF DISTRIBUTIONS

K_6	Cumulative distribution function $p(x)$ or $p(r)$	Hazard function $z(x)$
$\dfrac{\bar{M}_6}{(\bar{M}_2)^3}$	$\displaystyle\int_{-\infty}^{x} f(x)\,\mathrm{d}x$	$\dfrac{f(x)}{1-p(x)}$
15	$\dfrac{1}{\sigma\sqrt{(2\pi)}}\displaystyle\int_{-\infty}^{x} \exp[-(x-\mu)^2/2\sigma^2]\,\mathrm{d}x$	$\dfrac{f(x)}{1-p(x)}$
$\dfrac{\bar{M}_6}{(\bar{M}_2)^3}$	$\dfrac{1}{\sigma\sqrt{(2\pi)}}\displaystyle\int_{0}^{\log x} \exp[-(y-\mu)^2/2\sigma^2]\,\mathrm{d}y$	$\dfrac{f(x)}{1-p(x)}$
$\dfrac{\bar{M}_6}{(\bar{M}_2)^3}$	$1-\exp[-(x/\lambda)^\beta]$	$\dfrac{\beta x^{\beta-1}}{\lambda^\beta}$
$-\dfrac{130}{\beta}+\dfrac{120}{\beta^2}$	$1-\exp(-x/\lambda)\displaystyle\sum_{j=0}^{\beta-1} \left(\dfrac{x}{\lambda}\right)^j \dfrac{1}{\Gamma(j+1)}$	$\dfrac{x^{\beta-1}}{\lambda^\beta\,\Gamma(\beta)\displaystyle\sum_{j=0}^{\beta-1}\left(\dfrac{x}{\lambda}\right)^j \dfrac{1}{\Gamma(j+1)}}$
110	$1-\left(1+\dfrac{x}{\lambda}\right)\exp(-x/\lambda)$	$\dfrac{x}{\lambda(x+\lambda)}$
265	$1-\exp(-x/\lambda)$	$\dfrac{1}{\lambda}$
$\dfrac{27}{7}$	$\dfrac{x-a}{b-a}$	$\dfrac{1}{b-x}$
$+\dfrac{25}{\alpha}+\dfrac{1}{\alpha^2}$	$\displaystyle\sum_{j=0}^{r} \dfrac{\alpha^j}{j!}\exp(-\alpha)$	—
$\dfrac{\bar{M}_6}{(\bar{M}_2)^3}$	$\displaystyle\sum_{j=0}^{r}\binom{n}{j}p^j(1-p)^{n-j}$	—

Example 7.2

Given that the Laplace transform of an exponential distribution is $1/(1+\lambda s)$, find the expression for the mth moment about the origin. Expanding $1/(1+\lambda s)$ by the binomial theorem gives:

$$\frac{1}{1+\lambda s} = (1 - \lambda s + \lambda^2 s^2 - \lambda^3 s^3 + \ldots)$$

$$= \sum_{j=0}^{\infty} (-1)^j s^j \lambda^j$$

Comparing this with Equation (7.31) shows that:

$$\lambda^m = \frac{M_m}{m!}$$

or:

$$M_m = m! \, \lambda^m$$

Other examples of general formulae for moments about the origin for various distributions are given in Table 7.1.

If Equation (7.31) is differentiated with respect to s and the complex variable s is then set equal to zero, this gives:

$$\left\{\frac{\mathrm{d}L[f(x)]}{\mathrm{d}s}\right\}_{s=0} = -M_1 = -\mu \qquad (7.32)$$

Repeating the process, leads to the general relationship:

$$\left\{\frac{\mathrm{d}^m L[f(x)]}{\mathrm{d}s^m}\right\}_{s=0} = (-1)^m M_m \qquad (7.33)$$

The Laplace transform of a distribution can, therefore, be used as a function for generating the mth moment about the origin.

Applying Equation (7.32) to the Laplace transform of the normal p.d.f. which was calculated in Example 7.1 gives:

$$-\frac{\mathrm{d}L[f(x)]}{\mathrm{d}s} = (\mu - \sigma^2 s) \exp\left[-(s\mu - \tfrac{1}{2}\sigma^2 s^2)\right]$$

hence:

$$M_1 = -\left\{\frac{\mathrm{d}L[f(x)]}{\mathrm{d}s}\right\}_{s=0} = \mu \qquad (7.34)$$

In the same way, it can be seen that the next three moments for the normal

PROPERTIES OF DISTRIBUTIONS

p.d.f. are:

$$M_2 = \mu^2 + \sigma^2 \tag{7.35}$$

$$M_3 = \mu^3 + 3\mu\sigma^2 \tag{7.36}$$

$$M_4 = \mu^4 + 6\mu^2\sigma^2 + 3\sigma^4 \tag{7.37}$$

The same results can also be obtained from the general formula for the moments of a normal p.d.f. given in Table 7.1.

Example 7.3

Taking the Laplace transform of the gamma and Poisson distributions as given in Table 7.1, find the mean value of these two distributions.

For the gamma distribution:

$$-\frac{dL[f(x)]}{ds} = \frac{\beta\lambda}{(1+\lambda s)^{\beta+1}}$$

hence:

$$\mu = -\left\{\frac{dL[f(x)]}{ds}\right\}_{s=0} = \beta\lambda$$

For the Poisson distribution:

$$-\frac{dL[f(r)]}{ds} = \alpha\exp(-s)\exp[-\alpha(1-e^{-s})]$$

hence:

$$\mu = -\left\{\frac{dL[f(r)]}{ds}\right\}_{s=0} = \alpha$$

Another class of moments, which are of interest, may be obtained by setting $x_0 = M_1 = \mu$ in Equation (7.23). This gives the *m*th moment about the mean as:

$$\bar{M}_m = \int_{-\infty}^{\infty} (x - M_1)^m f(x)\,dx \tag{7.38}$$

and:

$$\bar{M}_m = \sum_{r=0}^{\infty} (r - M_1)^m f(r) \tag{7.39}$$

It is obvious that:

$$\bar{M}_1 = 0 \tag{7.40}$$

and, in particular, that:

$$\bar{M}_2 = \int_{-\infty}^{\infty} (x - M_1)^2 f(x)\,dx \tag{7.41}$$

or:

$$\bar{M}_2 = \sum_{r=0}^{\infty}(r-M_1)^2 f(r) \qquad (7.42)$$

$\bar{M}_2$, or the second moment about the mean, is known as the variance of a distribution and is generally denoted by the symbol σ^2 or σ_x^2, so:

$$\bar{M}_2 = \sigma^2 \qquad (7.43)$$

This is equivalent, in the physical sense, to the 'radius of gyration' of the distribution shape. In the statistical sense, it is a measure of the dispersion or deviation of the variate about the mean value. The square root of the variance, namely σ, is known as the 'standard deviation' of the distribution.

By expanding Equation (7.41), the variance may be expressed in terms of the first and second moments about the origin, that is:

$$\sigma^2 = \bar{M}_2 = M_2 - M_1^2 = M_2 - \mu^2 \qquad (7.44)$$

More generally, by expanding Equation (7.38), it is seen that:

$$\bar{M}_m = \sum_{j=0}^{m}(-1)^j \binom{m}{j} M_{m-j} M_1^j \qquad (7.45)$$

So that any moment about the mean may be expressed in terms of the moments about the origin. Some typical forms of this expression are given in Table 7.1.

Considering again the Laplace transform of a distribution, where:

$$L[f(x)] = \int_{-\infty}^{\infty} e^{-sx} f(x)\,dx$$

this may be written in the form:

$$L[f(x)] = e^{-M_1 s} \int_{-\infty}^{\infty} e^{-(x-M_1)s} f(x)\,dx$$

which gives:

$$L[f(x)] = e^{-M_1 s} \int_{-\infty}^{\infty} \left[1 - (x-M_1)s + \frac{(x-M_1)^2}{2!}s^2 - \ldots\right] f(x)\,dx$$

or:

$$L[f(x)] = e^{-M_1 s} \sum_{m=0}^{\infty}(-1)^m \frac{s^m}{m!} \bar{M}_m \qquad (7.46)$$

Here the Laplace transform of a distribution is expressed in terms of the mean and all the moments about the mean. This may be compared with Equation (7.31) where the transform was expressed in terms of all the moments about the origin.

PROPERTIES OF DISTRIBUTIONS

Example 7.4

Find the standard deviation of the lognormal distribution given that the mth moment about the origin is $\exp(m\mu + \tfrac{1}{2}m^2\sigma^2)$. Here:

$$M_1 = \exp(\mu + \tfrac{1}{2}\sigma^2)$$

and:

$$M_2 = \exp(2\mu + 2\sigma^2)$$

Therefore, from Equation (7.44):

$$\sigma_x^2 = M_2 - M_1^2$$
$$= \exp(2\mu + 2\sigma^2) - \exp(2\mu + \sigma^2)$$

and hence:

$$\sigma_x = \exp(\mu + \tfrac{1}{2}\sigma^2)\sqrt{[\exp(\sigma^2) - 1]}$$

The moments about the mean, $\bar{M}_m$, can be expressed in a more dimensionless form if a coefficient K_m is defined where:

$$K_m = \frac{\bar{M}_m}{(\bar{M}_2)^{m/2}} \qquad (7.47)$$

The usefulness of this coefficient may be demonstrated with reference to the normal p.d.f. From the general equation for the mth moment about the mean the expression for the normal p.d.f. becomes (see Table 7.1):

$$\bar{M}_{2m} = \frac{(2m)!}{m!}\left(\frac{\sigma^2}{2}\right)^m$$

and:

$$\bar{M}_{2m-1} = 0$$

and since $\bar{M}_2 = \sigma^2$, then:

$$K_{2m} = \frac{(2m)!}{2^m m!}$$

and:

$$K_{2m-1} = 0$$

Particular values of K for the normal p.d.f. are then:

$$K_2 = 1$$
$$K_3 = 0$$
$$K_4 = 3$$
$$K_5 = 0$$

The coefficient K_3 is generally known as the coefficient of skewness for a distribution as it gives an indication of the degree of asymmetry. A distribution which is symmetrical about the mean leads to a value of K_3 equal to zero as has just been shown to apply to the normal distribution. Negative values of K_3 show a distribution peak to the right of the mean and positive values a peak to the left of the mean.

The quantity K_4 is known as the coefficient of 'kurtosis' and describes the extent to which the distribution is peaked. The value of $K_4 = 3$ which applies for the normal distribution is taken as the standard measure of kurtosis. Distributions where K_4 is less than 3 have a flatter top than a normal distribution and are called 'platykurtic'. When K_4 is greater than 3, the distribution is more peaked than the normal and is said to be 'leptokurtic'. In the case where K_4 is equal to 3 the distribution is described as 'mesokurtic'.

As has been mentioned earlier, a distribution may be completely described by all its moments. However, in practice, an adequate description can normally be obtained by considering only the moments up to the fourth. From this limited set of moments the particular characteristics usually derived in order to describe a distribution are the mean (M_1 or μ), the variance ($\bar{M}_2$ or σ^2), the coefficient of skewness (K_3) and the coefficient of kurtosis (K_4).

From the previous working the reader may readily verify that:

$$\mu = M_1 = -\left\{\frac{dL[f(x)]}{ds}\right\}_{s=0} \tag{7.48}$$

$$\sigma^2 = \bar{M}_2 = \left\{\frac{d^2 L[f(x)]}{ds^2}\right\}_{s=0} - \mu^2 \tag{7.49}$$

$$K_3 = -\frac{1}{\sigma^3}\left\{\frac{d^3 L[f(x)]}{ds^3}\right\}_{s=0} - \frac{\mu(\mu^2 + 3\sigma^2)}{\sigma^3} \tag{7.50}$$

$$K_4 = \frac{1}{\sigma^4}\left\{\frac{d^4 L[f(x)]}{ds^4}\right\}_{s=0} - \frac{\mu^2(\mu^2 + 6\sigma^2)}{\sigma^4} - \frac{4\mu K_3}{\sigma} \tag{7.51}$$

Example 7.5

Find the coefficients of skewness and kurtosis for the gamma distribution given that the Laplace transform of its density function is $1/(1+\lambda s)^\beta$.

It has already been shown in Example 7.3 that for this case:

$$\mu = \beta\lambda$$

PROPERTIES OF DISTRIBUTIONS

Further differentiation of the Laplace transform leads to:

$$\left\{\frac{d^2 L[f(x)]}{ds^2}\right\}_{s=0} = \lambda^2 \beta(\beta+1)$$

$$-\left\{\frac{d^3 L[f(x)]}{ds^3}\right\}_{s=0} = \lambda^3 \beta(\beta+1)(\beta+2)$$

and:

$$\left\{\frac{d^4 L[f(x)]}{ds^4}\right\}_{s=0} = \lambda^4 \beta(\beta+1)(\beta+2)(\beta+3)$$

Hence, from Equation (7.49):

$$\sigma^2 = \lambda^2 \beta$$

From Equation (7.50):

$$K_3 = \frac{2}{\sqrt{\beta}}$$

and from Equation (7.51):

$$K_4 = 3 + \frac{6}{\beta}$$

It may be noted that as $\beta \to \infty$, then $K_3 \to 0$ and $K_4 \to 3$. Which is an indication that the gamma distribution approaches the form of the normal distribution as β becomes large.

Values of the coefficients K_3 and K_4 for a range of typical distributions are shown in Table 7.1. It will be seen, by studying these coefficients, that the lognormal, Weibull, gamma, Poisson and binomial distributions all approach the form of a normal distribution when certain of their parameters approach certain limits.

As shown in Equations (5.40) and (5.41) in Chapter 5, the normal distribution function may be written in the standardized form:

$$f(t) = \frac{1}{\sqrt{(2\pi)}} e^{-t^2/2} \qquad (7.52)$$

where:

$$t = \frac{x - \mu}{\sigma} \qquad (7.53)$$

For the gamma distribution, as seen in Example 7.5, the normal limit occurs as β becomes large. In this case the equivalent 'normal' mean is

$\lambda\beta$ and the equivalent 'normal' variance is $\lambda^2\beta$, hence:

$$t = \frac{x - \lambda\beta}{\lambda\sqrt{\beta}} \tag{7.54}$$

The normal distribution may also be used as an approximation to discrete distributions such as the Poisson and the binomial. Some care needs to be taken in this approach since, strictly speaking, a continuous distribution cannot represent a discrete distribution. However, with this reservation in mind, the Poisson approaches the normal form as $\alpha \to \infty$ and then:

$$t = \frac{r - \alpha}{\sqrt{\alpha}} \tag{7.55}$$

For the binomial distribution, the limit occurs as $np \to \infty$ and here:

$$t = \frac{r - np}{\sqrt{[np(1-p)]}} \tag{7.56}$$

In this latter case, it can be shown that if $(np)^{3/2} > 1.07$, the error in using the normal distribution function instead of the binomial never exceeds 0.05 for any r.

Since, as was shown in Equations (7.11) to (7.13), the Poisson distribution is a limiting case of the binomial these two distributions can be interchanged by the relationship $\alpha = np$ if α and np are sufficiently large.

Example 7.6

The probability of an event occurring per trial is 0.1. What is the probability of exactly 12 events occurring out of 100 trials? How would this calculated probability be affected by using the normal and Poisson approximations to the binomial distribution?

Here, $p = 0.1$, $n = 100$ and $r = 12$. Hence the binomial distribution function gives:

$$f(12) = \binom{100}{12} p^{12}(1-p)^{88}$$

which, from tables gives:

$$f(12) = 0.099$$

Using the Poisson approximation:

$$\alpha = np = 10$$

therefore:

$$f(12) = \frac{10^{12}}{12!} e^{-10}$$

$$= 0.095$$

PROPERTIES OF DISTRIBUTIONS

Making the normal substitution for the binomial:

$$t = \frac{r-np}{\sqrt{[np(1-p)]}}$$

$$= \frac{12-10}{\sqrt{(10 \times 0.9)}}$$

then, from the standard form of the normal p.d.f.:

$$f(t) = 0.106$$

So, even though np is not very large in this example, both the Poisson and normal distributions give a reasonable approximation to the binomial.

The Cumulative Distribution Function

As mentioned at the beginning of this chapter, the p.d.f. for a continuous variate represents the probability $f(x)\,\delta x$ that the variate lies between the limits $(x-\tfrac{1}{2}\delta x)$ and $(x+\tfrac{1}{2}\delta x)$. Obviously, this probability tends to zero in the limit $\delta x \to 0$. Interest often lies, therefore, in the probability of x falling in some finite range from, say, a to b. This probability was represented in Equation (7.1). A more particular form of Equation (7.1) arises when the range considered is from the lower limit of the variate, which is often zero, to some specific value, x, of the variate. Here the probability is:

$$p(x) = \int_0^x f(x)\,\mathrm{d}x \qquad (7.57)$$

The lower limit is typically zero, but may be some other value such as $-\infty$ as in the case of the normal distribution.

Equation (7.57) is defined as the cumulative distribution function of the probability density function $f(x)$. This function has already been briefly referred to in connexion with Equation (2.3) in Chapter 2 and in connexion with Equation (5.38) in Chapter 5.

Differentiating Equation (7.57) obviously leads to the relationship:

$$f(x) = \frac{\mathrm{d}p(x)}{\mathrm{d}x} \qquad (7.58)$$

If the variate has a lower limit of zero and an upper limit of $+\infty$ then from Equation (7.57):

$$p(0) = 0 \qquad (7.59)$$

and:

$$p(\infty) = 1 \qquad (7.60)$$

The complement of the cumulative function, $\overline{p(x)}$, may be defined where:

$$\overline{p(x)} = 1 - p(x) \tag{7.61}$$

or:

$$\overline{p(x)} = \int_x^\infty f(x)\,dx \tag{7.62}$$

It follows then that:

$$f(x) = -\frac{d\overline{p(x)}}{dx} \tag{7.63}$$

and also that:

$$\overline{p(0)} = 1 \tag{7.64}$$

$$\overline{p(\infty)} = 0 \tag{7.65}$$

The first moment about the origin of a p.d.f., or the mean μ, is:

$$\mu = \int_0^\infty x f(x)\,dx$$

which, from Equation (7.63), may be written as:

$$\mu = -\int_0^\infty x\,d\overline{p(x)} \tag{7.66}$$

Integrating by parts gives:

$$\mu = -[x\overline{p(x)}]_0^\infty + \int_0^\infty \overline{p(x)}\,dx$$

The infinite limit of the function in the brackets is normally zero for most distribution functions, hence:

$$\mu = \int_0^\infty \overline{p(x)}\,dx \tag{7.67}$$

which is expressing the mean of the p.d.f. in terms of the complement of the corresponding cumulative distribution function.

The cumulative distribution function of the normal p.d.f. is not capable of simple analytical expression and tables, such as Table A.1, are usually used for its evaluation as mentioned in Chapter 5. Many other p.d.f.'s do, however, lead to relatively simple analytical expressions for their corresponding cumulative functions. For instance, in the case of the Weibull distribution:

$$f(x) = \frac{\beta x^{\beta-1}}{\lambda^\beta} e^{-(x/\lambda)^\beta}$$

PROPERTIES OF DISTRIBUTIONS

and:
$$p(x) = \int_0^x \frac{\beta x^{\beta-1}}{\lambda^\beta} e^{-(x/\lambda)^\beta} dx$$
$$= [-e^{-(x/\lambda)^\beta}]_0^x$$

or:
$$p(x) = 1 - e^{-(x/\lambda)^\beta} \quad (7.68)$$

Example 7.7

Find the cumulative distribution function for the gamma distribution and hence deduce the cumulative functions for the special Erlangian and exponential distributions.

The gamma distribution function is:
$$f(x) = \frac{x^{\beta-1}}{\lambda^\beta \Gamma(\beta)} e^{-x/\lambda}$$

Therefore:
$$p(x) = \int_0^x \frac{x^{\beta-1}}{\lambda^\beta \Gamma(\beta)} e^{-x/\lambda} dx$$

and integrating by parts leads to the recurrence formula:
$$p(x)_\beta = -\frac{x^{\beta-1}}{\lambda^{\beta-1} \Gamma(\beta)} e^{-x/\lambda} + p(x)_{\beta-1}$$

whence it is seen that:
$$p(x) = 1 - e^{-x/\lambda} \sum_{j=0}^{\beta-1} \left(\frac{x}{\lambda}\right)^j \frac{1}{\Gamma(j+1)} \quad (7.69)$$

In the case of the special Erlangian distribution, $\beta = 2$, hence:
$$p(x) = 1 - \left(1 + \frac{x}{\lambda}\right) e^{-x/\lambda} \quad (7.70)$$

and for the exponential distribution, $\beta = 1$, hence:
$$p(x) = 1 - e^{-x/\lambda} \quad (7.71)$$

Values of Equation (7.71) for the cumulative exponential distribution are tabulated in Table A.4.

Example 7.8

From observations made on a large number of nominally identical systems it is found that the times to the first system failure are normally distributed with a mean value of 1000 h and a standard deviation of 100 h. What is the probability that a system failure occurs between time zero and a time of 750 h? What would this probability be if the distribution of times to failure followed an exponential distribution of the same mean time to failure as the normal?

The standard substitution for the normal distribution is:

$$t = \frac{x-\mu}{\sigma}$$

For this case:

$$t = \frac{750 - 1000}{100}$$

$$= -2.5$$

Looking up the cumulative probability for a standardized normal value of 2.5 in Table A.1 gives a value of 0.4938. This value, of course, refers to the range 0 to x. The range required in the present example is from $-\infty$ to $-x$, therefore:

$$p(x) = 0.5 - 0.4938$$

$$= 0.0062$$

For the exponential case:

$$p(x) = 1 - e^{-t/\lambda}$$

$$= 1 - e^{-750/1000}$$

$$= 1 - e^{-0.75}$$

From Table A.4, this gives:

$$p(x) = 0.528$$

which illustrates the large difference between the characteristics of the normal and exponential distribution functions.

Cumulative distribution functions also exist for discrete probability functions and this has already been indicated with reference to the binomial cumulative distribution function in Equations (5.54) to (5.62) of Chapter 5. Generally the cumulative function, $p(r)$, for discrete distributions is given by:

$$p(r) = \sum_{j=0}^{r} f(j) \tag{7.72}$$

The lower limit of $f(r)$ is normally zero, but the upper limit may be some value n, as in the case of the binomial function, or $+\infty$, as in the case of the

PROPERTIES OF DISTRIBUTIONS

Poisson distribution. The corresponding limits of $p(r)$ are:

$$p(0) = f(0) \tag{7.73}$$

$$p(\infty) = 1 \tag{7.74}$$

and these equations are, of course, the discrete equivalents of Equations (7.59) and (7.60). The complementary function $\overline{p(r)}$ is defined in the same way as $\overline{p(x)}$, namely:

$$\overline{p(r)} = 1 - p(r) \tag{7.75}$$

and hence:

$$\overline{p(r)} = \sum_{j=r+1}^{\infty} f(j) \tag{7.76}$$

The cumulative distribution functions for the binomial and Poisson distributions are:
Binomial:

$$p(r) = \sum_{j=0}^{r} \binom{n}{j} p^j (1-p)^{n-j} \tag{7.77}$$

Poisson:

$$p(r) = \sum_{j=0}^{r} \frac{\alpha^j}{j!} e^{-\alpha} \tag{7.78}$$

The lower limit of the Poisson cumulative function, $p(0)$, and its complement, $\overline{p(0)}$, are very often of interest. It can be seen that:

$$p(0) = e^{-\alpha} \tag{7.79}$$

and:

$$\overline{p(0)} = 1 - e^{-\alpha} \tag{7.80}$$

from which it will be noticed that $\overline{p(0)}$ for the discrete Poisson distribution has the same form as $p(x)$ for the continuous exponential distribution given in Equation (7.71). An equivalence between these two particular functions can therefore be postulated where:

$$\alpha = x/\lambda \tag{7.81}$$

In Example 7.8, the value of $p(x)$ for the exponential distribution was calculated for a mean time to failure of 1000 h and a cumulative range of 0 to 750 h. For this example, then:

$$\alpha = \tfrac{750}{1000} = 0.75$$

From the dimensions of λ, the reciprocal $1/\lambda$ can be looked upon as a mean failure rate and the value of the corresponding α can be considered as a mean number of failures in a given time. So, in this case, the mean failure rate is 1 failure per 1000 h and the mean number of failures, α, in a 750-h period is 0.75. The cumulative exponential function, $p(x)$, is the probability that the time to first failure lies in the range 0 to 750 h for a mean failure rate of 1 failure per 1000 h. On the other hand, the cumulative Poisson function, $p(0)$, is the probability that 0 failures occur in the 750-h period when the mean number of failures for this period is 0.75. The complement, $\overline{p(0)}$, is the probability of 1 or more failures under the same conditions.

$p(x)$ and $\overline{p(0)}$, then, although they lead to the same numerical conclusions in certain cases, are in fact based upon different concepts and strictly a continuous function can never be exactly equated to a discrete function.

· Values for both the cumulative Poisson distribution and the cumulative binomial distribution can be obtained approximately from the curves in Figures A.28 and A.1 to A.27 respectively. For more precise values of these distributions, the reader is referred to the various standard sets of tables which exist. Under certain conditions the cumulative normal distribution may provide an adequate approximation in a similar way to that mentioned earlier in connexion with Example 7.6.

The general forms of both the continuous and discrete cumulative distribution functions that have been discussed so far are illustrated in Figure 7.1 together with the forms of the corresponding density and probability functions.

Conditional Distribution Functions

The cumulative distribution function expresses the probability of the value of a variate lying in the range from the lower limit of the variate to some specific value of the variate. Very often the probability of the variate value lying in some more general range from, say, a to b is of interest as indicated by Equation (7.1). This equation can be written in terms of the cumulative distribution function as follows:

$$p(a \leqslant x \leqslant b) = \int_a^b f(x)\,dx$$
$$= \int_0^b f(x)\,dx - \int_0^a f(x)\,dx$$

or:

$$p(a \leqslant x \leqslant b) = p(b) - p(a) \qquad (7.82)$$

Example 7.9

It is found that the random variations with respect to time in the output voltage signal of a particular system are exponentially distributed with a mean value of 100 V. What is the probability that the output voltage will be found at any time to lie in the range (a) from 90 V to 110 V? and (b) from 70 V to 90 V?

PROPERTIES OF DISTRIBUTIONS

From Equation (7.82):

$$p(90 \leqslant V \leqslant 110) = p(110) - p(90)$$

which, from Table A.4, gives:

$$p(90 \leqslant V \leqslant 110) = 0.667 - 0.593$$
$$= 0.074$$

In the same way it can be seen that:

$$p(70 \leqslant V \leqslant 90) = p(90) - p(70)$$
$$= 0.593 - 0.503$$
$$= 0.090$$

The probability function given by Equation (7.82) represents the limiting ratio of the number of variate values that lie in the range a to b to the total number of all possible variate values. This follows from the definitions of probability, the density function and the cumulative distribution function.

Suppose now that from the total possible range of all variate values some range of values was eliminated and a probability, such as Equation (7.82), was required on the basis of this elimination. An example of the way in which this sort of requirement could arise might be as follows. A manufactured item leaving the production line is found to have random variations in size which follow a particular p.d.f. The inspection process which follows then rejects all sizes which are below a certain value of, say, c. A customer might then be interested in the probability of the size of an item which he receives falling within a certain range of values of, say, from a to b. A probability for the range a to b is therefore required on the basis that or on the condition that the range 0 to c of the total population has been eliminated.

It is seen that this 'conditional probability' $p_c(x)$, must represent the limiting ratio of the number of variate values that lie in the range a to b to the total number of variate values that are in the new population range of c to ∞. Hence:

$$p_c(a \leqslant x \leqslant b) = \frac{\int_a^b f(x)\,dx}{\int_c^\infty f(x)\,dx} \qquad (7.83)$$

where $c \leqslant a \leqslant b$.

Obviously, if c is the lower limit of the variate which may be zero, then:

$$p_c(a \leqslant x \leqslant b) = p(a \leqslant x \leqslant b)$$

Equation (7.83) may be written in terms of the cumulative distribution

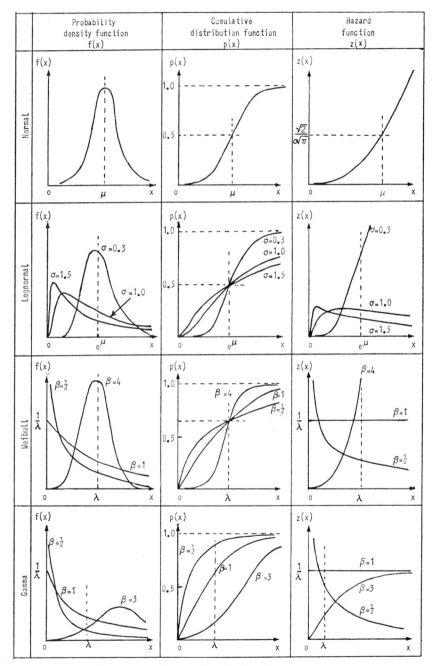

Figure 7.1 Probability distribution functions

PROPERTIES OF DISTRIBUTIONS

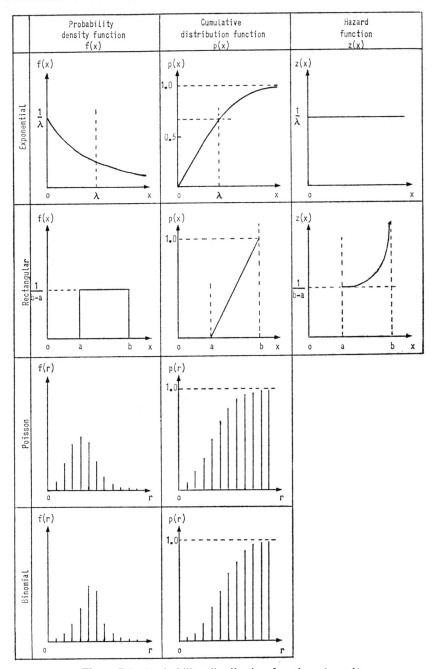

Figure 7.1 Probability distribution functions (*contd.*)

functions, $p_c(x)$ and $p(x)$, as follows:

$$p_c(b) - p_c(a) = \frac{p(b) - p(a)}{1 - p(c)} \quad (7.84)$$

In the particular case where $a = c$, Equation (7.84) becomes:

$$p_c(x) = \frac{p(x) - p(c)}{1 - p(c)} \quad (7.85)$$

since $p_c(c) = 0$ and since the value b may be expressed as the value x of the variate x.

$p_c(x)$ as defined by Equation (7.85) is the conditional cumulative distribution function. It is valid in the range $c \leqslant x \leqslant \infty$ for the case where the original p.d.f. has an upper limit of ∞ and lower limit $\leqslant c$. In particular, it is seen from Equation (7.85) that:

$$p_c(c) = 0 \quad (7.86)$$

$$p_c(\infty) = 1 \quad (7.87)$$

Any $p_c(x)$ has a corresponding density function, $f_c(x)$, which may be derived by differentiating Equation (7.85) with respect to x:

$$f_c(x) = \frac{f(x)}{1 - p(c)} \quad (7.88)$$

Example 7.10

The lifetimes of certain engineering devices are exponentially distributed with a mean lifetime of 10 years. What is the conditional probability of failure by the end of the succeeding year for devices which have already survived (a) 3 years and (b) 8 years?

For case (a), the value of $p(c)$, that is the cumulative probability of failure after 3 years, is given by:

$$p(c) = 1 - e^{-3/10}$$

$$= 0.2592$$

and the value of $p(x)$, that is the cumulative probability of failure after 4 years is:

$$p(x) = 1 - e^{-4/10}$$

$$= 0.3297$$

Hence, from Equation (7.85),

$$p_c(x) = \frac{0.3297 - 0.2592}{1 - 0.2592}$$

$$= 0.0951$$

PROPERTIES OF DISTRIBUTIONS

For case (b), $p(c)$ is evaluated at 8 years and $p(x)$ at 9 years giving values of 0.5507 and 0.5934 respectively. Hence:

$$p_c(x) = \frac{0.5934 - 0.5507}{1 - 0.5507}$$

$$= 0.0951$$

From the preceding example it appears, within the limits of calculation accuracy, that the conditional probability of end of life in a given period may be independent of the expired lifetime up to the beginning of the period for lifetimes which are exponentially distributed. This may be confirmed as follows.

For the exponential distribution:

$$p(x) = 1 - e^{-x/\lambda}$$
$$p(c) = 1 - e^{-c/\lambda}$$

Hence:

$$p_c(x) = \frac{(1 - e^{-x/\lambda}) - (1 - e^{-c/\lambda})}{e^{-c/\lambda}}$$

$$= 1 - e^{-(x-c)/\lambda}$$

$$= 1 - e^{-T/\lambda} \quad (7.89)$$

where $T = (x - c)$ is the period or range over which the conditional probability is calculated. Equation (7.89), therefore, shows that, for the exponential distribution, the conditional probability of lying in a certain range depends only upon the value of the range and not upon its position. This is a peculiar and significant property of the exponential distribution and applies to no other distribution. For instance, the reader may like to verify that in the case of the special Erlangian distribution:

$$p_c(x) = 1 - \frac{(\lambda + x)}{(\lambda + c)} e^{-(x-c)/\lambda} \quad (7.90)$$

and in the case of the Weibull distribution that:

$$p_c(x) = 1 - e^{-(x^\beta - c^\beta)/\lambda^\beta} \quad (7.91)$$

where it will be seen in both cases that $p_c(x)$ is dependent upon the individual values of x and c.

The corresponding value of $f_c(x)$ for any distribution may be obtained by differentiating such equations as Equations (7.89), (7.90) and (7.91) or, in some cases, more directly from Equation (7.88). For example:

Weibull:

$$f_c(x) = \frac{\beta x^{\beta-1}}{\lambda^\beta} e^{-(x^\beta - c^\beta)/\lambda^\beta}, \quad c \leqslant x \leqslant \infty \quad (7.92)$$

Special Erlangian:

$$f_c(x) = \frac{x}{\lambda(\lambda+c)} e^{-(x-c)/\lambda}, \quad c \leqslant x \leqslant \infty \tag{7.93}$$

Exponential:

$$f_c(x) = \frac{1}{\lambda} e^{-(x-c)/\lambda}, \quad c \leqslant x \leqslant \infty \tag{7.94}$$

The conditional probability density function may, in fact, be looked upon as a truncated form of the original distribution where the truncation takes place at a value c and applies for all values greater than c. Distributions may, of course, be truncated at the top end of their range and for these cases similar considerations to those just discussed will apply. Particular types of truncated distributions will be considered in later chapters which deal with some of the forms of distributions which arise under various combinational conditions.

The mean value of $f_c(x)$, or of a distribution truncated below a value of c, is:

$$\mu = \frac{1}{p(c)} \int_c^\infty x f(x) \, dx \tag{7.95}$$

$$= -\frac{1}{p(c)} \int_c^\infty x \, d\overline{p(x)}$$

$$= -\frac{1}{p(c)} [x\overline{p(x)}]_c^\infty + \frac{1}{p(c)} \int_c^\infty \overline{p(x)} \, dx$$

or:

$$\mu = c + \frac{1}{p(c)} \int_c^\infty \overline{p(x)} \, dx \tag{7.96}$$

If the value of the mean is measured from the point c instead of from the origin where:

$$\mu_c = \mu - c \tag{7.97}$$

then:

$$\mu_c = \frac{1}{p(c)} \int_c^\infty \overline{p(x)} \, dx \tag{7.98}$$

For the exponential distribution, this value of μ_c becomes:

$$\mu_c = \frac{1}{e^{-c/\lambda}} \int_c^\infty e^{-x/\lambda} \, dx$$

$$= \lambda \tag{7.99}$$

PROPERTIES OF DISTRIBUTIONS   269

So that the mean value of the exponential, measured from the point of truncation, is a constant irrespective of the point of truncation. This is another property which is peculiar to the exponential distribution.

It can be readily verified, as another case, that for the truncated special Erlangian distribution:

$$\mu = 2\lambda + \frac{c^2}{(c+\lambda)} \qquad (7.100)$$

and:

$$\mu_c = \lambda + \frac{\lambda^2}{(c+\lambda)} \qquad (7.101)$$

Hazard Distribution Functions

It will be noticed from Equation (7.88) that the value of $f_c(x)$ for a distribution is the corresponding value of $f(x)$ multiplied by a constant reduction factor of $1/[1-p(c)]$. For a particular x, c can take on any value in the range 0 to x. When $c = 0$:

$$f_c(x)_{c=0} = f(x) \qquad (7.102)$$

In other words, the conditional p.d.f. becomes equal to the original p.d.f. If, on the other hand, c is considered as having a value at the other extreme of its range such that $c = x$, then:

$$f_c(x)_{c=x} = \frac{f(x)}{1-p(x)} \qquad (7.103)$$

Here, c is no longer a constant but has taken on a value identical with the value of the variate, x. Equation (7.103) now represents some entirely new function of x which may be denoted by $z(x)$, where:

$$z(x) = \frac{f(x)}{1-p(x)} \qquad (7.104)$$

The question now arises as to what meaning can be attached to the function $z(x)$. The first thing to note, from the method of its postulation, is that it has some conditional probability connotation associated with it. Since $f(x)\,\delta x$ is the probability of x lying in the range $(x-\tfrac{1}{2}\delta x)$ to $(x+\tfrac{1}{2}\delta x)$, then $z(x)\,\delta x$ could be defined as the conditional probability of x lying in the range $(x-\tfrac{1}{2}\delta x)$ to $(x+\tfrac{1}{2}\delta x)$ where the condition is that all values of x less than x have been removed from the original distribution.

Looked at another way, if x is put equal to c in Equation (7.88) then:

$$z(c) = \frac{f(c)}{1-p(c)} \qquad (7.105)$$

and this is the initial value of the conditional probability density function, $f_c(c)$. So $z(x)$ is then a function which describes the initial conditional density as the initial variate value of such a density is varied.

A particular example may help to illustrate the general nature of $z(x)$. If x is a random variate which represents, say, the times to failure of a system, then $z(x) \delta x$ is the probability of a system, which has survived up to a time x, failing in the next increment of time x to $(x+\delta x)$. Because of this aspect of describing conditional failures in the time domain, the function $z(x)$ has been called by various authors the 'failure-rate function', the 'age specific failure-rate' and, in the case of human life expectancy, as the 'force of mortality'. In more general statistical parlance, however, it is often referred to as the 'hazard-rate function' or, simply, as the 'hazard function'.

Even though $z(x) \delta x$ may be regarded as an 'incremental' or 'instantaneous' probability, $z(x)$ itself is not a density function like $f(x)$. The density function, $f(x)$, may, however, be expressed entirely in terms of $z(x)$. From Equation (7.104):

$$\int_0^x z(x)\,dx = \int_0^x \frac{f(x)\,dx}{1-p(x)}$$

$$= \int_0^x \frac{dp(x)}{1-p(x)}$$

$$= -\log[1-p(x)]$$

Therefore:

$$p(x) = 1 - \exp\left[-\int z(x)\,dx\right] \qquad (7.106)$$

and differentiating with respect to x:

$$f(x) = z(x) \exp\left[-\int z(x)\,dx\right] \qquad (7.107)$$

These Equations (7.106) and (7.107) may be obtained by a different method which makes the prior supposition that a hazard function exists. For instance, the probability that a variate has a value less than or equal to $(x+\delta x)$, $p(x+\delta x)$, must equal the probability that it has a value less than or equal to x, $p(x)$, plus the probability that it has a value between x and $(x+\delta x)$, $f(x)\delta x$. Or:

$$p(x+\delta x) = p(x) + f(x)\,\delta x \qquad (7.108)$$

whence, in the limit:

$$\frac{dp(x)}{dx} = f(x) \qquad (7.109)$$

PROPERTIES OF DISTRIBUTIONS

which is already known and has now been deduced on the prior supposition of a density function.

A similar statement can now be made using the hazard function. The probability that a variate has a value less than or equal to $(x+\delta x)$, $p(x+\delta x)$, must equal the probability that it has a value less than or equal to x, $p(x)$, plus the probability that it has no value less then x, $\overline{p(x)}$, where this latter probability is then multiplied by the probability of x lying between x and $(x+\delta x)$ on the condition that x has no value less than x, $z(x)\,\delta x$. Or:

$$p(x+\delta x) = p(x) + \overline{p(x)}\,z(x)\,\delta x$$
$$= p(x) + [1-p(x)]\,z(x)\,\delta x \qquad (7.110)$$

Whence, in the limit:

$$\frac{1}{[1-p(x)]}\frac{dp(x)}{dx} = z(x)$$

or:

$$p(x) = 1 - \exp\left[-\int z(x)\,dx\right] \qquad (7.111)$$

as before.

It is now apparent that the hazard function, $z(x)$, can be written in the following alternative forms:

$$z(x) = \frac{f(x)}{1 - \int_0^x f(x)\,dx} \qquad (7.112)$$

$$z(x) = \frac{1}{1-p(x)}\frac{dp(x)}{dx} \qquad (7.113)$$

$$z(x) = -\frac{1}{\overline{p(x)}}\frac{d\overline{p(x)}}{dx} \qquad (7.114)$$

Example 7.11

Find the form of the hazard function for the Weibull and exponential distributions.

In the case of the Weibull:

$$z(x) = \frac{f(x)}{\overline{p(x)}}$$
$$= \frac{(\beta x^{\beta-1}/\lambda^\beta)\,e^{-(x/\lambda)^\beta}}{e^{-(x/\lambda)^\beta}}$$

or:
$$z(x) = \frac{\beta x^{\beta-1}}{\lambda^\beta} \qquad (7.115)$$

and for the exponential:
$$z(x) = \frac{(1/\lambda)e^{-x/\lambda}}{e^{-x/\lambda}}$$

or:
$$z(x) = 1/\lambda \qquad (7.116)$$

Once again, the example shows that the exponential distribution exhibits a characteristic which is not given by any other distribution. In this case it is seen that, for the exponential distribution, the hazard function is a constant and equal to the reciprocal of the mean value. If the times to failure of a system, therefore, are exponentially distributed, the hazard function has a constant value which has the dimensions of a failure-rate and which is equal to the reciprocal of the mean time to failure. This fact will be discussed further in the later chapters which deal with the analysis of system failures.

For any other distribution, the hazard function is not a constant but depends upon the variate x. Where system failures are being represented, this means that the effective failure-rate of the system is changing with time. Some typical expressions for the hazard functions of various distributions are given in Table 7.1 and the general forms of the functions are illustrated graphically in Figure 7.1.

Since a hazard function can represent the pattern of the rate of a system's failure, interest also lies in the ways in which different hazard functions may be combined. This aspect will be discussed in the next chapter which deals, in general, with the combinations of distributions.

Questions
1. Derive the Laplace transforms for the lognormal, gamma and Poisson distribution functions and compare your answers with those given in Table 7.1. From these transforms show that the total probability of all values of the original distributions is unity and explain your method.
2. The times to open for an electro-magnetically operated switch in a system are found to always lie between a lower limit of 6 s and an upper limit of 12 s. The distribution of times between these values is random and such that the probability of a time lying in any 1-s interval is the same wherever the interval is taken in the range. What is
    (a) the expression for the p.d.f.,
    (b) the Laplace transform of the p.d.f., and
    (c) what is the mean time to open?

3. Find the values for the second and fourth moments about the origin for
   (a) a Weibull distribution with parameters $\beta = 4$ and $\lambda = 2$,
   (b) a gamma distribution with $\beta = \frac{1}{2}$ and $\lambda = 2$, and
   (c) an exponential distribution with $\lambda = 1$.
4. A bank of nominally identical floodlights are used to illuminate a working area. The total number of lights in the bank is n and the probability that any one light is in the successful state at any time is p. What is the mean or expected number of lights that will be in operation at any one time if
   (a) $n = 20$, $p = 0.9$,
   (b) $n = 100$, $p = 0.9$, and
   (c) $n = 100$, $p = 0.5$?
5. It is found that the number of system breakdowns occurring in a given length of time follows a Poisson distribution with a mean value of 2 breakdowns. What are the probabilities, in the same length of time, of the system having exactly
   (a) no breakdowns,
   (b) 1 breakdown,
   (c) 2 breakdowns, and
   (d) 10 breakdowns?
6. Find
   (a) the variance,
   (b) the standard deviation,
   (c) the coefficient of skewness, and
   (d) the coefficient of kurtosis,
   for the times to open of the switch defined in Question 2. Explain what can be deduced about the distribution from these values.
7. The variations in size of a component part in a system follow a continuous random distribution of the variate x where the p.d.f., $f(x)$, is given by:

$$f(x) = \frac{x}{a^2} U(x) + \frac{2}{a^2}(a-x)\,U(x-a) - \frac{1}{a^2}(2a-x)\,U(x-2a)$$

Draw the shape of this distribution and calculate
   (a) the mean,
   (b) the variance,
   (c) the coefficient of skewness, and
   (d) the coefficient of kurtosis,
   if the value of the parameter a is 1 length unit.
8. Random variations in the power output, x, of an engine are found to follow a gamma distribution with values of $\beta = 3$ and $\lambda = 100$ kW. What is the probability that the power output is less than 200 kW? How would this probability be changed if it was calculated using a normal approximation to the gamma distribution?

9. The times that an operator takes to shut down a certain section of a chemical plant on the receipt of an alarm stimulus are found to be lognormally distributed. The parameters of the lognormal distribution, expressed in terms of the natural logarithm of the response times measured in seconds, are $\mu = 2.0723$ and $\sigma^2 = 0.4605$. What is
   (a) the mean time the operator takes, and
   (b) the probability that he succeeds in shutting down the plant in less than 20 s?
10. A resistor component in an electronic amplifier is subject to random and catastrophic failure. Over a large number of such components it has been deduced that the times to catastrophic failure are exponentially distributed with a mean time to failure of 10^7 h. What is the probability that such a resistor will fail within periods of
    (a) 100 h,
    (b) 2500 h,
    (c) 10,000 h, and
    (d) 50,000 h?
11. A large number of valve mechanisms which are produced are found to have times to first failure which follow a Weibull distribution. The parameters of this distribution are $\lambda = 10$ years and $\beta = 0.5$. What is the probability that such a mechanism will survive
    (a) 1 year,
    (b) 5 years, and
    (c) 10 years without failure and what is
    (d) the mean time to first failure?
12. An illuminated mimic diagram in a plant control room has 150 nominally identical bulbs which are required to be permanently illuminated. If the probability of any one bulb being out at any one time is 0.01, what is the probability of
    (a) at least 5 bulbs being out,
    (b) not more than 3 bulbs being out,
    (c) exactly 4 bulbs being out?
    Discuss the relationship between these three calculated answers.
13. Recalculate the three answers to Question 12 using the Poisson approximation for the binomial. On the same basis find the change in the probability values if the probability of bulbs being in the failed state is reduced to 0.001.
14. The lifetime of a bearing is subject to random variations which are normally distributed. The mean lifetime is 15 years and the standard deviation is 3 years. What is the estimated probability that such a bearing will come to the end of its life between the end of the fifteenth and sixteenth year of operation if

PROPERTIES OF DISTRIBUTIONS 275

(a) the estimate is made at time zero,
(b) the estimate is made for a bearing which has already survived 10 years, and
(c) the estimate is made for a bearing which has already survived 15 years?

15. A range of control units have times to their first failure which are exponentially distributed. The mean of this distribution is 5000 h. What is the probability that a unit which has survived
    (a) 1000 h,
    (b) 5000 h, and
    (c) 10,000 h,
    will survive a further 1000 h? What would be the probabilities for these three cases if the distribution was a special Erlangian with the parameter $\lambda = 5000$ h.

16. The hazard function for a device which has a performance characteristic x is $z(x) = 1/(2\sqrt{x})$. Derive the expressions for
    (a) the probability density function,
    (b) the cumulative distribution function, and
    (c) calculate the mean of the p.d.f.

17. If the hazard function for a variate x is a constant, show, from the definition of a hazard function and from simple probability principles, that this leads to a probability density function for x which is of the simple exponential form. Show also the type of distribution that may be derived if the hazard function is no longer constant but is of the form $z(x) = x/(1+x)$.

18. Using tables, plot the hazard function curve for the normal distribution as a function of the variate x. Show that for $x = \mu$, $z(x) = \sqrt{2}/(\sigma\sqrt{\pi})$.

CHAPTER 8
Combinations of Distributions

The Need for Combinations

Consider a complete system where, for every system performance parameter, the systematic variation and distribution of random variation is known. This set of characteristics, expressed either as an array of equations or in the form of some transfer characteristic matrix, would define the complete achieved performance of the overall system. In other words, the need for an overall expression of achieved performance, as discussed in Chapters 4 and 5, would have been met. If the required performance, as described in Chapter 3, is formulated in a similar form, then the reliability of the overall situation is just a matter of carrying out the correlation as indicated in Figure 2.1 of Chapter 2.

In practice, a complete statement of system performance can only be obtained in this way if adequate data have been collected on the overall system behaviour. This would mean that a large enough number of nominally identical systems would have to have been operated for a long enough period of time in order to deduce the distributions of variation in all the relevant performance parameters. With the larger and more complex technological systems this is rarely possible since the numbers of systems involved and/or their time of operation do not form a large enough statistical sample from which to estimate parameters of performance.

Most technological systems, however, are made up of a number of sub-systems. Each sub-system may have accumulated a sufficient number of unit-hours of operation from which an adequate estimate of its variational characteristics can be obtained. In order to arrive at the complete system's achieved performance from these data, it is necessary to combine the known variational characteristics of each sub-system in the appropriate logical way. If, again, a sub-system does not form a sufficient statistical sample of variational data, it may be split down into sub-sub-systems or, perhaps, individual items of equipment. This process can obviously be repeated until the stage or level is reached where all the necessary data exist. In the ultimate, although this is rarely necessary throughout the system, the system may be broken down into consideration of all its smallest and individual component parts.

COMBINATIONS OF DISTRIBUTIONS   277

As mentioned at the beginning of Chapter 5 under 'Achievement Variations', a knowledge of the variational performance of component parts of a system may exist where the similar characteristics of the complete system are totally unknown.

In order to evaluate the achieved performance of a complete system then, it is often necessary to synthesize the system behaviour on the basis of the behaviour of its constituent parts. This means that there is a need for a method of combining the variational characteristics of parts in order to arrive at the variational characteristics of the whole.

The parts of a system may be functionally combined in a variety of different ways. For instance, two items may be functionally connected in 'series' where, say, the delay time in the operation of the second item is added to that of the first item. Here, the two distributions of variations in delay times would require to be 'summed' in order to obtain the overall distribution of delay times for the two items. In other cases, the corresponding performance parameters of two items may combine in a subtractive or difference fashion. In a simple electrical circuit or analogue, the variational characteristics of a resistor and the current flowing through it may be known. The corresponding voltage distribution would then be derived from the 'product' of the appropriate resistor and current value distributions. The overall distribution of stress in a mechanical member may be required to be synthesized from the 'quotient' of the distributions of force and strength.

In addition to these 'arithmetic' functional combinations of items such as 'sum', 'difference', 'product' and 'quotient', there may also exist in a system certain other combinations of functions. These other types of combinations could typically be of the auctioneering type. That is, an overall function may be produced from the largest or, perhaps, the smallest of a range of sub-functions. The power output of a system could be a function of the product of, say, the output voltage and the output current. If, however, both the current and voltage were subject to variations in delay time, then the output power delay time would depend upon the largest delay time of either the current or the voltage.

The need has been established, then, for methods of combining distributions of random variates when the quantities represented by these variates are functionally linked in various arithmetic and selected ways. This chapter now examines these methods for various typical types of combinations.

Sums and Differences of Distributions

Consider, first of all, two independent items where their functional performance characteristics are represented by two corresponding random and independent variates x and y. The probability density functions for these two variates are $f_x(x)$ and $f_y(y)$ respectively. If some new dependent variate,

z, exists where z is derived from the sum of x and y, that is:

$$z = x+y \tag{8.1}$$

then it can be taken that z is distributed according to some other density function which will be designated as $f_z(z)$.

The probability that the sum of x and y is less than or equal to some particular z, $p_z(z)$, is given by the probability of occurrence of a particular x, $f_x(x)\,\delta x$, together with the probability that y is less than or equal to $(z-x)$, $p_y(z-x)$, for all x between the lower limit of x (taken as zero, say) and z. This relationship may be expressed as follows:

$$p_z(z) = \int_0^z f_x(x) p_y(z-x)\,\mathrm{d}x \tag{8.2}$$

or, in terms of the convolution notation used in Chapter 6:

$$p_z = f_x * p_y \tag{8.3}$$

The Laplace transform of this convolution, from Equation (6.44), is:

$$P_z(s) = F_x(s) P_y(s) \tag{8.4}$$

and multiplying both sides by the operator s gives:

$$sP_z(s) = F_x(s)\,sP_y(s)$$

or:

$$L\left[\frac{\mathrm{d}}{\mathrm{d}z}p_z(z)\right] = F_x(s) L\left[\frac{\mathrm{d}}{\mathrm{d}y}p_y(y)\right]$$

and, therefore:

$$F_z(s) = F_x(s) F_y(s) \tag{8.5}$$

This relationship can also be expressed in terms of the moments of the three distributions $M_m(z)$, $M_m(x)$ and $M_m(y)$ by using Equation (7.31). Hence:

$$\sum_{m=0}^{\infty}(-1)^m \frac{s^m}{m!} M_m(z) = \left[\sum_{m=0}^{\infty}(-1)^m \frac{s^m}{m!} M_m(x)\right]\left[\sum_{m=0}^{\infty}(-1)^m \frac{s^m}{m!} M_m(y)\right] \tag{8.6}$$

whence it is apparent that:

$$M_m(z) = \sum_{j=0}^{m}\binom{m}{j} M_{m-j}(x) M_j(y) \tag{8.7}$$

and, in a similar way from Equation (7.46), that:

$$\bar{M}_m(z) = \sum_{j=0}^{m}\binom{m}{j} \bar{M}_{m-j}(x) \bar{M}_j(y) \tag{8.8}$$

COMBINATIONS OF DISTRIBUTIONS

Two results are immediately apparent from Equations (8.7) and (8.8). First:

$$M_1(z) = M_1(x) + M_1(y)$$

or:

$$\mu_z = \mu_x + \mu_y \tag{8.9}$$

which is stating, as would be expected, that the mean of the resultant distribution of z is the sum of the means of the two original distributions of x and y. Secondly:

$$\bar{M}_2(z) = \bar{M}_2(x) + \bar{M}_2(y)$$

or:

$$\sigma_z^2 = \sigma_x^2 + \sigma_y^2 \tag{8.10}$$

So that the variance of the resultant distribution of z is the sum of the individual variances of the x and y distributions.

These concepts may be extended to the case where n independent random variates $x_1, x_2, ..., x_n$ are summated such that:

$$z = x_1 + x_2 + ... + x_n \tag{8.11}$$

The Laplace transform in Equation (8.5) now becomes:

$$F_z(s) = \prod_{j=1}^{n} F_{x_j}(s) \tag{8.12}$$

and the repeated application of Equations (8.7) and (8.8) yield:

$$\mu_z = \mu_{x_1} + \mu_{x_2} + ... + \mu_{x_n} \tag{8.13}$$

and:

$$\sigma_z^2 = \sigma_{x_1}^2 + \sigma_{x_2}^2 + ... + \sigma_{x_n}^2 \tag{8.14}$$

Example 8.1

Three resistors, R_1, R_2 and R_3, are connected in series. The mean resistance values are $R_1 = 300\ \Omega$, $R_2 = 400\ \Omega$ and $R_3 = 500\ \Omega$ and the standard deviations correspond to 5% of the mean value in each case. What is the mean value of the resistance chain and the standard deviation expressed as a percentage of the total mean value?

The overall mean value of resistance is:

$$\mu_R = \mu_{R_1} + \mu_{R_2} + \mu_{R_3}$$
$$= 300 + 400 + 500$$
$$= 1200\ \Omega$$

The overall variance is:

$$\sigma_R^2 = \sigma_{R_1}^2 + \sigma_{R_2}^2 + \sigma_{R_3}^2$$
$$= (15)^2 + (20)^2 + (25)^2$$

Hence:
$$\sigma_R = 35.36 \ \Omega$$
$$= 2.95\%$$

and it is interesting to note that this is a smaller standard deviation than that of each of the individual resistors. Putting elements in series in this way can therefore be said to produce a 'tolerance advantage'.

A particular case of interest arises when each of the variates $x_1, x_2, \ldots, x_n$ is taken from an identical distribution, say, $f_x(x)$. In this case, Equation (8.12) becomes:

$$F_z(s) = \{F_x(s)\}^n \tag{8.15}$$

Differentiating this with respect to s gives:

$$\frac{d}{ds} F_z(s) = n\{F_x(s)\}^{n-1} \frac{d}{ds} F_x(s) \tag{8.16}$$

and letting $s = 0$ yields:

$$\mu_z = n\mu_x \tag{8.17}$$

A result which could, of course, have been obtained directly from Equation (8.13).

Repeating the differentiation process on Equation (8.16)

$$\frac{d^2}{ds^2} F_z(s) = n(n-1)\{F_x(s)\}^{n-2} \left\{\frac{d}{ds} F_x(s)\right\}^2 + n\{F_x(s)\}^{n-1} \frac{d^2}{ds^2} F_x(s) \tag{8.18}$$

and again letting $s = 0$ gives:

$$\sigma_z^2 + \mu_z^2 = n(n-1)\mu_x^2 + n(\mu_x^2 + \sigma_x^2) \tag{8.19}$$

and combining Equations (8.17) and (8.19) yields:

$$\sigma_z^2 = n\sigma_x^2 \tag{8.20}$$

which again would have also followed from Equation (8.14).

A third differentiation process leads to:

$$K_{3_z} = \frac{K_{3_x}}{\sqrt{n}} \tag{8.21}$$

and differentiating Equation (8.15) four times and setting $s = 0$ gives:

$$K_{4_z} = 3\left(1 - \frac{1}{n}\right) + \frac{K_{4_x}}{n} \tag{8.22}$$

As $n \to \infty$, $K_{3_z} \to 0$ and $K_{4_z} \to 3$. Hence the coefficients of skewness and kurtosis of the combined distribution $f_z(z)$ tend to the values appropriate to

COMBINATIONS OF DISTRIBUTIONS

a normal distribution as the number of variates which are summed from identical distributions becomes large. So whatever the form of the particular distribution, $f_x(x)$, from which the variates $x_1, x_2, \ldots, x_n$ are taken, the summed distribution approaches a normal distribution of mean value $n\mu_x$ and variance $n\sigma_x^2$ as n becomes large. This represents a particular concept which arises from what is generally called the 'central limit theorem'.

It should be noted, of course, that when the distribution, $f_x(x)$, is itself normal, any number of variates taken and summed from this distribution will always produce a resultant distribution which is normal. For instance, in the case of the normal distribution:

$$F_x(s) = \exp[-(s\mu - \tfrac{1}{2}\sigma^2 s^2)]$$

and hence:

$$F_z(s) = \{F_x(s)\}^n = \exp[-sn\mu - \tfrac{1}{2}n\sigma^2 s^2)] \qquad (8.23)$$

and Equation (8.23) is always a normal distribution with mean value $n\mu$ and variance $n\sigma^2$ whatever the value of n.

Example 8.2

A chain system is made up of n nominally identical links. Each link is made by a manufacturing process which produces random changes in length which are found to follow a rectangular distribution. The density function for this distribution is shown in Figure 8.1. It has a mean value of $(a+b)/2$ and lower and upper limits of a and b respectively. What is the form of the distribution for the random variations in the overall length of the chain if
(a) there are n links,
(b) $n = 2$, and
(c) n is very large?

The density function, $f_x(x)$, for each link, as has been shown in previous chapters, is:

$$f_x(x) = \frac{1}{(b-a)}[U(x-a) - U(x-b)] \qquad (8.24)$$

and the Laplace transformation of this p.d.f., from Table 7.1 or Table A.2 is:

$$F_x(s) = \frac{e^{-as} - e^{-bs}}{(b-a)s} \qquad (8.25)$$

If $f_z(z)$ is the p.d.f. for the variations in length of the complete chain, then:

$$F_z(s) = \{F_x(s)\}^n = \frac{e^{-nas}}{(b-a)^n s^n}(1 - e^{-(b-a)s})^n \qquad (8.26)$$

Expanding this expression and taking the appropriate inverse transforms from Table A.2 gives:

$$f_z(x) = \frac{1}{(b-a)^n(n-1)!} \sum_{j=0}^{n}(-1)^j \binom{n}{j} [x-a-j(b-a)]^{n-1} U[x-a-j(b-a)] \quad (8.27)$$

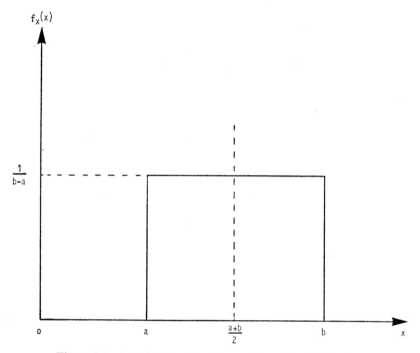

Figure 8.1 Density function for a rectangular distribution

If $n = 2$, this yields the following series of equations:

$$\left.\begin{aligned} f_z(x) &= 0 & \text{for } 0 \leqslant x \leqslant a \\ f_z(x) &= \frac{x-a}{(b-a)^2} & \text{for } a \leqslant x \leqslant b \\ f_z(x) &= \frac{(2b-a)-x}{(b-a)^2} & \text{for } b \leqslant x \leqslant (2b-a) \\ f_z(x) &= 0 & \text{for } (2b-a) \leqslant x \leqslant \infty \end{aligned}\right\} \quad (8.28)$$

and this result is shown graphically in Figure 8.2 where it is seen that the effect of summing two variates from a rectangular distribution is to produce a triangular distribution.

COMBINATIONS OF DISTRIBUTIONS 283

If n is large, Equation (8.26) can be written as:

$$F_z(s) = \exp(-nas)\exp\{-n[\tfrac{1}{2}(b-a)s - \tfrac{1}{24}(b-a)^2 s^2]\}$$
$$= \exp\{-[\tfrac{1}{2}(a+b)ns - \tfrac{1}{24}(b-a)^2 ns^2]\} \quad (8.29)$$

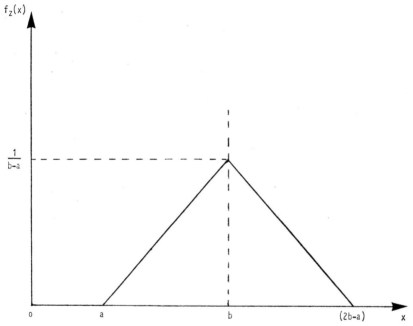

Figure 8.2 Result of summing two variates from the rectangular distribution of Figure 8.1.

and Equation (8.29) will be recognized as a normal distribution where:

$$\left.\begin{array}{c}\mu_z = \dfrac{n(a+b)}{2} = n\mu_x \\[2mm] \sigma_z^2 = \dfrac{n(b-a)^2}{12} = n\sigma_x^2\end{array}\right\} \quad (8.30)$$

since, as shown in Table 7.1, for a rectangular distribution:

$$\left.\begin{array}{c}\mu_x = \dfrac{a+b}{2} \\[2mm] \sigma_x^2 = \dfrac{(b-a)^2}{12}\end{array}\right\} \quad (8.31)$$

If now differences between variates taken from distributions are considered, Equation (8.1) may become of the form:

$$z = x - y \tag{8.32}$$

or:

$$z = x + u \tag{8.33}$$

where:

$$u = (-y) \tag{8.34}$$

Because of the particular relationship of Equation (8.34), it is worth, at this stage, considering the more general relationship:

$$u = ny \tag{8.35}$$

where n is any positive or negative constant.

If u is a variate distributed according to $f_u(u)$ and y is a variate distributed according to $f_y(y)$, then because of the relationship of Equation (8.35):

$$p_u(k) = p_y\left(\frac{k}{n}\right) \tag{8.36}$$

where k is some variate value of u and y.

Differentiating Equation (8.36) with respect to k:

$$\frac{d}{dk} p_u(k) = \frac{d(k/n)}{dk} \frac{d}{d(k/n)} p_y\left(\frac{k}{n}\right)$$

or:

$$f_u(k) = \frac{1}{n} f_y\left(\frac{k}{n}\right) \tag{8.37}$$

and conversely:

$$f_y(k) = n f_u(nk) \tag{8.38}$$

This relationship and other similar relationships of this type are tabulated for ease of reference in Table 8.1.

Referring to Table A.2 and taking the Laplace transforms of Equations (8.37) and (8.38):

$$F_u(s) = F_y(ns) \tag{8.39}$$

and:

$$F_y(s) = F_u\left(\frac{s}{n}\right) \tag{8.40}$$

So, in particular, if $n = -1$ as in Equation (8.34), then:

$$F_u(s) = F_y(-s)$$

and the difference of variates represented by Equation (8.32) leads to a

distribution $f_z(z)$ where:
$$F_z(s) = F_x(s) F_y(-s) \qquad (8.41)$$
which is also given in Table 8.1.

Table 8.1 Distribution Relationships

1. If $u = ny$		(u and y random variates, n a constant)
	then:	$f_u(x) = (1/n) f_y(x/n)$
	or:	$f_y(x) = nf_u(nx)$
	and, in terms of Laplace transformations:	
		$L[f_u(x)] = (1/n) L[f_y(x/n)] = F_y(ns)$
	or:	$L[f_y(x)] = nL[f_u(nx)] = F_u(s/n)$
2. If $u = v+w$		(u, v and w random variates)
	then:	$F_u(s) = F_v(s) F_w(s)$
3. If $u = v-w$		(u, v and w random variates)
	then:	$F_u(s) = F_v(s) F_w(-s)$
4. If $w = \log y$		(w and y random variates)
	then:	$f_w(x) = e^x f_y(e^x)$
	or:	$f_y(x) = (1/x) f_w(\log x)$
5. If $u = y+n$		(u and y random variates, n a constant)
	then:	$f_u(x) = f_y(x-n)$
	or:	$f_y(x) = f_u(x+n)$
	and, in terms of Laplace transformations:	
		$F_u(s) = e^{-ns} F_y(s)$
6. If $u = (n/y)$		(u and y random variates, n a constant)
	then:	$f_u(x) = (n/x^2) f_y(n/x)$
7. If $u = y^2$		(u and y random variates, y having both positive and negative values)
	then:	$f_u(x) = (1/\sqrt{x}) f_y(\sqrt{x})$
8. If $u = y^2$		(u and y random variates, y having only positive values)
	then:	$f_u(x) = (1/2\sqrt{x}) f_y(\sqrt{x})$

Example 8.3

Information on the temperature state of a plant is presented to an operator on a recorder which measures the difference between the actual plant temperature and some reference temperature. The plant temperature signal is subject to random variations which follow a rectangular distribution of mean value $(a_1+b_1)/2$ and variance $(b_1-a_1)^2/12$, and the reference temperature signal has a similar variational pattern with a mean value of $(a_2+b_2)/2$ and a variance of $(b_2-a_2)^2/12$. What is the form of the resultant distribution of variation as presented to the operator?

If $f_x(x)$ and $f_y(y)$ represent the p.d.f.'s for the plant and reference temperature signals respectively, then:

$$f_x(x) = \frac{1}{b_1-a_1}[U(x-a_1)-U(x-b_1)]$$

$$f_y(y) = \frac{1}{b_2-a_2}[U(x-a_2)-U(x-b_2)]$$

and:

$$F_x(s) = \frac{e^{-a_1 s}}{(b_1-a_1)s}[1-e^{-(b_1-a_1)s}]$$

$$F_y(-s) = \frac{-e^{+a_2 s}}{(b_2-a_2)s}[1-e^{+(b_2-a_2)s}]$$

Hence:

$$F_z(s) = \frac{e^{(a_2-a_1)s}}{(b_1-a_1)(b_2-a_2)s^2}[1-e^{-(b_1-a_1)s}][e^{(b_2-a_2)s}-1]$$

From which the reader may like to verify that the inverse transform leads to the following set of equations when $(a_1+b_1)>(a_2+b_2)$ and $(b_1-a_1)>(b_2-a_2)$

$$f_z(z) = 0 \qquad \text{for } 0 \leqslant z \leqslant (a_1-b_2)$$

$$f_z(z) = \frac{z+(b_2-a_1)}{(b_1-a_1)(b_2-a_2)} \qquad \text{for } (a_1-b_2) \leqslant z \leqslant (a_1-a_2)$$

$$f_z(z) = \frac{1}{(b_1-a_1)} \qquad \text{for } (a_1-a_2) \leqslant z \leqslant (b_1-b_2)$$

$$f_z(z) = \frac{(b_1-a_2)-z}{(b_1-a_1)(b_2-a_2)} \qquad \text{for } (b_1-b_2) \leqslant z \leqslant (b_1-a_2)$$

$$f_z(z) = 0 \qquad \text{for } (b_1-a_2) \leqslant z \leqslant \infty$$

and the mean value of $f_z(z)$ is:

$$\mu_z = \frac{(a_1+b_1)-(a_2+b_2)}{2}$$

The shape of this distribution for $f_z(z)$ is shown in Figure 8.3.

Cases may arise in practice where the Laplace transformation method leads to difficulties in solution. Here it should be remembered that the basic analytical equation for the resultant distribution of the sums of two variates

COMBINATIONS OF DISTRIBUTIONS 287

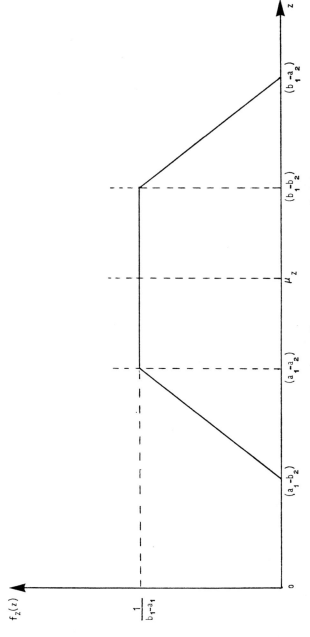

Figure 8.3 Resultant distribution for Example 8.3

is that given in Equation (8.2). This equation may, of course, be written as:

$$p_z(z) = \int_0^z f_x(x)\,dx \int_0^{z-x} f_y(y)\,dy \qquad (8.42)$$

and the corresponding p.d.f., $f_z(z)$, may be found by differentiating Equation (8.42) with respect to z since:

$$f_z(z) = \frac{d}{dz} p_z(z) \qquad (8.43)$$

Equation (8.42), however, only applies if both $f_x(x)$ and $f_y(y)$ are continuous distributions in the range 0 to ∞. For this case, $f_z(z)$ is then also a continuous distribution in the range 0 to ∞. If either $f_x(x)$ or $f_y(y)$ has finite upper or lower bounds, then, as has been seen in the recent examples, $f_z(z)$ contains points of discontinuity. To meet this case, a series of integral equations of the type of Equation (8.42) are required between each successive pair of discontinuous points. Suppose that $f_x(x)$ and $f_y(y)$ have lower and upper limits of x_l and x_u, and y_l and y_u, respectively. Suppose also that $(x_u - x_l) \geqslant (y_u - y_l)$, which is, in fact, a general case. Then the lower limit of $f_z(z)$ is $(x_l + y_l)$ and the first point of discontinuity is $(x_l + y_u)$.

Therefore:

$$\text{for} \quad (x_l + y_l) \leqslant z \leqslant (x_l + y_u)$$

$$p_z(z) = \int_{x_l}^{z-y_l} f_x(x)\,dx \int_{y_l}^{z-x} f_y(y)\,dy \qquad (8.44)$$

Provided now that $(x_u - x_l) \geqslant 2(y_u - y_l)$, then for the range:

$$(x_l + y_u) \leqslant z \leqslant [x_l + y_l + 2(y_u - y_l)]$$

$$p_z(z) = \int_{x_l}^{z-y_u} f_x(x)\,dx + \int_{z-y_u}^{z-y_l} f_x(x)\,dx \int_{y_l}^{z-x} f_y(y)\,dy \qquad (8.45)$$

Equation (8.45) now applies for every range of the type:

$$[x_l + y_l + n(y_u - y_l)] \leqslant z \leqslant [x_l + y_l + (n+1)(y_u - y_l)]$$

provided that:

$$(x_u - x_l) \geqslant (n+1)(y_u - y_l)$$

When this latter condition is no longer true, the last range becomes:

$$(x_u + y_l) \leqslant z \leqslant [x_u + y_l + n(y_u - y_l)]$$

and the corresponding integral equation is:

$$p_z(z) = \int_x^{z-y_u} f_x(x)\,dx + \int_{z-y_u}^{x_u} f_x(x)\,dx \int_{y_l}^{z-x} f_y(y)\,dy \qquad (8.46)$$

COMBINATIONS OF DISTRIBUTIONS

The reader may now like to re-work Examples 8.2 and 8.3 using Equations (8.44) to (8.46) instead of the Laplace transformation method. It should be remembered, of course, that where differences are involved Equation (8.37) yields:

$$f_u(k) = -f_y(-k) \tag{8.47}$$

with corresponding sign changes in the differentials and the limits.

What has been said so far about the sum of random variates taken from continuous distributions applies also to variates taken from discrete distributions. Suppose that t is a discrete variate which is given by the sum of two discrete variates r and q, where:

$$t = r + q \tag{8.48}$$

Then the equivalent summation equation to the integral equation of Equation (8.2) is:

$$p_t(t) = \sum_{r=0}^{t} f_r(r) \sum_{q=0}^{t-r} f_q(q) \tag{8.49}$$

or:

$$p_t(t) = \sum_{r=0}^{t} f_r(r) p_q(t-r) \tag{8.50}$$

For instance, if $f_r(r)$ and $f_q(q)$ both represent Poisson distribution functions where:

$$f_r(r) = \frac{\alpha_r^r}{r!} e^{-\alpha_r} \quad \text{and} \quad f_q(q) = \frac{\alpha_q^q}{q!} e^{-\alpha_q}$$

then Equation (8.49) becomes:

$$p_t(t) = \exp[-(\alpha_r + \alpha_q)] \sum_{r=0}^{t} \frac{\alpha_r^r}{r!} \sum_{q=0}^{t-r} \frac{\alpha_q^q}{q!}$$

which leads to:

$$p_t(t) = \exp[-(\alpha_r + \alpha_q)] \sum_{j=0}^{t} \frac{(\alpha_r + \alpha_q)^j}{j!} \tag{8.51}$$

and hence:

$$f_t(t) = \frac{(\alpha_r + \alpha_q)^t}{t!} \exp[-(\alpha_r + \alpha_q)] \tag{8.52}$$

which is also a Poisson distribution with a mean value α_t where:

$$\alpha_t = \alpha_r + \alpha_q \tag{8.53}$$

The Laplace transformation method would also lead to the same result since if it is taken that:

$$F_t(s) = F_r(s) F_q(s) \tag{8.54}$$

then, for the Poisson distributions:

$$F_t(s) = \exp[-\alpha_r(1-e^{-s})] \exp[-\alpha_q(1-e^{-s})]$$
$$= \exp[-(\alpha_r+\alpha_q)(1-e^{-s})] \tag{8.55}$$

which once again shows a resultant Poisson distribution with a mean value of $(\alpha_r+\alpha_q)$.

Example 8.4

A system consists of three parts, A, B and C. In each part the number of failures in a given length of time are found to follow a Poisson distribution. If part A has a mean number of 2 failures in a year, part B a mean of 1.5 failures in a year and part C a mean of 0.5 failures in a year, what is the distribution of failures in a year and the mean number of failures per year for the total system?

From the work just done, the distribution of total system failures is a Poisson distribution and the mean is given by:

$$\alpha_t = 2+1.5+0.5$$
$$= 4 \text{ failures in a year}$$

The combinational laws just discussed can be applied to any discrete distribution. However, further discussion on discrete distributions which consist of just two possible states will be left to the later chapters since this represents quite a large and pertinent subject on its own.

Products of Distributions

Suppose that a random variate, z, is dependent upon the product of two independent and random variates x and y, where:

$$z = xy \tag{8.56}$$

In the particular case where $f_x(x)$ and $f_y(y)$ are continuous from 0 to ∞, the cumulative distribution function of z is given by:

$$p_z(z) = \int_0^\infty f_x(x) \, dx \int_0^{z/x} f_y(y) \, dy \tag{8.57}$$

or:

$$p_z(z) = \int_0^\infty f_x(x) p_y\left(\frac{z}{x}\right) dx \tag{8.58}$$

COMBINATIONS OF DISTRIBUTIONS

However, more generally, $f_x(x)$ and $f_y(y)$ may be bounded by lower and upper limits. In this case, as with the sum of two variates just discussed, there are three ranges over which the integral equation applies. Assuming a general condition that:

$$x_l y_u \leqslant y_l x_u$$

then the first range is:

$$x_l y_l \leqslant z \leqslant x_l y_u$$

and here:

$$p_z(z) = \int_{x_l}^{z/y_l} f_x(x)\,dx \int_{y}^{z/x} f_y(y)\,dy \qquad (8.59)$$

and for the second range of:

$$x_l y_u \leqslant z \leqslant y_l x_u$$

$$p_z(z) = \int_{x_l}^{z/y_u} f_x(x)\,dx + \int_{z/y_u}^{z/y_l} f_x(x)\,dx \int_{y_l}^{z/x} f_y(y)\,dy \qquad (8.60)$$

and for the final range of:

$$y_l x_u \leqslant z \leqslant x_u y_u$$

$$p_z(z) = \int_{x_l}^{z/y_u} f_x(x)\,dx + \int_{z/y_u}^{x_u} f_x(x)\,dx \int_{y_l}^{z/x} f_y(y)\,dy \qquad (8.61)$$

Example 8.5

The value of a resistor is subject to random variations according to a rectangular distribution of mean value R_0 and variance $R^2/12$. The current flowing through the resistor also follows a rectangular distribution of variation with a mean of I_0 and a variance of $I^2/12$. What is the form of the distribution of voltage variation across the resistor?
Assume that $R_l I_u \leqslant R_u I_l$, then the first range of V is:

$$R_l I_l \leqslant V \leqslant R_l I_u \quad \text{or} \quad \left(R_0 - \frac{R}{2}\right)\left(I_0 - \frac{I}{2}\right) \leqslant V \leqslant \left(R_0 - \frac{R}{2}\right)\left(I_0 + \frac{I}{2}\right)$$

and for this range, Equation (8.59) leads to:

$$p(V) = \frac{1}{RI}\left[\left(R_0 - \frac{R}{2}\right)\left(I_0 - \frac{I}{2}\right) + V\left(\log\frac{V}{[R_0 - (R/2)][I_0 - (I/2)]} - 1\right)\right]$$

from which:

$$f(V) = \frac{1}{RI}\log\frac{V}{[R_0 - (R/2)][I_0 - (I/2)]} \qquad (8.62)$$

in a similar way, for:

$$\left(R_0 - \frac{R}{2}\right)\left(I_0 + \frac{I}{2}\right) \leqslant V \leqslant \left(R_0 + \frac{R}{2}\right)\left(I_0 - \frac{I}{2}\right)$$

$$f(V) = \frac{1}{RI} \log \frac{[I_0 + (I/2)]}{[I_0 - (I/2)]} \tag{8.63}$$

and for:

$$\left(R_0 + \frac{R}{2}\right)\left(I_0 - \frac{I}{2}\right) \leqslant V \leqslant \left(R_0 + \frac{R}{2}\right)\left(I_0 + \frac{I}{2}\right)$$

$$f(V) = \frac{1}{RI} \log \frac{[R_0 + (R/2)][I_0 + (I/2)]}{V} \tag{8.64}$$

Integrals of the type of Equation (8.57) do not form a generally soluble Laplace transformation as they did in the case of the sums of variates. However, it is apparent that Equation (8.56) can be converted from an expression of products into an expression of sums by taking logarithms, whence:

$$\log z = \log x + \log y \tag{8.65}$$

or:

$$u = v + w \tag{8.66}$$

where:

$$u = \log z \tag{8.67}$$

$$v = \log x \tag{8.68}$$

$$w = \log y \tag{8.69}$$

Considering an expression of the general type of Equation (8.69), the cumulative distribution functions for w and y are obviously related by the equation:

$$\int_0^k f_w(w)\,dw = \int_1^{e^k} f_y(y)\,dy \tag{8.70}$$

and differentiating both sides with respect to k, yields:

$$f_w(k) = e^k f_y(e^k) \tag{8.71}$$

and conversely:

$$f_y(k) = \frac{1}{k} f_w(\log k) \tag{8.72}$$

as tabulated in Table 8.1.

COMBINATIONS OF DISTRIBUTIONS

It will be immediately apparent, from Equation (8.71) and from the form of the lognormal and normal distributions previously discussed, that if $f_y(y)$ is lognormal then $f_w(w)$ is normal. Conversely, from Equation (8.72), if $f_w(w)$ is normal then $f_y(y)$ is lognormal.

Now, in connexion with Equation (8.66), it has already been seen in the previous section that:

$$L[f_u(u)] = L[f_v(v)]L[f_w(w)] \tag{8.73}$$

Therefore, making use of the relationship of Equation (8.71):

$$L[f_u(u)] = L[e^u f_x(e^u)]L[e^u f_y(e^u)]$$

or:

$$f_u(u) = L^{-1}\{L[e^u f_x(e^u)]L[e^u f_y(e^u)]\}$$

Now $f_z(z)$ is related to $f_u(u)$ by an expression of the type given in Equation (8.72). Therefore:

$$f_z(z) = \frac{1}{z}[\![L^{-1}\{L[e^z f_x(e^z)]L[e^z f_y(e^z)]\}]\!]_{z=\log z} \tag{8.74}$$

Applying this equation to Example 8.5 will lead to the same answer as obtained previously.

Generally, however, the integral Equation (8.57) and the transform Equation (8.74) can be difficult to evaluate unless the component p.d.f.'s are simple linear functions of the variate or logarithmic functions of the variate. It is often prudent, therefore, to look for some reasonable approximation of the functions which can lead to an easier solution.

An approximation worthy of consideration arises from the fact that, in practice, the random variations of a system's performance parameter generally represent proportionately small variations about some mean value. Suppose, therefore, that the variate x has a mean value x_0 about which random variations δx exist. Similarly, suppose that the variate y has small random variations about a mean value of y_0. Then:

$$x = x_0 + \delta x \tag{8.75}$$

$$y = y_0 + \delta y \tag{8.76}$$

and hence:

$$z = x_0 y_0 + y_0 \delta x + x_0 \delta y \tag{8.77}$$

neglecting the second order of small quantities.

Equation (8.77) now represents a simple summation so that the Laplace transformation becomes:

$$F_z(s) = F_{x_0 y_0}(s) F_{y_0 \delta x}(s) F_{x_0 \delta y}(s) \tag{8.78}$$

Now, from Relationship No. 5 in Table 8.1, Equation (8.75) leads to the distributional relationship:

$$f_{\delta x}(k) = f_x(k+x_0) \tag{8.79}$$

Therefore, from Table A.3:

$$F_{\delta x}(s) = e^{x_0 s} F_x(s) \tag{8.80}$$

and making use of the result of Equation (8.39):

$$F_{y_0 \delta x}(s) = e^{x_0 y_0 s} F_x(y_0 s) \tag{8.81}$$

Similarly:

$$F_{x_0 \delta y}(s) = e^{x_0 y_0 s} F_y(x_0 s) \tag{8.82}$$

and since $x_0 y_0$ is a single-valued quantity this is equivalent to the Dirac function of z, $\delta(z - x_0 y_0)$. Hence:

$$F_{x_0 y_0}(s) = e^{-x_0 y_0 s} \tag{8.83}$$

Substituting from Equations (8.81) to (8.83) in Equation (8.78):

$$F_z(s) = e^{x_0 y_0 s} F_x(y_0 s) F_y(x_0 s) \tag{8.84}$$

Equation (8.84) therefore applies to the product of two variates taken from distributions which represent small variations about their mean values. The principle of Equation (8.84) can obviously be applied to the product of any number of variates provided that each of the variates are contained within small deviations from the mean value.

By differentiating Equation (8.84) with respect to s and setting $s = 0$:

$$\mu_z = x_0 y_0$$

or:

$$\mu_z = \mu_x \mu_y \tag{8.85}$$

as would be expected.

A second process of differentiation yields the result that:

$$\sigma_z^2 = y_0^2 \sigma_x^2 + x_0^2 \sigma_y^2$$

or:

$$\sigma_z^2 = \mu_y^2 \sigma_x^2 + \mu_x^2 \sigma_y^2 \tag{8.86}$$

or, from Equations (8.85) and (8.86):

$$\frac{\sigma_z^2}{\mu_z^2} = \frac{\sigma_x^2}{\mu_x^2} + \frac{\sigma_y^2}{\mu_y^2} \tag{8.87}$$

More generally, it follows that if z is the product of n independent variates $x_1, x_2, ..., x_n$ then:

$$\mu_z = \mu_{x_1} \mu_{x_2} \cdots \mu_{x_n} \tag{8.88}$$

COMBINATIONS OF DISTRIBUTIONS

and:

$$\frac{\sigma_z^2}{\mu_z^2} = \frac{\sigma_{x_1}^2}{\mu_{x_1}^2} + \frac{\sigma_{x_2}^2}{\mu_{x_2}^2} + \ldots + \frac{\sigma_{x_n}^2}{\mu_{x_n}^2} \tag{8.89}$$

Example 8.6

A moving element in a control system of mean mass 2 kg is accelerated with a mean acceleration of 20 m s^{-2}. Both the mass of the element and its acceleration are subject to small random variations about their mean values. The standard deviations of these variations are 0.2 kg for the mass and 1.5 m s^{-2} for the acceleration. What is
(a) the mean force on the element and
(b) the standard deviation for the random variations in force?

The force is given by:

$$(\text{force}) = (\text{mass}) \times (\text{acceleration})$$

and the mean force is:

$$\mu_z = \mu_x \mu_y$$
$$= 2 \times 20$$
$$= 40 \text{ N}$$

The standard deviation of force is given by the relationship:

$$\sigma_z = \mu_z \left[\frac{\sigma_x^2}{\mu_x^2} + \frac{\sigma_y^2}{\mu_y^2} \right]^{\frac{1}{2}}$$

$$= 40 \left[\frac{(0.2)^2}{2^2} + \frac{(1.5)^2}{20^2} \right]^{\frac{1}{2}}$$

$$= 5 \text{ N}$$

Where the product is made up of n variates taken from the same distribution, then the extension of Equation (8.84) becomes:

$$F_z(s) = \exp\left[(n-1)\mu^n s\right] \{F_x(\mu^{n-1} s)\}^n \tag{8.90}$$

whence the usual process of differentiating with respect to s will show that:

$$\mu_z = \mu_x^n \tag{8.91}$$

$$\frac{\sigma_z^2}{\mu_z^2} = n \frac{\sigma_x^2}{\mu_x^2} \tag{8.92}$$

or:

$$\sigma_z^2 = n \sigma_x^2 \mu_x^{2n-2} \tag{8.93}$$

Also, as n becomes large, $K_{3_z} \to 0$ and $K_{4_z} \to 3$, so that $f_z(z)$ becomes asymptotic to a normal distribution as would be expected.

All of the preceding discussion upon the products of variates taken from continuous distributions can be adapted to apply to variates of discrete distributions. The reader may like to investigate this, using, say, the Poisson distribution as an example.

Quotients of Distributions

The case will now be considered where a random variate, z, is dependent upon the quotient of two independent and random variates x and y, where:

$$z = x/y \qquad (8.94)$$

In the particular case where $f_x(x)$ and $f_y(y)$ are continuous from 0 to ∞, the cumulative distribution function of z is given by:

$$p_z(z) = \int_0^\infty f_x(x)\,dx \int_{x/z}^\infty f_y(y)\,dy \qquad (8.95)$$

or:

$$p_z(z) = \int_0^\infty f_x(x)\,dx\,\overline{p_y\left(\frac{x}{z}\right)} \qquad (8.96)$$

In the more general case where $f_x(x)$ and $f_y(y)$ are bounded by finite limits, three ranges exist for $p_z(z)$ in a similar way to those which applied for the sums and products of distributions. Under the general assumption that:

$$x_l y_u \leqslant y_l x_u$$

the first range is:

$$\frac{x_l}{y_u} \leqslant z \leqslant \frac{x_l}{y_l}$$

for which:

$$p_z(z) = \int_{x_l}^{zy_u} f_x(x)\,dx \int_{x/z}^{y_u} f_y(y)\,dy \qquad (8.97)$$

and the second range is:

$$\frac{x_l}{y_l} \leqslant z \leqslant \frac{x_u}{y_u}$$

for which:

$$p_z(z) = \int_{x_l}^{zy_l} f_x(x)\,dx + \int_{zy_l}^{zy_u} f_x(x)\,dx \int_{x/z}^{y_u} f_y(y)\,dy \qquad (8.98)$$

COMBINATIONS OF DISTRIBUTIONS 297

and for the third range of:

$$\frac{x_u}{y_u} \leqslant z \leqslant \frac{x_u}{y_l}$$

$$p_z(z) = \int_{x_l}^{zy_l} f_x(x)\,dx + \int_{zy_l}^{x_u} f_x(x)\,dx \int_{x/z}^{y_u} f_y(y)\,dy \tag{8.99}$$

Example 8.7

The total force, f, acting over the total area, a, of a mechanical structure produces a pressure, p. If the random variations in the force are lognormally distributed with parameters μ_f and σ_f^2 and those in the area are similarly lognormally distributed with parameters μ_a and σ_a^2, what is the form of and the corresponding parameters for the distribution of the pressure?

The relationship between pressure, force and area is:

$$p = f/a$$

Taking logarithms of both sides:

$$\log p = \log f - \log a$$

Since each of the variates f and a is lognormally distributed, then both $\log f$ and $\log a$ are normally distributed. Hence, $\log p$ is also normally distributed as was seen in the discussion on sums and differences of distributions. With $\log p$ being normally distributed, the distribution of p is lognormal and the parameters of p are:

$$\mu_p = \mu_f - \mu_a$$

$$\sigma_p^2 = \sigma_f^2 + \sigma_a^2$$

This example now points the way towards dealing in general with quotients of distributions. The equivalent logarithmic expression of Equation (8.94) is:

$$\log z = \log x - \log y \tag{8.100}$$

Then by a similar process to that from which Equation (8.74) was derived from Equation (8.65), it can be seen that:

$$f_z(z) = \frac{1}{z}[[L^{-1}\{L[e^z f_x(e^z)](L[e^z f_y(e^z)])_{s=-s}\}]]_{z=\log z} \tag{8.101}$$

As with Equation (8.74), however, Equation (8.101) is not always easily soluble in terms of analytical functions. It is pertinent, therefore, to consider once again the case of small variations about the mean value as indicated in Equations (8.75) and (8.76). Substituting these equations in Equation (8.94) gives:

$$z = \frac{x_0 + \delta x}{y_0 + \delta y} \qquad (8.102)$$

or:

$$z = \frac{x_0}{y_0} + \frac{\delta x}{y_0} - \frac{x_0 \, \delta y}{y_0^2} \qquad (8.103)$$

and following similar steps to those used in the derivation of Equation (8.84), it can be seen that:

$$F_z(s) = \exp(-x_0 s/y_0) F_x\left(\frac{s}{y_0}\right) F_y\left(-\frac{x_0 s}{y_0^2}\right) \qquad (8.104)$$

whence:

$$\mu_z = \frac{x_0}{y_0}$$

or:

$$\mu_z = \frac{\mu_x}{\mu_y} \qquad (8.105)$$

and:

$$\frac{\sigma_z^2}{\mu_z^2} = \frac{\sigma_x^2}{\mu_x^2} + \frac{\sigma_y^2}{\mu_y^2} \qquad (8.106)$$

Example 8.8

Two electrical resistors, R_1 and R_2, of mean values 10 Ω and 30 Ω respectively are connected in series. A voltage V, of mean value 80 V, exists across the resistor combination. What is the mean value and standard deviation for the current through the total resistance if the standard deviations of resistance and voltage are $\sigma_{R_1} = 0.5$ Ω, $\sigma_{R_2} = 1$ Ω and $\sigma_V = 5$ V?

The current, I, is given by:

$$I = \frac{V}{R_1 + R_2}$$

The numerator has a mean value, μ_R, of:

$$\mu_R = 10 + 30 = 40$$

and a variance of:

$$\sigma_R^2 = 0.5^2 + 1.0^2 = 1.25$$

So, the mean value of the current, from Equation (8.105) is:

$$\mu_I = \frac{\mu_V}{\mu_R}$$

$$= \frac{80}{40} = 2 \text{ A}$$

COMBINATIONS OF DISTRIBUTIONS

and the variance of the current, from Equation (8.106) is:

$$\frac{\sigma_I^2}{\mu_I^2} = \frac{\sigma_V^2}{\mu_V^2} + \frac{\sigma_R^2}{\mu_R^2}$$

or:

$$\sigma_I^2 = 2^2 \left\{ \frac{5^2}{80^2} + \frac{1.25}{40^2} \right\}$$

or:

$$\sigma_I = 0.137 \text{ A}$$

General Combinations of Small Variations

The method employed in Example 8.8 to find the resultant parameters of a combined distribution may become very cumbersome where $f_z(z)$ is complex. It is worth noting, therefore, a simpler and more general method of approach based upon the theory of small quantities. Suppose that the variate z is some function of the variates x and y such that:

$$z = f(x, y) \tag{8.107}$$

then the mean value of z, μ_z, is given by:

$$\mu_z = f(\mu_x, \mu_y) \tag{8.108}$$

If the variates in Equation (8.107) are considered to represent only small variations about the mean, then:

$$x = \mu_x + \delta\mu_x \tag{8.109}$$
$$y = \mu_y + \delta\mu_y \tag{8.110}$$
$$z = \mu_z + \delta\mu_z \tag{8.111}$$

and hence:

$$\mu_z + \delta\mu_z = f(\mu_x + \delta\mu_x, \mu_y + \delta\mu_y) \tag{8.112}$$

or from Equations (8.108) and (8.112):

$$\delta\mu_z = f(\mu_x + \delta\mu_x, \mu_y + \delta\mu_y) - f(\mu_x, \mu_y) \tag{8.113}$$

Assuming that z is continuous and that the derivatives exist, then Equation (8.113) may be expanded as a Taylor series to the first order of small quantities to yield:

$$\delta\mu_z = \frac{\partial \mu_z}{\partial \mu_x} \delta\mu_x + \frac{\partial \mu_z}{\partial \mu_y} \delta\mu_y \tag{8.114}$$

or, squaring both sides:

$$(\delta\mu_z)^2 = \left(\frac{\partial \mu_z}{\partial \mu_x} \delta\mu_x + \frac{\partial \mu_z}{\partial \mu_y} \delta\mu_y \right)^2 \tag{8.115}$$

and since the variance, σ^2, is proportional to the square of the sum of the variations about the mean, $\delta\mu$, then Equation (8.115) becomes:

$$\sigma_z^2 = \left(\frac{\partial \mu_z}{\partial \mu_x}\right)^2 \sigma_x^2 + \left(\frac{\partial \mu_z}{\partial \mu_y}\right)^2 \sigma_y^2 \qquad (8.116)$$

The cross-product term being zero since the sum of the deviations from the mean is zero.

More generally, for the function:

$$z = f(x_1, x_2, x_3, \ldots, x_n) \qquad (8.117)$$

then:

$$\sigma_z^2 = \left(\frac{\partial \mu_z}{\partial \mu_{x_1}}\right)^2 \sigma_{x_1}^2 + \left(\frac{\partial \mu_z}{\partial \mu_{x_2}}\right)^2 \sigma_{x_2}^2 + \ldots + \left(\frac{\partial \mu_z}{\partial \mu_{x_n}}\right)^2 \sigma_{x_n}^2 \qquad (8.118)$$

Example 8.9

Find the standard deviation of the current in Example 8.8 by the method of Equation (8.118):

$$I = \frac{V}{R_1 + R_2}$$

Hence:

$$\frac{\partial I}{\partial V} = \frac{1}{R_1 + R_2} = \frac{1}{40}$$

evaluated at the mean values, and:

$$\frac{\partial I}{\partial R_1} = -\frac{V}{(R_1 + R_2)^2} = -\frac{1}{20}$$

and:

$$\frac{\partial I}{\partial R_2} = -\frac{V}{(R_1 + R_2)^2} = -\frac{1}{20}$$

Therefore, from Equation (8.118):

$$\sigma_I^2 = \left(\frac{1}{40}\right)^2 5^2 + \left(-\frac{1}{20}\right)^2 (0.5)^2 + \left(-\frac{1}{20}\right)^2 1^2$$

or:

$$\sigma_I = 0.137 \text{ A}$$

as before.

Selected Combinations of Distributions

Many systems, which are designed for high reliability, incorporate aspects of redundancy. That is, a particular function of the system may be capable

COMBINATIONS OF DISTRIBUTIONS

of being performed by more than one element of the system. Some typical types of redundant elements and redundant systems were described in Chapter 1. Redundancy may be carried out in a number of different ways. There is simple parallel redundancy where two nominally identical items work continually in parallel and where either item is capable of performing the required function on its own. There is standby redundancy which is similar to parallel redundancy except that only one of the items is normally in operation and the other is brought into operation when the function of the first item is lost or degraded. There are auctioneering systems where the system function is selected from the 'best' function of a number of nominally identical items. There are voting systems where the system function is taken from, say, the best m functional elements out of a total of n elements.

Each type of redundant system can lead to functional variations being combined in a number of different ways. The arithmetic type of combinations already dealt with will enter into the picture but often the idea of 'selected' or 'best' elements leads to different types of functional combinations. This frequently arises in connexion with response times or time delays. As an illustration, suppose that an emergency electrical supply consists essentially of two standby generating sets of which only one is actually required. On demand, both these sets are automatically started and run up to speed but only the set which reaches full output capability first is connected on to the distribution system. If there are variations in the times to full output, then the set with the shortest time is obviously the one which is selected. On the other hand, if both sets were required, then the time delay to the restoration of supplies would depend upon the machine that has the longest time to reach full output capability.

Consider, first of all therefore, some variate z which is dependent upon the smaller value of two independent variates x and y, that is:

$$z = \text{the smaller of either } x \text{ or } y \tag{8.119}$$

This is similar to a logical OR situation, and so applying the law for the combination of probabilities in this fashion derived in Equation (1.17):

$$p_z(z) = p_x(z) + p_y(z) - p_x(z)p_y(z) \tag{8.120}$$

which may be written as:

$$p_z(z) = 1 - \overline{p_x(z)}\,\overline{p_y(z)} \tag{8.121}$$

and differentiating with respect to z leads to:

$$f_z(z) = f_x(z)\overline{p_y(z)} + f_y(z)\overline{p_x(z)} \tag{8.122}$$

More generally, if z is selected from the smallest of a range of independent variates $x_1, x_2, ..., x_n$ then:

$$f_z(z) = \sum_{k=1}^{n} f_{x_k}(z) \left[\prod_{\substack{j=1 \\ k \neq j}}^{n} \overline{P_{x_j}(z)} \right] \quad (8.123)$$

If $x_1, x_2, ..., x_n$ are each taken from the same distribution, then Equation (8.123) becomes:

$$f_z(z) = n f_x(z) [\overline{P_x(z)}]^{n-1} \quad (8.124)$$

If Equation (8.123) is written in terms of the hazard functions described in Chapter 7 then by making use of Equations (7.106) and (7.107) it is seen that:

$$f_z(z) = [z_{x_1}(z) + ... + z_{x_n}(z)] \exp\{-\int [z_{x_1}(z) + ... + z_{x_n}(z)] dz\} \quad (8.125)$$

or:

$$f_z(z) = z_z(z) \exp[-\int z_z(z) dz] \quad (8.126)$$

where:

$$z_z(z) = \sum_{j=1}^{n} z_{x_j}(z) \quad (8.127)$$

So that the resultant distribution, which is obtained from the smallest of a set of variates from different distributions, is characterized by a hazard function which is simply equal to the sum of the hazard functions of each distribution. Obviously where the distributions are equal, as in Equation (8.124), then:

$$z_z(z) = n z_x(z) \quad (8.128)$$

Example 8.10

A system is made up of four sub-systems such that the failure of any sub-system results in complete system failure. The times to failure of each sub-system are exponentially distributed with parameters of $\lambda_1 = 10,000$ h, $\lambda_2 = 5000$ h, $\lambda_3 = 4000$ h and $\lambda_4 = 2000$ h. What is the form of the distribution for the times to failure for the complete system and the values of the parameters for this distribution?

The system fails at the smallest time to failure of the sub-systems. For the exponential distribution, as seen in Chapter 7.

$$z_x(z) = \frac{1}{\lambda_x}$$

Hence:

$$z_z(z) = \frac{1}{\lambda_z} = \frac{1}{\lambda_1} + \frac{1}{\lambda_2} + \frac{1}{\lambda_3} + \frac{1}{\lambda_4}$$

COMBINATIONS OF DISTRIBUTIONS

In this case then:

$$\lambda_z = \frac{20{,}000}{21} \simeq 952.4 \text{ h}$$

and the resultant distribution is:

$$f_z(z) = \frac{1}{\lambda_z} e^{-z/\lambda_z}$$

So the system distribution of times to failure is also exponential and has a mean time to failure, λ_z, of 952.4 h. It is often more convenient, because of the way in which the hazard functions are summated, to express the parameter λ of the exponential distribution in terms of its reciprocal θ where:

$$\theta = 1/\lambda \tag{8.129}$$

The parameter θ may then be termed a 'constant failure-rate'. It is seen, of course, in this example that:

$$\theta_1 = 10^{-4}\, f/h,\ \theta_2 = 2\times 10^{-4}\, f/h,\ \theta_3 = 2.5\times 10^{-4}\, f/h$$

and:

$$\theta_4 = 5\times 10^{-4}\, f/h$$

Hence:

$$\theta_z = \theta_1 + \theta_2 + \theta_3 + \theta_4$$
$$= 10.5 \times 10^{-4}\, f/h$$

and:

$$f_z(z) = \theta_z e^{-\theta_z z} \tag{8.130}$$

So far it has been assumed that all $f_{x_j}(x)$ are continuous in the range 0 to ∞. If, on the other hand, the $f_{x_j}(x)$ are bounded by finite lower and upper limits, then the corresponding $f_z(z)$ will contain points of discontinuity. Consider the case of two distributions $f_x(x)$ and $f_y(y)$ which have limits of x_l, x_u, y_l and y_u. Assume that $x_l \leqslant y_l$. Then, in the range:

$$0 \leqslant z \leqslant x_l$$
$$f_z(z) = 0 \tag{8.131}$$

If x_u is $\leqslant y_l$, then the next range is:

$$x_l \leqslant z \leqslant x_u$$

but, if x_u is $\geqslant y_l$, then this range becomes:

$$x_l \leqslant z \leqslant y_l$$

in either case:

$$f_z(z) = f_x(z) \tag{8.132}$$

With a second range of z of $x_l \leqslant z \leqslant x_u$, the third range of z is:

$$x_u \leqslant z \leqslant \infty$$

and here:

$$f_z(z) = 0 \qquad (8.133)$$

but with a second range of z of $x_l \leqslant z \leqslant y_l$, the third range becomes:

$$y_l \leqslant z \leqslant x_u$$

and then:

$$f_z(z) = f_x(z)\overline{p_y(z)} + f_y(z)\overline{p_x(z)} \qquad (8.134)$$

as in Equation (8.122). The fourth and final range is then:

$$x_u \leqslant z \leqslant \infty$$

where:

$$f_z(z) = 0 \qquad (8.135)$$

Example 8.11

Two systems, X and Y, are designed to perform a similar output function on receipt of an input signal. Each system is subject to random variations in time delays which follow rectangular distributions. For system X, the p.d.f. is

$$f_x(x) = \tfrac{1}{4}[U(x-3) - U(x-7)]$$

and for system Y:

$$f_y(y) = \tfrac{1}{2}[U(y-6) - U(y-8)]$$

What is the p.d.f. for an overall arrangement where the shortest time delay from system X or Y is always selected?

Let $f_z(z)$ be the required p.d.f., then for the range:

$$0 \leqslant z \leqslant 3$$

$$f_z(z) = 0$$

Following the result of Equation (8.132), the next range is:

$$3 \leqslant z \leqslant 6$$

and here:

$$f_z(z) = f_x(z)$$
$$= \tfrac{1}{4}[U(z-3) - U(z-6)]$$

The third range, from Equation (8.134), is:

$$6 \leqslant z \leqslant 7$$

COMBINATIONS OF DISTRIBUTIONS

where:

$$f_z(z) = \left(\frac{15-2z}{8}\right)[U(z-6) - U(z-7)]$$

and, finally, for:

$$7 \leqslant z \leqslant \infty$$
$$f_z(z) = 0$$

The shape of $f_z(z)$ for this example is shown in Figure 8.4.

An alternative method of selecting a variate is to take the largest instead of the smallest value. Suppose, therefore, that z is dependent upon the larger value of two independent variates x and y such that:

$$z = \text{the larger of either } x \text{ or } y \qquad (8.136)$$

This is similar to a logical AND situation, and so applying the law for the combination of probabilities of this type derived in Equation (1.14) gives:

$$P_z(z) = P_x(z) P_y(z) \qquad (8.137)$$

and therefore:

$$f_z(z) = f_x(z) P_y(z) + f_y(z) P_x(z) \qquad (8.138)$$

and generally for n variates $x_1, x_2, \ldots, x_n$:

$$f_z(z) = \sum_{k=1}^{n} f_{x_k}(z) \left[\prod_{\substack{j=1 \\ j \neq k}}^{n} P_{x_j}(z) \right] \qquad (8.139)$$

or, if all the variates are taken from the same distribution:

$$f_z(z) = n f_x(z) [P_x(z)]^{n-1} \qquad (8.140)$$

Equation (8.140) may be written in terms of the hazard function:

$$f_z(z) = n z_x(z) \exp[-\int z_x(z) dz] \{1 - \exp[-\int z_x(z) dz]\}^{n-1} \qquad (8.141)$$

but this does not lead to any simple relationship as it did in the case of selecting the smallest of a set of variates.

If $f_x(x)$ is a rectangular distribution of the form:

$$f_x(x) = \frac{1}{(b-a)} [U(x-a) - U(x-b)]$$

then, from Equation (8.140):

$$f_z(z) = \frac{n}{(b-a)^n} (z-a)^{n-1} \qquad (8.142)$$

the range of z being the same as x, namely a to b. The mean value of Equation

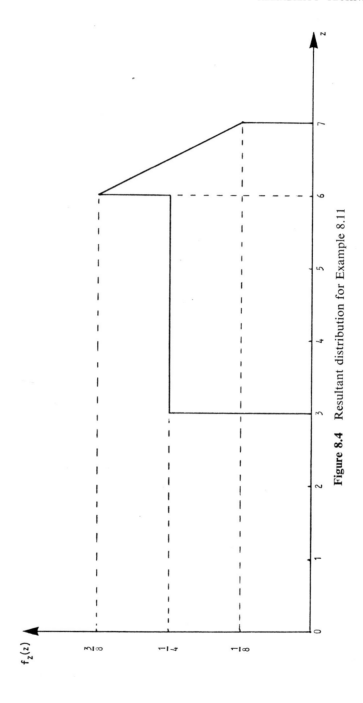

Figure 8.4 Resultant distribution for Example 8.11

COMBINATIONS OF DISTRIBUTIONS

(8.142) is seen to be:

$$\mu_z = a + \frac{n(b-a)}{n+1}$$

or:

$$\mu_z = \frac{a+b}{2} + \frac{(n-1)(b-a)}{2(n+1)} \tag{8.143}$$

So the mean of z is always greater than the mean of x, as would be expected. As n becomes large, μ_z approaches the upper limit of x. This will, of course, be true for any distribution combined in this way. If the upper limit of x is infinity, then the mean of z will also approach infinity as n becomes large. For large but finite n it is seen that Equation (8.143) may be written in the form:

$$\mu_z = b - \frac{(b-a)}{n} \tag{8.144}$$

so that the mean value is situated at a value of $(b-a)/n$ below the upper limit of x. The form of the limiting distribution of Equation (8.142) as $n \to \infty$ may be found by taking the Laplace transform as follows:

$$F_z(s) = n! \, e^{-bs} \sum_{j=0}^{\infty} \frac{[(b-a)s]^j}{(n+j)!} \tag{8.145}$$

As n becomes large, this approximates to:

$$F_z(s) = \frac{e^{-bs}}{1 - [(b-a)s]/n} \tag{8.146}$$

which on taking the inverse transform yields:

$$f_z(z) = \frac{n}{(b-a)} \exp\left[n(z-b)/(b-a)\right] U[b-z] \tag{8.147}$$

This is a negative exponential distribution with an upper limit of b and a mean value of $(b-a)/n$ below this limit.

Apart from taking the smallest or the largest variate out of a set of variates, many other forms of selection are possible. One other form of particular interest arises in connexion with majority-vote systems. As an illustration, suppose that three variates, x_1, x_2 and x_3, exist and a variate z is required such that it represents the larger of the two smallest values out of the possible three x values. Then, following the lines of the previous derivations, the cumulative distribution function for z will be given by:

$$p_z(z) = p_{x_1}(z) p_{x_2}(z) + p_{x_2}(z) p_{x_3}(z) + p_{x_3}(z) p_{x_1}(z) - 2 p_{x_1}(z) p_{x_2}(z) p_{x_3}(z) \tag{8.148}$$

whence, on differentiating with respect to z:

$$f_z(z) = f_{x_1}(z)[p_{x_2}(z)+p_{x_3}(z)-2p_{x_2}(z)p_{x_3}(z)]$$
$$+f_{x_2}(z)[p_{x_1}(z)+p_{x_3}(z)-2p_{x_1}(z)p_{x_3}(z)]$$
$$+f_{x_3}(z)[p_{x_1}(z)+p_{x_2}(z)-2p_{x_1}(z)p_{x_2}(z)] \quad (8.149)$$

and if all the x variates belong to the same distribution $f_x(x)$, then:

$$f_z(z) = 6f_x(z)p_x(z)[1-p_x(z)] \quad (8.150)$$

Example 8.12

The shut-down signal for a chemical process plant is derived from three separate but nominally identical measurement channels. In each channel, the time delay between the measured quantity reaching the shut-down value and the output signal to operate the shut-down mechanism is found to follow a rectangular distribution of the form:

$$f_x(x) = \tfrac{1}{4}[U(x-3)-U(x-7)]$$

where the parameter, x, is measured in seconds. The shut-down mechanism, however, only operates after it has received at least two shut-down signals. What is the distribution of delay times for complete plant shut-down and what is the mean and standard deviation of this distribution?

This is a majority-vote system acting on the largest delay time of any two smallest delay times from the three measurement channels. Hence, applying Equation (8.150) gives:

$$f_z(z) = \tfrac{3}{32}(z-3)(7-z) \quad \text{for } 3 \leqslant z \leqslant 7$$

The mean of the distribution is:

$$\mu_z = \int_3^7 \tfrac{3}{32} x(x-3)(7-x)\,dx$$
$$= 5 \text{ s}$$

and the variance is given by:

$$\sigma_z^2 = \int_3^7 \tfrac{3}{32}(x-\mu_z)^2 (x-3)(7-x)\,dx$$
$$= \frac{4}{5}$$

hence:

$$\sigma_z = \frac{2}{\sqrt{5}} \text{ s}$$

The distributions for other types of majority-vote systems may be derived and used in a similar way to that shown in Equation (8.150) and illustrated

COMBINATIONS OF DISTRIBUTIONS

in Example 8.12. Some further examples of selected combinations of distributions will be given in some of the following chapters which deal with system analysis. This chapter has aimed at laying the foundations for the techniques that are needed in order to combine distributions in a variety of different ways. It is hoped, on this basis, that the reader may now be able to deal with most system problems of this nature which may be presented to him.

Questions

1. A control surface is operated by two hydraulic actuators, X and Y, working in parallel. The distribution of force produced by actuator X is represented by the p.d.f.:

$$f_x(x) = \tfrac{1}{20}[U(x-90) - U(x-110)]$$

   where the variate x is measured in Newtons. The corresponding p.d.f. for the distribution of force produced by actuator Y is:

$$f_y(y) = \left(\frac{y-80}{800}\right)[U(y-80) - U(y-120)]$$

   Calculate:
   (a) the formula for the combined distribution of force acting upon the control surface,
   (b) the mean of this distribution,
   (c) the standard deviation of this distribution,
   and sketch:
   (d) the form of the distribution.

2. Twenty-five nominally identical resistors, each of mean value 10 Ω, are connected in series. The variation in each resistor value follows a distribution with a standard deviation of 10% of the mean value. What is the approximate probability that the overall resistance value is less than 240 Ω?

3. A control system is actuated on a signal z derived from the difference between two measured quantities x and y. The quantities x and y are subject to random variations which follow an exponential distribution with mean values of 4 units and 3 units respectively. What is:
   (a) the form of the p.d.f. for the distribution of z,
   (b) its mean value, and
   (c) its standard deviation?

4. Four motor car manufacturing companies produce an output of a number of cars per hour which follows a Poisson process. The mean outputs per hour are 2, 1.5, 1 and 0.5 cars. What is the probability in any

one hour that the combined car output is:
(a) 4 cars,
(b) 5 cars,
(c) less than or equal to 5 cars, and
(d) more than 5 cars?

5. A machine shop consists of 10 grinders, 60 lathes and 20 milling machines. The probabilities of any one of these three types of machines being out of service are 0.1, 0.05 and 0.25 respectively. What is the mean number of machines out of the total of 90 that are likely to be in service at any one time? What is the chance that the number of machines in service is less than or equal to this mean number? (Assume the Poisson approximation to the binomial distribution.)

6. A piece of apparatus supplied with a mean d.c. voltage of 110 V consumes a mean d.c. current of 5 A. If the random variations in voltage are rectangularly distributed over a total range of 10 V and the variations in current similarly distributed over a total range of 1 A, what is the probability that the power consumed is:
(a) less than 500 W, and
(b) less than 550 W?

7. The pressure upon the piston head of a hydraulic actuator has a mean value of 800 kN m^{-2} and a standard deviation of 8.0 kN m^{-2}. If the area of the piston head has a mean value of 0.01 m^2 and a standard deviation of 0.0001 m^2, what is the mean value and standard deviation of the force acting upon the piston? (Assume small variations about the mean values in each case.)

8. An electric motor with a mean electrical resistance of 2 Ω is driven by a mean d.c. current of 55 A. Both the resistance value of the motor and the drive current are subject to small random variations about their mean values. The standard deviations of these variations are 0.1 Ω for the resistance and 3 A for the current. What is:
(a) the mean value of the voltage drop across the motor, and
(b) the standard deviation of this voltage drop?

9. An aeroplane, when about to take off on a regular daily service, is found to have a mean mass of 50,000 kg and a mean initial thrust produced by the engines of 40 kN. The variations in aeroplane mass and engine thrust follow rectangular distributions of total ranges 10,000 kg and 2 kN respectively. What is:
(a) the form of the p.d.f. for the resultant acceleration,
(b) the mean acceleration,
(c) the probability that the acceleration is less than the mean value, and
(d) the probability that the acceleration is less than (41/55) m s^{-2}?

COMBINATIONS OF DISTRIBUTIONS   311

10. The power output of an engine is subject to small variations about a mean value of 75 kW. The variations follow a normal distribution with a standard deviation of 1 kW. The speed output of the engine is also subject to a normal distribution of small changes about the mean. The mean and standard deviation of the speed characteristics are 500 radians per second and 5 radians per second respectively. What is:
    (a) the mean value of the engine's torque output,
    (b) the standard deviation of the torque characteristic, and
    (c) the probability that the torque lies between 142.5 N m and 155 N m?
11. Five resistors are connected between points A and B as follows:

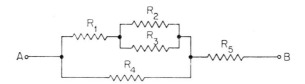

Each resistor is subject to small variations about its mean value, the characteristic of the variations being defined by this mean value and a standard deviation as shown in the following table:

Resistor	Mean Value (Ω)	Standard Deviation (Ω)
R_1	26	1
R_2	40	2
R_3	60	3
R_4	50	1
R_5	25	0.5

What is the mean value and standard deviation for the total resistance variation between points A and B?

12. A mean d.c. current of 5 A flows through two resistors in series of mean values 4 Ω and 6 Ω. If the standard deviation of the current variation is 0.05 A, that of the 4 Ω resistor is 0.1 Ω and that of the 6 Ω resistor is 0.2 Ω, what is the mean value and standard deviation of the power dissipated?

13. A unit is made up of three main components. At the start of the unit's life it is known that the times to failure of each of the three components follow a Weibull distribution. The parameters of this distribution for each of the components are $\lambda_1 = 10^2$ h, $\beta_1 = 3$, $\lambda_2 = 10^3/\sqrt{3}$ h, $\beta_2 = 2$, $\lambda_3 = 10^6/3$ h and $\beta_3 = 1$. If the unit is considered to have failed at the

time when the first component fails, what is:
(a) the mean time to failure of the unit, and
(b) the type of distribution of the unit times to failure?

14. Four standby generating sets have times to start and take up load which follow an exponential distribution with a mean value of 40 s. What is the mean time to the take up of load if the load is supplied from:
(a) the first set to start, and
(b) the last set to start?

15. A room is lit by five nominally identical lamps. All lamps are switched on together and left switched on. It is known that the times to lamp failures after they have been switched on is rectangularly distributed between a lower limit of 8000 h and an upper limit of 12,000 h. What is the mean time to the room being in darkness? How would this mean time be affected if the number of lamps was increased to a total of 15?

16. The shut-down signal for a nuclear reactor is derived from four separate but nominally identical temperature measurement channels. In each channel, the time delay between the reactor temperature reaching the required shut-down value and the output signal from the channel is found to follow an exponential distribution with a mean value of 12 s. The shut-down mechanism operates directly it has received three output signals. What is:
(a) the form of the distribution of delay times for reactor shut-down, and
(b) the mean value of this distribution?

CHAPTER 9

Sampling, Estimation and Confidence

Sampling

The probability functions and distributions discussed in the preceding chapters will render a mathematically exact value of the relevant probability which is calculated provided that the form of the distribution and the values of its associated parameters are known with certainty.

As an example, consider a control mechanism which has a probability of 0.5 of performing a certain function in a certain time. The probability of the mechanism performing 8 functions within the time and 12 functions outside the time in a sequence of 20 operations is, from the binomial distribution:

$$f(8) = \binom{20}{8} (0.5)^8 (1-0.5)^{20-8}$$

$$= \binom{20}{8} (0.5)^{20}$$

$$= 0.1201$$

This means that there is an exactly 12.01% chance of achieving the result postulated. Expressed another way, for a very large number of such performance functions, 1201 functions on average out of every 10,000 will yield a result of 8 functions within the time and 12 outside.

Suppose, as another example, that it is known that the response times of electronic amplifiers in a large population follow a normal distribution such that the mean, μ, is 6 s and the standard deviation, σ, is 0.5 s. Then, the probability of an amplifier selected at random having a response time less than 7 s (which is 2 standard deviations above the mean) is, from Table A.1:

$$p(7) = 0.5 + 0.4772$$

$$= 0.9772$$

This means that there is an exactly 97.72% chance of selecting an amplifier with a response time less than 7 s. Expressed another way, for a very large number of selections an amplifier with a response time less then 7 s will be selected on average every 9772 times out of every 10,000 attempts.

As a final example, suppose a device exists which exhibits a constant average failure-rate pattern and the value of this failure-rate is known to be precisely 0.1 faults per year. Then, from the exponential distribution, the probability of failure of this device by the end of a 1-year period is:

$$p(t) = 1 - e^{-0.1}$$

$$= 0.095$$

This means that there is an exactly 9.5% chance of failure in the time interval quoted. Expressed another way, for a very large number of identical devices, 950 devices on average out of every 10,000 will fail in the annual period.

There is no doubt about the results calculated in each of the preceding examples provided that they are interpreted in the manner described. Unfortunately, in most practical cases, the parameters of the distribution or even the distribution itself may not be precisely known. The values of the parameters and the appropriate characteristics may then have to be 'estimated' or 'inferred'. The question now arises not only as to the 'accuracy' of such estimates but also to the effect of this estimate on any predicted probabilities. Associated with any statistical estimation, therefore, is the concept of the 'accuracy' of or the 'confidence' in the values obtained.

Estimation is necessary because, in general, the characteristics of the complete population are not known. The estimation then has to be based upon the characteristics of a proportion or 'sample' of the complete population. The process is concerned with drawing inferences about the whole population based upon the evidence produced by a sample. Two common-sense aspects with regard to the sample itself are immediately apparent. First, the larger the sample, the more likely is the 'estimate' to be close to the 'true' value with complete correspondence, of course, being reached when the sample embraces the whole population. Secondly, whatever sample size is used it must be 'representative' of the parent population. These two aspects lead to the requirement for any estimation to be 'consistent' and 'unbiased' and this is discussed further in the next section.

Samples may be of two main types. On the one hand, there are those which represent a collection of data or results which is already in existence or which has been previously derived for some other purpose. Here, the problem is to obtain the best estimate of the required values from what is available. On the other hand, samples may be deliberately created or tests set up in order to make the specific estimate that is required. In this case, the problem is to select the sample or formulate the series of tests such that the best estimate can be made.

With both types of samples a number of general points arise in connexion with the following issues:
(a) There is a need for a method or methods of obtaining a representative sample. This is often referred to as random sampling techniques.
(b) Variations obviously exist between nominally identical samples and hence the sample estimates are likely to follow some form of statistical distribution.
(c) Assuming the form of the parent population, estimates may be made of the population distributional parameters. There is then a need to define the degree of trustworthiness in or the confidence in such estimates.
(d) If the form of the parent population is not known then it may be required to test the 'goodness of fit' with some standard distribution and obtain a 'level of significance' in relation to such an hypothesis.

In connexion with issue (a), a representative or random sample is often defined in the following general way:

'A random sample is obtained by selecting items from the population in such a way that each item in the population has an equal chance of being selected.'

The reader will appreciate from the above statement some of the difficulties which arise in defining 'random sample'. Terms such as 'random' and 'equal chance' appear to have similar meanings and the whole definition has the appearance of circularity. However, it is beyond the scope of this book to define or discuss random sampling in any detail and the reader is referred to other standard works on the subject. The simple definition given does, however, highlight the important characteristic that the word 'random' applies to the method of selection of the sample and not to the properties of the sample after it has been chosen.

The remaining issues listed in (b), (c) and (d) will now be discussed further in the following sections of this chapter.

Estimation and Confidence

General concepts

One of the most common types of estimation that is required to be made from a sample is that of a parameter value of the parent population. This parameter could be the mean, the variance or any other of the characteristics which describe a statistical distribution.

The estimate of a parameter θ, say, is generally denoted by the symbol $\hat{\theta}$. The caret ($\hat{\ }$) above the symbol for the parameter being used to denote an estimator of that parameter. This estimator, $\hat{\theta}$, is normally calculated from the sample information by the use of some formula. The choice of formula and method of estimation should fulfil a number of conditions of which the

following two are important:
(a) the estimation should be consistent, that is the probability of $\hat{\theta}$ approaching θ should tend to unity as the sample size approaches the whole population, and
(b) the estimation should be unbiased, that is the mean value of $\hat{\theta}$ over the number of possible estimations should be equal to θ.

This latter point leads to considerations related to the distribution of $\hat{\theta}$. It is obvious that if a large number of separate and independent estimates are made of θ that the individual values of $\hat{\theta}$ so obtained are likely to follow some particular distribution. This distribution is generally termed a 'sampling distribution' and is normally related in some way to the underlying distribution of the parent population. It is from the sampling distribution that the accuracy of, or the confidence in, the estimation in relation to the true value of the parameter can be obtained. Such an exercise will normally define a range of values within which the true value is believed to lie for a particular 'confidence level'. Set up in this way, the confidence level expresses the probability that the assertion made about the true value of the parameter from the estimates is true. The estimated range or interval for the parameter is termed the 'confidence interval' and the end points of the interval are called 'confidence limits'. The following symbolism is used:

$$\theta_l = \text{lower confidence limit of parameter}$$
$$\theta_u = \text{upper confidence limit of parameter}$$
$$\theta_u - \theta_l = \text{confidence interval of parameter}$$
$$c = \text{confidence level expressed as a probability}$$

Suppose now that a value of $\hat{\theta}$, designated as $\hat{\theta}_j$, is calculated for a particular sample using an appropriate formula for the estimator. Suppose also that the form of the sampling distribution, $f(\hat{\theta})$, is known and can be represented as shown in Figure 9.1.

The sampling distribution, $f(\hat{\theta})$, will have a mean value of $\mu_{\hat{\theta}}$ which, if the correct estimator is used, will be equal to the true value of θ. However, the particular value $\hat{\theta}_j$ which has been calculated may lie anywhere in the range of $\hat{\theta}$ from zero to infinity and could be either above or below the mean value $\mu_{\hat{\theta}}$. There is nothing on the basis of a single sample estimate to say which may be true. Deductions may be made, however, by assuming that each condition is true in turn. First, therefore, let it be assumed that $\hat{\theta}_j$ lies below the mean value $\mu_{\hat{\theta}}$. This condition is illustrated in Figure 9.2. If the shaded area in Figure 9.2 represents a probability value of $(1-c)$, then this chosen position for $\hat{\theta}_j$ means that on average, over a large number of nominally identical and independent samples, the calculated value of $\hat{\theta}$ would exceed $\hat{\theta}_j$ in $100c\%$ of all cases. Hence, it follows, if a $\mu_{\hat{\theta}}$ is calculated for

SAMPLING, ESTIMATION AND CONFIDENCE

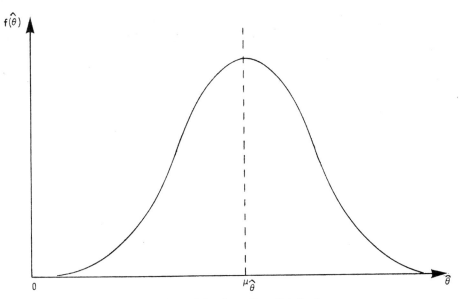

Figure 9.1 Sampling distribution

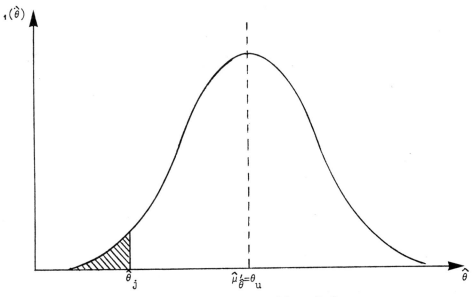

Figure 9.2 Upper confidence limit

this selected position of $\hat{\theta}_j$ and on the basis of this $\hat{\theta}_j$ position, that such a value, $\hat{\mu}'_\theta$ say, will be greater than the true value of the parameter θ in $100c\%$ of all samples which are analysed in this way. This calculated value, $\hat{\mu}'_\theta$, may now be equated to the upper confidence limit θ_u.

Secondly, let it be assumed that $\hat{\theta}_j$ lies above the mean value μ_θ as shown in Figure 9.3. If the shaded area in Figure 9.3 is again chosen to represent a

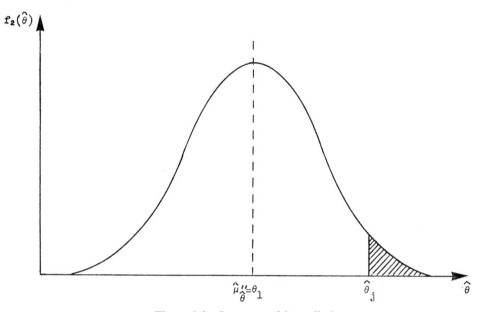

Figure 9.3 Lower confidence limit

probability of $(1-c)$ then the corresponding value of μ_θ which may be calculated from the position of $\hat{\theta}_j$ represents a value, $\hat{\mu}''_\theta$ say, which will be lower than θ in $100c\%$ of all samples which are analysed in this way. This second calculated value, $\hat{\mu}''_\theta$, can be equated to the lower confidence limit θ_l.

Two statements can now be made as follows:

$$\theta < \theta_u \text{ with a } 100c\% \text{ confidence level}$$

and:

$$\theta > \theta_l \text{ with a } 100c\% \text{ confidence level}$$

Each of these statements represents what is generally termed a 'single-ended' confidence limit. For the upper limit only case, it is obvious that the following relationship holds:

$$c = \int_{\hat{\theta}_j}^{\infty} f_1(\hat{\theta}) \, d\hat{\theta} \qquad (9.1)$$

SAMPLING, ESTIMATION AND CONFIDENCE

and also that:
$$\theta_u = \hat{\mu}'_\theta \tag{9.2}$$

For the lower limit only case:
$$c = \int_{-\infty}^{\hat{\theta}_j} f_2(\hat{\theta}) \, d\hat{\theta} \tag{9.3}$$

and:
$$\theta_l = \hat{\mu}''_\theta \tag{9.4}$$

It follows, in the more general case of 'double-ended' confidence limits, that a statement may be made in the following form:

$$\theta_l \leqslant \theta \leqslant \theta_u \text{ with } 100c\% \text{ confidence level}$$

where the lower shaded area of Figure 9.2 and the upper shaded area of Figure 9.3 now have a combined or summated probability value of $(1-c)$ and hence:

$$1 + c = \int_{\hat{\theta}_j}^{\infty} f_1(\hat{\theta}) \, d\hat{\theta} + \int_{-\infty}^{\hat{\theta}_j} f_2(\hat{\theta}) \, d\hat{\theta} \tag{9.5}$$

The appropriate values of θ_u and θ_l are still given by the type of relationships given in Equations (9.2) and (9.4) respectively.

In finding such 'double-ended' confidence limits, the confidence interval is often calculated with a view to symmetry, that is:

$$\frac{1+c}{2} = \int_{\hat{\theta}_j}^{\infty} f_1(\hat{\theta}) \, d\hat{\theta} \tag{9.6}$$

and:

$$\frac{1+c}{2} = \int_{-\infty}^{\hat{\theta}_j} f_2(\hat{\theta}) \, d\hat{\theta} \tag{9.7}$$

but such a procedure is not necessary and the confidence interval can be calculated for any range, symmetrical or not.

The general procedure for parameter estimation and for determining the corresponding confidence limits associated with such an estimation may now be summarized as follows:

(i) Determine or assume the form of the appropriate sampling distribution, $f(\hat{\theta})$, for the particular parameter θ whose estimation is required.

(ii) From the sample information calculate a value $\hat{\theta}$ which represents a consistent and unbiased estimator of θ.

(iii) Fit this particular value of $\hat{\theta}$, namely $\hat{\theta}_j$, to the sampling distribution to give the required level of confidence.

(iv) Calculate the corresponding value or values of μ_θ which then represent the required confidence limits.

Having carried out this exercise, the results need careful interpretation because the final confidence limits have been calculated on the basis of certain assumptions. The principal assumptions are concerned with where the calculated value, $\hat{\theta}_j$, actually lies in the sampling distribution. This is still not known and all that can be said is that if enough nominally identical and independent samples were analysed then $\hat{\theta}_j$ would lie in the range or confidence interval assumed $100c\%$ of the time. The same type of remarks apply, therefore, to the deduced confidence limits. In other words, if such confidence limits were calculated separately and independently from a large number of nominally identical samples then a statement made from the cases so analysed that the actual value of θ lay within the particular confidence interval expressed would be true on $100c\%$ occasions. Confidence, then, in this context is not a probability of what the true value of θ may be but rather the probability that a statement concerning the range of values in which θ may fall is true as a statement.

It is now obvious that the whole process of parameter estimation depends upon the shape and form of the appropriate sampling distribution and upon the use of the correct estimator. These aspects will be dealt with in the next two sub-sections which are concerned with two of the most important population parameters which normally require to be estimated, namely the mean and variance of the parent population.

Sample means

One of the first parameters of a parent population which normally requires estimating from a given sample is that of the mean. A sample will, of course, contain a discrete number, n, of variate values, $x_1, x_2, \ldots, x_n$, taken from the parent population. The mean of the sample can then provide an estimate, $\hat{\mu}$, of the population mean where this estimator is given by:

$$\hat{\mu} = \frac{1}{n} \sum_{j=1}^{n} x_j \tag{9.8}$$

or:

$$\hat{\mu} = \frac{z}{n} \tag{9.9}$$

where:

$$z = x_1 + x_2 + \ldots + x_n \tag{9.10}$$

Since each x_j is taken from the same parent distribution, $f_x(x)$ say, then the distribution of z, $f_z(z)$, representing the sum of all x_j's can be obtained from Equation (8.15) in Chapter 8, namely:

$$L[f_z(z)] = \{L[f_x(x)]\}^n \tag{9.11}$$

SAMPLING, ESTIMATION AND CONFIDENCE

Also, from Equation (9.9) and Relationship No. 1 in Table 8.1:

$$L[f_{\hat{\mu}}(\hat{\mu})] = nL[f_z(nz)] \quad (9.12)$$

where $f_{\hat{\mu}}(\hat{\mu})$ is the distribution of sample means.

Combining Equation (9.11) with Equation (9.12) gives:

$$L[f_{\hat{\mu}}(x)] = \{nL[f_x(nx)]\}^n \quad (9.13)$$

from which the distribution of sample means may be found in terms of the variate x and the parent distribution of x.

Differentiating Equation (9.13) with respect to s:

$$\frac{dL[f_{\hat{\mu}}(x)]}{ds} = n\{nL[f_x(nx)]\}^{n-1}\frac{dL[nf_x(nx)]}{ds} \quad (9.14)$$

Now, from Equation (7.48):

$$\mu_{\hat{\mu}} = -\left\{\frac{dL[f_{\hat{\mu}}(x)]}{ds}\right\}_{s=0}$$

therefore:

$$\mu_{\hat{\mu}} = -\left[n\{L[nf_x(nx)]\}^{n-1}\frac{dL[nf_x(nx)]}{ds}\right]_{s=0} \quad (9.15)$$

Also from Relationship No. 7 in Table A.3 or Relationship No. 1 in Table 8.1:

$$\{L[nf_x(nx)]\}_{s=0} = \{L[f_x(x)]\}_{s=s/n=0}$$
$$= 1 \quad (9.16)$$

Similarly:

$$\{L[nf_x(nx)]\}^{n-1}_{s=0} = 1 \quad (9.17)$$

In addition:

$$\frac{dL[nf_x(nx)]}{ds} = \frac{d}{ds}\{L[f_x(x)]\}_{s=s/n} \quad (9.18)$$

and therefore:

$$\left\{\frac{dL[nf_x(nx)]}{ds}\right\}_{s=0} = \frac{1}{n}\left\{\frac{dL[f_x(x)]}{ds}\right\}_{s=0}$$

$$= -\frac{\mu}{n} \quad (9.19)$$

where μ is the mean value of the parent population of x.

Hence, substituting Equations (9.16), (9.17) and (9.19) in to Equation (9.15) gives:

$$\hat{\mu} = \mu \quad (9.20)$$

So that the mean of the sample means is equal to the mean of the parent population and the estimator of Equation (9.8) has produced an unbiased estimate. This is a result which is generally true irrespective of the form of $f_x(x)$ or the value of n.

Now, returning to Equation (9.14) and carrying out a further differentiation with respect to s gives:

$$\frac{d^2 L[f_{\hat{\mu}}(x)]}{ds^2} = n(n-1)\{L[nf_x(nx)]\}^{n-2}\left\{\frac{dL[nf_x(nx)]}{ds}\right\}^2$$

$$+ n\{L[nf_x(nx)]\}^{n-1}\left\{\frac{d^2 L[nf_x(nx)]}{ds^2}\right\} \quad (9.21)$$

and letting $s = 0$ in Equation (9.21) and applying the results of Equations (9.17) and (9.19) yields:

$$\left\{\frac{d^2 L[f_{\hat{\mu}}(x)]}{ds^2}\right\}_{s=0} = \frac{(n-1)\mu^2}{n} + n\left\{\frac{d^2 L[nf_x(nx)]}{ds^2}\right\}_{s=0} \quad (9.22)$$

Now, as shown in Equation (7.49):

$$\left\{\frac{d^2 L[f_{\hat{\mu}}(x)]}{ds^2}\right\}_{s=0} = \sigma_{\hat{\mu}}^2 + \mu_{\hat{\mu}}^2$$

$$= \sigma_{\hat{\mu}}^2 + \mu^2 \quad (9.23)$$

In a similar way it can be seen that:

$$\left\{\frac{d^2 L[nf_x(nx)]}{ds^2}\right\}_{s=0} = \frac{1}{n^2}\left\{\frac{d^2 L[f_x(x)]}{ds^2}\right\}_{s=0}$$

$$= \frac{1}{n^2}(\sigma^2 + \mu^2) \quad (9.24)$$

Therefore, substituting from Equations (9.23) and (9.24) in Equation (9.22) gives:

$$\sigma_{\hat{\mu}}^2 + \mu^2 = \frac{(n-1)\mu^2}{n} + \frac{1}{n}(\sigma^2 + \mu^2)$$

or:

$$\sigma_{\hat{\mu}}^2 = \frac{\sigma^2}{n} \quad (9.25)$$

So that the variance of the sample means is equal to the variance of the parent population divided by the sample size n. This result is also generally true for any $f_x(x)$ and for any value of $n \geq 1$.

SAMPLING, ESTIMATION AND CONFIDENCE

By a third process of differentiation with respect to s on Equation (9.13) it can be shown that:

$$K_{3_{\hat{\mu}}} = \frac{K_3}{\sqrt{n}} \qquad (9.26)$$

where $K_{3_{\hat{\mu}}}$ and K_3 are the coefficients of skewness for the distribution of the sample means and for the parent distribution respectively. Obviously, as n approaches infinity:

$$K_{3_{\hat{\mu}}} = 0 \qquad (9.27)$$

which, as seen in Chapter 7, is a condition for normality.

Differentiating Equation (9.13) four times and making the appropriate substitutions, it is found that:

$$K_{4_{\hat{\mu}}} = 3\left(1 - \frac{1}{n}\right) + \frac{K_4}{n} \qquad (9.28)$$

or, as n becomes large:

$$K_{4_{\hat{\mu}}} = 3 \qquad (9.29)$$

which, once again, is a condition of normality.

In general, then, for any parent population, as the sample size n becomes large, the distribution of sample means approaches a normal distribution with mean μ and variance σ^2/n where μ and σ^2 are the mean and variance of the parent population respectively. This result, which is also a particular aspect of the 'central limit theorem' may be compared with the previous discussion in Chapter 8 dealing with the sums of variates. In particular, Equations (9.20), (9.25), (9.26) and (9.28) may be compared with Equations (8.17), (8.20), (8.21) and (8.22).

Example 9.1

The lengths of a large number of pivot pins produced on a production line are known to be rectangularly distributed with lower and upper limits of a units and b units respectively. If a random sample of n pivot pins is taken from the production line and an estimate made of the mean length, what is the form of the distribution of such sample means and what form does this distribution take as n becomes large?

The Laplace transform of a rectangular distribution, from Table 7.1, is:

$$F_x(s) = \frac{e^{-as} - e^{-bs}}{(b-a)s} \qquad (9.30)$$

Now, from Equation (9.13) and Relationship No. 1 in Table 8.1:

$$F_{\hat{\mu}}(s) = \left[F_x\left(\frac{s}{n}\right)\right]^n \qquad (9.31)$$

Hence, from Equations (9.30) and (9.31):

$$F_{\hat{\mu}}(s) = \left[\frac{n(e^{-as/n} - e^{-bs/n})}{(b-a)s}\right]^n \qquad (9.32)$$

Expanding the exponential terms in Equation (9.32) and neglecting powers of $(1/n)$ above $(1/n)^3$, that is allowing n to become large, it is seen that Equation (9.32) approximates to:

$$F_{\hat{\mu}}(s) = [\![\exp\{-[(b+a)s/2n - (b-a)^2 s^2/24n^2]\}]\!]^n$$
$$= \exp\{-[(b+a)s/2 - (b-a)^2 s^2/24n]\} \qquad (9.33)$$

This will be recognized from the work done in Chapters 7 and 8 as the transform of a normal distribution where:

$$\mu_{\hat{\mu}} = \frac{(b+a)}{2} = \mu$$

and

$$\sigma_{\hat{\mu}}^2 = \frac{(b-a)^2}{12n} = \frac{\sigma^2}{n}$$

and where μ and σ^2 are the mean and variance of the parent rectangular distribution respectively.

This example may be compared with Example 8.2 where similar, although not identical, relationships were found for the case of simple sums of variates taken from a rectangular distribution.

It should be noted, of course, that when the parent population is normal, the distribution of the sample means is also normal irrespective of the value of n. Hence, for a normal distribution:

$$f_{\hat{\mu}}(x) = \frac{\sqrt{n}}{\sigma\sqrt{(2\pi)}} \exp[-n(x-\mu)^2/2\sigma^2] \qquad (9.34)$$

and from a similar process to the derivation of Equation (7.19):

$$L[f_{\hat{\mu}}(x)] = \exp[-(\mu s - \tfrac{1}{2}\sigma^2 s^2/n)] \qquad (9.35)$$

Some slight restrictions on the conclusions reached so far may be necessary when the parent population consists of a finite number of members. Suppose, therefore, that a parent population consists of N individual members

SAMPLING, ESTIMATION AND CONFIDENCE

$x_1, x_2, \ldots, x_N$, then the population mean is given by:

$$\mu = \frac{1}{N} \sum_{j=1}^{N} x_j \qquad (9.36)$$

and the population variance by:

$$\sigma^2 = \frac{1}{N} \sum_{j=1}^{N} (x_j - \mu)^2$$

$$= \frac{1}{N} \sum_{j=1}^{N} x_j^2 - \mu^2 \qquad (9.37)$$

If samples are taken from the population where each sample is of size n, then the total number of possible independent samples is restricted to:

$$\binom{N}{n}$$

The mean of all possible sample means is then given by:

$$\mu_{\hat{\mu}} = \frac{1}{\binom{N}{n}} (\hat{\mu}_1 + \hat{\mu}_2 + \ldots + \hat{\mu}_{\binom{N}{n}}) \qquad (9.38)$$

where $\hat{\mu}_1, \hat{\mu}_2, \ldots$ are the means calculated from each sample of size n. Therefore:

$$\mu_{\hat{\mu}} = \frac{1}{\binom{N}{n}} \left(\frac{1}{n} \sum_{j=1}^{n} x_j + \frac{1}{n} \sum_{k=1}^{n} x_k + \ldots \right) \qquad (9.39)$$

where each summation in Equation (9.39) represents a different sample of size n.

Hence, combining all the terms in each summation:

$$\mu_{\hat{\mu}} = \frac{1}{\binom{N}{n}} \frac{1}{n} \binom{N-1}{n-1} \sum_{j=1}^{N} x_j$$

$$= \frac{1}{N} \sum_{j=1}^{N} x_j$$

$$= \mu \qquad (9.40)$$

which shows that the mean of a sample still represents the best and unbiased estimate of the population mean as has already been shown in Equation (9.20) for the case of an infinite population.

The variance of the sample means can now be expressed in the form:

$$\sigma_{\hat{\mu}}^2 = \frac{1}{\binom{N}{n}} [(\hat{\mu}_1-\mu)^2 + (\hat{\mu}_2-\mu)^2 + \ldots + (\hat{\mu}_{\binom{N}{n}}-\mu)^2]$$

$$= \frac{1}{\binom{N}{n}} \left[(\hat{\mu}_1^2 + \hat{\mu}_2^2 + \ldots) - 2\mu(\hat{\mu}_1 + \hat{\mu}_2 + \ldots) + \binom{N}{n}\mu^2 \right]$$

$$= \frac{1}{\binom{N}{n}} \left[\frac{1}{n^2}\left(\sum_{j=1}^{n} x_j\right)^2 + \frac{1}{n^2}\left(\sum_{k=1}^{n} x_k\right)^2 + \ldots \right] - \mu^2$$

$$= \frac{1}{\binom{N}{n}} \frac{1}{n^2} \binom{N-2}{n-1} \sum_{j=1}^{N} x_j^2 - \frac{(N-n)}{n(N-1)}\mu^2$$

$$= \frac{(N-n)}{n(N-1)} \left[\frac{1}{N}\sum_{j=1}^{N} x_j^2 - \mu^2\right]$$

$$= \frac{(N-n)}{n(N-1)} \sigma^2 \qquad (9.41)$$

From which it may be seen that the variance of the sample means is no longer given simply by σ^2/n except in the case where N approaches infinity.

It is now pertinent to conclude this section by seeing how the derived distributions and parameters for the sample means of a population can be used to ascribe confidence limits to the estimate of the mean for the parent population.

If double-ended confidence limits are required for the estimated mean, then Equations (9.6) and (9.7) become in this case:

$$\frac{1+c}{2} = \int_{\hat{\mu}_j}^{\infty} f_{\hat{\mu}}(x) \, dx \qquad (9.42)$$

and:

$$\frac{1+c}{2} = \int_{-\infty}^{\hat{\mu}_j} f_{\hat{\mu}}(x) \, dx \qquad (9.43)$$

Assuming now that the distribution of sample means is normal, then substituting from Equation (9.34) in Equations (9.42) and (9.43):

$$\frac{1+c}{2} = \int_{\hat{\mu}_j}^{\infty} \frac{\sqrt{n}}{\sigma\sqrt{(2\pi)}} \exp[-n(x-\mu)^2/2\sigma^2] \, dx \qquad (9.44)$$

and:

$$\frac{1+c}{2} = \int_{-\infty}^{\hat{\mu}_j} \frac{\sqrt{n}}{\sigma\sqrt{(2\pi)}} \exp[-n(x-\mu)^2/2\sigma^2]\,dx \qquad (9.45)$$

The two values of μ which satisfy Equations (9.44) and (9.45) represent the upper and lower confidence limits, μ_u and μ_l, respectively.

Suppose, however, that Equations (9.44) and (9.45) are written in standardized form by making the standard substitution:

$$y = \frac{(x-\mu)\sqrt{n}}{\sigma} \qquad (9.46)$$

then:

$$\frac{1+c}{2} = \int_{t'}^{\infty} \frac{1}{\sqrt{(2\pi)}} e^{-\frac{1}{2}y^2}\,dy \qquad (9.47)$$

and:

$$\frac{1+c}{2} = \int_{-\infty}^{t'} \frac{1}{\sqrt{(2\pi)}} e^{-\frac{1}{2}y^2}\,dy \qquad (9.48)$$

where:

$$t' = \frac{(\hat{\mu}_j - \mu)\sqrt{n}}{\sigma} \qquad (9.49)$$

The values of t' which satisfy Equations (9.47) and (9.48) lead to two values of μ for the upper and lower confidence limits repectively. Due to symmetry, these two values of μ are given by:

$$\mu = \hat{\mu}_j \pm t'\frac{\sigma}{\sqrt{n}} \qquad (9.50)$$

Also, it is seen from Equations (9.47) and (9.48) that t' corresponds to the number of standard deviations of a standardized normal distribution which are equivalent to a value of $c/2$.

Example 9.2

A very large population of electro-magnetic relays are known to have response times which follow a distribution whose standard deviation is 10 ms. From this population, measurements are made of the response times on a relatively large sample of 100 relays and these measurements yield an estimated mean value of 100 ms. Assuming that the distribution of the sample means is normal what are the symmetrical lower and upper limits on the mean value for a 90% confidence level?

For this example:

$$c = 0.9$$

Hence, from Table A.1, the number of standard deviations, t', for this value of c is:

$$t' = 1.645$$

Therefore, from Equation (9.50):

$$\left(\hat{\mu}_j - 1.645 \frac{\sigma}{\sqrt{n}}\right) \leqslant \mu \leqslant \left(\hat{\mu}_j + 1.645 \frac{\sigma}{\sqrt{n}}\right)$$

which, on inserting the example values gives:

$$\left(100 - 1.645 \frac{10}{\sqrt{100}}\right) \leqslant \mu \leqslant \left(100 + 1.645 \frac{10}{\sqrt{100}}\right)$$

or:

$$98.355 \text{ ms} \leqslant \mu \leqslant 101.645 \text{ ms} \quad \text{for a 90\% confidence level}$$

It will be noticed in the preceding example that the method adopted for calculating the confidence limits for the mean presupposes a knowledge of the value of the standard deviation, σ, for the parent population. Where this value is not known, the standard deviation may have to be estimated from some estimator $\hat{\sigma}_c$ and Equation (9.50) then becomes:

$$t = \frac{(\hat{\mu}_j - \mu)\sqrt{n}}{\sigma_c} \tag{9.51}$$

t is now a distribution function which depends not only upon the distribution of the sample means but also on the distribution of the sample variances. The forms and parameters of these latter distributions are discussed in the next sub-section.

Sample variances

Having derived relationships for estimating the mean value of a parent population from a sample of variates, the next criterion of general interest is a method of estimating the variance of the population. Following the lines of Equation (9.37), the variance of a sample of size n taken from an infinite population of the variate x may be written in the form:

$$\hat{\sigma}^2 = \frac{1}{n} \sum_{j=1}^{n} x_j^2 - \hat{\mu}^2 \tag{9.52}$$

Hence, knowing the distribution of $\hat{\mu}$ and x, the problem is to find the distribution of $\hat{\sigma}^2$ and its associated parameters. It is seen that the sample variance is a function related to the squares of the individual variates such as x_j and $\hat{\mu}$. It is pertinent, therefore, to consider first of all a relationship

SAMPLING, ESTIMATION AND CONFIDENCE

between random variates of the type:

$$u = y^2 \qquad (9.53)$$

Then, from Relationship No. 8 in Table 8.1:

$$f_u(x) = \frac{1}{\sqrt{x}} f_y(\sqrt{x}) \qquad (9.54)$$

or, if x has only positive values, then:

$$f_u(x) = \frac{1}{2\sqrt{x}} f_y(\sqrt{x}) \qquad (9.55)$$

and the case represented in Equation (9.55) may now be further considered without loss of generality.

The Laplace transform of $f_u(x)$ is:

$$L[f_u(x)] = \int_0^\infty e^{-sx} \frac{f_y(\sqrt{x})}{2\sqrt{x}} dx \qquad (9.56)$$

and writing x in the form $(\sqrt{x})^2$, Equation (9.56) becomes:

$$L[f_u(x)] = \int_0^\infty e^{-s(\sqrt{x})^2} f_y(\sqrt{x}) \, d(\sqrt{x}) \qquad (9.57)$$

therefore:

$$\frac{dL[f_u(x)]}{ds} = -\int_0^\infty (\sqrt{x})^2 e^{-s(\sqrt{x})^2} f_y(\sqrt{x}) \, d(\sqrt{x}) \qquad (9.58)$$

and:

$$\left\{ \frac{dL[f_u(x)]}{ds} \right\}_{s=0} = -\int_0^\infty (\sqrt{x})^2 f_y(\sqrt{x}) \, d(\sqrt{x})$$

$$= -M_2 \qquad (9.59)$$

Hence from the relationships given in Equations (7.44) and (7.48), Equation (9.59) becomes:

$$\mu_u = \sigma^2 + \mu^2 \qquad (9.60)$$

where μ_u is the mean of the distribution of u and σ^2 and μ are the variance and mean of the distribution of y.

Differentiating Equation (9.57) twice with respect to s and setting s to zero leads to the relationship:

$$\left\{ \frac{d^2 L[f_u(x)]}{ds^2} \right\}_{s=0} = \left\{ \frac{d^4 L[f_y(x)]}{ds^4} \right\}_{s=0} \qquad (9.61)$$

and making use of the relationships in Equations (7.49) and (7.51) to express

Equation (9.61) in terms of the respective parameters of the distributions:

$$\sigma_u^2 + \mu_u^2 = \sigma^4 K_4 + \mu^2(\mu^2 + 6\sigma^2) + 4\mu K_3 \sigma^3 \tag{9.62}$$

and substituting for μ_u from Equation (9.60):

$$\sigma_u^2 = \sigma^4(K_4 - 1) + 4\sigma^2 \mu^2 + 4\mu K_3 \sigma^3 \tag{9.63}$$

The values of μ_u and σ_u^2 given by Equations (9.60) and (9.63) represent the mean and variance of the distribution of y^2 in terms of the parameters for the distribution of y. The next step in establishing an estimate for the variance of the parent population of x is to consider the distribution of the means of a number of x^2 values in a sample of size n as given by the first term in Equation (9.52). Hence, let:

$$v = \frac{1}{n} \sum_{j=1}^{n} x_j^2 \tag{9.64}$$

or:

$$v = \frac{1}{n} \sum_{j=1}^{n} u_j \tag{9.65}$$

where:

$$u_j = x_j^2 \tag{9.66}$$

Equation (9.65) is of the same form as Equation (9.8). Hence, following the same discussion as the last sub-section it is seen that:

$$\mu_v = \mu_u \tag{9.67}$$

and:

$$\sigma_v^2 = \frac{\sigma_u^2}{n} \tag{9.68}$$

and because of the relationship of Equation (9.66), the values of μ_u and σ_u^2 may be obtained from Equations (9.60) and (9.63) respectively, so that:

$$\mu_v = \sigma^2 + \mu^2 \tag{9.69}$$

and:

$$\sigma_v^2 = \frac{1}{n}[\sigma^4(K_4 - 1) + 4\sigma^2 \mu^2 + 4\mu K_3 \sigma^3] \tag{9.70}$$

In connexion now with the second term of Equation (9.52) let:

$$w = \hat{\mu}^2 \tag{9.71}$$

whence, in the same way that Equation (9.60) was derived from Equation (9.53):

$$\mu_w = \sigma_{\hat{\mu}}^2 + \mu_{\hat{\mu}}^2 \tag{9.72}$$

SAMPLING, ESTIMATION AND CONFIDENCE

and from a corresponding similarity with Equation (9.63):

$$\sigma_w^2 = \sigma_{\hat{\mu}}^4(K_{4\hat{\mu}}-1)+4\sigma_{\hat{\mu}}^2\mu_{\hat{\mu}}^2+4\mu_{\hat{\mu}}K_{3\hat{\mu}}\sigma_{\hat{\mu}}^3 \quad (9.73)$$

and substituting for $\mu_{\hat{\mu}}$, $\sigma_{\hat{\mu}}^2$, $K_{3\hat{\mu}}$ and $K_{4\hat{\mu}}$ from Equations (9.20), (9.25), (9.26) and (9.28) in Equations (9.72) and (9.73) gives:

$$\mu_w = \frac{\sigma^2}{n}+\mu^2 \quad (9.74)$$

and:

$$\sigma_w^2 = \frac{1}{n^2}\left\{\sigma^4\left[\frac{1}{n}(K_4-3)+2\right]+4n\sigma^2\mu^2+4\mu K_3\sigma^3\right\} \quad (9.75)$$

Now, from Equations (9.52), (9.64) and (9.71), it is seen that:

$$\hat{\sigma}^2 = v-w \quad (9.76)$$

which from Relationship No. 3 in Table 8.1 yields:

$$F_{\hat{\sigma}^2}(s) = F_v(s)F_w(-s) \quad (9.77)$$

and differentiating Equation (9.77) with respect to s and equating s to zero gives from Equations (7.16) and (7.48):

$$-\mu_{\hat{\sigma}^2} = -\mu_v+\mu_w \quad (9.78)$$

and substituting for μ_v and μ_w from Equations (9.69) and (9.74):

$$\mu_{\hat{\sigma}^2} = (\sigma^2+\mu^2)-\left(\frac{\sigma^2}{n}+\mu^2\right)$$

$$= \frac{(n-1)}{n}\sigma^2 \quad (9.79)$$

So that the mean of the sample variance is not equal to the variance of the parent population except in the case where n tends to infinity. On the other hand, if a new sample variance, $\hat{\sigma}_c^2$, is constructed such that:

$$\hat{\sigma}_c^2 = \frac{1}{(n-1)}\sum_{j=1}^{n}(x_j-\hat{\mu})^2 \quad (9.80)$$

then a similar process to the derivation of Equation (9.79) will show that:

$$\mu_{\hat{\sigma}_c^2} = \sigma^2 \quad (9.81)$$

Hence, Equation (9.80) represents the best or unbiased estimate of the population variance and this expression is sometimes known as the Bessel correction. A certain modification, however, is necessary to the Bessel correction when the parent population consists of a finite number of members. In this case, the mean and variance of the parent population are given by

Equations (9.36) and (9.37) respectively and the sample mean is equal to the population mean as already shown in Equation (9.40). The mean of all possible sample variances is given by:

$$\mu_{\hat{\sigma}^2} = \frac{1}{\binom{N}{n}}(\hat{\sigma}_1^2 + \hat{\sigma}_2^2 + \ldots + \hat{\sigma}_{\binom{N}{n}}^2)$$

$$= \frac{1}{\binom{N}{n}} \frac{1}{n} \left[\binom{N-1}{n-1} \sum_{j=1}^{N} x_j^2 - n \sum_{j=1}^{\binom{N}{n}} \mu_j^2 \right]$$

$$= \frac{1}{\binom{N}{n}} \frac{1}{n} \left\{ \binom{N-1}{n-1} \sum_{j=1}^{N} x_j^2 - \frac{1}{n} \left[\binom{N-1}{n-1} \sum_{j=1}^{N} x_j^2 + 2\binom{N-2}{n-2} \sum_{j \neq k} x_j x_k \right] \right\}$$

$$= \frac{1}{\binom{N}{n}} \frac{1}{n} \left\{ \binom{N-1}{n-1} \sum_{j=1}^{N} x_j^2 - \frac{1}{n} \left[\binom{N-1}{n-1} \sum_{j=1}^{N} x_j^2 \right. \right.$$

$$\left. \left. + \binom{N-2}{n-2} N^2 \mu^2 - \binom{N-2}{n-2} \sum_{j=1}^{N} x_j^2 \right] \right\}$$

$$= \frac{1}{\binom{N}{n}} \frac{N}{n^2} \binom{N-2}{n-2} \left(\sum_{j=1}^{N} x_j^2 - N\mu^2 \right)$$

$$= \frac{N(n-1)}{n(N-1)} \sigma^2 \quad (9.82)$$

The best and unbiased estimate of the variance of a finite population is therefore given by:

$$\hat{\sigma}_c^2 = \frac{(N-1)}{N(n-1)} \sum_{j=1}^{n} (x_j - \hat{\mu})^2 \quad (9.83)$$

and Equation (9.83) only approaches the relationship for the Bessel correction given in Equation (9.80) as N tends to infinity.

Returning now to the expression for the distribution of variances given by Equation (9.77), differentiating the expression twice with respect to s, equating s to zero and making use of the relationships given in Equations (7.16), (7.48) and (7.49) yields:

$$\sigma_{\hat{\sigma}^2}^2 + \mu_{\hat{\sigma}^2}^2 = (\sigma_v^2 + \mu_v^2) - 2\mu_v \mu_w + (\sigma_w^2 + \mu_w^2) \quad (9.84)$$

and substituting for the various parameters in Equation (9.84) in terms of the parameters of the parent population as given by Equations (9.69), (9.70),

SAMPLING, ESTIMATION AND CONFIDENCE

(9.74), (9.75) and (9.79) leads to:

$$\sigma_{\hat{\sigma}^2}^2 = \frac{1}{n^3}[\sigma^4(K_4-3)+\sigma^4 n^2(K_4-1)+2n\sigma^4+8\sigma^2\mu^2 n^2+4\mu K_3 \sigma^3 n(n+1)] \quad (9.85)$$

However, the relationship of Equation (9.85) has been derived on the basis of the biased estimator for variance, $\hat{\sigma}^2$, as given by Equation (9.52). From Equation (9.80), it is seen that the relationship between this biased estimator and the unbiased estimator, $\hat{\sigma}_c^2$, is:

$$\hat{\sigma}_c^2 = \frac{n}{(n-1)}\hat{\sigma}^2 \quad (9.86)$$

therefore, from Relationship No. 1 in Table 8.1:

$$F_{\hat{\sigma}_c^2}(s) = F_{\hat{\sigma}^2}\left(\frac{ns}{n-1}\right) \quad (9.87)$$

The usual process of differentiating with respect to s and setting s equal to zero then leads to:

$$\mu_{\hat{\sigma}_c^2} = \frac{n}{(n-1)}\mu_{\hat{\sigma}^2} \quad (9.88)$$

and:

$$(\sigma_{\hat{\sigma}_c^2}^2+\mu_{\hat{\sigma}_c^2}^2) = \frac{n^2}{(n-1)^2}(\sigma_{\hat{\sigma}^2}^2+\mu_{\hat{\sigma}^2}^2) \quad (9.89)$$

whence, from Equations (9.79), (9.81) and (9.85):

$$\sigma_{\hat{\sigma}_c^2}^2 = \frac{1}{n(n-1)^2}[\sigma^4(K_4-3)+\sigma^4 n^2(K_4-1)+2n\sigma^4+8\sigma^2\mu^2 n^2+4\mu K_3 \sigma^3 n(n+1)] \quad (9.90)$$

Example 9.3

For samples of size n taken from a parent normal distribution, find the expressions for the variance of the sample variances when using both the biased and unbiased estimator of variance.

For a normal distribution, the following parameter values apply:

$$\mu = \mu \quad K_3 = 0$$
$$\sigma^2 = \sigma^2 \quad K_4 = 3$$

Substituting these values in Equation (9.85) gives the expression for the variance of the sample variances based on the biased estimator, $\hat{\sigma}^2$:

$$\sigma_{\hat{\sigma}^2}^2 = \frac{2\sigma^2}{n^2}[(n+1)\sigma^2+4n\mu^2] \quad (9.91)$$

Similarly, in the case of the unbiased estimator, from Equation (9.90):

$$\sigma_{\hat{\sigma}_e^2}^2 = \frac{2\sigma^2}{(n-1)^2}[(n+1)\sigma^2 + 4n\mu^2] \quad (9.92)$$

It is seen from the preceding example and from Equation (9.90) that the variance of sample variances depends both upon the parameters of the parent population and on the sample size. This means that any calculations of the confidence limits associated with the sample variances must also depend upon the form of the parent population and the sample size. For this reason, some typical standard parent distributions are now examined in order to derive the specific forms of sampling distributions and related parameters that may apply.

Sampling from the normal distribution

For a normal distribution, the distribution of variates in the parent population is given by:

$$f_x(x) = \frac{1}{\sigma\sqrt{(2\pi)}} \exp[-(x-\mu)^2/2\sigma^2] \quad (9.93)$$

Now, from Equations (9.77) and (9.87), it is seen that the transform of the distribution of the unbiased estimator of variance can be generally expressed in the form:

$$F_{\hat{\sigma}_e^2}(s) = F_v\left(\frac{ns}{n-1}\right) F_w\left(\frac{-ns}{n-1}\right) \quad (9.94)$$

and from the relationships for the variates v and w given in Equations (9.64) and (9.71), it follows from Equation (9.31) and the previous working that:

$$F_{\hat{\sigma}_e^2}(s) = \left[F_{x^2}\left(\frac{s}{n-1}\right)\right]^n \left[F_{\mu^2}\left(\frac{-ns}{n-1}\right)\right] \quad (9.95)$$

Also, from Equations (9.54) and (9.93):

$$f_{x^2}(x) = \frac{1}{\sigma\sqrt{(2\pi)}} \frac{1}{\sqrt{x}} \exp[-(\sqrt{x}-\mu)^2/2\sigma^2] \quad (9.96)$$

where the variate x now lies in the range 0 to ∞. Hence, taking the Laplace transform of Equation (9.96):

$$F_{x^2}(s) = \frac{1}{\sigma\sqrt{(2\pi)}} \int_0^\infty \exp(-sx)\frac{1}{\sqrt{x}} \exp[-(\sqrt{x}-\mu)^2/2\sigma^2]\,dx \quad (9.97)$$

SAMPLING, ESTIMATION AND CONFIDENCE

and making the substitution $y = \sqrt{x}$, Equation (9.97) becomes:

$$F_{x^2}(s) = \frac{2}{\sigma\sqrt{(2\pi)}} \int_0^\infty \exp(-sy^2) \exp[-(y-\mu)^2/2\sigma^2] \, dy$$

$$= \frac{2\sqrt{(2\sigma^2 s + 1)} \exp[-\mu^2 s/(2\sigma^2 s + 1)]}{\sigma\sqrt{[2\pi(2\sigma^2 s + 1)]}}$$

$$\times \int_0^\infty \exp\{-(2\sigma^2 s + 1)[y - \mu/(2\sigma^2 s + 1)]^2/2\sigma^2\} \, dy$$

$$= \frac{\exp[-\mu^2 s/(2\sigma^2 s + 1)]}{\sqrt{(2\sigma^2 s + 1)}} \tag{9.98}$$

Hence:

$$\left[F_{x^2}\left(\frac{s}{n-1}\right)\right]^n = \left[1 + \frac{2\sigma^2 s}{(n-1)}\right]^{-n/2} \exp\{-\mu^2 ns/[2\sigma^2 s + (n-1)]\} \tag{9.99}$$

Now, in addition, from Equations (9.31) and (9.93):

$$f_{\hat\mu}(x) = \frac{\sqrt{n}}{\sigma\sqrt{(2\pi)}} \exp[-n(x-\mu)^2/2\sigma^2] \tag{9.100}$$

and, therefore, following a similar process to that used in the derivation of Equation (9.98):

$$F_{\hat\mu^2}(s) = \frac{\exp[-\mu^2 s/(2\sigma^2 s/n + 1)]}{\sqrt{(2\sigma^2 s/n + 1)}} \tag{9.101}$$

and hence:

$$F_{\hat\mu^2}\left(\frac{-ns}{n-1}\right) = \left[1 - \frac{2\sigma^2 s}{(n-1)}\right]^{-1/2} \exp\{-\mu^2 ns/[2\sigma^2 s - (n-1)]\} \tag{9.102}$$

Therefore, substituting from Equations (9.99) and (9.102) in Equation (9.95):

$$F_{\hat\sigma_e^2}(s) = \left[1 + \frac{2\sigma^2 s}{(n-1)}\right]^{-n/2} \left[1 - \frac{2\sigma^2 s}{(n-1)}\right]^{-1/2} \exp\{-4n\mu^2 \sigma^2 s^2/[4\sigma^4 s^2 - (n-1)^2]\} \tag{9.103}$$

and differentiating Equation (9.103) with respect to s and letting s equal zero gives:

$$\mu_{\hat\sigma_e^2} = \sigma^2 \tag{9.104}$$

which may be compared with the general result obtained previously in Equation (9.81). Also, the reader may like to verify that differentiating Equation (9.103) twice with respect to s and letting s equal zero gives the same result as obtained in Equation (9.92).

As n becomes large, the exponential term in Equation (9.103) rapidly approaches unity, hence:

$$F_{\hat{\sigma}_c^2}(s) \simeq \left[1 + \frac{2\sigma^2 s}{(n-1)}\right]^{-n/2} \left[1 - \frac{2\sigma^2 s}{(n-1)}\right]^{-1/2}$$

$$\simeq \left[1 + \frac{2\sigma^2 s}{(n-1)}\right]^{-(n-1)/2} \quad (9.105)$$

and Equation (9.105) may be written in the form:

$$F_{\hat{\sigma}_c^2}(s) \simeq (f/2\sigma^2)^{f/2} [s + f/2\sigma^2]^{-f/2} \quad (9.106)$$

where:

$$f = (n-1) \quad (9.107)$$

From Table A.2, the inverse transform of Equation (9.106) becomes:

$$f_{\hat{\sigma}_c^2}(x) = \frac{(f/2\sigma^2)^{f/2}}{\Gamma(f/2)} x^{(f/2-1)} e^{-fx/2\sigma^2} \quad (9.108)$$

Consider now a variate, χ^2, which is defined by:

$$\chi^2 = \sum_{j=1}^{n} \left(\frac{x_j - \mu}{\sigma}\right)^2 \quad (9.109)$$

then substituting from Equations (9.80) and (9.107) in Equation (9.109):

$$\chi^2 = \frac{f\hat{\sigma}_c^2}{\sigma^2} \quad (9.110)$$

Hence, from Relationship No. 1 in Table 8.1:

$$f_{\chi^2}(x) = \frac{\sigma^2}{f} f_{\hat{\sigma}_c^2}\left(\frac{\sigma^2 x}{f}\right) \quad (9.111)$$

and combining Equations (9.108) and (9.111):

$$f_{\chi^2}(x) = \frac{1}{2^{f/2} \Gamma(f/2)} x^{(f/2-1)} e^{-x/2} \quad (9.112)$$

Equation (9.112) is known as the 'χ^2' or 'chi-square' distribution and the parameter f in the distribution is known as the 'degree of freedom'. Values of the cumulative χ^2 distribution for various percentage probabilities and various values of f are tabulated in Table A.5.

Using Table A.5, estimates of the true value of the variance for certain percentage probabilities can be obtained from the sample variance from the relationship given in Equation (9.110). In other words the confidence limits

SAMPLING, ESTIMATION AND CONFIDENCE

for σ^2 can be expressed in the form:

$$\frac{\hat{\sigma}_c^2 f}{\chi_2^2} \leqslant \sigma^2 \leqslant \frac{\hat{\sigma}_c^2 f}{\chi_1^2} \qquad (9.113)$$

where χ_1^2 and χ_2^2 are the lower and upper values of the cumulative χ^2 distribution which correspond to the particular confidence limits required.

Example 9.4

The power outputs of a large population of electronic amplifiers are known to follow a normal distribution. From this population, the power outputs of a small sample of five amplifiers are measured and found to be 5.1 mW, 5.5 mW, 5.7 mW, 6.3 mW and 6.9 mW. What are the symmetrical 90% confidence limits on the estimate of the population variance?

The first step is to calculate the sample mean which is the best estimate of the population mean:

$$\mu = \hat{\mu}$$

$$= \frac{1}{5} \sum_{j=1}^{5} x_j$$

$$= \tfrac{1}{5}(5.1 + 5.5 + 5.7 + 6.3 + 6.9)$$

$$= 5.9 \text{ mW}$$

The best estimate of the population variance is then given by the Bessel correction of Equation (9.80):

$$\hat{\sigma}_c^2 = \frac{1}{4} \sum_{j=1}^{5} (x_j - \mu)^2$$

$$= \tfrac{1}{4}(0.8^2 + 0.4^2 + 0.2^2 + 0.4^2 + 1.0^2)$$

$$= 0.5$$

The number of degrees of freedom, f, is given by Equation (9.107):

$$f = (n-1)$$

$$= 4$$

For a 90% symmetrical confidence level, χ_1^2 will represent the 5% probability and χ_2^2 the 95% probability of the cumulative χ^2 distribution for $f = 4$ degrees of freedom. Therefore, from Table A.5:

$$\chi_1^2 = 0.711$$

$$\chi_2^2 = 9.488$$

Substituting all the appropriate values in Equation (9.113) then yields:

$$\frac{0.5 \times 4}{9.488} \leqslant \sigma^2 \leqslant \frac{0.5 \times 4}{0.711}$$

or:

$$0.21 \leqslant \sigma^2 \leqslant 2.8 \quad \text{for a 90\% confidence level}$$

The mean and variance of the χ^2 distribution may be obtained from the Laplace transformation of Equation (9.112) which, from Table A.2, is seen to be:

$$F_{\chi^2}(s) = (2s+1)^{-f/2} \tag{9.114}$$

Therefore:

$$\mu_{\chi^2} = -\left[\frac{dF_{\chi^2}(s)}{ds}\right]_{s=0}$$

$$= f \tag{9.115}$$

and:

$$\sigma^2_{\chi^2} = \left[\frac{d^2 F_{\chi^2}(s)}{ds^2}\right]_{s=0} - \mu^2_{\chi^2}$$

$$= 2f \tag{9.116}$$

Successive processes of differentiation then show that:

$$K_{3\chi^2} = \left(\frac{8}{f}\right)^{\frac{1}{2}} \tag{9.117}$$

and:

$$K_{4\chi^2} = \frac{3}{f}(f+4) \tag{9.118}$$

So that for large f, the χ^2 distribution converges to the normal distribution with a mean of f and a variance of $2f$.

Now, the degree of confidence that can be expressed in the estimated variance affects the degree of confidence that can be expressed in the estimate of the mean as shown by the relationship given in Equation (9.51), from which:

$$\frac{t^2}{n} = \frac{y^2}{\hat{\sigma}^2_c} \tag{9.119}$$

where:

$$y = (\hat{\mu} - \mu) \tag{9.120}$$

The variate $\hat{\mu}$ is normally distributed with a mean μ and a variance of σ^2/n.

SAMPLING, ESTIMATION AND CONFIDENCE 339

Therefore, y is normally distributed with a zero mean and a variance of σ^2/n, or:

$$f_y(x) = \frac{\sqrt{n}}{\sigma\sqrt{(2\pi)}} e^{-nx^2/2\sigma^2} \qquad (9.121)$$

and:

$$f_{y^2}(x) = \frac{\sqrt{n}}{\sigma\sqrt{(2\pi)}} \frac{1}{\sqrt{x}} e^{-nx/2\sigma^2} \qquad (9.122)$$

The distribution of the variate t^2/n is therefore dependent upon the ratio of the two distributions given in Equations (9.122) and (9.108). It can be shown that this leads to a distribution of t given by the following expression:

$$f_t(x) = \frac{1}{\sqrt{(\pi f)}} \frac{\Gamma[(f+1)/2]}{\Gamma(f/2)} (1 + x^2/f)^{-(f+1)/2} \qquad (9.123)$$

Equation (9.123) is known as the 'Student t' distribution and it is symmetrical about the mean converging to the normal distribution as f approaches infinity. Values of the cumulative t distribution for various percentage probabilities and various values of f are tabulated in Table A.6. Because of the symmetrical nature of the distribution, the values in Table A.6 are only tabulated for percentage probabilities in the range 50% to 100%.

The t distribution is related to the χ^2 distribution as can be seen by eliminating $\hat{\sigma}_c$ from Equations (9.51) and (9.110):

$$t = \frac{\sqrt{(nf)}(\hat{\mu}-\mu)}{\sigma\chi} \qquad (9.124)$$

Returning now to Equation (9.50), the confidence limits on the mean, which take account of both the variations in the sample means and the sample variances, can be expressed, for the normal distribution, in the form:

$$\left(\hat{\mu} - t\frac{\hat{\sigma}_c}{\sqrt{n}}\right) \leqslant \mu \leqslant \left(\hat{\mu} + t\frac{\hat{\sigma}_c}{\sqrt{n}}\right) \qquad (9.125)$$

Example 9.5

For the same sample of electronic amplifiers described in Example 9.4, calculate the symmetrical confidence limits on the estimated mean value for a 90% confidence level.

The values of $\hat{\mu}$ and $\hat{\sigma}_c$ previously calculated for this example are:

$$\hat{\mu} = 5.9 \text{ mW}$$

$$\hat{\sigma}_c = \sqrt{0.5} \text{ mW}$$

For a symmetrical 90% confidence level, the appropriate value of t is that corresponding to 5% and 95% cumulative probability at $f = (n-1) = 4$

degrees of freedom. From Table A.6, this leads to:

$$t = 2.132$$

Hence, inserting the example values in Equation (9.125) yields:

$$\left(5.9 - 2.132\frac{\sqrt{0.5}}{\sqrt{5}}\right) \leq \mu \leq \left(5.9 + 2.132\frac{\sqrt{0.5}}{\sqrt{5}}\right)$$

or:

$$5.23 \leq \mu \leq 6.57 \quad \text{for a 90\% confidence level}$$

Sampling from the exponential distribution

Suppose that the probability density for the times to failure of elements of a system in a large population of such elements is known to follow the exponential distribution. The density function may then be expressed in the form:

$$f_t(t) = \frac{1}{\lambda}e^{-t/\lambda} \tag{9.126}$$

where t is the time variable or variate and the parameter λ represents the mean time to failure. Equation (9.126) may also be written as:

$$f_t(t) = \theta e^{-\theta t} \tag{9.127}$$

where:

$$\theta = \frac{1}{\lambda} \tag{9.128}$$

is a parameter representing the average failure-rate.

From Table A.2 or Table 7.1 in Chapter 7, the Laplace transform of Equation (9.126) is:

$$F_t(s) = \frac{1}{1+s\lambda} \tag{9.129}$$

For a sample of times to failure, $(t_1, t_2, ..., t_n)$, taken from the population of elements, the sample mean time to failure can be calculated according to:

$$\hat{\mu}_t = \hat{\lambda}$$

$$= \frac{1}{n}\sum_{j=1}^{n} t_j \tag{9.130}$$

and, from Equation (9.31), the distribution of sample mean times to failure is given by:

$$F_{\hat{\lambda}}(s) = \left[F_t\!\left(\frac{s}{n}\right)\right]^n \tag{9.131}$$

SAMPLING, ESTIMATION AND CONFIDENCE

which, on combining Equations (9.129) and (9.131), leads to:

$$F_{\hat{\lambda}}(s) = \frac{(n/\lambda)^n}{[s+(n/\lambda)]^n} \qquad (9.132)$$

and taking the inverse transform from Table A.2:

$$f_{\hat{\lambda}}(t) = \frac{(n/\lambda)^n}{\Gamma(n)} t^{n-1} e^{-nt/\lambda} \qquad (9.133)$$

Now, from Equation (9.6), the upper confidence limit of the mean time to failure is given by the solution of the following expression with respect to t:

$$\frac{1+c}{2} = \int_{\hat{\lambda}}^{\infty} f_{\hat{\lambda}}(t) \, dt \qquad (9.134)$$

whence:

$$\frac{1+c}{2} = \int_{\hat{\lambda}}^{\infty} \frac{(n/\lambda)^n}{\Gamma(n)} t^{n-1} e^{-nt/\lambda} \, dt \qquad (9.135)$$

The similarity of this expression with the cumulative χ^2 distribution may be seen by making the following substitutions in Equation (9.135):

$$\chi^2 = \frac{2nt}{\lambda} \qquad (9.136)$$

$$n = \frac{f}{2} \qquad (9.137)$$

whence:

$$\frac{1+c}{2} = \int_{\chi_1^2}^{\infty} f_{\chi^2}(\chi^2) \, d(\chi^2) \qquad (9.138)$$

where:

$$\chi_1^2 = \frac{2n\hat{\lambda}}{\lambda}$$

$$= \frac{f\hat{\lambda}}{\lambda} \qquad (9.139)$$

Similarly, the lower confidence limit expression of Equation (9.7) leads to a value of λ given by:

$$\chi_2^2 = \frac{f\hat{\lambda}}{\lambda} \qquad (9.140)$$

In Equation (9.139), χ_1^2 is evaluated at $100[(1-c)/2]\%$ probability at $f = 2n$ degrees of freedom and in Equation (9.140) χ_2^2 is evaluated at $100[(1+c)/2]\%$

probability at $f = 2n$ degrees of freedom. With this interpretation, Equations (9.139) and (9.140) can be written in the form:

$$\frac{\chi_1^2}{f\hat{\lambda}} \leqslant \frac{1}{\lambda} \leqslant \frac{\chi_2^2}{f\hat{\lambda}} \tag{9.141}$$

or, in terms of the failure-rate, θ:

$$\frac{\hat{\theta}\chi_1^2}{f} \leqslant \theta \leqslant \frac{\hat{\theta}\chi_2^2}{f} \tag{9.142}$$

Example 9.6

The times to failure of a motor are known to follow an exponential distribution. The motor is put on test and the test is run until the first failure occurs. The test is then stopped, the motor repaired and the test continued until the second failure occurs. This process is repeated until the fifth failure has occurred when the test is terminated. In each phase of the test the time is measured from the commencement of correct operation until failure. These five measured times are found to be 632 h, 3450 h, 816 h, 928 h and 150 h. What are the symmetrical upper and lower confidence limits for the estimate of the mean time to failure of the motor at a 90% confidence level?

Applying Equation (9.130):

$$\hat{\lambda} = \frac{632 + 3450 + 816 + 928 + 150}{5}$$

$$= 1195.2 \text{ h}$$

It should be noted, of course, that $\hat{\lambda}$ can be derived from just a knowledge of the total actual running time and the total number of faults and that it is not necessary to know the individual times to failure.

The corresponding value of the estimated mean failure-rate, $\hat{\theta}$, is:

$$\hat{\theta} = \frac{1}{\hat{\lambda}}$$

$$= 837 \text{ faults per } 10^6 \text{ h}$$

Since the sample size in this example is 5, the number of degrees of freedom from Equation (9.137) is 10. Hence, for the required 90% confidence level, χ_1^2 is evaluated at 5% cumulative probability at 10 degrees of freedom and χ_2^2 is evaluated at 95% cumulative probability at 10 degrees of freedom. From Table A.5 this gives:

$$\chi_1^2 = 3.940$$

and:

$$\chi_2^2 = 18.307$$

SAMPLING, ESTIMATION AND CONFIDENCE

Substituting the derived values into Equation (9.141) yields:

$$\frac{3.940}{10 \times 1195.2} \leqslant \frac{1}{\lambda} \leqslant \frac{18.307}{10 \times 1195.2}$$

or:

$$652 \text{ h} \leqslant \lambda \leqslant 3030 \text{ h} \quad \text{at 90\% confidence level}$$

or again, in terms of the failure-rate parameter:

$$330 \text{ faults}/10^6 \text{ h} \leqslant \theta \leqslant 1530 \text{ faults}/10^6 \text{ h} \quad \text{at 90\% confidence level}$$

The application of the exponential distribution assumes that the running time of the device does not affect its failure-rate. On this basis, the same results would have been achieved in the example just given if five separate, but nominally identical, motors had each been run up to the time of their first failure. With some reservations then and provided that the failure-rate of devices is constant with respect to time, equivalent samples may be taken from a small number of devices running for a long time or from a large number of devices running for a relatively shorter time.

An assumption made in the derivation of Equations (9.141) and (9.142) and also in Example 9.6 is that the sample test is terminated at the time of the last failure of the device or of each device. Very often in practice, however, a sample test or a selection of field data may be terminated at a time which does not correspond to the time of the last failure. Two assumptions are now possible, either that the n failures which have actually occurred are related to the total running time or that $(n+1)$ failures are related to the total running time, i.e. that another failure was about to occur when the test or data was terminated. The former assumption may be an optimistic one and the latter assumption may be a pessimistic one. The two assumptions will, however, lead to two sample estimates of the mean time to failure given by:

$$\hat{\lambda}_1 = \frac{T}{n} \tag{9.143}$$

and:

$$\hat{\lambda}_2 = \frac{T}{n+1} \tag{9.144}$$

where T is the total actual running time of the device or devices.

Equations (9.143) and (9.144) lead to two expressions for a confidence interval, which are derived in a similar way to Equation (9.141), namely:

$$\frac{\chi_{11}^2}{f_1 \hat{\lambda}_1} \leqslant \frac{1}{\lambda} \leqslant \frac{\chi_{12}^2}{f_1 \hat{\lambda}_1} \tag{9.145}$$

and:

$$\frac{\chi_{21}^2}{f_2 \hat{\lambda}_2} \leqslant \frac{1}{\lambda} \leqslant \frac{\chi_{22}^2}{f_2 \hat{\lambda}_2} \qquad (9.146)$$

where $f_1 = 2n$ and $f_2 = 2(n+1)$ degrees of freedom.

Hence, χ_{11}^2 and χ_{12}^2 are evaluated at $2n$ degrees of freedom and χ_{21}^2 and χ_{22}^2 are evaluated at $2(n+1)$ degrees of freedom.

It is normal practice in cases of this type to choose an overall confidence interval which gives the widest spread in values. This can be achieved by taking the lower limit term in Equation (9.145) together with the upper limit term in Equation (9.146) and expressing the confidence interval as:

$$\frac{\chi_{11}^2}{f_1 \hat{\lambda}_1} \leqslant \frac{1}{\lambda} \leqslant \frac{\chi_{22}^2}{f_2 \hat{\lambda}_2} \qquad (9.147)$$

Now, from the definitions of f_1 and f_2 and from Equations (9.143) and (9.144):

$$f_1 \hat{\lambda}_1 = f_2 \hat{\lambda}_2 \qquad (9.148)$$

Hence:

$$\frac{\chi_{11}^2}{f_1 \hat{\lambda}_1} \leqslant \frac{1}{\lambda} \leqslant \frac{\chi_{22}^2}{f_1 \hat{\lambda}_1} \qquad (9.149)$$

or, in terms of the failure-rate parameter, θ:

$$\frac{\hat{\theta}_1 \chi_{11}^2}{f_1} \leqslant \theta \leqslant \frac{\hat{\theta}_1 \chi_{22}^2}{f_1} \qquad (9.150)$$

Example 9.7

A relatively large number of nominally identical pumps are installed on a process plant and it is known that the times to failure for the pumps follow an exponential distribution. Over a 6-month period it is found that 5 pump failures have occurred in a sample of 20 of the pumps. If an estimate of the population mean failure-rate is made from this sample, what are the symmetrical 90% confidence limits on the estimate?

Applying Equations (9.128) and (9.143):

$$\hat{\theta}_1 = \frac{n}{T} \qquad (9.151)$$

where $n = 5$ and $T = 10$ pump-years. Hence:

$$\hat{\theta}_1 = 0.5 \text{ faults per year}$$

SAMPLING, ESTIMATION AND CONFIDENCE

Also, from Equation (9.137):

$$f_1 = 2n$$
$$= 10 \text{ degrees of freedom}$$

and:

$$f_2 = 2(n+1)$$
$$= 12 \text{ degrees of freedom}$$

Now, for the required 90% confidence level, χ_{11}^2 is evaluated at 5% cumulative probability at $f_1 = 10$ degrees of freedom and χ_{22}^2 is evaluated at 95% cumulative probability at $f_2 = 12$ degrees of freedom. Hence, from Table A.5:

$$\chi_{11}^2 = 3.940$$

and:

$$\chi_{22}^2 = 21.026$$

Then, substituting the appropriate values into Equation (9.150):

$$\frac{0.5 \times 3.940}{10} \leqslant \theta \leqslant \frac{0.5 \times 21.026}{10}$$

or:

$$0.197 \text{ faults/year} \leqslant \theta \leqslant 1.05 \text{ faults/year} \quad \text{at a 90\% confidence level}$$

Example 9.8

If, for the same sample as that described in Example 9.7, no faults had occurred in the sample instead of the 5 faults originally postulated, what can be said about the estimated mean failure-rate?

Obviously, the straightforward estimate of the mean failure-rate for this case must be zero. However, it is interesting to note that Equation (9.147) may be used to give an expression for an upper confidence limit on failure-rate even though no faults have developed in the sample taken. For instance, an estimate of $\hat{\theta}_2$ exists where:

$$\hat{\theta}_2 = \frac{n+1}{T} \qquad (9.152)$$

which, for the example values and even though $n = 0$, yields:

$$\hat{\theta}_2 = 0.1 \text{ faults/year}$$

and:

$$f_2 = 2(n+1)$$
$$= 2 \text{ degrees of freedom}$$

The value of χ^2_{22} at 95% cumulative probability and 2 degrees of freedom is then, from Table A.5:

$$\chi^2_{22} = 5.991$$

from which it follows that:

$$\theta \leqslant \frac{0.1 \times 5.991}{2}$$

or:

$$\theta \leqslant 0.3 \text{ faults/year} \quad \text{at a 95\% confidence level}$$

So, on the basis of no faults having occurred in the 10 pump-year sample of operation it may be stated that the estimated actual failure-rate is less than 0.3 faults per year and this statement will be true for 95% of all sample estimates of failure-rate made in this way.

Sampling from the Poisson distribution

The Poisson distribution function as discussed in Chapter 7 and given, in particular, in Table 7.1 and Equation (7.12) may be expressed in the form:

$$f(r) = \frac{\alpha^r}{r!} e^{-\alpha} \tag{9.153}$$

In Equation (9.153), $f(r)$ represents the probability of exactly r events taking place and α is the average occurrence of events relating to the whole population of the situation where the r events are considered.

In a sampling process where r events are observed, these number of events may be taken as an estimate of the true value of α, or:

$$r = \hat{\alpha} \tag{9.154}$$

This estimate of α will lead, from Equation (9.153), to a distribution of sample estimates given by:

$$f(\hat{\alpha}) = \frac{\alpha^{\hat{\alpha}}}{\hat{\alpha}!} e^{-\alpha} \tag{9.155}$$

The mean of this distribution from Table 7.1 is:

$$\mu_{\hat{\alpha}} = \alpha \tag{9.156}$$

so that Equation (9.154) represents an unbiased estimate of α.

An upper confidence limit may now be obtained for α from a particular sample estimate $\hat{\alpha}$ by applying to Equation (9.155) the principle involved in the derivation of Equation (9.6). Whence:

$$\frac{1+c}{2} = \sum_{j=\hat{\alpha}}^{\infty} \frac{\alpha^j}{j!} e^{-\alpha} \tag{9.157}$$

SAMPLING, ESTIMATION AND CONFIDENCE

It is seen in Equation (9.157) that a summation process instead of an integration process is involved since the Poisson function is a discrete and not a continuous distribution. In a similar way, a lower confidence limit may be expressed in the form:

$$\frac{1+c}{2} = \sum_{j=0}^{\hat{\alpha}-1} \frac{\alpha^j}{j!} e^{-\alpha} \qquad (9.158)$$

The upper limit of summation is taken as $(\hat{\alpha}-1)$ in Equation (9.158) since the value at $\hat{\alpha}$ has already been included in the summation process of Equation (9.157) and it cannot be legitimately included twice in a distribution of discrete values. A slightly different confidence interval could, of course, have been set up by making the lower limit of summation in Equation (9.157) equal to $(\hat{\alpha}+1)$ and the corresponding upper limit in Equation (9.158) equal to $\hat{\alpha}$. This will be discussed further later in this sub-section.

Equation (9.158) can be written in a slightly different way by subtracting both sides from unity and making use of the fact that the summation of all terms in the Poisson distribution is unity:

$$\frac{1-c}{2} = \sum_{j=\hat{\alpha}}^{\infty} \frac{\alpha^j}{j!} e^{-\alpha} \qquad (9.159)$$

Where the events considered in Equations (9.157) and (9.159) are failures of devices in the time domain then, as shown when discussing the implications of Equations (7.78) to (7.81), α may be expressed in the form:

$$\alpha = \frac{T}{\lambda} = \theta T \qquad (9.160)$$

where θ is the mean failure-rate of the device or devices and T is the actual overall time interval during which the failures occurred.

Combining Equations (9.157), (9.159) and (9.160) leads to a confidence interval expression of the following form:

$$\left[\sum_{j=\hat{\theta}T}^{\infty} \frac{(\theta T)^j}{j!} e^{-\theta T}\right]_{(1-c)/2} \leqslant \theta T \leqslant \left[\sum_{j=\hat{\theta}T}^{\infty} \frac{(\theta T)^j}{j!} e^{-\theta T}\right]_{(1+c)/2} \qquad (9.161)$$

where the lower limit in Equation (9.161) is evaluated at a probability value of $(1-c)/2$ and the upper limit at a probability value of $(1+c)/2$.

Example 9.9

Suppose that the motor failures considered in Example 9.6 are taken to follow a Poisson distribution when evaluated over some time scale, then the sample taken involves 5 failures over a 5976-h period. Under these conditions, what is the estimate of the Poisson parameter and corresponding mean failure-rate at a symmetrical 90% confidence level?

In this case, the sample number of events is 5 or:

$$\hat{\alpha} = \hat{\theta}T = 5$$

By the use of standard tables of the cumulative Poisson distribution or the plot of the cumulative distribution given in Figure A.28, the limits in Equation (9.161) can be evaluated at probability values of 0.05 and 0.95 for $\hat{\alpha} = 5$. Using Figure A.28, involves setting $r = 5$ and taking the two values of α corresponding to $[1-p(r-1)] = 0.05$ and $[1-p(r-1)] = 0.95$ since:

$$1 - p(r-1) = \sum_{j=r}^{\infty} \frac{\alpha^j}{j!} e^{-\alpha} \tag{9.162}$$

This then gives:

$$1.97 \leqslant \theta T \leqslant 9.15$$

and since, in the example being considered, T is the total running time of 5976 h, then:

330 faults/10^6 h $\leqslant \theta \leqslant$ 1530 faults/10^6 h   at a 90% confidence level

and it is seen that this is the same result as that obtained in Example 9.6. This is another aspect of the certain similarities between the exponential and Poisson distributions which were originally discussed in Chapter 7.

It is now pertinent to consider the case, already discussed in connexion with the exponential distribution in the previous sub-section, where the sample test or sample data on the failure of devices is terminated somewhere between the time of the nth failure and the $(n+1)$th failure. Under these conditions and in a similar way to the remarks that led to the derivation of Equations (9.143) and (9.144), $\hat{\alpha}$ now has two possible values:

$$\hat{\alpha}_1 = n \tag{9.163}$$

and:

$$\hat{\alpha}_2 = (n+1) = \hat{\alpha}_1 + 1 \tag{9.164}$$

For the case of Equation (9.163), the confidence limit Equation (9.161) becomes:

$$\left[\sum_{j=\theta_1 T}^{\infty} \frac{(\theta T)^j}{j!} e^{-\theta T}\right]_{(1-c)/2} \leqslant \theta T \leqslant \left[\sum_{j=\theta_1 T}^{\infty} \frac{(\theta T)^j}{j!} e^{-\theta T}\right]_{(1+c)/2} \tag{9.165}$$

and, in the case of Equation (9.164), another confidence interval from Equation (9.161) is:

$$\left[\sum_{j=(\theta_1 T+1)}^{\infty} \frac{(\theta T)^j}{j!} e^{-\theta T}\right]_{(1-c)/2} \leqslant \theta T \leqslant \left[\sum_{j=(\theta_1 T+1)}^{\infty} \frac{(\theta T)^j}{j!} e^{-\theta T}\right]_{(1+c)/2} \tag{9.166}$$

SAMPLING, ESTIMATION AND CONFIDENCE   349

As was mentioned in discussing Equations (9.145) and (9.146), it is normal practice in cases of this sort to take the combined confidence interval which gives the widest spread of values. This is achieved by taking the lower limit of Equation (9.165) in combination with the upper limit of Equation (9.166). Hence, doing this and writing:

$$\theta T = n \qquad (9.167)$$

where n is the number of failures, yields:

$$\left[\sum_{j=\hat{n}}^{\infty} \frac{n^j}{j!} e^{-n}\right]_{(1-c)/2} \leqslant n \leqslant \left[\sum_{j=(\hat{n}+1)}^{\infty} \frac{n^j}{j!} e^{-n}\right]_{(1+c)/2} \qquad (9.168)$$

Example 9.10

Assuming that the pump failures described in Example 9.7 follow a Poisson distribution, what are the symmetrical 90% confidence limits on the estimate of the mean number of failures for the sample of 10 pump-years and also the corresponding limits for the estimated mean failure-rate?
In this case:

$$c = 0.9$$

and:

$$\hat{n} = 5$$

Therefore, from standard tables or Figure A.28, Equation (9.168) may be evaluated as follows:

$$1.97 \leqslant n \leqslant 10.5 \quad \text{at a 90\% confidence level}$$

Hence the estimated number of pump failures lies in the band 1.97 failures to 10.5 failures at the 90% confidence level. Since the sample size corresponds to 10 pump-years, the corresponding estimate for the pump failure-rate is:

$$0.197 \text{ faults/year} \leqslant \theta \leqslant 1.05 \text{ faults/year}$$

which, as would now be expected, is the same as the result calculated in Example 9.7.

It is often useful when a sample test or a collection of sample data yields zero or a small number of failures to estimate simply the upper confidence limit for the number of failures or failure-rate. Such an estimate is given by the upper limit only of Equation (9.168) evaluated at the single upper confidence probability, c, that is required, that is:

$$n \leqslant \left[\sum_{j=(\hat{n}+1)}^{\infty} \frac{n^j}{j!} e^{-n}\right]_{c} \qquad (9.169)$$

It will be noticed that in terms of the cumulative Poisson distribution:

$$\sum_{j=(\hat{n}+1)}^{\infty} \frac{n^j}{j!} e^{-n} = 1 - p(\hat{n}) \qquad (9.170)$$

In Figure A.28, the cumulative probability $[1-p(r-1)]$ is plotted against α. Hence, the summation of Equation (9.170) or (9.169) can be solved from Figure A.28 by reading $\alpha = n$ and $r = \hat{n}+1$. Only the top right-hand part of Figure A.28, however, is generally of use in this context and so this part of the figure is replotted on an expanded scale in Figure A.29. In this case the plot is expressed directly in terms of $[1-p(\hat{n})]$ against n.

Example 9.11

Calculate, assuming the Poisson distribution of failures, the single 95% upper confidence limit on the estimate of pump failures as would pertain to the situation of Example 9.8.

In this case:

$$\hat{n} = 0$$

and, from Figure A.29, the corresponding value of n for 95% probability is:

$$n = 3.0$$

So, the upper 95% limit for the estimate of failures in the sample of 10 pump-years, when no failures have actually occurred in one such sample, is 3 failures. The corresponding limit for the failure-rate estimate is:

$$\theta \leqslant 0.3 \text{ faults/year} \quad \text{at a 95\% confidence level}$$

and, again, this corresponds with the previous working in Example 9.8.

Generally, then, for the type of cases considered in this sub-section, the confidence intervals calculated using the Poisson function as a sampling distribution yield the same results as those calculated in connexion with the exponential distribution where the sampling distribution is χ^2.

Sampling from the binomial distribution

In a sample where, say, r successes are observed out of n trials, an estimate of the probability of success may be given by:

$$\hat{p} = \frac{r}{n} \qquad (9.171)$$

This estimator will lead, from the binomial distribution function of Equation (7.11), to a distribution of sample estimators given by:

$$f(n\hat{p}) = \binom{n}{n\hat{p}} p^{n\hat{p}} (1-p)^{n-n\hat{p}} \qquad (9.172)$$

SAMPLING, ESTIMATION AND CONFIDENCE

The mean of this distribution, from Table 7.1, is:

$$\mu_{n\hat{p}} = np \tag{9.173}$$

so that Equation (9.171) represents an unbiased estimate of p.

An upper and lower confidence limit may now be obtained for p from a particular sample estimate $\hat{p}$ by applying to Equation (9.172) the principle involved in the derivation of Equations (9.6) and (9.7), whence:

$$\frac{1+c}{2} = \sum_{j=n\hat{p}}^{n} \binom{n}{j} p^j (1-p)^{n-j} \tag{9.174}$$

and:

$$\frac{1+c}{2} = \sum_{j=0}^{n\hat{p}-1} \binom{n}{j} p^j (1-p)^{n-j} \tag{9.175}$$

and similar remarks apply to the upper limit of $(n\hat{p}-1)$ as were made with regard to the upper limit of $(\hat{a}-1)$ in Equation (9.158).

Equations (9.174) and (9.175) could, of course, be written in the form where the particular value of $n\hat{p}$ is included in the expression for the lower confidence limit instead of in the expression for the upper confidence limit. This would lead to two expressions for confidence limits of a slightly different form, namely:

$$\frac{1+c}{2} = \sum_{j=n\hat{p}+1}^{n} \binom{n}{j} p^j (1-p)^{n-j} \tag{9.176}$$

and:

$$\frac{1+c}{2} = \sum_{j=0}^{n\hat{p}} \binom{n}{j} p^j (1-p)^{n-j} \tag{9.177}$$

The choice between the uses of Equations (9.174) and (9.175) on the one hand and Equations (9.176) and (9.177) on the other hand may depend upon the particular application being considered. However, one criterion, as was adopted in the derivation of Equation (9.168), is to take the combined confidence interval which gives the widest spread of values. This is achieved by taking the lower limit of Equation (9.175) in combination with the upper limit of Equation (9.176), that is:

$$\left[\sum_{j=0}^{n\hat{p}-1} \binom{n}{j} p^j (1-p)^{n-j}\right]_{(1+c)/2} \leq p \leq \left[\sum_{j=n\hat{p}+1}^{n} \binom{n}{j} p^j (1-p)^{n-j}\right]_{(1+c)/2} \tag{9.178}$$

or, if both sides of Equation (9.175) are subtracted from unity, then Equation (9.178) becomes:

$$\left[\sum_{j=n\hat{p}}^{n} \binom{n}{j} p^j (1-p)^{n-j}\right]_{(1-c)/2} \leq p \leq \left[\sum_{j=n\hat{p}+1}^{n} \binom{n}{j} p^j (1-p)^{n-j}\right]_{(1+c)/2} \tag{9.179}$$

and this may also be written in terms of the binomial cumulative probability function, whence:

$$[1-p(n\hat{p}-1)]_{(1-c)/2} \leqslant p \leqslant [1-p(n\hat{p})]_{(1+c)/2} \qquad (9.180)$$

Values of $[1-p(r-1)]$ for a range of r and n values, as has already been mentioned in Chapter 7, are plotted for the binomial cumulative distribution function in Figures A.1 to A.27 inclusive. These figures may therefore be used to solve the limit expressions in Equation (9.180). If more accuracy is required, then the reader is referred to the various publications of tables of the binomial function which are in existence.

Example 9.12

From a large population of control mechanisms, a sample of 20 mechanisms is taken and a test carried out on each to see whether they perform correctly. In this sample test, only 8 are found to perform correctly and the remaining 12 perform outside the requirement. What are the 90% confidence limits on the estimated probability of any mechanism in the population performing correctly?

Here,

$$\hat{p} = \tfrac{8}{20} = 0.4$$

Hence:

$$n\hat{p} = 8$$

The lower limit of Equation (9.180) is then obtained by reading $r = n\hat{p} = 8$ and $[1-p(r-1)] = (1-c)/2 = 0.05$ in Figure A.11 and this gives:

$$p = 0.217$$

The upper limit of Equation (9.180) is obtained by reading $r = n\hat{p}+1 = 9$ and $[1-p(r-1)] = (1+c)/2 = 0.95$ in Figure A.11 and this yields:

$$p = 0.606$$

Hence:

$$0.217 \leqslant p \leqslant 0.606 \quad \text{at a 90\% confidence level}$$

Both the Poisson distribution and the normal distribution can be used as an approximation to the binomial distribution as was originally mentioned in Chapter 7. This fact can therefore be used to estimate confidence limits for the binomial with some lesser accuracy than the method just described. In some cases this lesser accuracy may be adequate.

Let it be assumed, therefore, that the estimates of the binomial mean, $n\hat{p}$, follow a normal distribution. Then, expressing the normal mean and variance in terms of the binomial mean and variance as given in Table 7.1

SAMPLING, ESTIMATION AND CONFIDENCE 353

and Equation (7.56), it may be stated that:

$$\mu = np \tag{9.181}$$

$$\hat{\mu} = n\hat{p} \tag{9.182}$$

and:

$$\sigma = \sqrt{[n\hat{p}(1-\hat{p})]} \tag{9.183}$$

These expressions may now be substituted in Equation (9.50), whence:

$$\{n\hat{p} - t'\sqrt{[n\hat{p}(1-\hat{p})]}\} \leqslant np \leqslant \{n\hat{p} + t'\sqrt{[n\hat{p}(1-\hat{p})]}\} \tag{9.184}$$

Example 9.13

Recalculate the confidence limits required in Example 9.12 using the normal approximation of Equation (9.184).

Using the values of Example 9.12 in Equations (9.182) and (9.183) gives:

$$\hat{\mu} = 8$$

$$\sigma = 2.19$$

For a 90% symmetrical confidence level, the value of t' from Table A.1 is 1.645 since, due to symmetry, this represents a cumulative probability of 0.45. Hence, substituting the requisite values in Equation (9.184) yields:

$$0.22 \leqslant p \leqslant 0.58 \quad \text{at a 90\% confidence level}$$

which, considering the assumptions, is in fair agreement with the result obtained in Example 9.12.

Weighted combination of estimates

So far, in the sampling processes described, estimates have been made of population parameters from the results of a single 'representative' sample. It is possible, in practice, that evidence may be available from the results of several different samples taken under, perhaps, different conditions and at different times. Obviously, each separate sample of this nature is likely to yield a different estimate. The question then arises as to how the estimates from a range of such samples may be combined in order to yield a 'best' estimate derived from all the information available.

The first and simplest case to consider is where, in a range of samples of sizes $n_1, n_2, ..., n_k$, each sample is completely representative of the parent population in exactly the same way and under exactly the same conditions. Here every member of one sample has the same merit or weight compared with any member of any other sample. In this case, all the members of all the samples may be 'lumped' together to provide, in effect, just one overall sample of size n where

$$n = \sum_{j=1}^{k} n_j \qquad (9.185)$$

This single overall sample will then obviously produce a better estimate than any of the original and individual samples in the range n_j since $n > n_j$ for all j. In addition, the overall sample can be treated in exactly the same way as shown in the previous discussions of this section.

The second and more relevant case in the present discussion is where all the individual samples, although representative in some way of the parent population, do not possess the same degree of representation. This may arise due to different errors in the sampling process itself or to different methods of measuring the variate values in each sample. For instance, in a certain sample of certain size, the power output of a population of engines may be measured by the use of a dynamic brake. In another sample of another size, the power outputs may be computed from a measurement of engine efficiency and fuel input. Apart from the differences in sample sizes, there may be differences in the accuracy of the measurement process between the two cases. One might place, for instance, far more reliance on the dynamic brake method of measurement than on the efficiency method. It could possibly be wrong, therefore, to lump both samples together and make an estimate from the directly combined results. Some form of 'weighting' is therefore required in order to combine results from different samples in a more acceptable fashion.

A general case will now be considered where the infinite parent population of the required quantity is sampled by, say, n different methods. This general process is illustrated in Figure 9.4. In this figure, method 1 concerns itself with the whole population associated with this particular method. The corresponding mean for this population is μ_1 and the standard deviation is σ_1. A particular sample examined by method 1 may have a size of n_{11} made up of individual measurements $x_{1n_{11}}$. The estimate of the method 1 mean, μ_1, from this particular sample is then designated $\hat{\mu}_{11}$. Following this technique for other samples in the method 1 series such as n_{1k}, a population of sample means, $\hat{\mu}_{1k}$'s, is considered which has a mean of $\mu_{\hat{\mu}_1}$ and a standard deviation of $\sigma_{\hat{\mu}_1}$. This process can be repeated for other methods such as method m where $\mu_{\hat{\mu}_m}$ and $\sigma_{\hat{\mu}_m}$ apply. From this information the best estimate of the mean of the required quantity, designated $\hat{\mu}$, and the standard deviation of this estimate, shown as $\hat{\sigma}_{\hat{\mu}}$, may be derived.

It has already been seen in this chapter that a sample mean is generally the best estimate of the population mean. On this basis $\mu_{\hat{\mu}_1}$ is the best estimate of μ and the same can be said about any of the other individual methods of sampling. Also, in the limit, it has been shown that the distribution of sample

SAMPLING, ESTIMATION AND CONFIDENCE 355

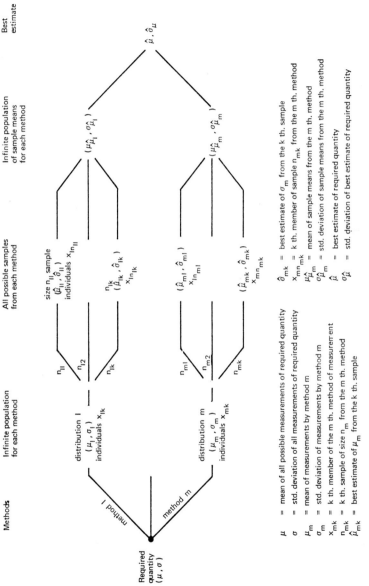

Figure 9.4 Sampling process and best estimate of a required quantity

means is normal. Hence, making this assumption of normality and considering the representation of sample means in Figure 9.5, the probability of $\hat{\mu}_{1k}$ occurring may be written as:

$$dP_1 = \frac{\sqrt{n_1}}{\sigma_1 \sqrt{(2\pi)}} \exp\left[-n_1(\hat{\mu}_{1k}-\mu)^2/2\sigma_1^2\right] d\hat{\mu}_1$$

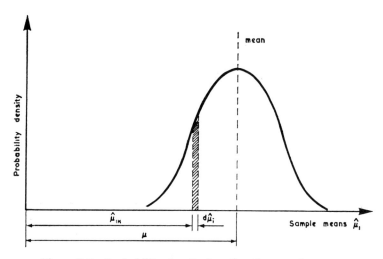

Figure 9.5 Probability density function for sample means

Similarly, it may be argued for $\hat{\mu}_{2k}$ that:

$$dP_2 = \frac{\sqrt{n_2}}{\sigma_2 \sqrt{(2\pi)}} \exp\left[-n_2(\hat{\mu}_{2k}-\mu)^2/2\sigma_2^2\right] d\hat{\mu}_2$$

and for $\hat{\mu}_{mk}$:

$$dP_m = \frac{\sqrt{n_m}}{\sigma_m \sqrt{(2\pi)}} \exp\left[-n_m(\hat{\mu}_{mk}-\mu)^2/2\sigma_m^2\right] d\hat{\mu}_m$$

If dP is the compound probability that the set of m estimates of the type $\hat{\mu}_{jk}$ will occur, then:

$$\begin{aligned}
dP &= dP_1 dP_2 dP_3 \ldots dP_m \\
&= \prod_{j=1}^{m} \frac{\sqrt{n_j}}{\sigma_j \sqrt{(2\pi)}} \exp\left[-n_j(\hat{\mu}_{jk}-\mu)^2/2\sigma_j^2\right] d\hat{\mu}_j \\
&= \exp\left[-\sum_{j=1}^{m} n_j(\hat{\mu}_{jk}-\mu)^2/2\sigma_j^2\right] \prod_{j=1}^{m} \frac{\sqrt{(n_j)} d\hat{\mu}_j}{\sigma_j \sqrt{(2\pi)}}
\end{aligned} \quad (9.186)$$

A basic statistical assumption may now be made that the best estimate of μ

SAMPLING, ESTIMATION AND CONFIDENCE

is obtained when the compound probability of Equation (9.186) is a maximum. Differentiating Equation (9.186) with respect to μ and equating to zero then yields:

$$\sum_{j=1}^{m} \frac{n_j(\hat{\mu}_{jk}-\mu)}{\sigma_j^2} = 0 \tag{9.187}$$

hence, the best estimate of μ is given by:

$$\hat{\mu} = \sum_{j=1}^{m} \frac{n_j \hat{\mu}_{jk}}{\sigma_j^2} \bigg/ \sum_{j=1}^{m} \frac{n_j}{\sigma_j^2} \tag{9.188}$$

and by a second derivative test Equation (9.188) may be shown to give a minimum condition.

If Equation (9.188) is written with:

$$\frac{n_j}{\sigma_j^2} = w_j \tag{9.189}$$

then:

$$\hat{\mu} = \sum_{j=1}^{m} w_j \hat{\mu}_{jk} \bigg/ \sum_{j=1}^{m} w_j \tag{9.190}$$

Equation (9.190) is sometimes called the 'weighted mean' where w is the 'weighting factor'. It will be noted that the data, in this respect, are weighted inversely proportional to the variance. Also, since the condition for maximum compound probability of the estimates is:

$$\sum_{j=1}^{m} \frac{n_j(\hat{\mu}_{jk}-\mu)^2}{2\sigma_j^2} \text{ is a minimum}$$

then this may be written:

$$\sum_{j=1}^{m} w_j(\hat{\mu}_{jk}-\mu)^2 \text{ is a minimum}$$

which means that the best estimate from a number of sample measurements having different weights is that which makes the weighted sum of the residuals, $(\hat{\mu}_{jk}-\mu)$, a minimum. This is the 'Principle of Least Squares'.

Since $\hat{\sigma}_{\hat{\mu}}^2$ is the variance of the best estimate of the mean or of the weighted mean, then from Equations (9.190) and (8.118):

$$\hat{\sigma}_{\hat{\mu}}^2 = \frac{1}{\left(\sum_{j=1}^{m} w_j\right)^2} (w_1^2 \sigma_{\hat{\mu}_1}^2 + w_2^2 \sigma_{\hat{\mu}_2}^2 + \ldots + w_m^2 \sigma_{\hat{\mu}_m}^2) \tag{9.191}$$

Now from Equation (9.25):

$$\sigma_{\hat{\mu}_j}^2 = \frac{\sigma_j^2}{n_j} \tag{9.192}$$

Hence, substituting Equations (9.189) and (9.192) in to Equation (9.191):

$$\hat{\sigma}_{\hat{\mu}}^2 = \sum_{j=1}^{m} w_j \Big/ \left(\sum_{j=1}^{m} w_j\right)^2$$

$$= 1 \Big/ \sum_{j=1}^{m} w_j \qquad (9.193)$$

Equation (9.190), therefore, leads to a best estimate of the population mean, or to a weighted mean, which can provide a more acceptable estimate in those cases where the evidence of a number of samples is not equally representative. The corresponding variance associated with this estimate is given by Equation (9.193). It should be noted, however, that the rigorous justification of the existence of normality may lead to assumptions requiring careful examination in certain cases.

Example 9.14

An estimate is required of the mean response time performance of a large population of thermal switches. It is found that evidence exists from sample tests which have been carried out by six different methods. For each method, the sample size and the mean estimate of response time is known. Also known, from long experience of employing the particular methods, are the standard deviations associated with each method. The available information is then of the following form:

Method	Standard deviation σ_j (s)	Sample size n_j	Sample mean $\hat{\mu}_j$ (s)
1	0.37	16	18.5
2	0.205	5	20.5
3	0.57	14	19.0
4	1.05	15	21.0
5	1.17	17	19.5
6	3.00	25	20.0

whence, applying Equation (9.190), the weighted mean is given by:

$$\hat{\mu} = \sum_{j=1}^{6} w_j \hat{\mu}_j \Big/ \sum_{j=1}^{6} w_j$$

$$= 6003.3/307.7$$

$$= 19.51 \text{ s}$$

SAMPLING, ESTIMATION AND CONFIDENCE 359

and the variance of this weighted mean, from Equation (9.193), is:

$$\hat{\sigma}_{\hat{\mu}}^2 = 1 \bigg/ \sum_{j=1}^{6} w_j$$

$$= 1/307.7$$

$$= 0.00325$$

or:

$$\hat{\sigma}_{\hat{\mu}} = 0.057 \text{ s}$$

Goodness of Fit

General

In the first section of this chapter on 'Sampling', some of the general problems which may require solution were discussed and these were listed under the points (a), (b), (c) and (d). The first three of these points have now been discussed and, in particular, the last section has dealt at some length with the estimation of parameters from a parent population. Both from a general point of view and from the requirements which arise out of forming a sampling distribution there is a need to know something about the form of the parent population distribution. It is important, therefore, to have some type of test which can establish the 'goodness of fit' between some postulated distribution and the evidence that exists as to the nature of the population distribution. Such evidence, of course, is likely to exist in the same basic form of sample data as has been used to estimate population parameters.

The tests which may be performed to determine the best fit of the sample data to some hypothesis relating to the distribution of the parent population fall into two main types. These types are graphical or analytical and the following sub-sections give a brief review of the ways in which both methods may be applied.

Graphical methods

Suppose, by way of illustration, it is required to test the data given in Table 2.2 of Chapter 2 to see whether it is likely to have been derived from a parent distribution which is normal. The first step, in the graphical approach, is to obtain values of the cumulative probability or cumulative relative frequency in relation to various variate values in the range of the data. To do this it is necessary to divide the data values into certain group intervals or class intervals. Such a division of data was made previously in Table 2.3 of Chapter 2. It is now important, though, to decide for the purposes of test the optimum number of intervals into which the data should be divided. A useful rule in this context is that proposed by Sturges, namely:

$$K = 1 + 3.322 \log n \tag{9.194}$$

where n is the sample size and K is the number of class intervals into which the range between the two extreme values of the data should be divided. Applying this rule to the data of Table 2.2 gives a value of approximately $K = 7$. Since the rule is only a guide, however, the six intervals that were originally used in Table 2.3 will still be maintained for this present illustration. In this table only the frequency corresponding to each interval was tabulated, hence the corresponding cumulative frequencies and cumulative relative frequencies (or percentage probabilities) are shown additionally in Table 9.1. As before, three decimal places have been used to define the intervals in order to leave no ambiguity as to the interval in which any of the data falls. The data values themselves being given to two decimal places.

The cumulative relative frequency or cumulative probability may now be plotted for each group interval against the upper variate value of each interval. If this plot is made on 'normal probability paper' of the type shown in Figure 9.6, then a straight line indicates that the data are derived from a normal distribution. The reader will notice the non-linearity of the cumulative probability scale on this type of paper in order to achieve the straight-line relationship. The actual data points for the first five intervals in Table 9.1

Table 9.1 Derivation of cumulative relative frequency for the data of Table 2.2

Group interval	Frequency of response times	Cumulative frequency	Cumulative relative frequency (%)
1.205–1.305	7	7	8.24
1.305–1.405	14	21	24.71
1.405–1.505	21	42	49.40
1.505–1.605	23	65	76.47
1.605–1.705	16	81	95.30
1.705–1.805	4	85	100.00

are shown by ×'s on Figure 9.6 and the straight line drawn through these points shows the closeness of the data fit to a normal distribution. It is generally expected that the 'end' points in such a data plot may deviate slightly from the general pattern due to the small numbers of events that are normally recorded at the 'tails' of the distribution.

Figure 9.6 shows that the mean value (corresponding to the 50% cumulative probability point) is 1.504 s. One standard deviation from the mean is shown by the variate values corresponding to the cumulative probability points of either 15.87% or 84.13%. These variate values are 1.365 s and 1.643 s respectively, hence for instance:

$$\hat{\sigma} = 1.643 - 1.504$$
$$= 0.139 \text{ s}$$

SAMPLING, ESTIMATION AND CONFIDENCE

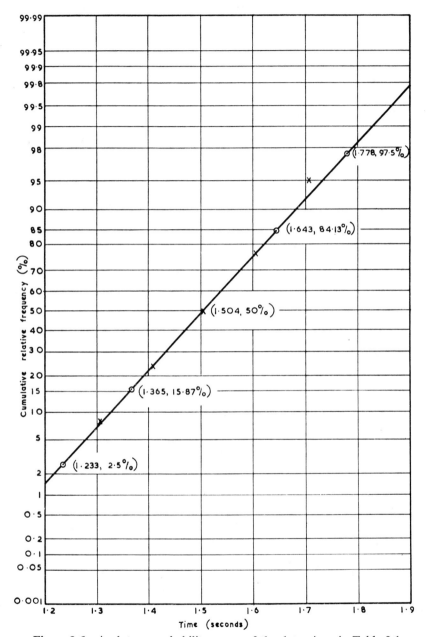

Figure 9.6 A plot on probability paper of the data given in Table 9.1

With reference to the mean value and the standard deviation for the 85 readings, these could, of course, have been calculated directly from the sample data without resorting to a normal probability plot. For instance, from Equation (9.8):

$$\hat{\mu} = \frac{1}{85} \sum_{j=1}^{85} x_j$$

$$= 1.506 \text{ s}$$

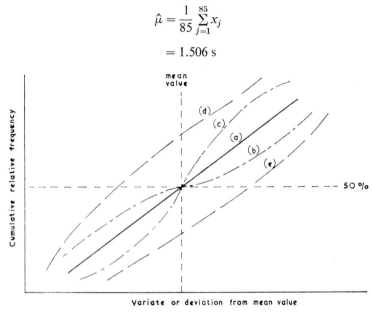

Figure 9.7 Plots on probability paper

and, from Equation (9.80):

$$\hat{\sigma}_c^2 = \frac{1}{84} \sum_{j=1}^{85} (x_j - 1.506)^2$$

or:

$$\hat{\sigma}_c = 0.130 \text{ s}$$

and these two answers may be compared with the more approximate graphical technique. However, the main purpose of the graphical technique has been to demonstrate the goodness of fit of the data to a normal distribution and this, in a subjective way, it has achieved.

A plot of data on normal probability paper may not, of course, produce a straight line but certain types of deviations from a straight line have certain significances. Some typical characteristic plots which may be produced from various data are shown in Figure 9.7. Plot (a), being a straight line, represents, once again, the characteristics of a normal distribution. If the distribution is skewed then the data plots may take the form of either (d) or (e). Positive values of K_3 or a distribution skewed so that its peak is to the left of the mean is represented by plot (d). Similarly, plot (e) represents distributions with a negative value of K_3. Plots (b) and (c) show the effects, on a symmetrical distribution, of different values of kurtosis. A distribution which is flat-topped or platykurtic is represented by plot (b) and plot (c) represents a peaked or leptokurtic distribution.

Apart from deducing, from the type of plots illustrated in Figure 9.7, the sort of deviation from a normal distribution that may exist, the relevant sample data may be replotted on alternative forms of probability paper. The cumulative probability scale in Figure 9.6 is scaled so that a normal distribution leads to a straight-line plot. Obviously, the scaling of such a cumulative probability scale can be adjusted so that a straight-line plot is obtained for some other type of distribution. Typical probability papers produced in this way are those for exponential, Weibull and lognormal functions. With the latter function, of course, the cumulative probability scale is the same as on normal probability paper but the variate scale is logarithmic instead of linear.

Obviously, with the graphical methods so far described, the degree of goodness of fit that is obtained between the sample data and the distribution that is hypothesized is subject to personal judgement. There is no direct way, by these methods, of obtaining a quantitative value for the goodness of fit. This leads on to considerations of alternative analytical techniques which are discussed in the next sub-section.

Analytical methods

In the graphical methods just described, the assessment of goodness-of-fit depends upon how nearly the sample data point for each class interval approaches the corresponding point for the class interval calculated from the hypothesized distribution function. The analytical methods are similar but attempt to quantify, in some way, the degree of difference between these respective points.

Consider, as before, that some sample data of size n may be divided into K class or group intervals. The choice of K may be determined from some relationship such as Equation (9.194). For the jth class interval, the variate x of the sample data may take on a value x_j where x_j represents the mid-value of this interval. If the incremental range of the jth class interval is δx_j then the interval extends from $(x_j - \tfrac{1}{2}\delta x_j)$ to $(x_j + \tfrac{1}{2}\delta x_j)$. Suppose also that the number of variate values falling within the jth interval is n_j, then it follows that:

$$\sum_{j=1}^{K} n_j = n \qquad (9.195)$$

Let $f(x)$ be the density function of the hypothesized distribution to which it is required to fit the data, then the hypothesized number of variate values of x_j that would be expected to fall in the interval $(x_j - \tfrac{1}{2}\delta x_j)$ to $(x_j + \tfrac{1}{2}\delta x_j)$ is:

$$n \int_{x_j - \tfrac{1}{2}\delta x_j}^{x_j + \tfrac{1}{2}\delta x_j} f(x)\,dx = np_j \quad (9.196)$$

where p_j represents the integral of the hypothesized density function over the range of the jth interval.

The question now to be asked is that of how nearly the values of n_j approach the corresponding values of np_j for all j. The straightforward differences between n_j and np_j could be examined, but, as will be seen, it is useful to consider a statistic, X_j, based upon the following difference relationship:

$$X_j = \frac{(n_j - np_j)}{\sqrt{(np_j)}} \quad (9.197)$$

The number of times, n_j, that the variate x takes on the value of x_j may be looked upon as a particular result out of a total of n trials. On this basis, n_j will be binomially distributed with a mean value of np_j and a variance of $np_j(1 - p_j)$. As n becomes large and p_j becomes small, this binomial distribution will, as shown in Equation (7.12) of Chapter 7, approximate to the Poisson distribution with a mean of np_j and a variance of np_j, that is:

$$\mu_{n_j} = np_j \quad (9.198)$$

$$\sigma^2_{n_j} = np_j \quad (9.199)$$

For a particular sample of size n and a given hypothesized distribution, $f(x)$, the quantity np_j is a constant. Hence, the distribution of X_j, from the work done earlier in this chapter and from the appropriate relationships in Table 8.1, is given by:

$$F_{X_j}(s) = e^{\sqrt{(np_j)}} F_{n_j}\left[\frac{s}{\sqrt{(np_j)}}\right] \quad (9.200)$$

The mean of the distribution of X_j may be found in the usual way by differentiating Equation (9.200) with respect to s and setting s equal to zero. This process yields:

$$\mu_{X_j} = \frac{\mu_{n_j}}{\sqrt{(np_j)}} - \sqrt{(np_j)} \quad (9.201)$$

and substituting for μ_{n_j} from Equation (9.198) gives:

$$\mu_{X_j} = 0 \quad (9.202)$$

SAMPLING, ESTIMATION AND CONFIDENCE

From the second differential of Equation (9.200) it will be seen that:

$$\sigma^2_{X_j} = \frac{\sigma^2_{n_j}}{np_j} \tag{9.203}$$

or, from Equations (9.199) and (9.203):

$$\sigma^2_{X_j} = 1 \tag{9.204}$$

A similar analysis of the higher moments of the distribution of X_j will show that it approaches the normal distribution as np_j becomes relatively large. In practice this amounts to np_j being greater than about 5. The variates X_j, therefore, can be taken as being normally distributed with zero mean and unit variance and hence expressed by the distribution function:

$$f_{X_j}(x) = \frac{1}{\sqrt{(2\pi)}} e^{-x^2/2} \tag{9.205}$$

The variate X_j represents only a difference relationship for the jth interval. In determining a goodness of fit it is obviously necessary to consider the difference relationship associated with all the class intervals taken together in some manner. A possible way of doing this is to consider an overall relationship based on the 'sums of the squares'. A variate, y, can be constructed then where:

$$y = \sum_{j=1}^{K} X_j^2 \tag{9.206}$$

Now, from the work previously done in deriving Equation (9.96), if X_j is distributed according to Equation (9.205) then X_j^2 is distributed according to:

$$f_{X_j^2}(x) = \frac{1}{\sqrt{(2\pi)}} \frac{1}{\sqrt{x}} e^{-x/2} \tag{9.207}$$

Whence, from Table A.2 or by a similar process to the derivation of Equation (9.98):

$$F_{X_j^2}(s) = \frac{1}{\sqrt{(2s+1)}} \tag{9.208}$$

The distribution of y in Equation (9.206) which represents the sum of all X_j^2 can then be obtained from Equation (8.15) in Chapter 8, namely:

$$F_y(s) = [F_{X_j^2}(s)]^K \tag{9.209}$$

or:

$$F_y(s) = \frac{1}{2^{K/2}(s+\frac{1}{2})^{K/2}} \tag{9.210}$$

and the inverse transform of Equation (9.210) from Table A.2 is:

$$f_y(x) = \frac{x^{(K/2-1)} e^{-x/2}}{2^{K/2} \Gamma(K/2)} \qquad (9.211)$$

which, by comparison with Equation (9.112), will be recognized as the form of the χ^2 distribution where, in this case:

$$\chi^2 = y$$

or:

$$\chi^2 = \sum_{j=1}^{K} \frac{(n_j - np_j)^2}{np_j} \qquad (9.212)$$

The overall relationship of Equation (9.212) which represents the differences between the sample data values for each class interval and the corresponding values from the hypothesized distribution is therefore distributed according to a χ^2 distribution with K degrees of freedom. However, it has been assumed in the derivation of Equation (9.211) that all the X_j's are independent. In fact, because of the relationship of Equation (9.195) only $(K-1)$ of the X_j's, at the most, are independent. If it is assumed, more generally, in any sample data that only $(K-m)$ of the values for each class interval are independent then Equation (9.209) becomes:

$$F_y(s) = [F_{X_j^2}(s)]^{K-m} \qquad (9.213)$$

which leads to:

$$f_y(x) = \frac{x^{(f/2-1)} e^{-x/2}}{2^{f/2} \Gamma(f/2)} \qquad (9.214)$$

where f, the number of degrees of freedom, is now given by:

$$f = K - m \qquad (9.215)$$

It is now apparent that, if for any set of sample data and any hypothesized distribution the value of χ^2 is calculated from Equation (9.212) then this value of χ^2 falls somewhere on the sampling χ^2 distribution with f degrees of freedom represented by Equations (9.214) and (9.215). These results are based, of course, on the assumption made throughout the working so far that the hypothesized distribution is the correct one. The χ^2 distribution of Equation (9.214), therefore, represents the variation that will take place, from sample to sample, in the χ^2 value of Equation (9.212) even when the hypothesis is correct.

Suppose now that a decision is taken to accept the hypothesis as correct if the calculated value of χ^2 from Equation (9.212) falls below a cumulative probability value of, say, p on the χ^2 distribution of Equation (9.214). This value of p is then the probability of accepting the hypothesis as correct when

SAMPLING, ESTIMATION AND CONFIDENCE

it is, in fact, correct and is given by:

$$p = \int_0^{\chi^2} f_y(x)\,dx \qquad (9.216)$$

Alternatively, a value α given by:

$$\alpha = 1 - p \qquad (9.217)$$

represents the probability of rejecting a correct hypothesis. This value of α is often known as the 'level of significance' and like the corresponding value of p may be used as a quantitative measure of the goodness of fit.

Because of the distributional relationships involved the test for goodness of fit on the basis of Equations (9.212), (9.214) and (9.217) is known as the 'χ^2 test'.

Example 9.15

Using the χ^2 test, determine whether the sample data of Table 9.1 would be accepted as coming from a normal distribution when using a level of significance, α, of 0.1.

If the data are divided into the same class intervals as used in Table 9.1, then the frequency for each interval or the number, n_j, of variate values falling in each interval are shown in column two of Table 9.1. These are repeated, for this example, in column two of Table 9.2.

Table 9.2 χ^2 test applied to data of Table 9.1

$n = 85$

Group interval	Frequency or number of variate values per interval n_j	Hypothesized probability per interval p_j	Hypothesized frequency np_j	X_j^2
1.205–1.305	7	0.061	5.17	0.647
1.305–1.405	14	0.158	13.40	0.027
1.405–1.505	21	0.278	23.68	0.303
1.505–1.605	23	0.280	23.81	0.028
1.605–1.705	16	0.160	13.60	0.420
1.705–1.805	4	0.063	5.34	0.338
Totals:	85	1.000	85.00	1.763

Having determined n_j from the data, the next parameter to determine is the value of p_j calculated from the hypothesized distribution. This distribution is normal and the best estimate of the mean and variance of this normal distribution is given, as shown in the previous sub-section, by:

$$\hat{\mu} = \frac{1}{85} \sum_{j=1}^{85} x_j$$

$$= 1.506$$

and:

$$\hat{\sigma}_c^2 = \frac{1}{84} \sum_{j=1}^{85} (x_j - 1.506)^2$$

or:

$$\hat{\sigma}_c = 0.130$$

Using these values of $\hat{\mu}$ and $\hat{\sigma}_c$ for $f(x)$ in Equation (9.196), the interval values of p_j may be calculated from Table A.1 and these are shown in the third column of Table 9.2. It should be noted that because the normal distribution has a range from $-\infty$ to $+\infty$ the end intervals of the hypothesized probability must be taken from $-\infty$ to 1.305 and from 1.705 to $+\infty$ respectively. Multiplying the p_j values by the total sample size of $n = 85$ gives the hypothesized frequency, np_j, for each interval as shown in the fourth column of Table 9.2. The final column of the table shows the corresponding values of X_j^2 calculated from Equation (9.197). The appropriate value of χ^2 is then derived from the sum of column five as shown in Equation (9.212). This leads to:

$$\chi^2 = 1.763$$

There are 6 class intervals. One restriction arises from the fact that the total probability or total sample size of the hypothesized distribution has been forced to agree with the corresponding sample size of the actual data. Also, both the mean and the variance of the hypothesized distribution have been estimated from the sample data. There are thus 3 constraints and consequently there are $6-3 = 3$ degrees of freedom. Also, with $\alpha = 0.1$ the value of the probability by which the hypothesis is accepted is given from Equation (9.217) by:

$$p = 0.9$$

The corresponding value of χ^2 in Equation (9.216) is now required for $p = 0.9$ and 3 degrees of freedom. This may be found from Table A.5 and yields:

$$\chi^2 = 6.251$$

The calculated value of χ^2 from the data of 1.763 is significantly less than this value of 6.251 and hence the hypothesis of a normal distribution would be accepted at the level of significance postulated.

SAMPLING, ESTIMATION AND CONFIDENCE

Apart from the χ^2 test, various other statistical tests may be used to evaluate the correctness or otherwise of a hypothesis. For instance a test known as the Kolmogorov–Smirnov test exists which depends upon the evaluation of the maximum absolute difference between the actual cumulative probability derived from the data and the hypothesized cumulative probability. The derivation of the Kolmogorov–Smirnov distribution for this maximum difference variate and the tabulated values of its cumulative probability function will be found in a number of standard works on the subject. Typically, though, for large sample sizes, which in practice means sample sizes greater than about 30, the Kolmogorov–Smirnov cumulative function approaches the following simplified relationship:

$$D = \frac{a}{\sqrt{n}} \qquad (9.218)$$

where D is the difference variate, n is the sample size and a is a parameter depending upon the cumulative probability p or level of significance α. Some values of a for a few typical values of α are shown in Table 9.3.

Table 9.3 Typical values of 'a' for the simplified Kolmogorov–Smirnov relationship of Equation (9.218)

Level of significance α	Parameter a
0.20	1.07
0.15	1.14
0.10	1.22
0.05	1.36
0.01	1.63

Example 9.16

Using the simplified Kolmogorov–Smirnov test, determine whether the sample data of Table 9.1 would be accepted as coming from a normal distribution when using a level of significance of $\alpha = 0.1$.

The actual cumulative probabilities from the sample data have already been calculated, as a percentage, in the fourth column of Table 9.1. The corresponding hypothesized cumulative probabilities can be obtained by summing the values in column three of Table 9.2. These two sets of cumulative probabilities together with the corresponding absolute differences are shown in Table 9.4.

Table 9.4 The simplified Kolmogorov–Smirnov test applied to the data of Table 9.1

Actual cumulative probability	Hypothesized cumulative probability	Absolute difference
0.082	0.061	0.021
0.247	0.219	0.028
0.494	0.497	0.003
0.765	0.777	0.012
0.953	0.937	0.016
1.000	1.000	0.000

It is now seen that the maximum absolute difference is:

$$D_{max} = 0.028$$

From Equation (9.218) and Table 9.3, the Kolmogorov–Smirnov critical value for a level of significance of 0.1 and a sample size of 85 becomes:

$$D = \frac{1.22}{\sqrt{85}}$$

$$= 0.132$$

Since D_{max} is significantly less than the critical value of D, the hypothesis of a normal distribution would again be accepted on the basis of this test and a level of significance of 0.1.

The reader will appreciate that differences arise between the various types of analytical tests for goodness of fit. These differences are related to the assumptions on which the tests are based and the assumptions associated with their method of application to any particular set of sample data. A review of all the assumptions involved and of the relative merits of one type of test compared with another could be quite extensive and is outside the scope of this present book. In general, however, the reader will find that the adoption of the χ^2 test will give a reasonable and quantitative measure of goodness of fit for quite a wide range of sample data problems which are likely to arise in the technological field.

Sample Testing

The general issues listed under (a), (b), (c) and (d) in the opening section of this chapter dealing with the problems of 'sampling' have now been discussed. These issues, it will be recalled, were associated with representative samples, sample variations, estimation and its associated confidence and goodness of fit. It was also mentioned in the same opening section that

SAMPLING, ESTIMATION AND CONFIDENCE 371

samples may be considered as falling into two main types. First, there are those samples which already exist and from which some inferences may be drawn. Secondly, there are those samples which are deliberately created in order to check some predetermined inference. The basic properties of the samples are obviously the same in both cases, but whereas most of the preceding discussion in this chapter has been concerned with samples of the first type it is now the intention to briefly review some of the particular characteristics of samples of the second type.

As an illustration of the difference between the two sample types the case may be considered where the mean time to failure of an item of equipment is of interest. In the first sample type, some information may already exist in the form of a series of recorded times to failure as was instanced for the motor in Example 9.6. For this case, some 'after-the-fact' inferences may be made and an estimate obtained of a mean failure-rate at a certain confidence level. On the other hand, with samples of the second type, it may be required to choose or set up a sample such that some level of confidence in the mean failure-rate of the overall population of items of interest is substantiated. This would normally involve, in the present illustration, the testing of a certain selection of the items in order to obtain their time-to-failure characteristics. Before starting such a sample test it is necessary to lay down certain test rules by which the failure characteristics of the overall population are accepted or rejected on the basis of the sample evidence. In the case of a time-to-failure parameter, this may be done by selecting a value of time-to-failure, λ_t, such that:

The overall population of items is rejected if:

$$\hat{\lambda} < \lambda_t \qquad (9.219)$$

and, the overall population of items is accepted if:

$$\hat{\lambda} \geq \lambda_t \qquad (9.220)$$

where $\hat{\lambda}$ is the mean time-to-failure of the items estimated from the sample under test.

The value of λ_t will be related in some way to the actual required value of the population mean time-to-failure, λ_u, which may also be regarded as some upper limit of acceptable mean time-to-failure. In addition, λ_t will also be related to some lowest acceptable value of population mean time-to-failure which may be designated as λ_l. Some elements of this relationship are illustrated in Figure 9.8. The right-hand density function, $f_2(\hat{\lambda})$, in Figure 9.8 shows the distribution of sample mean times-to-failure, $\hat{\lambda}$, for a sample of size n which is taken from a population whose true mean time-to-failure is λ_u. In the same way, the left-hand density function, $f_1(\hat{\lambda})$, is the distribution of sample mean times-to-failure for a similiar size sample taken from a population whose true mean time-to-failure is λ_1. The test criterion value of λ_t is taken to lie somewhere between λ_1 and λ_u although this is not necessarily the case.

It can now be seen that a sample which is taken from a population whose true mean is at the lower limit of λ_1 will be accepted with a probability of β where:

$$\beta = \int_{\lambda_t}^{\infty} f_1(\hat{\lambda}) \, d\hat{\lambda} \qquad (9.221)$$

This value of β can therefore be looked upon as the probability of accepting a reject item and is sometimes called the 'consumer's risk'.

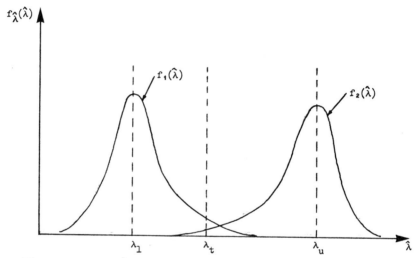

Figure 9.8 Plot of density functions for sample mean times to failure

In a similar way, a sample which is taken from a population whose true mean is at the upper required value of λ_u will be rejected with a probability α where:

$$\alpha = \int_{-\infty}^{\lambda_t} f_2(\hat{\lambda}) \, d\hat{\lambda} \qquad (9.222)$$

This value of α is then the probability of rejecting an acceptable item and is sometimes called the 'producer's risk'. It will be noticed that the value of α defined by Equation (9.222) is identical in concept to the value of α defined by Equations (9.217) and (9.216). In other words, the probability of rejecting an acceptable item is the same in concept as the probability of rejecting a correct hypothesis.

SAMPLING, ESTIMATION AND CONFIDENCE

If the sample size n is relatively large, then, as shown earlier in this chapter, the distributions of sample means as given by $f_1(\lambda)$ and $f_2(\lambda)$ will both approach normal distributions and Equation (9.221), for instance, can be written in the form:

$$\beta = \tfrac{1}{2} - \int_0^{t_1} \frac{1}{\sqrt{(2\pi)}} e^{-t^2/2} dt \qquad (9.223)$$

where:

$$t_1 = \frac{\lambda_t - \lambda_l}{\sigma_1/\sqrt{n}} \qquad (9.224)$$

and σ_1 is the standard deviation of the parent population of mean times-to-failure whose mean value is λ_l. The value t_1 obviously corresponds to the number of standard deviations that λ_t lies above λ_l in the distribution of $f_1(\lambda)$.

Similarly:

$$\alpha = \tfrac{1}{2} - \int_0^{t_2} \frac{1}{\sqrt{(2\pi)}} e^{-t^2/2} dt \qquad (9.225)$$

where:

$$t_2 = \frac{\lambda_u - \lambda_t}{\sigma_2/\sqrt{n}} \qquad (9.226)$$

Here, σ_2 is the standard deviation of the parent population of mean times-to-failure whose mean value is λ_u and t_2 represents the number of standard deviations that λ_t lies below λ_u in the distribution of $f_2(\lambda)$.

If the values of the consumer's risk, β, and the producer's risk, α, are predecided then this fixes the corresponding values of t_1 and t_2 from Equations (9.223) and (9.225) respectively. Combining Equations (9.224) and (9.226) then yields:

$$\frac{(\lambda_u - \lambda_t)}{(\lambda_t - \lambda_l)} = \frac{\sigma_2 t_2}{\sigma_1 t_1} \qquad (9.227)$$

whence:

$$\lambda_t = \frac{(\sigma_2 t_2/\sigma_1 t_1)\lambda_l + \lambda_u}{(\sigma_2 t_2/\sigma_1 t_1) + 1} \qquad (9.228)$$

So that the decision value for the acceptability of the sample test, λ_t, may be obtained from the predetermined values of λ_l, λ_u, α and β provided that the standard deviations of the corresponding parent populations, σ_1 and σ_2, are also known.

In determining the mean value of times-to-failure it is very often assumed that the parent population of distributions of times-to-failure is exponential.

For this distribution, as shown in Chapter 7, the standard deviation is equal to the mean, hence:

$$\sigma_1 = \lambda_1 \tag{9.229}$$

and:

$$\sigma_2 = \lambda_u \tag{9.230}$$

Substituting these values of σ_1 and σ_2 in Equation (9.228) yields:

$$\lambda_t = \frac{K(1+k)\lambda_u}{(1+kK)} \tag{9.231}$$

where:

$$K = \frac{\lambda_1}{\lambda_u} \tag{9.232}$$

and:

$$k = \frac{t_1}{t_2} \tag{9.233}$$

Hence, Equation (9.231) gives the sample test acceptance value of λ_t solely in terms of the predetermined values of λ_1, λ_u, α and β.

The required sample size, n, for the test may be derived, in terms of the same parameters, from either Equation (9.224) or (9.226). From the former, it is seen that the sample size may be expressed in the form:

$$\sqrt{n} = \frac{(1+kK)t_1}{k(1-K)} \tag{9.234}$$

or, from the latter:

$$\sqrt{n} = \frac{(1+kK)t_2}{(1-K)} \tag{9.235}$$

Example 9.17

It is required to test a sample of electro-magnetic relays from a large population of such relays in order to see whether the population is acceptable from the point of view of the mean time-to-failure of the relays. The required value of mean time-to-failure is 2×10^5 h and the lowest acceptable value is 4×10^4 h. It is required to achieve a consumer's risk of 0.1 and a producer's risk of 0.1. What is:
(a) the value of the estimated mean time-to-failure on which the decision to accept or reject the population should be based, and
(b) the number of failures that would need to be recorded in the sample?
(Assume that the distribution of relay times-to-failure is exponential and that the distribution of sample means is normal.)

SAMPLING, ESTIMATION AND CONFIDENCE 375

Putting the example values of λ_l and λ_u in Equation (9.232) gives:

$$K = \frac{4 \times 10^4}{2 \times 10^5} = 0.2$$

For a consumer's risk of $\beta = 0.1$, the value of t_1 from Equation (9.223) and Table A.1 is:

$$t_1 = 1.282$$

Similarly, for $\alpha = 0.1$, from Equation (9.225):

$$t_2 = 1.282$$

Hence, from Equation (9.233):

$$k = 1.0$$

Substituting the appropriate values of λ_u, K and k in Equation (9.231) yields:

$$\lambda_t = \frac{\lambda_u}{3}$$

$$= \frac{2 \times 10^5}{3} \text{ h}$$

Also, from Equation (9.235):

$$\sqrt{n} = 1.923$$

or:

$$n \simeq 3.7$$

Hence the sample test would have to proceed until 4 failures had occurred. On the basis of the acceptable mean time-to-failure, λ_u, this would require a test covering 8×10^5 relay-hours of operation.

It is often of interest to consider how the probability of accepting a population of items varies as the true mean time-to-failure of the population varies. The probability, p, of accepting a population when the sample criterion of acceptance is λ_t is given, on the same lines as Equation (9.221), by:

$$p = \int_{\lambda_t}^{\infty} f(\hat{\lambda}) \, d\hat{\lambda} \tag{9.236}$$

where $f(\hat{\lambda})$ is the distribution of sample mean times-to-failure, $\hat{\lambda}$, from a population whose true mean time-to-failure is λ. If $f(\hat{\lambda})$ is normal then the corresponding number of standard deviations, t, from the integral of Equation (9.236) is:

$$t = \frac{\lambda_t - \lambda}{\sigma/\sqrt{n}} \tag{9.237}$$

where $\sigma = \lambda$ is the population standard deviation for a population whose distribution of times-to-failure is exponential. If λ_t is decided by the criterion of Equation (9.231) and $\sqrt{n}$ by the corresponding criterion of Equation (9.235), then:

$$t = \frac{[K(1+k)-(1+kK)x]t_2}{(1-K)x} \qquad (9.238)$$

where:

$$x = \frac{\lambda}{\lambda_u} \qquad (9.239)$$

By way of illustration, the case may be typically considered where:

$$K = \frac{\lambda_l}{\lambda_u} = 0.2$$

$$\alpha = 0.05$$

and:

$$\beta = 0.1$$

Under these conditions, from Equations (9.223) and (9.225) and from Table A.1:

$$t_1 = 1.282$$

and:

$$t_2 = 1.645$$

and hence:

$$k = 0.7794$$

Equation (9.238) therefore becomes:

$$t = \frac{0.732}{x} - 2.378 \qquad (9.240)$$

which, for certain particular values of x, leads to the following values of p:

x	t	p
1.0	−1.646	0.95
0.8	−1.463	0.93
0.6	−1.158	0.88
0.5	−0.914	0.82
0.4	−0.548	0.71
0.3	+0.062	0.48
0.2	+1.282	0.10

The corresponding probability of acceptance curve is shown by the broken line curve in Figure 9.9.

SAMPLING, ESTIMATION AND CONFIDENCE

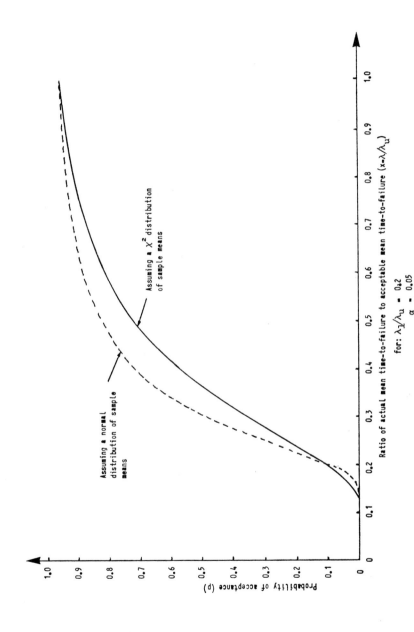

Figure 9.9 Typical acceptance probability curves

For small sample sizes, the distribution of sample means from an exponential population more nearly approaches a χ^2 distribution rather than a normal distribution. For this case, the values of α and β are still given by Equations (9.222) and (9.221) respectively but $f_2(\hat{\lambda})$ and $f_1(\hat{\lambda})$ are now both χ^2 distributions. For a given value of β and using the relationship of Equation (9.139), the corresponding value of χ_1^2 is:

$$\chi_1^2 = \frac{2n\lambda_t}{\lambda_1} \qquad (9.241)$$

Similarly, for a given value of α, the corresponding value of χ_2^2 is:

$$\chi_2^2 = \frac{2n\lambda_t}{\lambda_u} \qquad (9.242)$$

Hence, from Equations (9.232), (9.241) and (9.242):

$$K = \frac{\chi_2^2}{\chi_1^2} \qquad (9.243)$$

If now, by way of example, $\alpha = 0.05$, $\beta = 0.1$ and $K = 0.2$, then it is seen from Table A.5 that the nearest fit to Equation (9.243) is given by a value of:

$$f = 2n = 8 \qquad (9.244)$$

For 8 degrees of freedom and $\alpha = 0.05$, the 5% probability value yields:

$$\chi_2^2 = 2.733$$

Also, for 8 degrees of freedom and $\beta = 0.1$ (that is a percentage cumulative probability of 90%), the result is:

$$\chi_1^2 = 13.362$$

whence:

$$K \simeq 0.2$$

From either Equation (9.241) or (9.242) this leads to an approximate value of λ_t of:

$$\lambda_t = 0.335\,\lambda_u \qquad (9.245)$$

and also, of course, from Equation (9.244):

$$n = 4 \qquad (9.246)$$

Example 9.18

Rework Example 9.17 assuming that the distribution of sample means is χ^2 rather than normal.

For this example, $\alpha = 0.1$, $\beta = 0.1$ and $K = 0.2$. From Table A.5, the best fit of Equation (9.243) for these values of α and β is given by 6 degrees of

SAMPLING, ESTIMATION AND CONFIDENCE   379

freedom. For this fit:
$$\chi_1^2 = 10.645$$
$$\chi_2^2 = 2.204$$

From Equations (9.241) and (9.242) this leads to:
$$\lambda_t \simeq 0.36 \lambda_u$$
or:
$$\lambda_t \simeq 7.2 \times 10^4 \text{ h}$$
Also:
$$n = \frac{f}{2}$$
$$= 3$$

and these results may be compared with those previously obtained in Example 9.17.

The acceptance probability function for the case where the sample means are χ^2 distributed is still given by Equation (9.236) except that $f(\hat{\lambda})$ is now a χ^2 density function. The value of the acceptance probability can therefore be written as:
$$p = 1 - p_{\chi^2}(\hat{\lambda}) \quad (9.247)$$
and the corresponding value of χ^2 becomes:
$$\chi^2 = \frac{2n\lambda_t}{\lambda} \quad (9.248)$$
which on using the values derived, by way of example, in Equations (9.245) and (9.246) becomes:
$$\chi^2 = \frac{2.68}{x} \quad (9.249)$$

The value of p from Equation (9.247) is shown by the full-line curve in Figure 9.9 for a range of x values given by Equation (9.249). It is seen, for the relatively small sample size selected, that there is some difference between the curve based on the assumption of a normal distribution and that based on the assumption of a χ^2 distribution.

An assumption made so far is that the sample test is terminated at the time of the last failure. It is possible, however, to terminate the sample test at some time after the last failure. This leads to two possible estimates for the sample mean time-to-failure as shown in the sub-section dealing with 'Sampling from the exponential distribution'. One way of dealing with this situation is to take the more pessimistic of the two estimates or the lower sample mean time-to-failure as given, for instance, by Equation (9.144). In this case, the corresponding χ^2 distribution may be evaluated at $(2n+1)$ degrees of freedom instead of $2n$ degrees of freedom.

It will have been seen that if λ_l, λ_u, α and β are all specified then this leads to corresponding unique values of the test criterion λ_t and the sample size n. It may be decided, however, to continue the test or increase the sample size to some value greater than the unique value of n. Such a larger value of sample size, n_t say, leads to two values for λ_t. One is a lower limit, λ_{t_1}, given by, for instance, Equations (9.223) and (9.224) and the other an upper limit, λ_{t_2}, given by, for instance, Equations (9.225) and (9.226) where:

$$\lambda_{t_1} = \frac{t_1 \lambda_l}{\sqrt{n_t}} + \lambda_l \tag{9.250}$$

and:

$$\lambda_{t_2} = \lambda_u - \frac{t_2 \lambda_u}{\sqrt{n_t}} \tag{9.251}$$

The test may then be so conducted that if at any time the estimated mean time-to-failure falls between the values of λ_{t_1} and λ_{t_2} then the test is continued. If, on the other hand, the estimated value falls either below λ_{t_1} or above λ_{t_2} then the population is either rejected or accepted respectively and the test concluded. If no decision is reached by the time that the sample size reaches n_t, then the test is concluded any way and a decision made on the basis of the final estimate. This is generally known as a 'truncated sequential test' and various test rules may be formulated for such a test on the basis of formulae such as those given in Equations (9.250) and (9.251) or their equivalent for other distributions.

Example 9.19

Under the conditions stipulated in Example 9.17 it is required to carry out a truncated sequential test such that the test is terminated when the sample size reaches 12 failures. What are the two values of test criterion which determine the reject zone, continue testing zone and acceptance zone under the assumption that the sample mean times-to-failure are:
(a) normally distributed, and
(b) χ^2 distributed?
(a) For this example:

$$\lambda_l = 4 \times 10^4 \text{ h}$$
$$\lambda_u = 2 \times 10^5 \text{ h}$$
$$n_t = 12$$

SAMPLING, ESTIMATION AND CONFIDENCE 381

and, for $\alpha = 0.1$ and $\beta = 0.1$, then:

$$t_1 = t_2 = 1.282$$

Substituting these values in Equations (9.250) and (9.251) yields:

$$\lambda_{t_1} = 5.48 \times 10^4 \text{ h}$$

$$\lambda_{t_2} = 1.26 \times 10^5 \text{ h}$$

(b) For the χ^2 distribution with $2n = 24$ degrees of freedom the values of χ_1^2 and χ_2^2 corresponding to $\beta = 0.1$ (i.e. 90% cumulative probability) and $\alpha = 0.1$ (i.e. 10% cumulative probability) respectively are, from Table A.5:

$$\chi_1^2 = 33.196$$

$$\chi_2^2 = 15.659$$

Now from Equations (9.241) and (9.242):

$$\lambda_{t_1} = \frac{\chi_1^2 \lambda_l}{2n} \qquad (9.252)$$

and:

$$\lambda_{t_2} = \frac{\chi_2^2 \lambda_u}{2n} \qquad (9.253)$$

Hence, for the example values:

$$\lambda_{t_1} = 5.53 \times 10^4 \text{ h}$$

and:

$$\lambda_{t_2} = 1.31 \times 10^5 \text{ h}$$

which may be compared with the corresponding values just calculated in part (a) of the example.

The example just considered represents only one of a number of possible ways of deriving test criteria values for truncated sequential tests. The reader is referred to other works on the subject for a more extensive review of the procedures that may be adopted and of their relative merits.

This then concludes the general outline in this book of some of the theory behind sampling, estimation and confidence and some of the ways in which such theory may be applied to practical technological situations. The methods are referred to again in Chapter 13, but the immediate succeeding chapters return to the general analytical problems associated with the reliability analysis of technological systems.

Questions

1. When samples of a particular size are drawn from a large population of a certain type of hydraulic jack it is known that the sample estimates of the mean jack travel follow a rectangular distribution whose density function is of the form $f(x) = 1/(b-a)$ with a standard deviation of $5/\sqrt{12}$ mm. If, from one of these samples, the estimate made of mean jack travel is 12 mm, what are the symmetrical confidence limits of the estimate at a 90% confidence level?

2. The power outputs of a large population of a particular type of engine are rectangularly distributed between a lower limit of 100 kW and an upper limit of 120 kW. From this population a sample of two engines is taken and an estimate made of the mean power output. What is the form of the sampling distribution of the estimates of mean power output made in this way?

3. If, in the previous question, the sample size is increased to 20 engines, what is:
   (a) the new approximate form of the sampling distribution of mean power outputs,
   (b) the mean of this distribution, and
   (c) the standard deviation of this distribution?

4. A sample of 30 electronic amplifiers is taken from a total batch of 50 amplifiers and an estimate made of the mean amplifier voltage gain by measurements carried out on the 30 sample amplifiers. If this estimate is found to have a value of 40, what is the upper confidence limit on this estimate at a confidence level of 99% in relation to the batch of 50 amplifiers? The standard deviation in voltage gain for the complete batch is known to be 5.

5. A very large population of transistorized devices are known to have time constants which follow a distribution whose standard deviation is 20 μs. From this population, measurements are made of the time constants on a relatively large population of 49 devices and these measurements yield a mean value of 62 μs. Assuming that the distribution of sample means is normal what are the symmetrical lower and upper confidence limits on the estimate for a 95% confidence level?

6. From a production line which is producing electrical resistors a sample of 6 are withdrawn and measured for their resistance value. This yields measurements of 9.6 Ω, 9.9 Ω, 10.0 Ω, 10.1 Ω, 10.2 Ω and 10.2 Ω. What is the best and unbiased estimate of the standard deviation in resistance value in the production batch if:
   (a) the batch is very large, and
   (b) the batch consists of only 10 resistors?

7. The variations in length of a certain type of pivot pin are known to follow a normal distribution. From this population of pivot pins a small sample of 8 pins are measured and found to have lengths of 17.9 mm, 18.4 mm, 18.8 mm, 19.6 mm, 19.9 mm, 20.6 mm, 21.8 mm and 23.0 mm. What are the symmetrical 95% confidence limits on the estimate of the population standard deviation?
8. For the same sample data as that given in Question 7, calculate the 95% symmetrical confidence limits on the sample estimate of the mean pivot-pin length.
9. The times-to-failure of a particular type of transformer when operating under its normal design conditions are known to follow an exponential distribution. A sample of four such transformers are put on a test which simulates their normal running conditions and the test is continued until all four transformers have failed. The measured times-to-failure in this situation are found to be 11,500 h, 13,000 h, 14,760 h and 16,500 h. What are the symmetrical lower and upper confidence limits for the estimate of mean time-to-failure of the population of transformers at a 90% confidence level?
10. When used in rolling-mill applications it is known that the times-to-failure of a particular type of electric motor follow an exponential distribution. A total of 15 such motors are installed in one particular mill and over a period of 10,000 h it is found that 5 motor failures have occurred somewhere within the overall period. What are the symmetrical 90% confidence limits on the estimated population mean time-to-failure that may be derived from this sample?
11. A sample of 6 newly designed control systems have been fitted on a process plant and run for a period of 5000 h. During this time no failures have occurred. What is the 95% upper confidence limit on the estimate of the mean time-to-failure that might be expected from a large population of such control systems assuming that their times-to-failure will be exponentially distributed?
12. The number of failures occurring in a given length of time in the case of a certain type of aircraft engine are found to follow a Poisson distribution. What is the upper 95% confidence limit on the estimate of the mean number of failures in this length of time if for a particular engine sample it is found for this period that:
    (a) no failures have occurred,
    (b) one failure has occurred, and
    (c) four failures have occurred?
13. In a sample of 10 electronic amplifiers it is found that 9 are within specification and that 1 is outside. What are the 90% symmetrical confidence limits on the estimated probability that any amplifier chosen from the overall population will perform within specification?

14. The power output of a particular type of motor is measured by four different sampling methods. The characteristics of each of the four samples and the results they produce are as follows:

Method	Standard deviation of method (kW)	Sample size	Sample mean (kW)
1	7.4	8	370
2	4.1	4	410
3	11.4	6	380
4	21.0	7	420

What is the value of the weighted mean estimate and the standard deviation on this estimate?

15. The response times of an operator in shutting down a process plant on the receipt of an alarm signal were measured on 100 different occasions with the grouped results:

Group interval (s)	Frequency
0.40–0.80	1
0.80–1.20	6
1.20–1.60	37
1.60–2.00	34
2.00–2.40	16
2.40–2.80	5
2.80–3.20	1

Plot these results on lognormal probability paper and comment on any conclusions that you may draw.

16. The times-to-failure are recorded from a sample of 100 temperature measuring instruments and are found to fall in the following group intervals with the frequencies shown:

Group interval (h)	Frequency
0– 50	38
50–100	25
100–150	16
150–200	7
200–250	5
250–300	4
300–350	3
350–400	2

Assuming the hypothesis that the times-to-failure of the instruments are exponentially distributed, what is:
(a) the value of χ^2 from a χ^2 test, and
(b) does this value yield an acceptance of the hypothesis at a 0.1 level of significance?

17. Check the sample data of Question 16 using the simplified Kolmogorov–Smirnov test, and
    (a) calculate the maximum difference, D_{max}, and
    (b) compare this value with the critical Kolmogorov–Smirnov value at a 0.1 level of significance.

18. A large batch of component parts whose times-to-failure are known to be exponentially distributed are produced on a production line. The basic criteria for accepting this batch of parts are that a true value of mean time-to-failure of 1000 h should be rejected with a probability of 0.05 (producer's risk) and a true value of mean time-to-failure of 250 h should be accepted with a probability of 0.10 (consumer's risk). If a sample test is set up on this basis what is:
    (a) the value of mean time-to-failure against which the estimated mean should be compared for an acceptance–rejection criterion, and
    (b) the corresponding number of failures that would have to be recorded in the sample?

CHAPTER 10
Reliability Considerations for Systems

Overall Concepts

As the reader will appreciate, the overall method of defining and calculating the reliability of a system has now been established. The method is based upon the quantified correlation between the concepts of performance requirement and performance achievement which was originally introduced in Chapter 2. In the immediate succeeding chapters, this correlation was expressed in the form of a functional equation of the type:

$$R = f(Q, H) \qquad (10.1)$$

where Q represented the required performance and H the achieved performance. Methods of defining and obtaining Q and H in a suitable and compatible form were discussed in Chapters 3, 4 and 5. One such form was that of a transfer characteristic or matrix which was made up of elements representing transfer functions. The meanings of these functions were also discussed and their properties enlarged upon in Chapter 6.

It was shown, for any linear system, that the requirement and achievement could be represented by transfer matrices of similar forms. The random and systematic differences between these matrices with respect to space or time could then be expressed in terms of the variational characteristics in the coefficients of the corresponding transfer functions. This was examined specifically in Chapters 5 and 6. The meanings of deterministic and probabilistic concepts associated with these coefficient characteristics have also been explained.

Variational aspects in the coefficients for any system have been shown to be important and these can often be represented by suitable statistical distributions. The properties of such distributions and the methods by which they may be manipulated and combined have been outlined in Chapters 7 and 8.

The techniques so far discussed will have enabled the reader to see how both Q and H can be expressed in the form of an array of representative values and their associated statistical distributions. Where this array is not directly obtainable at the overall system level, methods have been indicated for

RELIABILITY CONSIDERATIONS FOR SYSTEMS

building up the system functions from a knowledge of the component part characteristics.

The next step is to examine in some more detail the correlation process by which the reliability function is produced from the knowledge of Q and H. This is the process represented by Equation (10.1) and indicated at the beginning of the book by Figure 2.1.

The Correlation Process

Simple correlations

It is obvious, for a large and complex system, that the correlation between requirement and achievement represents a comparison between two transfer matrices in n-dimensional space. Before considering such a general case,

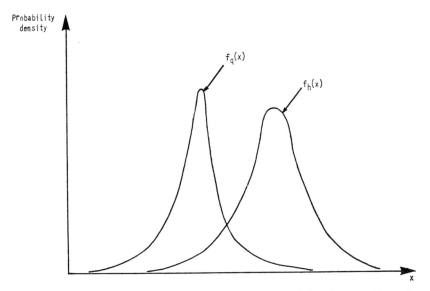

Figure 10.1 Requirement and achievement characteristics in one dimension

however, it is worth studying the simplest possible situation. Here both the requirement and achievement are each completely represented by a single variational characteristic of just one performance parameter or coefficient. Let this performance parameter be expressed by a variate x, then Figure 10.1 shows the way in which this simple case may be represented.

Suppose that the success of the system, whose characteristics are depicted in Figure 10.1, depends upon the achieved value of the performance parameter x being greater than the required value of x. Then the total measure of success will be found from the integration of a particular value of the achieved variate together with the probability that the required value is less than this particular achieved value. The corresponding integral expression for this total probability of success or reliability is:

$$R = \int_0^\infty f_h(x) \left\{ \int_0^x f_q(y) \, dy \right\} dx \tag{10.2}$$

$$= \int_0^\infty f_h(x) p_q(x) \, dx$$

or:

$$R = \int_0^\infty p_q(x) \, dp_h(x) \tag{10.3}$$

In the very particular case where the distribution of achieved values is equal to the distribution of required values, that is $p_q(x) = p_h(x) = p(x)$, the overall reliability becomes:

$$R = \int_0^\infty p(x) \, dp(x)$$

$$= \left\{ \frac{[p(x)]^2}{2} \right\}_0^\infty$$

$$= \tfrac{1}{2} \tag{10.4}$$

and this is a result which would have been intuitively expected under the postulated conditions, that is a 50 : 50 chance of success.

Very often, in practical situations, the required density function, $f_q(x)$, may be approximated to a single valued quantity. This is saying, in fact, that the requirement is deterministic with no variation. If this single valued requirement is x_q, then:

$$R = \int_{x_q}^\infty f_h(x) \, dx$$

$$= 1 - p_h(x_q) \tag{10.5}$$

Example 10.1

The requirement for the output power of an engine is subject to variations which follow a rectangular distribution with a mean value of 75 kW and a variance of 3 kW². The achieved power outputs of the engine are found to follow a similar distribution with a mean and variance of 80 kW and 12 kW² respectively. What is:

(a) the reliability for the power output of the engine being greater than the requirement, and

RELIABILITY CONSIDERATIONS FOR SYSTEMS

(b) how would this reliability be affected if the requirement became single valued at 75 kW?

The density functions for requirement and achievement are:

$$f_q(x) = \tfrac{1}{6}[U(x-72) - U(x-78)]$$
$$f_h(x) = \tfrac{1}{12}[U(x-74) - U(x-86)]$$

where the variate x is measured in kilowatts.

The corresponding cumulative distribution functions are:

$$p_q(x) = \tfrac{1}{6}(x-72)[U(x-72) - U(x-78)]$$
$$p_h(x) = \tfrac{1}{12}(x-74)[U(x-74) - U(x-86)]$$

Applying Equation (10.2) gives:

$$R = \int_{72}^{86} \tfrac{1}{12}[U(x-74) - U(x-86)] \frac{(x-72)}{6}[U(x-72) - U(x-78)]\,dx$$

$$= \int_{74}^{78} \frac{(x-72)}{72}\,dx + \int_{78}^{86} \frac{dx}{12}$$

$$= \tfrac{2}{9} + \tfrac{2}{3}$$

$$= \tfrac{8}{9}$$

and for the case where the requirement is single valued at 75 kW, then from Equation (10.5):

$$R = 1 - p_h(75)$$
$$= 1 - \tfrac{1}{12}(75 - 74)$$

or:

$$R = \tfrac{11}{12}$$

Another case of interest is when, in Equation (10.2), both $f_h(x)$ and $f_q(y)$ are normally distributed. The lower limits of the integrals must then be $-\infty$ and the reliability expression becomes:

$$R = \int_{x=-\infty}^{\infty} \int_{y=-\infty}^{x} f_h(x) f_q(y)\,dy\,dx \qquad (10.6)$$

Considering a rectangular co-ordinate system of x, y, the range of integration for the double integral of Equation (10.6) is shown by the shaded area in Figure 10.2. Changing the rectangular co-ordinates to the new axes v, w which are at 45° to x, y, the limits of the double integral become fixed as follows:

$$R = \int_{w=-\infty}^{0} \int_{v=-\infty}^{+\infty} f_h\!\left(\frac{v-w}{\sqrt{2}}\right) f_q\!\left(\frac{v+w}{\sqrt{2}}\right) dv\,dw \qquad (10.7)$$

If both f_h and f_q are normal with means and variances of μ_h, σ_h^2 and μ_q, σ_q^2 respectively, then:

$$R = \frac{1}{2\pi\sigma_h\sigma_q} \int_{w=-\infty}^{0} \int_{v=-\infty}^{+\infty} \exp\left[-\frac{1}{2\sigma_h^2}\left(\frac{v-w}{\sqrt{2}} - \mu_h\right)^2\right]$$

$$\times \exp\left[-\frac{1}{2\sigma_q^2}\left(\frac{v+w}{\sqrt{2}} - \mu_q\right)^2\right] dv\, dw \qquad (10.8)$$

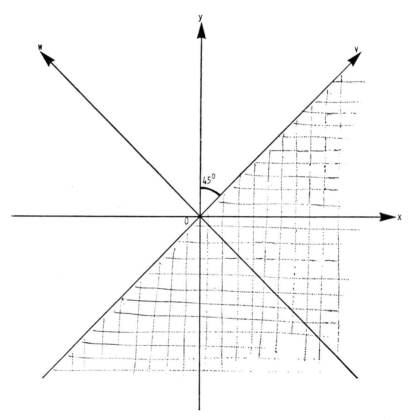

Figure 10.2 Transform of co-ordinates for Equation (10.6)

making the substitutions:

$$v = \sqrt{(2)}\sigma_h z + w + \sqrt{(2)}\mu_h \qquad (10.9)$$

$$dv = \sqrt{(2)}\sigma_h\, dz \qquad (10.10)$$

and rearranging the exponential terms leads to:

$$R = \frac{1}{\sqrt{(2)}\pi\sigma_q} \int_{w=-\infty}^{0} \int_{z=-\infty}^{\infty} \exp\left[-\frac{\sigma_h^2+\sigma_q^2}{2\sigma_q^2}\left(z+\frac{\sigma_h(\sqrt{(2)}w+\mu_h-\mu_q)}{\sigma_h^2+\sigma_q^2}\right)^2\right]$$

$$\times \exp\left[-\frac{(\sqrt{(2)}w+\mu_h-\mu_q)^2}{2(\sigma_h^2+\sigma_q^2)}\right] dz\, dw \qquad (10.11)$$

The variate z is now in the normal form and hence the integration of z between $-\infty$ and $+\infty$ gives:

$$R = \frac{1}{\sqrt{\pi}\sqrt{(\sigma_h^2+\sigma_q^2)}} \int_{-\infty}^{0} \exp\left[-\frac{(\sqrt{(2)}w+\mu_h-\mu_q)^2}{2(\sigma_h^2+\sigma_q^2)}\right] dw \qquad (10.12)$$

Making the substitution:

$$t = \frac{\sqrt{(2)}w+\mu_h-\mu_q}{\sqrt{(\sigma_h^2+\sigma_q^2)}} \qquad (10.13)$$

$$dt = \frac{\sqrt{(2)}\,dw}{\sqrt{(\sigma_h^2+\sigma_q^2)}} \qquad (10.14)$$

yields:

$$R = p\left[\frac{\mu_h-\mu_q}{\sqrt{(\sigma_h^2+\sigma_q^2)}}\right] \qquad (10.15)$$

where the cumulative distribution function of Equation (10.15) is of the normal form.

Example 10.2

The ability of a particular component to withstand a power dissipation follows a normal distribution with a mean of 100 W and a standard deviation of 4 W. The actual power developed in the component is also normally distributed with a mean of 90 W and a standard deviation of 3 W. What is the reliability for the power developed being less than the capability of power dissipation?
Here:

$$\mu_q = 100, \quad \sigma_q^2 = 16$$
$$\mu_h = 90, \quad \sigma_h^2 = 9$$

The reliability in this case is concerned with the achievement being less than the requirement. This involves a simple sign reversal in Equation (10.15), that is:

$$R = p\left[\frac{\mu_q-\mu_h}{\sqrt{(\sigma_h^2+\sigma_q^2)}}\right]$$
$$= p(2)$$

From Table A.1 this gives:

$$R = 0.9772$$

The cases so far considered have been concerned with only a single requirement function, $f_q(x)$. More generally, as previously discussed in Chapter 3, the requirement function is in the form of a lower and upper limit between which the achieved performance is required to fall. Suppose that the lower limit of required performance can be expressed by the density function $f_{ql}(x)$ and the corresponding upper limit by $f_{qu}(x)$, then the general characteristics of the situation may be represented as shown in Figure 10.3.

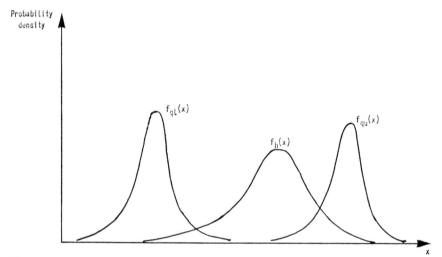

Figure 10.3 One-dimensional lower requirement, achievement and upper requirement characteristics

The reliability is now the probability that the achieved performance value is both greater than the lower required value and less than the upper required value. The corresponding integral equation is therefore:

$$R = \int_0^\infty f_h(x) \left[\int_0^x f_{ql}(y) \, dy \right] \left[\int_x^\infty f_{qu}(y) \, dy \right] dx \qquad (10.16)$$

or:

$$R = \int_0^\infty f_h(x) p_{ql}(x) [1 - p_{qu}(x)] \, dx \qquad (10.17)$$

or:

$$R = \int_0^\infty p_{ql}(x) [1 - p_{qu}(x)] \, dp_h(x) \qquad (10.18)$$

RELIABILITY CONSIDERATIONS FOR SYSTEMS

and for the case where both the lower and upper requirement distributions are single valued at x_{ql} and x_{qu} respectively:

$$R = \int_{x_{ql}}^{x_{qu}} f_h(x)\,dx$$
$$= P_h(x_{qu}) - P_h(x_{ql}) \qquad (10.19)$$

Example 10.3

An electronic amplifier has a distribution of times to failure which is of an exponential form with a mean value of 4000 h. The lower and upper required limits on the time to failure are randomly variable and also follow exponential distributions with means of 2500 h and 10,000 h respectively. What is the reliability of the amplifier in meeting the requirement in this respect?

The parameters of the three exponential distributions are:

$$\lambda_h = 4000 \text{ h} \quad \text{or} \quad \theta_h = 250 \times 10^{-6} \ f/h$$
$$\lambda_{ql} = 2500 \text{ h} \quad \text{or} \quad \theta_{ql} = 400 \times 10^{-6} \ f/h$$
$$\lambda_{qu} = 10{,}000 \text{ h} \quad \text{or} \quad \theta_{qu} = 100 \times 10^{-6} \ f/h$$

Substituting the appropriate exponential expressions in Equation (10.17):

$$R = \int_0^\infty [1 - \exp(-\theta_{ql} x)] \exp(-\theta_{qu} x)\, \theta_h \exp(-\theta_h x)\, dx$$

$$= \frac{\theta_h\, \theta_{ql}}{(\theta_h + \theta_{qu})(\theta_h + \theta_{qu} + \theta_{ql})}$$

whence, for the example values:

$$R = \tfrac{8}{21} = 0.381$$

If the density functions in Equation (10.16) are each of the normal form, then an approximate solution may be found for R by assuming that the variances of f_{ql} and f_{qu} are relatively small. Suppose that f_{ql} is vanishingly small for all values of x greater than μ_h, then:

$$P_{ql}(x) = P_{ql}(x) \quad \text{for} \quad -\infty \leqslant x \leqslant \mu_h \qquad (10.20)$$

and:

$$P_{ql}(x) = 1 \quad \text{for} \quad \mu_h \leqslant x \leqslant \infty \qquad (10.21)$$

In a similar way, assume that f_{qu} is vanishingly small for all x less than μ_h, then:

$$P_{qu}(x) = 0 \quad \text{for} \quad -\infty \leqslant x \leqslant \mu_h \qquad (10.22)$$
$$P_{qu}(x) = P_{qu}(x) \quad \text{for} \quad \mu_h \leqslant x \leqslant \infty \qquad (10.23)$$

If these particular conditions are inserted in Equation (10.17), then:

$$R = \int_{-\infty}^{\mu_h} f_h(x) p_{ql}(x)\, dx + \int_{\mu_h}^{\infty} f_h(x) [1 - p_{qu}(x)]\, dx$$

$$= \int_{-\infty}^{\infty} f_h(x) p_{ql}(x)\, dx - \int_{-\infty}^{\infty} f_h(x) p_{qu}(x)\, dx \quad (10.24)$$

which, from Equation (10.15), yields:

$$R = p\left[\frac{\mu_h - \mu_{ql}}{\sqrt{(\sigma_h^2 + \sigma_{ql}^2)}}\right] - p\left[\frac{\mu_h - \mu_{qu}}{\sqrt{(\sigma_h^2 + \sigma_{qu}^2)}}\right] \quad (10.25)$$

or:

$$R = p\left[\frac{\mu_h - \mu_{ql}}{\sqrt{(\sigma_h^2 + \sigma_{ql}^2)}}\right] + p\left[\frac{\mu_{qu} - \mu_h}{\sqrt{(\sigma_h^2 + \sigma_{qu}^2)}}\right] - 1 \quad (10.26)$$

Multiple correlations

In many systems there is more than one performance parameter of interest. So, for instance, an aircraft may be required to achieve a certain minimum speed, fly at a certain minimum height and carry a certain minimum payload. Or, again, a process plant may be required to convert a range of input products $a_1, a_2, a_3, \ldots$ into a range of output products $b_1, b_2, b_3, \ldots$. As was shown in Chapter 6, one method of expressing the multiple conversion characteristics of a system is by an expression of the form:

$$\{G_i(s)\} = [\Phi_{ij}(s)]\{F_j(s)\} \quad (10.27)$$

where $\{G_i(s)\}$ is the column matrix of the functions $G_1(s), G_2(s)$, etc., $G_i(s)$ is the Laplace transform of the ith output function $g_i(x)$, $\{F_j(s)\}$ is the column matrix of $F_j(s)$ which in its turn is the Laplace transform of the jth input function $f_j(x)$, $[\Phi_{ij}(s)]$ is the transfer matrix made up of elements $\Phi_{ij}(s)$ and each $\Phi_{ij}(s)$ is the transfer function relating the ith output function to the jth input function.

For a system with two output functions $g_1(x)$ and $g_2(x)$ related to two input functions $f_1(x)$ and $f_2(x)$, Equation (10.27) becomes:

$$\begin{bmatrix} G_1(s) \\ G_2(s) \end{bmatrix} = \begin{bmatrix} \Phi_{11}(s) & \Phi_{12}(s) \\ \Phi_{21}(s) & \Phi_{22}(s) \end{bmatrix} \begin{bmatrix} F_1(s) \\ F_2(s) \end{bmatrix} \quad (10.28)$$

which was the form originally derived in Equation (6.148).

A particular case of interest is where the matrix $[\Phi_{ij}(s)]$ is diagonal such that a particular output function is related only to a particular input function

RELIABILITY CONSIDERATIONS FOR SYSTEMS   395

and no other. This was shown, in Equation (6.149), to lead to the relationship:

$$\begin{bmatrix} G_1(s) \\ G_2(s) \end{bmatrix} = \begin{bmatrix} \Phi_{11}(s) & 0 \\ 0 & \Phi_{22}(s) \end{bmatrix} \begin{bmatrix} F_1(s) \\ F_2(s) \end{bmatrix} \quad (10.29)$$

for the two-parameter case.

Equation (10.29) is the matrix expression for two entirely independent equations:

$$G_1(s) = \Phi_{11}(s) F_1(s) \quad (10.30)$$

and:

$$G_2(s) = \Phi_{22}(s) F_2(s) \quad (10.31)$$

For an n-parameter system of the same type, there will be n independent equations of the form of Equations (10.30) and (10.31). The corresponding n independent transfer functions are $\Phi_{ij}(s)$ where $i = j$. For this case, each transfer function may be treated separately when carrying out the appropriate reliability analysis. This means that the correlation between the required values for the coefficients of any $\Phi_{ij}(s)$ and the achieved values can be carried out using the correlation process previously described. If each $\Phi_{ij}(s)$, or more simply $\Phi_i(s)$, has a single variable coefficient a_i, then this will lead to a single reliability value R_i which correlates the achieved value of the ith coefficient with its required value. Because of the assumed independence between the functions, the R_i's are independent of each other and hence the overall reliability becomes simply:

$$R = \prod_{i=1}^{n} R_i \quad (10.32)$$

Example 10.4

An alarm system associated with a process plant is required to perform two simple and independent functions. When one of the process variables exceeds a predetermined value, the alarm system is required to produce an audible tone with an equivalent power output of not less than 4 W and not more than 6 W. At the same time, the system is required to light a lamp within 5.0 s of the process variable limit being exceeded. The installed alarm system converts the input signal from the process variable limit switch into an alarm tone with a transfer function of k_1 W per switch operation. The corresponding transfer function for lighting the lamp is of the form:

$$\frac{k_2}{1 + \tau_2 s}$$

The value of k_1 is found to be randomly distributed according to a normal distribution with a mean value of 5 W per operation and a standard deviation of 0.5 W. The conversion factor, k_2, for the lamp system is invariable but the time constant, τ_2, of the simple lag function follows a Weibull distribution with parameters $\beta = 2.0$ and $\lambda = 2.5$ s. The lamp lights after a time delay equivalent to the time constant. All other aspects of the alarm system are assumed perfect. What is the reliability of the system in sounding the alarm and lighting the lamp to the requirements stated for each operation of the process variable limit switch?

For the audible alarm system, the reliability from Equation (10.19) is:

$$R_1 = \int_4^6 f_h(x)\,dx$$

where $f_h(x)$ is a normal distribution of mean value 5 and standard deviation 0.5. The integral, therefore, represents a spread of two standard deviations on either side of the mean and from Table A.1 this is seen to give:

$$R_1 = 2 \times 0.4772$$
$$= 0.9544$$

For the light system, again from Equation (10.19):

$$R_2 = p_h(x_{qu}) - p_h(x_{ql})$$

where here $x_{qu} = 5$ s and $x_{ql} = 0$ s and $p_h(x)$ is the cumulative Weibull distribution function of the form:

$$p_h(x) = 1 - \exp[-(x/\lambda)^\beta]$$

which for the example values of λ and β becomes:

$$p_h(x) = 1 - \exp[-(x/2.5)^2]$$

and putting $x = 5$ gives:

$$R_2 = p_h(5) = 1 - \exp(-2^2)$$
$$= 0.9817$$

The overall alarm system reliability, from Equation (10.32), is then:

$$R = R_1 R_2$$
$$= 0.9544 \times 0.9817$$
$$= 0.937$$

The system characteristic function is:

$$\begin{bmatrix} G_1(s) \\ G_2(s) \end{bmatrix} = \begin{bmatrix} k_1 & 0 \\ 0 & k_2/(1+\tau_2 s) \end{bmatrix} \begin{bmatrix} F(s) \\ F(s) \end{bmatrix}$$

RELIABILITY CONSIDERATIONS FOR SYSTEMS   397

and if $F(s)$ is assumed to be the transform of an input switch step function, then this leads, as would be expected, to:

$$g_1(x) = k_1$$
$$g_2(x) = k_2[1 - \exp(-t/\tau_2)]$$

where $g_1(x)$ is the output audible alarm and $g_2(x)$ is the output light.

The method illustrated in the preceding example can obviously be applied to a system with any number of independent parameters as it is just a simple extension of the concepts for a single-parameter system. Equation (10.32) does, however, assume that all parameters are required to function within the appropriate limits for the correct functioning of the overall system. This is not always the case and a redundant system, for example, may only require one of a selected number of parameters to function correctly for overall system success. A completely independent redundant system would lead to an overall reliability relationship of the form:

$$R = 1 - \prod_{i=1}^{n} \bar{R}_i \qquad (10.33)$$

For partly redundant systems other relationships can be derived from the probabilistic theory originally discussed in Chapter 1 and enlarged upon in Chapter 8. This aspect will, however, be treated further in the subsequent chapters.

Dependent correlations

The discussion in the previous section and the validity of equations such as Equations (10.32) and (10.33) depend upon the system parameters or the coefficients of all transfer functions being independent of each other. Very often, however, the coefficients in any one transfer function or the coefficients in different transfer functions may be cross related or dependent upon each other in some way. If such a dependence exists then the reliability factors in the series R_i cannot be directly combined by a simple multiplication process. For instance, suppose that two reliability factors for a system are given by the following types of relationship:

$$R_1 = p_a p_b \qquad (10.34)$$
$$R_2 = p_a p_c \qquad (10.35)$$

where p_a, p_b, p_c are independent probabilities but the p_a appearing in Equation (10.34) is identical with the p_a appearing in Equation (10.35). Applying the laws for the combination of independent and dependent probabilities discussed in Chapter 1 it is seen that the overall reliability, R, is:

$$R = R_1 R_2 = p_a p_b p_c \qquad (10.36)$$

and it should be noted that:

$$R \neq p_a^2 p_b p_c \tag{10.37}$$

Hence, if R_1 and R_2 were evaluated separately, the overall reliability R cannot be obtained from the product of R_1 and R_2 because of the mutually dependent term, p_a, which exists in both original relationships. The reliability, R, therefore needs to be evaluated directly from p_a, p_b and p_c and not via the individual reliabilities R_1 and R_2.

Consider two parameter variates z_1 and z_2 each of which are given by the sum of two other variates such that:

$$z_1 = x_1 + y_1 \tag{10.38}$$

$$z_2 = x_2 + y_2 \tag{10.39}$$

Suppose that the system requirement is such that parameter z_1 must be less than z_{1u} and parameter z_2 must be less than z_{2u}. Then from the work done in this chapter and in Chapter 8:

$$R_1 = p_{z_1}(z_{1u}) = \int_0^{z_{1u}} f_{x_1}(x) p_{y_1}(z_{1u} - x) \, dx \tag{10.40}$$

$$R_2 = p_{z_2}(z_{2u}) = \int_0^{z_{2u}} f_{x_2}(x) p_{y_2}(z_{2u} - x) \, dx \tag{10.41}$$

and:

$$R = \left\{ \int_0^{z_{1u}} f_{x_1}(x) p_{y_1}(z_{1u} - x) \, dx \right\} \left\{ \int_0^{z_{2u}} f_{x_2}(x) p_{y_2}(z_{2u} - x) \, dx \right\} \tag{10.42}$$

Provided x_1, y_1, x_2 and y_2 are all independent, Equation (10.42) is valid and may be evaluated by the methods previously described. Suppose now that Equation (10.39) is of the form:

$$z_2 = x_1 + y_2 \tag{10.43}$$

where the x_1 in Equation (10.38) is identical with the x_1 in Equation (10.43), then Equation (10.42) is no longer valid since:

$$f_{x_1}(x) \equiv f_{x_2}(x) \tag{10.44}$$

and the simple multiplication employed in Equation (10.42) cannot be carried out.

The overall requirement needs to be related directly to both Equations (10.38) and (10.43) taken together, whence it is seen that:

$$R = p_{z_1}(z_{1u}) p_{z_2}(z_{2u}) = \int_0^z f_{x_1}(x) p_{y_1}(z_{1u} - x) p_{y_2}(z_{2u} - x) \, dx \tag{10.45}$$

RELIABILITY CONSIDERATIONS FOR SYSTEMS

where:

$$z = z_{1u} \quad \text{if} \quad z_{1u} \leqslant z_{2u} \tag{10.46}$$

$$z = z_{2u} \quad \text{if} \quad z_{2u} \leqslant z_{1u} \tag{10.47}$$

Equation (10.45) could also have been derived from Equation (10.42) by applying the rules for the products of independent and dependent probabilities.

By way of illustration, consider the case where x_1, y_1 and y_2 are each rectangularly distributed with the following density functions:

$$f_{x_1}(x) = \tfrac{1}{2}[U(x-2) - U(x-4)] \tag{10.48}$$

$$f_{y_1}(x) = \tfrac{1}{2}[U(x-6) - U(x-8)] \tag{10.49}$$

$$f_{y_2}(x) = \tfrac{1}{2}[U(x-10) - U(x-12)] \tag{10.50}$$

and where the requirement is such that:

$$z_{1u} = 10 \tag{10.51}$$

$$z_{2u} = 14 \tag{10.52}$$

then Equation (10.45) becomes:

$$R = \int_2^4 \frac{1}{2}\left[\frac{10-x}{2} - 3\right]\left[\frac{14-x}{2} - 5\right]dx \tag{10.53}$$

$$= \tfrac{1}{3}$$

If R_1 and R_2 were calculated separately, for the same conditions, from Equations (10.40) and (10.41), then it is seen that:

$$R_1 = \tfrac{1}{2} \tag{10.54}$$

and:

$$R_2 = \tfrac{1}{2} \tag{10.55}$$

whence:

$$R_1 R_2 = \tfrac{1}{4} \tag{10.56}$$

Although Equations (10.54) and (10.55) are each valid in themselves, Equation (10.56) is not valid due to the presence of the dependent function x_1. Equation (10.56) gives the erroneous value of $\tfrac{1}{4}$ compared with the correct value of $\tfrac{1}{3}$ derived in Equation (10.53).

Equations of the type of Equation (10.45) may be readily solved for certain analytical functions such as the rectangular distributions just described. For other functions, a simple analytical solution may not be possible. In cases of this sort there are some alternative methods of approach which may provide an adequate solution.

In the first case it may be possible to express the requirement for Equations (10.38) and (10.43) in terms of x_1, y_1 and y_2 instead of in terms of z_1 and z_2. In this example, the overall reliability could then be readily derived from three independent functions instead of from two dependent functions.

In some instances, it may transpire that the dependent function, such as x_1, has relatively small variations or produces relatively small variations compared with the other functions. If this is so, then x_1 may be adequately treated as a constant represented by its mean value which leads to z_1 and z_2 becoming independent functions.

Where the above type of approaches are not valid, then the integral of Equation (10.45) may be solved by a step-by-step process. For instance, the integral may be written in the approximate form:

$$R = \sum_{j=1}^{n} \{p_{x_1}[j\delta x] - p_{x_1}[(j-1)\delta x]\} p_{y_1}[z_{1u} - (j-\tfrac{1}{2})\delta x] p_{y_2}[z_{2u} - (j-\tfrac{1}{2})\delta x] \quad (10.57)$$

where:

$$n = \frac{z}{\delta x} \quad (10.58)$$

Obviously, the smaller the choice of δx, or the larger the value of n, the more accurately will Equation (10.57) represent Equation (10.45).

If x_1 has restrictive upper and lower limits of x_{1u} and x_{1l} respectively as in the case of the rectangular distributions just discussed, then Equation (10.57) becomes:

$$R = \sum_{j=1}^{n} \{p_{x_1}[x_{1l} + j\delta x] - p_{x_1}[x_{1l} + (j-1)\delta x]\}$$
$$\times p_{y_1}[z_{1u} - (x_{1l} + j - \tfrac{1}{2})\delta x] p_{y_2}[z_{2u} - (x_{1l} + j - \tfrac{1}{2})\delta x] \quad (10.59)$$

where:

$$n = \frac{x_{1u} - x_{1l}}{\delta x} \quad (10.60)$$

If, in the case of the rectangular distribution example, $f_{x_1}(x)$ is divided into just four equal parts, that is $\delta x = \tfrac{1}{2}$, then:

$$R = \{p_{x_1}(2\tfrac{1}{2}) - p_{x_1}(2)\} p_{y_1}(10 - 2\tfrac{1}{4}) p_{y_2}(14 - 2\tfrac{1}{4})$$
$$+ \{p_{x_1}(3) - p_{x_1}(2\tfrac{1}{2})\} p_{y_1}(10 - 2\tfrac{3}{4}) p_{y_2}(14 - 2\tfrac{3}{4})$$
$$+ \{p_{x_1}(3\tfrac{1}{2}) - p_{x_1}(3)\} p_{y_1}(10 - 3\tfrac{1}{4}) p_{y_2}(14 - 3\tfrac{1}{4})$$
$$+ \{p_{x_1}(4) - p_{x_1}(3\tfrac{1}{2})\} p_{y_1}(10 - 3\tfrac{3}{4}) p_{y_2}(14 - 3\tfrac{3}{4})$$
$$= \tfrac{1}{4}(\tfrac{7}{8} \cdot \tfrac{7}{8}) + \tfrac{1}{4}(\tfrac{5}{8} \cdot \tfrac{5}{8}) + \tfrac{1}{4}(\tfrac{3}{8} \cdot \tfrac{3}{8}) + \tfrac{1}{4}(\tfrac{1}{8} \cdot \tfrac{1}{8})$$
$$= \tfrac{21}{64} \quad (10.61)$$

which may be compared with the exact answer of $\tfrac{1}{3}$ derived in Equation (10.53).

RELIABILITY CONSIDERATIONS FOR SYSTEMS

Example 10.5

The circuit depicted in Figure 10.4 has three components R_1, R_2 and C each of which are subject to random variations according to a normal distribution. The relevant parameters of these distributions are:

$$\mu_{R1} = 10 \text{ M}\Omega, \quad \sigma_{R1} = 1 \text{ M}\Omega$$
$$\mu_{R2} = 20 \text{ M}\Omega, \quad \sigma_{R2} = 2 \text{ M}\Omega$$
$$\mu_C = 2 \text{ }\mu\text{F}, \quad \sigma_C = 0.2 \text{ }\mu\text{F}$$

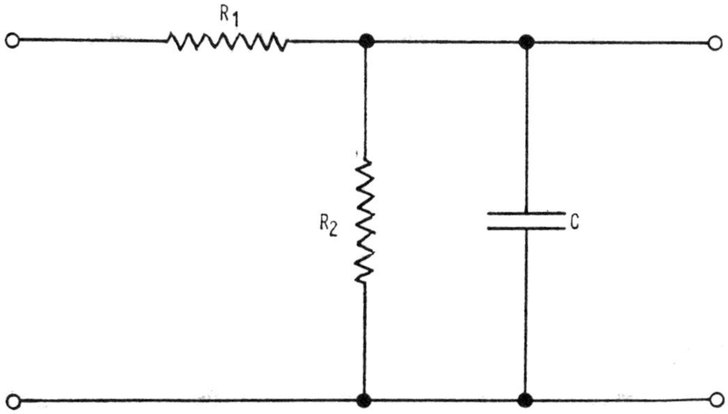

Figure 10.4 Example circuit with a dependent parameter

A requirement exists for the time constant, τ, of the circuit to lie between the limits of 18 s and 22 s and, at the same time, for the attenuation, a, to be between the limits 0.6 and 0.73. What is the reliability of the circuit against this requirement?

It will be readily seen that the transfer function for the circuit is:

$$\Phi(s) = \frac{1}{(1/a) + s\tau}$$

where the time constant, τ, is:

$$\tau = R_1 C$$

and the attenuation, a, is:

$$a = \frac{R_2}{R_1 + R_2}$$

Both τ and a, therefore, are dependent upon a common function R_1 and hence the overall reliability cannot be obtained from the reliabilities of τ and a derived separately. An approach can, however, be adopted using the approximate relationship of Equation (10.57).

For any particular R_1, C must lie between the limits:

$$\frac{18}{R_1} \leqslant C \leqslant \frac{22}{R_1}$$

and, for the same R_1, R_2 must lie between the limits:

$$\frac{0.6}{1-0.6} R_1 \leqslant R_2 \leqslant \frac{0.7\dot{3}}{1-0.7\dot{3}} R_1$$

or:

$$1\tfrac{1}{2} R_1 \leqslant R_2 \leqslant 2\tfrac{3}{4} R_1$$

Consider an incremental interval of R_1 equivalent to one standard deviation of R_1 and carry out a summation from minus three standard deviations of R_1 to plus three standard deviations of R_1, then:

$$R = \{p_{R_1}(8) - p_{R_1}(7)\}\{p_C(\tfrac{22}{7\frac{1}{2}}) - p_C(\tfrac{18}{7\frac{1}{2}})\}\{p_{R_2}(2\tfrac{3}{4}.7\tfrac{1}{2}) - p_{R_2}(1\tfrac{1}{2}.7\tfrac{1}{2})\}$$
$$+ \ldots +$$
$$\{p_{R_1}(13) - p_{R_1}(12)\}\{p_C(\tfrac{22}{12\frac{1}{2}}) - p_C(\tfrac{18}{12\frac{1}{2}})\}\{p_{R_2}(2\tfrac{3}{4}.12\tfrac{1}{2}) - p_{R_2}(1\tfrac{1}{2}.12\tfrac{1}{2})\}$$

Then, working out the various normal probabilities from Table A.1 with the example values given yields:

$$R = 0.021 \times 0.023 \times 0.622$$
$$+ 0.136 \times 0.277 \times 0.954$$
$$+ 0.341 \times 0.643 \times 0.997$$
$$+ 0.341 \times 0.607 \times 0.983$$
$$+ 0.136 \times 0.317 \times 0.915$$
$$+ 0.021 \times 0.112 \times 0.734$$
$$= 0.0003 + 0.0359 + 0.2190 + 0.2036 + 0.0394 + 0.0018$$
$$= 0.5000$$

Correlations associated with discrete states

The complete range of variations associated with any one performance parameter of a system may be represented by a single, but perhaps complex, statistical distribution. If this is possible, then the reliability calculations follow the methods of correlation just discussed. Very often, however, the performance parameters of a technological system are subject to discrete or 'catastrophic' changes in value. This concept was fully described in Chapter 5 under the heading of 'Combinations of continuous and discrete distributions'. The diagrammatic representation of this approach was also given in Figure 5.4 to which the reader is again referred.

RELIABILITY CONSIDERATIONS FOR SYSTEMS

Suppose, in connexion with Figure 5.4, that the probability of the performance parameter, x, being in the working state is p and also that $f_h(x)$ applies to the distribution of the variation of x in this state. Then the reliability of this situation as represented by an Equation of the type of Equation (10.17) is weighted by the probability p, that is:

$$R = p \int_0^\infty f_h(x)\, p_{ql}(x)\, [1 - p_{qu}(x)]\, dx \qquad (10.62)$$

it being assumed that $f_h(x)$, $f_{ql}(x)$ and $f_{qu}(x)$ do not overlap the value of x corresponding to the failed state.

For fixed requirement limits, the corresponding value of R obtained by a similar weighting of Equation (10.19) becomes:

$$R = p[p_h(x_{qu}) - p_h(x_{ql})] \qquad (10.63)$$

and the complementary unreliability, $\bar{R}$, is:

$$\bar{R} = \bar{p} + p[\overline{p_h(x_{qu}) + p_h(x_{ql})}] \qquad (10.64)$$

For the particular case where the variations of the parameter x in the working state always remain within the requirement limits, then:

$$p_h(x_{qu}) = 1 \qquad (10.65)$$
$$p_h(x_{ql}) = 0 \qquad (10.66)$$

and, therefore:

$$R = p \qquad (10.67)$$

and:

$$\bar{R} = \bar{p} \qquad (10.68)$$

which shows, as would be expected, that the reliability of the situation then only depends upon the catastrophic or discrete change of state in the performance parameter x.

An example of the application of the formula given in Equation (10.63) to a simple system problem has already been studied in Example 5.9 in Chapter 5. Other examples dealing with the variational part of Equation (10.63) have been examined earlier in this chapter. The only modification in these examples to the two-state situation is to multiply the final answer by the appropriate value of p.

The concept of discrete changes of state in a system or the component parts of the system can be extremely important and, in some systems, may be the predominant factor. Where they are the predominant factor, the reliability of the system may be adequately represented by the simple relationship given in Equation (10.67). It is pertinent, therefore, to consider the practical ways in which system discrete changes of state may take place and the implications of such changes on the value of p.

Changes of State
Practical implications

The simplest system model exists when only two discrete states are considered. These two states, as indicated in the earlier chapters, can be considered as the 'successful' or 'working state' on the one hand and the 'unsuccessful' or 'failed state' on the other. If no other variations exist, then the system reliability is simply the probability of the system being in the successful state as indicated by Equation (10.67).

This two-state model may, in many cases, be an adequate representation of system performance. The term 'failure' or 'failed state' is often taken to imply a catastrophic degradation in performance and whether or not the system is in this state may well be the main criterion of interest.

Where a system is made up of n component parts each of which can be in either the working state or failed state, then the system has effectively a total number of 2^n possible sub-states. The implications of this sort of situation were adequately discussed in Chapter 5 under the heading of 'Discrete distributions'. Where there are more than two system states, the system reliability is obviously given by the sum of the probabilities for the occurrence of all the states that lie between the requirement limits. Hence, if the discrete probability distribution of states for a parameter r of the system is $f(r)$ then:

$$R = \sum_{j=r_l}^{r_u} f(j) \qquad (10.69)$$

where r_l is the parameter state value immediately above the lower requirement limit and r_u is the parameter state value immediately below the upper requirement limit. The corresponding unreliability is:

$$\bar{R} = \sum_{j=0}^{r_l-1} f(j) + \sum_{j=r_u+1}^{n} f(j) \qquad (10.70)$$

where n represents the total number of possible states.

Unless there is some lack of inherent capability as discussed in Chapter 4, most systems can be assumed to be in the working state at the start of their intended operational period. For a two-state system, the first possible change of state is then from the working state into the failed state. This, for some systems, may be the only possible change and, having entered the failed state, it remains there. If this same restriction applies throughout a multi-component system then the possible changes are from all components in the working state, through successive component failures until all components are in the failed state. This type of process may be described as an irreversible change of state.

RELIABILITY CONSIDERATIONS FOR SYSTEMS

The other possibility is that, once a system or its component part has entered the failed state, some repair, replacement or restoration process is carried out to restore the system or its component part to the working state. This may be described as a reversible change of state or a renewal process.

Sometimes a series of random changes of state is known as a 'stochastic' process. However, the word 'stochastic' in its widest meaning implies any random variation of either the continuous or discrete type. Another term which is often used for system changes of state is to describe the process as a 'Markov process'. A Markov process may be defined as a stochastic process such that the conditional probability distribution for the system state at any future instant, given the present system state, is unaffected by any additional knowledge of the past history of the system. For some systems this situation may approximately apply so that the use of the term Markov process to describe the system's changes of state may well be justifiable in these cases.

Whatever the process description it is obviously necessary to be able to evaluate the chances of various changes of state taking place and also the probabilities of the system being in any particular state. The methods of deriving these probabilities are examined in the subsequent sections which deal with irreversible processes and reversible processes.

Irreversible changes of state

With most systems, interest centres around the chance of a particular system parameter failing or changing its state from the working state to the failed state at some point or during some interval in the time domain. If the distribution of times to failure for the parameter x of a system is $f_x(t)$, then the probability that the system has entered the failed state, $\bar{p}$, in the time interval 0 to t is:

$$\bar{p} = \int_0^t f_x(t) \, dt \qquad (10.71)$$

$$= p_x(t)$$

Since, for the moment it is assumed that a change of state into the failed state is irreversible, then Equation (10.71) also represents the probability of the system being in the failed state at time t.

It is useful at this stage to introduce an alternative method of reasoning which will find an application later on when considering reversible processes. For the system to be in the failed state at time $t + \delta t$, it must either have been in the failed state at time t or it must have been in the working state at time t and suffered an instantaneous failure in the time increment δt. The chance of an instantaneous failure occurring in the increment δt under these conditions is $z_x(t)\,\delta t$ from the definition of the hazard function previously discussed in Chapter 7. Hence:

$$\bar{p}(t+\delta t) = \bar{p}(t) + p(t)z_x(t)\,\delta t \tag{10.72}$$

whence, in the limit:

$$\frac{\mathrm{d}\bar{p}(t)}{\mathrm{d}t} + \bar{p}(t)z_x(t) = z_x(t) \tag{10.73}$$

This is a standard first-order differential equation the solution of which is:

$$\bar{p}(t) = 1 + K\exp\left[-\int_0^t z_x(t)\,\mathrm{d}t\right] \tag{10.74}$$

where K is an arbitrary constant. When $t = 0$, $\bar{p}(t) = 0$ and, therefore, $K = -1$. Whence:

$$\bar{p}(t) = 1 - \exp\left[-\int_0^t z_x(t)\,\mathrm{d}t\right]$$

$$= p_x(t) \tag{10.75}$$

as before.

The complete reliability expression for a single parameter, x, of a system which is subject to irreversible catastrophic changes of state is given by substituting the value of p obtained in Equation (10.75) for the p in Equation (10.63), that is:

$$R = \overline{p_x(t)}\,[p_{x_h}(x_{\mathrm{qu}}) - p_{x_h}(x_{\mathrm{ql}})] \tag{10.76}$$

Equation (10.76) is time dependent and, in fact, expresses the reliability obtained at a particular time, t. Very often the main contributor to this time dependency is the term $\overline{p_x(t)}$. The density function f_{x_h}, on which the terms p_{x_h} are based, may describe variations of performance in the time domain but the function itself may well be time independent. The particular case will be considered, therefore, where the variation of R with time depends only on $\overline{p_x(t)}$. Other more general cases can be dealt with in a similar way.

The reliability of a system at a particular time t, as given by Equation (10.76), may well be of interest. However, it is often required to know, in addition, the reliability over a period of time. The period of time of interest could be, for example, an operational period of the system or the complete life of the system. Consider, therefore, a period of time 0 to τ for which the reliability at any time t in this period is given by Equation (10.76). The reliability for the whole period is defined as the proportion of the total time that the system is in the working state. This is the concept of availability, A, which was originally introduced to the reader in connexion with Equation (1.34) in Chapter 1.

RELIABILITY CONSIDERATIONS FOR SYSTEMS   407

In order to evaluate the availability due to changes of state it is obviously of interest to know the lengths of time and the distribution of those lengths of time that the system is in the working or successful state. For the interval 0 to τ, two possible overall conditions can exist as shown in Figure 10.5.

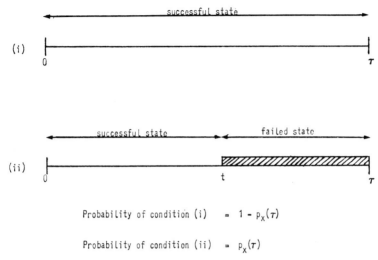

Figure 10.5 Successful times for irreversible changes of state

Condition (i) represents no failure, or no change to the failed state, occurring in the interval. This is a completely successful interval and the probability of its occurrence is the probability that the system is in the successful state at the time τ, hence from Equation (10.75):

$$\text{Probability of condition (i)} = 1 - p_x(\tau) \tag{10.77}$$

The length of successful time for this successful interval, t_{ss}, is obviously of a constant value τ:

$$t_{ss} = \tau \tag{10.78}$$

In the case of condition (ii), the system fails or changes from the successful to the failed state at some time, t, in the interval 0 to τ. This is a partly successful interval and its probability of occurrence is the same as the probability of the system being in the failed state at time τ.

$$\text{Probability of condition (ii)} = p_x(\tau) \tag{10.79}$$

The length of successful time for this interval which contains a change to the failed state, t_{sf}, is:

$$t_{sf} = t \tag{10.80}$$

and obviously t_{sf} is a variable which can take on any value between 0 and τ. Suppose that t_{sf} is distributed according to the density function $f_{t_{sf}}(t)$ then:

$$\int_0^\tau f_{t_{sf}}(t) \, dt = 1 \tag{10.81}$$

and:

$$p_{t_{sf}}(t) = \int_0^t f_{t_{sf}}(t) \, dt \tag{10.82}$$

In order to find an expression for $f_{t_{sf}}(t)$ in terms of $f_x(t)$ it is useful to consider the probability of a successful time being less than or equal to t for both the conditions depicted in Figure 10.5. The total probability of success up to a time t is $[1 - p_x(t)]$ and this is made up of the sum of the probabilities that first there is a completely successful interval, $[1 - p_x(\tau)]$ and secondly that there is a failed interval which contains a successful time greater than or equal to t, $p_x(\tau)[1 - p_{t_{sf}}(t)]$. Hence:

$$[1 - p_x(t)] = [1 - p_x(\tau)] + p_x(\tau)[1 - p_{t_{sf}}(t)] \tag{10.83}$$

whence:

$$p_{t_{sf}}(t) = \frac{p_x(t)}{p_x(\tau)} \tag{10.84}$$

and:

$$f_{t_{sf}}(t) = \frac{f_x(t)}{p_x(\tau)} \tag{10.85}$$

If the length of failed time in a failed interval is t_{ff}, then:

$$t_{ff} = \tau - t_{sf} \tag{10.86}$$

and so:

$$f_{t_{ff}}(t) = \frac{f_x(\tau - t)}{p_x(\tau)} \tag{10.87}$$

Example 10.6

The times to failure of a device are exponentially distributed with a mean failure-rate of θ. In an interval of time from 0 to τ any change of the device into the failed state is irreversible. For the intervals of time which contain a failure, what is the mean time in
(a) the successful state, and
(b) the failed state?
Assume that $\theta\tau \ll 1$.

RELIABILITY CONSIDERATIONS FOR SYSTEMS 409

Here:
$$f_x(t) = \theta e^{-\theta t}$$
$$= \theta \quad \text{since} \quad \theta \tau \ll 1$$

and:
$$p_x(t) = \theta t$$

whence, from Equation (10.85):
$$f_{t_{sf}}(t) = \frac{1}{\tau} \quad \text{for} \quad 0 \leqslant t \leqslant \tau$$

This is a rectangular distribution between the limits 0 and τ and hence the mean value of $f_{t_{sf}}(t)$, T_{sf}, is:

$$T_{sf} = \frac{\tau}{2} \tag{10.88}$$

and, in a similar way:

$$T_{ff} = \frac{\tau}{2} \tag{10.89}$$

The mean values of the successful times or failed times are of general interest since these can be used to derive the mean availabilities, A, and the mean fractional dead times, D. First, however, it is pertinent to consider the distributions of availabilities and fractional dead times.

For the period 0 to τ in which a change to the failed state occurs, the availability for such a period, A_f, is given by:

$$A_f = \frac{t_{sf}}{\tau} \tag{10.90}$$

from the definition of availability given in Chapter 1.

The distribution of t_{sf} was derived in Equation (10.85), hence the distribution of A_f from Relationship No. 1 in Table 8.1 is:

$$f_{A_f}(z) = \tau f_{t_{sf}}(\tau z)$$
$$= \frac{\tau f_x(\tau z)}{p_x(\tau)} \quad \text{for} \quad 0 \leqslant z \leqslant 1 \tag{10.91}$$

and the mean of A_f is given by:

$$\mu_{A_f} = \int_0^1 z f_{A_f}(z) \, dz$$

$$= \frac{\tau}{p_x(\tau)} \int_0^1 z f_x(\tau z) \, dz$$

and making the subsititution, $t = \tau z$:

$$\mu_{A_f} = \frac{1}{\tau p_x(\tau)} \int_0^\tau t f_x(t) \, dt$$

$$= 1 - \frac{1}{\tau p_x(\tau)} \int_0^\tau p_x(t) \, dt \qquad (10.92)$$

The availability for the period 0 to τ which is a successful period is:

$$A_s = 1 \qquad (10.93)$$

Hence, the total availability, A, for all periods is given by:

$$A = A_s[1 - p_x(\tau)] + A_f p_x(\tau) \qquad (10.94)$$

so that the distribution of A becomes:

$$f_A(z) = \frac{1}{p_x(\tau)} f_{A_f}\left\{\frac{z - [1 - p_x(\tau)]}{p_x(\tau)}\right\}$$

$$= \frac{\tau}{[p_x(\tau)]^2} f_x\left[\frac{\tau\{z - [1 - p_x(\tau)]\}}{p_x(\tau)}\right] \quad \text{for} \quad [1 - p_x(\tau)] \leqslant z \leqslant 1 \qquad (10.95)$$

and it is seen that the availability values lie between $[1 - p_x(\tau)]$ and 1.

The mean availability is given by:

$$\mu_A = \frac{\tau}{[p_x(\tau)]^2} \int_{1-p_x(\tau)}^1 z f_x\left[\frac{\tau\{z - [1 - p_x(\tau)]\}}{p_x(\tau)}\right] dz$$

which, on making the substitution:

$$t = \frac{\tau\{z - [1 - p_x(\tau)]\}}{p_x(\tau)}$$

yields:

$$\mu_A = \frac{1}{\tau} \int_0^\tau t f_x(t) \, dt + \frac{[1 - p_x(\tau)]}{p_x(\tau)} \int_0^\tau f_x(t) \, dt$$

$$= 1 - \frac{1}{\tau} \int_0^\tau p_x(t) \, dt \qquad (10.96)$$

The fractional dead time, D_f, for a period 0 to τ in which a failure occurs is:

$$D_f = \frac{t_{ff}}{\tau} \qquad (10.97)$$

from the definition given in Chapter 1. In a similar way to the availability

RELIABILITY CONSIDERATIONS FOR SYSTEMS

derivation, this leads to a distribution of D_f of the form:

$$f_{D_f}(z) = \frac{\tau f_x(\tau - \tau z)}{p_x(\tau)} \quad \text{for} \quad 0 \leqslant z \leqslant 1 \tag{10.98}$$

and a mean of:

$$\mu_{D_f} = \frac{1}{\tau p_x(\tau)} \int_0^\tau p_x(t)\,dt \tag{10.99}$$

The overall fractional dead time for all periods is:

$$D = D_s[1 - p_x(\tau)] + D_f p_x(\tau) \tag{10.100}$$

but, the fractional dead time for a successful period, D_s, is zero, hence:

$$D = D_f p_x(\tau) \tag{10.101}$$

so that:

$$f_D(z) = \frac{\tau}{[p_x(\tau)]^2} f_x\left[\tau - \frac{\tau z}{p_x(\tau)}\right] \quad \text{for} \quad 0 \leqslant z \leqslant p_x(\tau) \tag{10.102}$$

and the mean of D is given by:

$$\mu_D = \frac{\tau}{[p_x(\tau)]^2} \int_0^{p_x(\tau)} z f_x\left[\tau - \frac{\tau z}{p_x(\tau)}\right] dz$$

$$= \frac{1}{\tau} \int_0^\tau p_x(t)\,dt \tag{10.103}$$

Example 10.7

For the device described in Example 10.6, what are the distributions and means of the overall availability and fractional dead time?
The distribution functions for the times to failure of the device are:

$$f_x(t) = \theta \tag{10.104}$$

$$p_x(t) = \theta t \tag{10.105}$$

Substituting these values in Equation (10.95) gives:

$$f_A(z) = \frac{1}{\theta \tau}$$

so the distribution of availability is a rectangular distribution between the limits of $[1 - \theta \tau]$ and 1. The mean of this distribution is:

$$\mu_A = 1 - \frac{\theta \tau}{2} \tag{10.106}$$

The fractional dead time, from Equation (10.102), is also a rectangular distribution of the form:

$$f_D(z) = \frac{1}{\theta\tau} \qquad (10.107)$$

between the limits 0 and $\theta\tau$, and the mean is:

$$\mu_D = \frac{\theta\tau}{2} \qquad (10.108)$$

If the mean availability, μ_A, as just discussed represents the availability due to catastrophic changes of state then this value of μ_A can be substituted for the $\overline{p_x(t)}$ in Equation (10.76) to give an average reliability over the period 0 to τ of:

$$R = \mu_A [p_{x_h}(x_{qu}) - p_{x_h}(x_{ql})] \qquad (10.109)$$

provided that the term in brackets is time independent.

So far, only a completely irreversible process has been discussed. Most devices or systems, however, are capable of being repaired or restored to the successful state once they have entered the failed state. The model just discussed can, though, still be appropriate over certain specified time intervals. A system may operate for a period 0 to τ in an unmonitored or an unmanned condition where repairs are not possible or faults are not detected. At the end of the interval the system may then be thoroughly examined or checked and all necessary repairs carried out. The cycle of unmonitored or unmanned operation is then repeated for another period of length τ, and so on. Certain aircraft or other transportation systems fall into this type of category where repairs are not possible during active operation. A range of plant standby or protective systems also follow a similar pattern due to the fact that faults may occur over a limited period of time without being detected.

The simplest case for analysis is where each period of operation is of the same length and immediately follows the preceding period. For instance, the second period of operation is over the interval τ to 2τ. In addition, the repair process, which is taken to be instantaneous from the above assumption, is also assumed to restore the system to the time zero condition. Under these conditions, the reliability at any time is given by Equation (10.76) except that the value of t is measured from the beginning of the period in which it falls. So, for the jth period:

$$R = \overline{p_x[t-(j-1)\tau]} [p_{x_h}(x_{qu}) - p_{x_h}(x_{ql})] \qquad (10.110)$$

if t is measured from time zero.

Since each period is identical in length and starting conditions, the reliability for a period given by Equation (10.109) applies to each period and, therefore, equally applies to the total length of operation made up of n periods.

RELIABILITY CONSIDERATIONS FOR SYSTEMS

If the repair process, instead of being instantaneous, occupies a fixed and constant period of time τ_r, then the new total availability, A', becomes:

$$A' = \frac{\tau}{\tau+\tau_r} A \qquad (10.111)$$

whence:

$$f_{A'}(z) = \frac{(\tau+\tau_r)}{[p_x(\tau)]^2} f_x\left[\frac{z(\tau+\tau_r)-\tau[1-p_x(\tau)]}{p_x(\tau)}\right] \qquad (10.112)$$

for:

$$\frac{\tau[1-p_x(\tau)]}{p_x(\tau)} \leqslant z \leqslant \frac{\tau}{\tau+\tau_r}$$

and:

$$\mu_{A'} = \frac{\tau}{\tau+\tau_r} - \frac{1}{\tau+\tau_r}\int_0^\tau p_x(t)\,dt \qquad (10.113)$$

or:

$$\mu_{A'} = \frac{\tau}{\tau+\tau_r}\mu_A \qquad (10.114)$$

Similarly:

$$f_{D'}(z) = \frac{(\tau+\tau_r)}{[p_x(\tau)]^2} f_x\left[\frac{\tau p_x(\tau)+\tau_r-z(\tau+\tau_r)}{p_x(\tau)}\right] \qquad (10.115)$$

$$\text{for } \frac{\tau_r}{\tau+\tau_r} \leqslant z \leqslant \frac{\tau}{\tau+\tau_r}p_x(\tau)+\frac{\tau_r}{\tau+\tau_r}$$

and:

$$\mu_{D'} = \frac{\tau_r}{\tau+\tau_r} + \frac{\tau}{\tau+\tau_r}\mu_D \qquad (10.116)$$

With the approximate forms of the exponential distribution used in Examples 10.6 and 10.7, Equations (10.114) and (10.116) become:

$$\mu_{A'} = \frac{\tau}{\tau+\tau_r} - \frac{\theta\tau^2}{2(\tau+\tau_r)} \qquad (10.117)$$

$$\mu_{D'} = \frac{\tau_r}{\tau+\tau_r} + \frac{\theta\tau^2}{2(\tau+\tau_r)} \qquad (10.118)$$

If the checking and repair process which takes place at the end of each interval does not restore the system to the time zero condition, then the mean availability for each interval will be generally different. The situation, therefore, needs examining for each interval. A convenient assumption to make is that the checking and repair process does not alter the inherent failure characteristics of the system. This means that what takes place in the jth interval is unaffected by any events which have taken place up to the time $(j-1)\tau$. This is equivalent to saying that the changes to a failed state which can take place at any time follow a Markov process.

Consider, therefore, the jth consecutive interval which lies between the times $(j-1)\tau$ and $j\tau$. Equation (10.74) is still valid, but the constant, K, in this equation is no longer unity. The initial conditions for the interval are that when $t = (j-1)\tau$, $\bar{p}(t) = 0$. Hence:

$$K = \frac{-1}{\exp[-\int_0^{(j-1)\tau} z_x(t)\,dt]}$$

$$= \frac{-1}{1-p_x[(j-1)\tau]} \tag{10.119}$$

Therefore,

$$\bar{p}(t) = \frac{p_x(t) - p_x[(j-1)\tau]}{1 - p_x[(j-1)\tau]} \tag{10.120}$$

and:

$$p(t) = \frac{1 - p_x(t)}{1 - p_x[(j-1)\tau]} \tag{10.121}$$

Notice that Equations (10.120) and (10.121) could have been obtained from the conditional probability relationship of Equation (7.85), since the population of the events is being considered where times less than $(j-1)\tau$ are excluded from the original distribution.

The reliability at any time in the jth interval is:

$$R = \frac{\overline{p_x(t)}}{1 - p_x[(j-1)\tau]} \{p_{x_h}(x_{qu}) - p_{x_h}(x_{ql})\} \tag{10.122}$$

which may be compared with Equation (10.76).

To establish the distribution of successful times in the jth interval, Equation (10.83) needs to be modified to contain the appropriate conditional probabilities, whence:

$$\frac{1-p_x(t)}{1-p_x[(j-1)\tau]} = \frac{1-p_x(j\tau)}{\{1-p_x[(j-1)\tau]\}} + \frac{p_x(j\tau)-p_x[(j-1)\tau]}{\{1-p_x[(j-1)\tau]\}}\{1-p_{t_{st}}[t-(j-1)\tau]\} \tag{10.123}$$

which gives:

$$f_{t_{st}}(t) = \frac{f_x[t+(j-1)\tau]}{p_x(j\tau)-p_x[(j-1)\tau]} \tag{10.124}$$

RELIABILITY CONSIDERATIONS FOR SYSTEMS 415

Then, following similar steps to those taken between Equations (10.90) and (10.96), it is found that:

$$f_A(z) = \frac{\tau\{1-p_x[(j-1)\tau]\}}{\{p_x(j\tau)-p_x[(j-1)\tau]\}^2}$$

$$\times f_x\left(\frac{\tau[\![z\{1-p_x[(j-1)\tau]\}-[1-p_x(j\tau)]]\!]}{\{p_x(j\tau)-p_x[(j-1)\tau]\}}+(j-1)\tau\right)$$

$$\text{for } \frac{1-p_x(j\tau)}{1-p_x[(j-1)\tau]} \leqslant z \leqslant 1 \qquad (10.125)$$

and:

$$\mu_A = \frac{1}{1-p_x[(j-1)\tau]}\left[1-\frac{1}{\tau}\int_{(j-1)\tau}^{j\tau} p_x(t)\,dt\right] \qquad (10.126)$$

In a similar way, the fractional dead time for the jth period is given by:

$$f_D(z) = \frac{\tau\{1-p_x[(j-1)\tau]\}}{\{p_x(j\tau)-p_x[(j-1)\tau]\}^2} f_x\left(j\tau - \frac{\tau z\{1-p_x[(j-1)\tau]\}}{p_x(j\tau)-p_x[(j-1)\tau]}\right)$$

$$\text{for } 0 \leqslant z \leqslant \frac{p_x(j\tau)-p_x[(j-1)\tau]}{1-p_x[(j-1)\tau]} \qquad (10.127)$$

and:

$$\mu_D = \frac{1}{1-p_x[(j-1)\tau]}\left\{\frac{1}{\tau}\int_{(j-1)\tau}^{j\tau} p_x(t)\,dt - p_x[(j-1)\tau]\right\} \qquad (10.128)$$

For the exponential distribution, it can be readily verified that Equation (10.126) yields the same result as Equation (10.96) and that Equation (10.128) is the same as Equation (10.103). This means that with an exponential distribution of times to failure, the availabilities and fractional dead times are independent of the period chosen. This is a result which would have been expected from the characteristics of the exponential distribution previously discussed in Chapter 7. For any other distribution, μ_A depends upon the interval being examined. The overall mean availability for n consecutive periods of length τ is:

$$\mu_A = \frac{1}{n}\sum_{j=1}^{n} \mu_{A_j} \qquad (10.129)$$

where μ_{A_j} is given by Equation (10.126).

The probability that the jth interval contains a change into the failed state has been seen to be:

$$\frac{p_x(j\tau)-p_x[(j-1)\tau]}{1-p_x[(j-1)\tau]}$$

Therefore, the mean number of failures, μ_N, in a total time T made up of n consecutive intervals of length τ is:

$$\mu_N = \sum_{j=1}^{n} \frac{p_x(j\tau) - p_x[(j-1)\tau]}{1 - p_x[(j-1)\tau]} \qquad (10.130)$$

If the exponential distribution of times to failure applies or if, for any other distribution, the situation is restored to the time zero condition at the beginning of each interval, then Equation (10.130) simplifies to:

$$\mu_N = np_x(\tau) = \frac{T}{\tau} p_x(\tau) \qquad (10.131)$$

Example 10.8

An unmonitored protective system is run continuously and checked for faults every 1000 h. If a fault is found, it is instantaneously repaired without affecting the failure or age characteristics of the system. The times to system failure follow a special Erlangian distribution where the parameter λ has a value of 1000 h. Any other variations in system performance are insignificant. What is the mean availability of the system for the first, fifth and tenth period? What is the expected number of failures in the overall time from 0 to 10,000 h?

From Chapter 7, the cumulative distribution function for the special Erlangian is:

$$p_x(t) = 1 - \left(1 + \frac{t}{\lambda}\right) e^{-t/\lambda}$$

Substituting this expression in Equation (10.126) and integrating yields:

$$\mu_{A_j} = \frac{\lambda}{\tau[\lambda + (j-1)\tau]} \{[2\lambda + j\tau](1 - e^{-\tau/\lambda}) - \tau\} \qquad (10.132)$$

Since, in this example, τ and λ are numerically equal, then Equation (10.132) reduces to:

$$\mu_{A_j} = 0.6321 + \frac{0.2642}{j}$$

Hence:

$$\mu_{A_1} = 0.8963$$
$$\mu_{A_5} = 0.6849$$
$$\mu_{A_{10}} = 0.6585$$

Notice, that after a long period of time, the mean availability for any period is asymptotic to a value of 0.6321.

From Equation (10.130), the mean number of failures in 10,000 h or the first ten periods is:

$$\mu_N = \sum_{j=1}^{10}\left\{1 - \frac{\lambda+j\tau}{[\lambda+(j-1)\tau]}e^{-\tau/\lambda}\right\} \quad (10.133)$$

$$= \sum_{j=1}^{10}\left(1 - \frac{j+1}{j}\times 0.3679\right)$$

$$= 5.263$$

$$\simeq 5 \text{ failures}$$

Reversible changes of state

In this connexion it is now assumed that a system, or its component parts, are continuously monitored or manned such that any change into the failed state is immediately discovered. Following this discovery, a repair, replacement or renewal process is started in order to restore the system to the successful or working state again. It is obvious, in such a situation, that the times to repair the system may follow a random distribution in a similar way to the system's times to failure.

Let t_f be the time to failure and $f_{t_f}(t)$ the density function for the distribution of these times to failure. Then the probability that no failures have occurred by some time t is:

$$\overline{p_{t_f}(t)} = 1 - \int_0^t f_{t_f}(t)\,dt \quad (10.134)$$

Let t_r be the time to repair a failure and $f_{t_r}(t)$ the corresponding density function, then the probability that a repair has been completed in the time interval 0 to t is:

$$p_{t_r}(t) = \int_0^t f_{t_r}(t)\,dt \quad (10.135)$$

The main interest, though, generally lies in the chance of the system being in the successful or failed state at any particular time or over some time period. As in the previous section, the method of analysis for obtaining these probabilities depends upon the assumptions with regard to the effect of the repair process on the system's age. Suppose, in the first instance, that both the rates of failure and the rates of repair are functions of the passage of real time, t, from the initial time zero condition. This means that the instantaneous changes from success to failure and vice versa may be represented by the appropriate hazard functions for failure and repair, $z_{t_f}(t)$ and $z_{t_r}(t)$ respectively. The probability that the system is in the successful state at any time $(t+\delta t)$

is then given, in a similar way to Equation (10.72), by:

$$p_s(t+\delta t) = p_s(t)[1-z_{t_r}(t)\,\delta t] + p_f(t)z_{t_f}(t)\,\delta t$$

whence:

$$\frac{dp_s(t)}{dt} + z_{t_r}(t)p_s(t) - z_{t_f}(t)p_f(t) = 0 \qquad (10.136)$$

Similarly, the probability of the system being in the failed state at a time $(t+\delta t)$ is:

$$p_f(t+\delta t) = p_f(t)[1-z_{t_f}(t)\,\delta t] + p_s(t)z_{t_r}(t)\,\delta t$$

or:

$$\frac{dp_f(t)}{dt} + z_{t_f}(t)p_f(t) - z_{t_r}(t)p_s(t) = 0 \qquad (10.137)$$

In addition, since only one of the two possible states of success or failure is possible at any given time:

$$p_s(t) + p_f(t) = 1 \qquad (10.138)$$

From Equations (10.137) and (10.138):

$$\frac{dp_f(t)}{dt} + [z_{t_r}(t) + z_{t_f}(t)]p_f(t) = z_{t_r}(t) \qquad (10.139)$$

This is a standard first-order differential equation, the solution of which is:

$$p_f(t) = K \exp\left\{-\int_0^t [z_{t_r}(t) + z_{t_f}(t)]\,dt\right\}$$
$$+ \exp\left\{-\int_0^t [z_{t_r}(t) + z_{t_f}(t)]\,dt\right\} \int_0^t z_{t_r}(t) \exp\left\{+\int_0^t [z_{t_r}(t) + z_{t_f}(t)]\,dt\right\} dt$$

$$(10.140)$$

When $t = 0$, $p_f(t) = 0$ and hence $K = 0$. Then, writing the hazard functions in terms of their appropriate cumulative distribution functions:

$$p_f(t) = [1-p_{t_r}(t)][1-p_{t_f}(t)] \int_0^t \frac{f_{t_r}(t)\,dt}{[1-p_{t_r}(t)][1-p_{t_f}(t)]^2} \qquad (10.141)$$

and the corresponding $p_s(t)$ may be obtained from Equations (10.138) or (10.136).

For many distribution functions, Equation (10.141) does not lead to a particularly simple solution. One particular case of interest, however, is when the rates of failure and repair are constant. That is, when the distributions for the times to failure and repair are exponential. Let:

$$f_{t_f}(t) = \theta e^{-\theta t} \qquad (10.142)$$

and:

$$f_{l_r}(t) = \frac{1}{\tau} e^{-t/\tau} \qquad (10.143)$$

Then:

$$p_f(t) = e^{-l(\theta+1/\tau)} \int_0^t e^{-t/\tau} \theta e^{+\theta t} \, dt$$

$$= \frac{\theta \tau}{1+\theta \tau} [1 - e^{-l(\theta+1/\tau)}] \qquad (10.144)$$

An alternative set of assumptions that can be made is when, on the completion of each repair process, the system starts again in the 'as new' or the time zero condition. Similarly, at the time of each failure, this time is taken as the time zero condition for the repair process. Under these assumptions the whole process is sometimes known as an 'alternating renewal process' and, in many instances, this may represent a more realistic model of the practical situation.

In order to analyse the alternating renewal process, it is useful to define a conditional state probability of the form $p_{jk}(t)$. This is the probability of the system being in a state k at time t when it was in the state j at time zero. For the present application, j and k can have either of two values denoted by f for the failed state (or state of being under repair) and s for the successful or working state. This means, for instance, that $p_{sf}(t)$ is the probability of being in the failed state at time t when the system was in the successful state at time zero. Being in the successful state at time zero means that the system is starting from new at time zero. Conversely, being in the failed state at time zero means that the repair process is starting from new at time zero.

One condition for $p_{sf}(t)$ to pertain is that there must have been at least one failure in the time interval 0 to t. Suppose that the first of these failures took place at some time u where $0 \leq u \leq t$. The probability of a first failure at u is simply $f_{l_t}(u) \, du$. If the time u is then considered as a starting condition in the failed state, then the probability of being in the failed state at t is $p_{ff}(t-u)$. Hence, the following equality can be expressed.

$$p_{sf}(t) = \int_0^t p_{ff}(t-u) f_{l_t}(u) \, du \qquad (10.145)$$

or in terms of the convolutions discussed in Chapter 6:

$$p_{sf} = p_{ff} * f_{l_t} \qquad (10.146)$$

whence, on taking the Laplace transform:

$$P_{sf}(s) = P_{ff}(s) F_{l_t}(s) \qquad (10.147)$$

The conditional probability $p_{ff}(t)$ can be considered as the sum of two probabilities. The first is the probability that the repair which commenced at time zero is not completed by time t. This probability is simply:

$$[1-p_{t_r}(t)]$$

Secondly, the probability that the first repair is completed at time u, which is $f_{t_r}(u)\,du$, together with the conditional probability, $p_{sf}(t-u)$, that the system starting successful at u is in the failed state at t. This combined probability is given by:

$$\int_0^t p_{sf}(t-u)f_{t_r}(u)\,du$$

Hence:

$$p_{ff}(t) = [1-p_{t_r}(t)] + \int_0^t p_{sf}(t-u)f_{t_r}(u)\,du \qquad (10.148)$$

and taking the Laplace transform of Equation (10.148):

$$P_{ff}(s) = P_{sf}(s)F_{t_r}(s) + \frac{1}{s} - P_{t_r}(s) \qquad (10.149)$$

and multiplying through by the operator s gives:

$$sP_{ff}(s) = sP_{sf}(s)F_{t_r}(s) + 1 - F_{t_r}(s) \qquad (10.150)$$

Combining Equations (10.147) and (10.150) now yields:

$$sP_{sf}(s) = \frac{F_{t_r}(s)[1-F_{t_r}(s)]}{1-F_{t_r}(s)F_{t_r}(s)} \qquad (10.151)$$

But $p_{sf}(t)$ is the same as $p_f(t)$, the probability of being in the failed state at time t, if the condition of system success at time zero is always assumed. Hence, Equation (10.151) expresses $p_f(t)$ in terms of the density functions of failure and repair. A similar process of derivation will yield the expression for the probability of being in the successful state at time t, this is:

$$sP_s(s) = sP_{ss}(s) = \frac{1-F_{t_r}(s)}{1-F_{t_r}(s)F_{t_r}(s)} \qquad (10.152)$$

The Laplace transform of a density function may be expressed in terms of the mean and all the other moments about the origin as shown in Equation (7.31), for example:

$$F_{t_r}(s) = 1 - s\mu_{t_r} + O(s^2) \qquad (10.153)$$

where $O(s^2)$ represents all terms of the order of s^2 and higher. Substituting

RELIABILITY CONSIDERATIONS FOR SYSTEMS

Equation (10.153) in Equation (10.152) yields:

$$sP_s(s) = \frac{\mu_{t_t} + O(s)}{\mu_{t_t} + \mu_{t_r} + O(s)} \qquad (10.154)$$

Then, making use of the limiting relationship given in Table A.3 that:

$$\lim_{t \to \infty} f(t) = \lim_{s \to 0} sF(s) \qquad (10.155)$$

it follows that:

$$\lim_{t \to \infty} p_s(t) = \frac{\mu_{t_t}}{\mu_{t_t} + \mu_{t_r}} \qquad (10.156)$$

Similarly:

$$\lim_{t \to \infty} p_f(t) = \frac{\mu_{t_r}}{\mu_{t_t} + \mu_{t_r}} \qquad (10.157)$$

When $p_s(t)$ becomes a constant independent of time then it is equivalent to the mean availability. Hence, in the limit:

$$\mu_A = \frac{\mu_{t_t}}{\mu_{t_t} + \mu_{t_r}} \qquad (10.158)$$

and, in the same way:

$$\mu_D = \frac{\mu_{t_r}}{\mu_{t_t} + \mu_{t_r}} \qquad (10.159)$$

With certain assumptions regarding the existence of higher-order moments, Equations (10.158) and (10.159) apply for all types of distributions of $f_{t_t}(t)$ and $f_{t_r}(t)$ when t is large enough to make the limit valid. With many practical systems, such a value of t is reached fairly quickly so that the equations for mean availability and mean fractional dead time may be used for nearly the whole time period of the system.

Example 10.9

A system, starting in the working state at time zero, has times to failure which follow an exponential distribution where the parameter θ has a value of 10^{-4} faults per hour. Starting from the instant of each failure, the times to repair the faults are also of the exponential form with a mean value, τ, of 10 h. What is the probability of the system being in the failed state at
(a) 1.0 h,
(b) 10 h,
(c) 100 h, and
(d) ∞ h?

The appropriate density functions are given by Equations (10.142) and (10.143). From Table A.2 the corresponding Laplace transforms are:

$$F_{t_f}(s) = \frac{\theta}{s+\theta} \qquad (10.160)$$

and:

$$F_{t_r}(s) = \frac{1}{1+s\tau} \qquad (10.161)$$

Putting these values in Equation (10.151) yields:

$$P_f(s) = \frac{\theta}{s[s+(\theta+1/\tau)]} \qquad (10.162)$$

whence, from Table A.2:

$$p_f(t) = \frac{\theta\tau}{1+\theta\tau}[1-e^{-t(\theta+1/\tau)}] \qquad (10.163)$$

which for the example values gives approximately:

$$p_f(t) = \frac{1-e^{-t/10}}{1000}$$

Hence:

$$p_f(1) = 0.000095$$

$$p_f(10) = 0.000632$$

$$p_f(100) = 0.000999$$

$$p_f(\infty) = 0.001$$

The first thing to notice from the example is that after only the first 100 h of operation the system probability of being in the failed state is very close to its asymptotic value of 0.001. Since $1/\tau$ is generally very much greater than θ, this means that the asymptotic value is reached, for all practical purposes, after about 10 times the mean time for repair.

The second thing is that Equation (10.163) is identical with Equation (10.144). So for an exponential distribution of failure and repair, the same result is achieved whether a renewal process is assumed or not. In both cases, the constant mean value of the probability of being in the failed state, which is approached, or the mean fractional dead time is:

$$\mu_D = \lim_{t\to\infty} p_f(t) = \frac{\theta\tau}{1+\theta\tau} \qquad (10.164)$$

RELIABILITY CONSIDERATIONS FOR SYSTEMS

and the mean availability is:

$$\mu_A = \lim_{t \to \infty} p_s(t) = \frac{1}{1 + \theta\tau} \tag{10.165}$$

Another case, which is sometimes of interest, is when the repair process is always carried out in a fixed or constant time τ_r. The corresponding 'density function' for this 'distribution' is, of course, single valued and may be represented by the Dirac function:

$$f_{t_r}(t) = \delta(t - \tau_r) \tag{10.166}$$

and the Laplace transform of this, from Table A.2, is:

$$F_{t_r}(s) = e^{-s\tau_r} \tag{10.167}$$

If the times to failure are assumed to be exponentially distributed, then substituting Equations (10.167) and (10.160) in Equation (10.151) gives:

$$P_f(s) = \frac{1}{s} - \frac{1}{(s+\theta) - \theta e^{-s\tau_r}} \tag{10.168}$$

which may be written as:

$$P_f(s) = \frac{1}{s} - \left\{ \frac{1}{(s+\theta)} + \frac{\theta e^{-s\tau_r}}{(s+\theta)^2} + \frac{\theta^2 e^{-2s\tau_r}}{(s+\theta)^3} + \ldots \right\} \tag{10.169}$$

and which, from Table A.2, leads to an inverse transform of the form:

$$p_f(t) = 1 - \sum_{j=0}^{\infty} \frac{\theta^j (t - j\tau_r)^j U(t - j\tau_r) e^{-\theta(t - j\tau_r)}}{j!} \tag{10.170}$$

from which again it may be shown that:

$$\mu_D = \lim_{t \to \infty} p_f(t) = \frac{\theta\tau_r}{1 + \theta\tau_r} \tag{10.171}$$

A final aspect which may be of importance in reversible changes of state is the mean number of failures that are likely to take place in a given length of time. Associated with each failed state is a mean length of time in the successful state, μ_{t_f}, and a mean length of time under repair, μ_{t_r}. So, over a long length of time, T, the mean number of entries into the failed state will be:

$$\mu_N = \frac{T}{\mu_{t_f} + \mu_{t_r}} \tag{10.172}$$

which, with the appropriately defined parameters of the exponential distributions for failure and repair, becomes:

$$\mu_N = \frac{\theta T}{1 + \theta\tau} \tag{10.173}$$

The probability of being in the failed state and the availability over a given length of time has now been discussed for reversible changes of state and, in particular, for the alternating renewal process. In the more general case, where continuous random variations take place within the working state, the reliability at a particular time is given by substituting the value of $p_s(t)$ obtained from equations such as Equation (10.152) for the value of p in Equation (10.63). For example:

$$R = \left\{ L^{-1} \frac{1}{s} \left[\frac{1 - F_{t_f}(s)}{1 - F_{t_f}(s) F_{t_r}(s)} \right] \right\} [p_{x_h}(x_{qu}) - p_{x_h}(x_{ql})] \qquad (10.174)$$

and where the reliability is required over a specified interval of time for which the mean availability of the working state is μ_A as, for instance, given by Equation (10.158), then:

$$R = \mu_A [p_{x_h}(x_{qu}) - p_{x_h}(x_{ql})] \qquad (10.175)$$

Equations (10.174) and (10.175) may then be compared with the equivalent Equations (10.76) and (10.109) which were obtained for the irreversible changes of state.

Reliability Parameters of Interest

The correlation process discussed at the beginning of this chapter and the subsequent discussion on change of state models have shown that there are two main reliability characteristics of interest. These are the reliability, or the probability of H meeting Q, at some particular point in the time domain and the reliability, or probability of H meeting Q, over some specified period of time. The latter may be considered as a mean reliability.

In both cases, it is often convenient to define a two-state model consisting of a working or successful state on the one hand and a failed state on the other. For the working state it is necessary to know the distribution $f_{x_h}(x)$ for the variation of each achieved performance parameter of the system. This is then correlated with the upper and lower required performance limits, x_{qu} and x_{ql}, as shown typically in Equation (10.19). Ideally, the complete form of $f_{x_h}(x)$ is of interest, but some assumed analytical results are possible where only the mean and variance of this distribution can be specified.

For the change into the failed state it is necessary to know the distribution of system times to failure, $f_{t_f}(t)$. Where this distribution is of the exponential form, the parameter of interest is the mean time to failure, λ, sometimes abreviated to m.t.t.f. or the mean failure rate, θ. For a change which is irreversible over some length of time, the additional parameter of interest is the length of the period, τ, between the proof checks on the system which reveal and repair the failed condition.

RELIABILITY CONSIDERATIONS FOR SYSTEMS   425

In the case of reversible changes of state, the required distribution is that of the times to repair or restore the successful state from the failed state, $f_{t_r}(t)$. Once again, if this is of the exponential form, then the parameter of interest is the mean time to repair, τ.

Where exponential distributions do not apply to either the times to failure or the times to repair, then the specification of the distributions will require at least the variance as well as the mean values.

Using the type of parameters just discussed, the appropriate use of the two-state model then yields the probability of the system being in either the successful or working state at any particular time. When required, this may be developed to yield the further derived parameters of availability, A, or fractional dead time, D, with their appropriate distributions and mean values. Such derived probabilities or parameters of interest can then be used to modify the variational reliability of Equation (10.19) to give an overall reliability for the system.

In this process, other reliability parameters of interest may arise. A typical one is the mean number of failures, μ_N, or changes into the failed state, which are likely to occur in a specified period of time.

This chapter, then, has attempted to derive the general model for system reliability analysis and discuss its method of use and manipulation. In the process, specific reliability parameters of interest have arisen which are typical of any technological system. These parameters are not exhaustive, but they nevertheless give the reader an idea of the sort of system characteristic information which is required for a system reliability analysis.

As has been indicated previously in Chapters 5 and 8, the required characteristics may not be obtainable at the system level. This requires a method of synthesis to obtain the appropriate system parameters from a knowledge of the corresponding component part characteristics. Combinational synthesis of this type has already been discussed in Chapter 8 with principal reference to the combinations of continuous distributions. Some further methods are now necessary to deal with the change of state system model and these are described in the next chapter.

Questions

1. The variations in the output voltage of a control system are found to approximate to an exponential distribution with a mean value of 90 mV. The corresponding required minimum output voltage for the system is also exponentially distributed with a mean value of 10 mV. What is the reliability of the situation for

   (a) the distribution of requirement stated, and

   (b) if the requirement for a minimum voltage is single valued at 9.45 mV?

2. A range of engines are found to produce power outputs which follow a Weibull distribution with parameters of $\beta = 4$ and $\lambda = 100$ kW. What is the reliability of any engine meeting a minimum requirement for a power output of
   (a) 80 kW,
   (b) 90 kW, and
   (c) 100 kW?
3. The minimum required daily output from a process plant is for a mean value of 5000 kg of the output product. This requirement follows a normal distribution for the day-to-day variation with a standard deviation of 30 kg. The achieved output from the plant also follows a normal distribution in terms of variation of the product and this distribution has a mean and standard deviation of 5125 kg and 40 kg respectively. What is the reliability that the achievement meets the requirement on any one day?
4. A resistance component is produced by a factory in large quantities. The variation in resistance value produced by the manufacturing process leads to a rectangular distribution between the limits of 8 Ω and 12 Ω. A requirement exists for a particular resistor, selected at random, to have a value lying between a mean lower limit of 8 Ω and a mean upper limit of 12 Ω. If these lower and upper limits are also rectangularly distributed with standard deviations of $1/\sqrt{3}$ Ω, what is the reliability in meeting the requirement?
5. If all the rectangular distributions in Question 4 are replaced by normal distributions with the same means and variances and it is assumed that the requirement distributions do not overlap each other, what is the reliability in meeting the requirement in this case?
6. A shut-down system associated with a process plant is required to shut down the process when either of two process variables move outside certain limits. One of the process variables is temperature which has a random distribution of values over a daily period following a normal distribution with a mean of 250°C and a standard deviation of 10°C. The other process variable is pressure whose variations over the same daily period are rectangularly distributed between the values of 800 kN m^{-2} and 900 kN m^{-2}. The limits, outside which the shut-down system operates, are 220°C and 280°C for temperature and 801 kN m^{-2} and 899 kN m^{-2} for pressure. What is the reliability of the plant remaining in continuous operation for the daily period?
7. When an engine temperature exceeds a certain value, a fire protection system switches off the fuel supply and sprays the engine with foam. The protective sub-system for fuel supply has a mean delay of 4.0 s from

the moment of excess temperature to the cessation of fuel flow. This delay is normally distributed with a standard deviation of 0.5 s. The spray sub-system is also subject to random variations in time delay which are exponentially distributed with a mean of 2.0 s. What is the reliability of the protective system in performing both its functions in a time less than 5.0 s?

8. In the following resistance network:

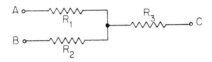

the three resistance elements are subject to random variations in value which follow rectangular distributions. The lower and upper limits of these distributions are 9 Ω and 11 Ω for R_1, 18 Ω and 22 Ω for R_2 and 4 Ω and 6 Ω for R_3. What is the reliability of the network if the resistance between points A and C is required to be less than 16 Ω at the same time as that between B and C is required to be less than 26 Ω?

9. For the same resistance network and variational characteristics of resistors defined in Question 8, it is now found that each resistor has a probability of moving catastrophically into either the complete open-circuit or the complete short-circuit state. The probabilities of being in the open-circuit state are 5×10^{-2} for R_1, 2×10^{-2} for R_2 and 2×10^{-2} for R_3. The corresponding probabilities for the short-circuit states are 10^{-2} for each resistor. What is the reliability of the network if
   (a) the resistance between A and C is required to lie between 0 Ω and 16 Ω at the same time as that between B and C is required to lie between 0 Ω and 26 Ω, and
   (b) the resistance between A and C is required to lie between 4 Ω and 17 Ω at the same time as that between B and C is required to lie between 4 Ω and 28 Ω?

10. The output of a digital device has a mean value of 5 units. Random variations cause integer changes in this output which are found to follow a Poisson distribution. If the required output is required to be greater than 2 units and less than 8 units, what is the reliability of the device?

11. A piece of equipment is to be operated in an unmanned situation for a period of $\tau = 1$ year. The times to failure of the equipment are exponentially distributed with a mean failure rate of $\theta = 0.1$ faults per year. Derive a formula for the distribution of the lengths of successful times for those yearly periods which contain a failure and find the formula and value for the mean of these successful times.

12. For the piece of equipment defined in Question 11, derive the formulae and values for the overall
    (a) distribution of availability,
    (b) mean availability,
    (c) distribution of fractional dead time, and
    (d) mean fractional dead time.
13. A protective system for a nuclear reactor is run over a number of years of operation. The system is prone to certain faults which, when they occur, remain undetected until the system is thoroughly proof checked. The times to occurrence of these faults follow an exponential distribution with a mean value of 10^5 h. The system is proof checked every 1000 h and the proof-checking procedure lasts 10 h during which time the system is inoperative. What is the mean fractional dead time of the system?
14. An aircraft control system operates for a period of 3 h at a time. Any failures which occur during this operational period are unrepairable until the end of the period when the repair is then assumed to be instantaneous but does not affect the failure/time characteristic of the system. Failures which do occur in the control system have times to occurrence which follow a special Erlangian distribution with a parameter, λ, of value 6 h. The system starts at time zero and operates through three consecutive periods of operation. What is
    (a) the mean availability, and
    (b) the mean number of failures for the control system over this 9 h period?
15. A steel rolling mill alternates between the two states of continuous on-line production and complete standstill. From time zero, the times to the mill entering the standstill state from the on-line production state and the times for the reverse restoration process are both exponentially distributed with mean values of 100 h and 5 h respectively. What is
    (a) the probability of the mill being in the standstill state after $2\frac{1}{2}$ h, and
    (b) the average annual revenue loss if the loss of production costs £2000 per hour and there are 8760 h in a year?
16. A process plant is subject to faults which cause stoppages of production. When such faults occur, a repair process is carried out which when complete restores the plant to the 'as new' or 'time zero' condition. The times to occurrence of the faults have a mean value of 200 h. The distribution of times to repair the faults, starting from the occurrence of each fault, has a mean value of 2 h. What is
    (a) the limiting value of the plant's mean availability, and
    (b) the average total stoppage time over an operating period of 100,000 h?

17. A motor vehicle is required to be in continuous use for 8 h a day. Every 20 days it is taken out of service for a complete day for routine maintenance. During its time of continuous use it is subject to disabling faults whose times to occurrence are exponentially distributed with a mean value of 100 h. These faults always take a constant time of 1 h to rectify. What is
    (a) the mean availability of the vehicle against the required conditions, and
    (b) the mean number of unscheduled faults that are likely to occur in 212 days starting from time zero?

CHAPTER 11
Synthesis of System Reliability

The Need for Synthesis

The reliability considerations discussed in the previous chapter have shown the need to establish various typical reliability parameters at the system level for each performance aspect of the system. If such parameters were directly obtainable, then there would be no need for any further discussion. However, to obtain the appropriate reliability parameters it is generally necessary to have collected data on the performance of a large number of nominally identical systems operating over an extended period of time. With most technological systems of any complexity the information population in this respect is rarely adequate and with new or relatively new systems the required information may be non-existent.

The data which are not available at the system level may, however, be available at some lower hierarchy in the system. A process of system synthesis may then be used to derive the data which are applicable to the complete system. This process, and the need for it, was discussed in Chapter 5 under the heading 'Achievement Variations' and in Chapter 8 under the heading of 'The Need for Combinations'.

In Chapter 8, the discussion was principally orientated to the synthesis or combining of continuously distributed variations in performance in the working state. The change of state models, involving more catastrophic effects, which have been dealt with in Chapter 10 now require some additional treatment with regard to the ways that they can be combined in an overall synthesis process. In the ultimate, the reliability model which combines both discrete changes of state and continuous variations in performance needs to be examined for methods of synthesis.

It is the purpose of this chapter to review these various methods. For the most part, it is convenient, in carrying out this review, to consider principally a system or sub-system comprised of just two elements. Generally, if a method is shown for combining the characteristics of two elements, then this same method can be repeated throughout a multi-element system in order to arrive at the overall synthesis required. There are some reservations to this approach which apply in systems with certain types of cross-connexions

SYNTHESIS OF SYSTEM RELIABILITY

or common elements. However, additional complexities of this type will be dealt with in the next chapter.

The Two-element System

Suppose that a system or sub-system can be considered as being made up of two elements. If each of the elements is capable of entering one of two states, then the system, as already shown in Chapter 5('Discrete distributions'), possesses $2^2 = 4$ possible sub-states. The two states of element A may imply a working state, W, of the element on its own and a failed state, F, of the element on its own. The same may apply to element B. The truth table approach described in Chapters 4 and 5, will then lead to a system sub-state representation as shown in Table 11.1.

Table 11.1 Truth table of two-element system

System sub-state	State of element A	State of element B
1	W	W
2	W	F
3	F	W
4	F	F

Which of the four system sub-states represents overall system success and which represents overall system failure now depends on the functional way in which the two elements of the system are combined. In general, if the system can be represented by just two states then n sub-states out of the 4 will combine for success and the remaining $(4-n)$ will combine for failure.

Whatever the system arrangement it is generally true, although not necessarily so, that sub-state No. 1 will lead to system success and sub-state No. 4 to system failure. It is the effects of sub-states 2 and 3 which are likely to lead to variations. For instance, if the system is so arranged that both elements A and B are required to function, then sub-states 2 and 3 join sub-state No. 4 in the failure category. On the other hand, if the system depends upon either element A or element B functioning then sub-states 2 and 3 join sub-state No. 1 in the success category.

It will be noticed that the logical method by which the system functions determines the appropriate combination of sub-states. Into this functioning description enter the logical terms such as 'AND' and 'OR' whose implications were discussed fully in Chapter 4. In that chapter the notational algebra of Boole was found to have certain conveniences for system representation. This same notation and its correspondence with state probabilities will be illustrated in this chapter with reference to system synthesis and the combinations of system sub-states.

Since the two-element system can take on a number of forms depending upon the logical arrangement of the elements, it is convenient to consider each of the forms separately. Where both elements are required for system success, this will be termed a 'non-redundant' arrangement. The term 'redundancy' will be used where only one element is necessary for system success and, in this case, there may exist either 'parallel redundancy' or 'standby redundancy'. The meaning of these terms was originally discussed and described in Chapter 1.

For each form of system arrangement there will be a need to establish such aspects as state probabilities, availabilities, fractional dead times and number of failures from the corresponding characteristics of each individual element. This will apply when the elements are subject to either irreversible or reversible changes of state or both. These aspects are discussed in the following sections.

Non-redundant Elements

Change of state representation

The non-redundant arrangement of two elements implies that both are required to function correctly for the system to be successful. This may be expressed by a logical statement of the type described in Chapter 4, namely:
If:

(Element A in working state) AND (Element B in working state)

then:

(System X in working state)

If the symbol, A, is used to denote the working state of element A with the corresponding symbol, $\bar{A}$, for the failed state and similar symbols are used for element B and system X, then, using the Boolean notation of Chapter 4:

$$X = AB \tag{11.1}$$

or, conversely, using De Morgan's laws:

$$\bar{X} = \bar{A} + \bar{B} \tag{11.2}$$

which is obvious from the corresponding logical statement:
If:

(Element A in failed state) OR (Element B in failed state)

then:

(System X in failed state)

The diagrammatic representation of the two-element non-redundant system can consist of the two elements connected in series as shown in

SYNTHESIS OF SYSTEM RELIABILITY 433

Figure 11.1(a). The series connexion is then taken to imply that both elements must function to achieve the overall system function. This could be considered as a reasonable representation in those cases where the 'signal' path through a system passes through first one element and then another. However, two non-redundant elements in a system need not necessarily function in this way and therefore the equivalent Boolean representation of Equation (11.1), as shown in Figure 11.1(b), is probably a more consistent notation. Notice, that just as Equation (11.2) is the complement of Equation (11.1) so Figure 11.1(c) is the corresponding complement to Figure 11.1(b). These

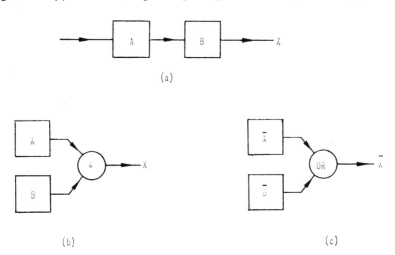

Figure 11.1 Diagrammatic representations of two-element non-redundant systems

figures employ some of the graphical symbols for logical functions introduced in Chapter 4. The representation in Figure 11.1(b) may be looked upon as drawn in terms of 'success', whereas Figure 11.1(c) represents a diagram in terms of 'failure'. Obviously, either diagram is equally valid and, in fact, various other diagrams can be drawn for the same situation where the blocks are mixtures of failed or successful states. Generally, however, the diagram which is drawn entirely in terms of success or entirely in terms of failure tends to be the most useful.

Although Boolean expressions are inherently deterministic as indicated in Chapter 4, they have equivalent probability relationships in connexion with discrete changes of state. Suppose that the probabilities of A, B and X being in the successful states are denoted by p_a, p_b and p_x respectively. Then, from the product law for the combination of independent probabilities in an 'AND' function as described in Chapter 1, the equivalent of Equation (11.1) becomes:

$$p_x = p_a p_b \tag{11.3}$$

which can be seen to have a similar form to Equation (11.1).

In the case of Equation (11.2), the events are combined in an 'OR' function and the equivalent probability expression for this, from Chapter 1, is:

$$\bar{p}_x = \bar{p}_a + \bar{p}_b - \bar{p}_a \bar{p}_b \tag{11.4}$$

which, in this case, is no longer quite similar to Equation (11.2) due to the additional product term which has appeared. The reader may easily verify, from Equations (11.3) and (11.4), that:

$$p_x + \bar{p}_x = 1 \tag{11.5}$$

as would be expected.

It is of interest to check the results of Equations (11.3) and (11.4) by means of the truth table given in Table 11.1. If a further column is added to the truth table giving the probabilities of each sub-state, then the result is as shown in Table 11.2.

Table 11.2 Truth table probabilities

System sub-state	State of element A	State of element B	Probability of sub-state
1	W	W	$p_a p_b$
2	W	F	$p_a \bar{p}_b$
3	F	W	$\bar{p}_a p_b$
4	F	F	$\bar{p}_a \bar{p}_b$

Each of the four system sub-states is a mutually exclusive event and, therefore, the sum of the probabilities in the last column of Table 11.2 is unity. For the non-redundant system being considered, the only sub-state condition which leads to system success is condition 1. Hence:

$$p_x = p_a p_b$$

as before.

Conversely, the overall condition for failure is the sum of the sub-state conditions 2, 3 and 4. Hence:

$$\bar{p}_x = p_a \bar{p}_b + \bar{p}_a p_b + \bar{p}_a \bar{p}_b$$
$$= \bar{p}_a + \bar{p}_b - \bar{p}_a \bar{p}_b$$

as before.

SYNTHESIS OF SYSTEM RELIABILITY

Equations (11.3) and (11.4) can be expressed in a number of different ways by using various appropriate combinations of the individual element probabilities of success and failure. For instance:

$$p_x = p_a p_b$$

or:

$$p_x = p_a - p_a \bar{p}_b \tag{11.6}$$

or:

$$p_x = 1 - \bar{p}_a - \bar{p}_b + \bar{p}_a \bar{p}_b \tag{11.7}$$

Similarly:

$$\bar{p}_x = \bar{p}_a + \bar{p}_b - \bar{p}_a \bar{p}_b$$

or:

$$\bar{p}_x = \bar{p}_a + p_a \bar{p}_b \tag{11.8}$$

or:

$$\bar{p}_x = 1 - p_a p_b \tag{11.9}$$

The probability equations of Equations (11.3) and (11.4) may also be represented in diagrammatic form in a similar way to the original Boolean equations. Such a representation is shown in Figures 11.2(a) and 11.2(b). It is important to remember now, that the '&' function implies multiplication of the probabilities and the 'OR' function implies the sum minus the product of the two probabilities. By writing the probabilities of system success or failure in terms of both success and failure of the individual elements, it is possible to express the overall probability in terms of just the sums and products of the individual probabilities as shown in Equation (11.8). This facet may be used in a diagrammatic representation as shown in Figures 11.2(c) and 11.2(d), where Figure 11.2(c) is equivalent to Equation (11.3) and Figure 11.2(d) is equivalent to Equation (11.8). As a further alternative, the series and parallel representation of products and sums as defined in Figure 4.9 of Chapter 4 may be used and this leads to the diagrams of Figures 11.2(c) and 11.2(d) becoming as shown in Figures 11.2(e) and 11.2(f) respectively.

It is now a simple matter to extend the two-element non-redundant system to an n-element system. The Boolean expressions of Equations (11.1) and (11.2) become:

$$X = ABCD\ldots \tag{11.10}$$

and:

$$\bar{X} = \bar{A} + \bar{B} + \bar{C} + \bar{D} + \ldots \tag{11.11}$$

and the corresponding probability expressions yield:

$$p_x = \prod_{j=1}^{n} p_j \tag{11.12}$$

and:

$$\bar{p}_x = 1 - \prod_{j=1}^{n} p_j \qquad (11.13)$$

or, in the case of the extension of the expression given in Equation (11.8):

$$\bar{p}_x = \bar{p}_a + p_a \bar{p}_b + p_a p_b \bar{p}_c + p_a p_b p_c \bar{p}_d + \ldots \qquad (11.14)$$

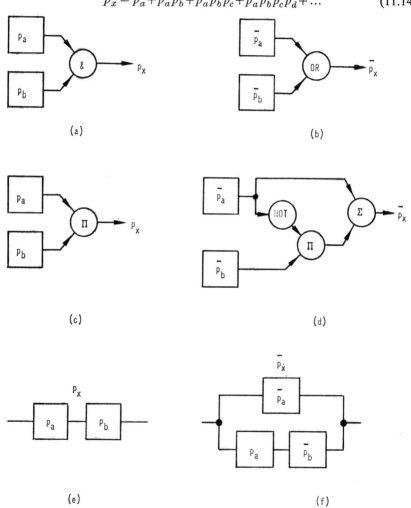

Figure 11.2 Probabilistic diagrams for two-element non-redundant systems

The reader will now be able to deduce equivalent extensions for the n-element non-redundant system in the diagrammatic representations of Figures 11.1 and 11.2.

SYNTHESIS OF SYSTEM RELIABILITY

It should be remembered that all the expressions derived so far depend upon the elements of the system being independent. Cases where dependency exists will be discussed in the next chapters. If, however, all the elements are nominally identical but still independent of each other, then:

$$p_x = p^n \tag{11.15}$$

and:

$$\bar{p}_x = 1 - p^n \tag{11.16}$$

or:

$$\bar{p}_x = \bar{p}(1 + p + p^2 + p^3 + \ldots) \tag{11.17}$$

Irreversible changes of state

These types of changes of state were discussed fully in Chapter 10. If the density function for times to failure of element A is $f_a(t)$ and that for element B is $f_b(t)$, then from Equation (10.71):

Probability element A in failed state at time t, $\bar{p}_a, = p_a(t)$ (11.18)

Probability element B in failed state at time t, $\bar{p}_b, = p_b(t)$ (11.19)

where $p_a(t)$ and $p_b(t)$ are the cumulative distribution functions of times to failure.

Consequently:

$$p_a = 1 - p_a(t) \tag{11.20}$$

or:

$$p_a = \exp\left[-\int z_a(t)\,dt\right] \tag{11.21}$$

and:

$$p_b = 1 - p_b(t) \tag{11.22}$$

or:

$$p_b = \exp\left[-\int z_b(t)\,dt\right] \tag{11.23}$$

Hence:

$$p_x = p_a p_b = \exp\left\{-\int [z_a(t) + z_b(t)]\,dt\right\} \tag{11.24}$$

and:

$$\bar{p}_x = 1 - \exp\left\{-\int [z_a(t) + z_b(t)]\,dt\right\} \tag{11.25}$$

This means that the probability of a two-element non-redundant system being in the failed state at time t is given by the integration up to t of the density function whose hazard function is the sum of the hazard functions of the individual elements. This type of hazard function relationship was first introduced to the reader in the derivations of Equations (8.125) to (8.127) in Chapter 8.

Obviously, for an n-element system:

$$p_x = \exp\left[-\int z_x(t)\,dt\right] \tag{11.26}$$

where:

$$z_x(t) = \sum_{j=1}^{n} z_j(t) \tag{11.27}$$

and:

$$f_x(t) = z_x(t)\exp\left[-\int z_x(t)\,dt\right] \tag{11.28}$$

The mean time to the first system failure, μ_f, is obviously the mean of the combined distribution of Equation (11.28). Hence, from Equation (7.27) or (7.67):

$$\mu_f = \int_0^\infty \exp\left[-\int z_x(t)\,dt\right] dt \tag{11.29}$$

If each element has times to failure which are exponentially distributed with a mean failure-rate of θ_j, then:

$$z_x(t) = \sum_{j=1}^{n} \theta_j \tag{11.30}$$

and:

$$\mu_f = 1 \bigg/ \sum_{j=1}^{n} \theta_j \tag{11.31}$$

and the probability of the system being in the successful state at time t is:

$$p_x = \exp\left\{-t\sum_{j=1}^{n} \theta_j\right\} \tag{11.32}$$

or:

$$p_x \simeq 1 - t\sum_{j=1}^{n} \theta_j \quad \text{if} \quad t\sum_{j=1}^{n} \theta_j \ll 1 \tag{11.33}$$

and the corresponding system probability of being in the failed state is:

$$\bar{p}_x \simeq t\sum_{j=1}^{n} \theta_j \tag{11.34}$$

Example 11.1

An electronic amplifier is made up of a total of 100 components. Each component has times to failure which follow an exponential distribution.

SYNTHESIS OF SYSTEM RELIABILITY

For each of 20 of the components the failure-rate parameter, θ, of the distribution has a value of 0.5 faults per 10^6 h. For another 50 components, $\theta = 0.1$ faults per 10^6 h per component. For a further 20 components, $\theta = 0.05$ faults per 10^6 h per component, and for the remaining 10 components, $\theta = 1.0$ fault per 10^6 h per component. Each component starts in the working state at time zero and is subject to irreversible changes of state. The amplifier fails if any one or more components fail. What is the probability of the amplifier being in the failed state after 500 h of operation?

Using the relationship of Equation (11.30), the total failure-rate parameter for the amplifier is given by:

$$\sum \theta = 20 \times 0.5 + 50 \times 0.1 + 20 \times 0.05 + 10 \times 1.0$$

$$= 26 \text{ faults per } 10^6 \text{ h}$$

Therefore, at a time $t = 500$

$$t \sum \theta = 500 \times 26 \times 10^{-6}$$

$$= 0.013$$

Here, $t \sum \theta$ is sufficiently small for the approximation of Equation (11.34) to apply, hence:

$$\bar{p}_x = 0.013$$

and this is the probability of the amplifier being in the failed state at 500 h.

The reader is referred back to Example 8.10 in Chapter 8 for a further example on the summation of hazard functions.

It has now been seen, for irreversible changes of state in non-redundant systems, how the system probability of being in the failed state may be synthesized from the element probability of being in the failed state. This synthesis may also be related to the element hazard functions or, in the case of the exponential distribution of times to failure, to the mean failure-rate of each element.

The next parameter of interest is that of availability. In order to attach any meaning to availability for irreversible changes of state some assumptions need to be made about the checking and restoration procedures which may take place at the end of every time interval τ. These aspects were fully discussed in the last chapter. For illustration in the present case it will be assumed that all periods are consecutive and of equal length τ, that no reversible change out of the failed state is possible except at each τ, and that the elements of the system are restored to the time zero condition at each τ.

The distribution of availability of the system for the above assumptions is given by Equation (10.95), namely:

$$f_{A_z}(z) = \frac{\tau}{[p_x(\tau)]^2} f_x\left[\frac{\tau\{z-[1-p_x(\tau)]\}}{p_x(\tau)}\right] \quad \text{for} \quad [1-p_x(\tau)] \leqslant z \leqslant 1 \quad (11.35)$$

where f_x is the density function for system times to failure and $p_x(\tau)$ is the corresponding cumulative function evaluated at τ. Now in terms of the system state probabilities for time τ, p_x and $\bar{p}_x$:

$$p_x(\bar{\tau}) = \bar{p}_x$$
$$= 1 - p_x$$
$$= 1 - p_a p_b$$
$$= 1 - \overline{p_a(\tau)}\,\overline{p_b(\tau)} \quad (11.36)$$

where $p_a(\tau)$ and $p_b(\tau)$ are the cumulative distribution functions for the times to failure of elements A and B respectively evaluated at τ.

Differentiating Equation (11.36) yields:

$$f_x(t) = f_a(t)\overline{p_b(t)} + f_b(t)\overline{p_a(t)} \quad (11.37)$$

which may be compared with the derivation of Equation (8.122) in Chapter 8.

Hence, for any failure distribution of elements A and B, the system density and cumulative functions may be derived from Equations (11.36) and (11.37). Then, substituting these functions in Equation (11.35) gives the distribution of availability. Similarly, the distribution of system fractional dead time may be obtained by the appropriate substitutions in Equation (10.102).

If the times to failure of the elements are exponentially distributed, then the process yields:

$$f_{A_z}(z) = \frac{\theta_x \tau}{[1-\exp(-\theta_x \tau)]^2} \exp\left\{\frac{-\theta_x \tau[z - \exp(-\theta_x \tau)]}{1 - \exp(-\theta_x \tau)}\right\} \quad (11.38)$$

and:

$$f_{D_z}(z) = \frac{\theta_x \tau}{[1-\exp(-\theta_x \tau)]^2} \exp\left\{\frac{-\theta_x \tau[1 - z - \exp(-\theta_x \tau)]}{1 - \exp(-\theta_x \tau)}\right\} \quad (11.39)$$

where:

$$\theta_x = \theta_a + \theta_b \quad (11.40)$$

In the case where $\theta_x \tau \ll 1$, then:

$$f_{A_z}(z) = \frac{1}{\theta_x \tau} = \frac{1}{(\theta_a + \theta_b)\tau} \quad \text{for} \quad (1 - \theta_x \tau) \leqslant z \leqslant 1 \quad (11.41)$$

SYNTHESIS OF SYSTEM RELIABILITY

and:

$$f_{D_x}(z) = \frac{1}{\theta_x \tau} = \frac{1}{(\theta_a + \theta_b)\tau} \quad \text{for} \quad 0 \leqslant z \leqslant \theta_x \tau \tag{11.42}$$

That is, the distributions of availability and fractional dead time are of the same form, but over different ranges of z.

The mean system availability and fractional dead time may be derived by substituting the value of $p_x(t)$, obtained from the relationship in Equation (11.36), in Equations (10.96) and (10.103) respectively. This leads to:

$$\mu_{A_x} = 1 - \frac{1}{\tau}\int_0^\tau p_a(t)\,dt - \frac{1}{\tau}\int_0^\tau p_b(t)\,dt + \frac{1}{\tau}\int_0^\tau p_a(t)p_b(t)\,dt \tag{11.43}$$

and it should be noted that this is not quite the same as the product of the individual mean availabilities, that is:

$$\mu_{A_x} \neq \mu_{A_a} \times \mu_{A_b} \tag{11.44}$$

For the exponential distribution:

$$\mu_{A_x} = \frac{1}{(\theta_a + \theta_b)\tau}\{1 - \exp[-(\theta_a + \theta_b)\tau]\} \tag{11.45}$$

or:

$$\mu_{A_x} = 1 - \frac{(\theta_a + \theta_b)\tau}{2} \quad \text{if} \quad (\theta_a + \theta_b)\tau \ll 1 \tag{11.46}$$

The corresponding relationships to Equations (11.43), (11.45) and (11.46) for fractional dead time are:

$$\mu_{D_x} = \frac{1}{\tau}\int_0^\tau p_a(t)\,dt + \frac{1}{\tau}\int_0^\tau p_b(t)\,dt - \frac{1}{\tau}\int_0^\tau p_a(t)p_b(t)\,dt \tag{11.47}$$

$$\mu_{D_x} = 1 - \frac{1}{(\theta_a + \theta_b)\tau}\{1 - \exp[-(\theta_a + \theta_b)\tau]\} \tag{11.48}$$

$$\mu_{D_x} = \frac{(\theta_a + \theta_b)\tau}{2} \quad \text{if} \quad (\theta_a + \theta_b)\tau \ll 1 \tag{11.49}$$

In an n-element non-redundant system which is subject to exponentially distributed times to failure and where:

$$\left\{\sum_{j=1}^n \theta_j\right\}\tau \ll 1 \tag{11.50}$$

then Equation (11.47) becomes:

$$\mu_{D_x} = \frac{1}{\tau}\int_0^\tau \left\{\sum_{j=1}^n \theta_j\right\} t\,dt \tag{11.51}$$

or:

$$\mu_{D_x} = \frac{1}{2}\left\{\sum_{j=1}^{n}\theta_j\right\}\tau \qquad (11.52)$$

or, if $\theta_j = \theta$ for all j, then:

$$\mu_{D_x} = \tfrac{1}{2}n\theta\tau \qquad (11.53)$$

Example 11.2

A non-redundant system is made up of two parts A and B such that both parts are required to function for system success. The system is restored to the 'as new' condition every 1000 h. Part A has times to failure following a special Erlangian distribution with a value of the parameter λ_a of 2000 h. Part B has times to failure following an exponential distribution where $\lambda_b = 1000$ h. What is the mean fractional dead time of the system?

From Equation (11.47):

$$\begin{aligned}\mu_{D_x} &= \frac{1}{\tau}\int_0^\tau\left[1-\left(1+\frac{t}{\lambda_a}\right)e^{-t/\lambda_a}\right]dt \\ &\quad+\frac{1}{\tau}\int_0^\tau(1-e^{-t/\lambda_b})\,dt \\ &\quad-\frac{1}{\tau}\int_0^\tau\left[1-\left(1+\frac{t}{\lambda_a}\right)e^{-t/\lambda_a}\right](1-e^{-t/\lambda_b})\,dt \\ &= 1-\frac{\lambda}{\tau}\left[\left(1+\frac{\lambda}{\lambda_a}\right)-\left(1+\frac{\lambda}{\lambda_a}+\frac{\tau}{\lambda_a}\right)e^{-\tau/\lambda}\right]\end{aligned}$$

where:

$$\lambda = \frac{\lambda_a\lambda_b}{\lambda_a+\lambda_b}$$

Then, substituting the appropriate values for λ_a, λ_b and τ, yields:

$$\mu_{D_x} = 0.3838$$

If at the end of each time interval τ, the elements of the system are out of action for a finite period of time τ_r during which time the renewal process takes place, then the mean availability and fractional dead time equations for the system are modified according to the relationships of Equations (10.114) and (10.116) respectively.

However, with an n-element system the testing and restoration of each element need not necessarily take place at the same time. If such staggered testing does take place and there is no overlap between the test periods for each individual element, then the dead time due to testing becomes $n\tau_r$ in any interval of length τ, where τ_r is the time that each element is out of action for test purposes. In this case:

$$\mu_{A'_x} = \frac{\tau}{\tau + n\tau_r} \mu_{A_x} \tag{11.54}$$

with a similar modification in the case of the expression for mean fractional dead time.

However, another effect of staggered testing is that the elements in the system are all at different ages at any one time. Hence, it is pertinent to see whether this age difference in itself affects the mean availability. Suppose that, in a two-element system, element A starts new at time zero and is then renewed after every time interval τ. Element B, on the other hand, starts new at a time $k\tau$ and is also then renewed every time interval τ measured from the starting time of $k\tau$. The fractional value of k can lie between 0 and 1.

Two sub-intervals now exist, one of length $k\tau$ and the other of length $(1-k)\tau$. The mean fractional dead time for the sub-interval of length $(1-k)\tau$ is:

$$\mu_{D_{x_1}} = \frac{1}{(1-k)\tau} \int_0^{(1-k)\tau} p_b(t)\,dt + \frac{1}{(1-k)\tau} \int_0^{(1-k)\tau} p_a(t+k\tau)\,dt$$

$$- \frac{1}{(1-k)\tau} \int_0^{(1-k)\tau} p_a(t+k\tau) p_b(t)\,dt \tag{11.55}$$

and the corresponding mean fractional dead time for the sub-interval of length $k\tau$ is:

$$\mu_{D_{x_2}} = \frac{1}{k\tau} \int_0^{k\tau} p_a(t)\,dt + \frac{1}{k\tau} \int_0^{k\tau} p_b[t+(1-k)\tau]\,dt$$

$$- \frac{1}{k\tau} \int_0^{k\tau} p_a(t) p_b[t+(1-k)\tau]\,dt \tag{11.56}$$

and the overall mean fractional dead time is given by:

$$\mu_{D_x} = (1-k)\mu_{D_{x_1}} + k\mu_{D_{x_2}} \tag{11.57}$$

In the case of the exponential distribution Equation (11.57) becomes:

$$\mu_{D_x} = 1 - \frac{1}{(\theta_a+\theta_b)\tau} [(1-e^{-\theta_a\tau})e^{-(1-k)\theta_b\tau} + (1-e^{-\theta_b\tau})e^{-k\theta_a\tau}] \tag{11.58}$$

It is now seen, by comparison with Equation (11.48), that the mean fractional dead time is different for the staggered testing compared with testing the elements at the same time. This is generally true for all distributions of times to failure. If, however, $(\theta_a+\theta_b)\tau \ll 1$, then:

$$\mu_{D_x} = \frac{(\theta_a+\theta_b)\tau}{2} \qquad (11.59)$$

and Equation (11.59) is now identical with Equation (11.49). So, provided that the times to failure of the elements are exponentially distributed and provided that the cumulative exponential distribution can be approximated to a linear function of time, then there is no difference in the mean fractional dead times or availabilities for any value of k in a staggered testing routine. This applies for an n-element system as well as for the two-element system.

Example 11.3

A non-redundant system is made up of two nominally identical but independent elements. Each element has times to failure which are exponentially distributed with a mean value of 1000 h. The interval between test and renewal for each element is 1000 h, but the tests are staggered by a difference of $1000k$ h. Find the mean fractional dead time of the system for the cases when:
(a) $k = \frac{1}{4}$,
(b) $k = \frac{1}{2}$,
(c) $k = \frac{3}{4}$, and
(d) $k = 1$.

For two nominally identical elements, Equation (11.58) reduces to:

$$\mu_{D_x} = 1 - \frac{1}{2\theta\tau}[(1-e^{-\theta\tau})(e^{-k\theta\tau}+e^{-(1-k)\theta\tau})] \qquad (11.60)$$

and substituting the example values for θ and τ gives:

$$\mu_{D_x} = 1 - \frac{1}{2}\left(1-\frac{1}{e}\right)(e^{-k}+e^{-(1-k)})$$

whence, for:

$$k = \tfrac{1}{4}, \quad \mu_{D_x} = 0.604$$
$$k = \tfrac{1}{2}, \quad \mu_{D_x} = 0.617$$
$$k = \tfrac{3}{4}, \quad \mu_{D_x} = 0.604$$
$$k = 1, \quad \mu_{D_x} = 0.568$$

The number of failures that are likely to occur in a system subject to irreversible changes of state was derived in Equations (10.130) and (10.131). These formulae apply to a non-redundant system of n elements where the system cumulative probability function $p_x(\tau)$ is given by Equation (11.36). Hence, for a two-element system which is renewed at the end of every time

SYNTHESIS OF SYSTEM RELIABILITY 445

interval τ, the mean number of failures over some total time T is:

$$\mu_{N_x} = \frac{T}{\tau}[1 - \overline{p_a(\tau)}\,\overline{p_b(\tau)}] \tag{11.61}$$

which, for the exponential distribution, becomes:

$$\mu_{N_x} = \frac{T}{\tau}[1 - e^{-(\theta_a + \theta_b)\tau}]$$

or:

$$\mu_{N_x} \simeq (\theta_a + \theta_b)T \quad \text{if} \quad (\theta_a + \theta_b)\tau \ll 1 \tag{11.62}$$

Example 11.4

An area is lit by two electric light bulbs and the light from both bulbs is required. The times to failure of the bulbs are normally distributed with a mean of 1000 h and a standard deviation of 100 h. Both bulbs are renewed every 800 h and not at any other time. What is the average number of light system failures that will occur in a total time of 20,000 h?

Here, the value of $p_a(\tau)$ from Table A.1 is:

$$p_a(\tau) = 0.0228$$

hence:

$$\overline{p_a(\tau)} = 0.9772$$

similarly:

$$\overline{p_b(\tau)} = 0.9772$$

and:

$$1 - \overline{p_a(\tau)}\,\overline{p_b(\tau)} = 0.0451$$

Therefore, from Equation (11.61):

$$\mu_{N_x} = \frac{20{,}000}{800} \times 0.0451$$

$$\simeq 1.13$$

Reversible changes of state

The reader is referred to Chapter 10 for the definition and appropriate derivation of formulae for this type of change of state. Considering first the probability of a system being in the failed state at some time t, then Equation (10.141) applies for time-dependent failures and repairs and Equation (10.151) applies for the alternating renewal process. Either of these equations can then be used to express the probability of element A being in the failed state, $\bar{p}_a$, and element B being in the failed state, $\bar{p}_b$, at any particular time. The system probability of being in the failed state, $\bar{p}_x$, is then derived from

Equation (11.4). Similarly, the chance of system success, for the two-element system, is derived from Equation (11.3).

Typically, then, the two-element system chance of success at time t for elements subject to the alternating renewal process is:

$$p_x = \left\{ L^{-1} \left[\frac{1 - F_{f_a}(s)}{s[1 - F_{f_a}(s) F_{r_a}(s)]} \right] \right\} \left\{ L^{-1} \left[\frac{1 - F_{f_b}(s)}{s[1 - F_{f_b}(s) F_{r_b}(s)]} \right] \right\} \quad (11.63)$$

where $f_{f_a}(t)$ and $f_{f_b}(t)$ are the density functions for the times to failure of elements A and B, and $f_{r_a}(t)$ and $f_{r_b}(t)$ are the corresponding repair functions. For exponential functions, Equation (11.63) becomes:

$$p_x = 1 - \frac{\theta_a \tau_a}{1 + \theta_a \tau_a} \{1 - \exp[-t(\theta_a + 1/\tau_a)]\} - \frac{\theta_b \tau_b}{1 + \theta_b \tau_b} \{1 - \exp[-t(\theta_b + 1/\tau_b)]\}$$

$$+ \frac{\theta_a \theta_b \tau_a \tau_b}{(1 + \theta_a \tau_a)(1 + \theta_b \tau_b)} \{1 - \exp[-t(\theta_a + 1/\tau_a)]\} \{1 - \exp[-t(\theta_b + 1/\tau_b)]\}$$

$$(11.64)$$

However, as pointed out in Chapter 10, such equations as Equations (11.63) and (11.64) very rapidly approach, in practice, a constant value which is equivalent to the mean availability. Therefore, invoking the same limiting condition as that of Equation (10.156), it is seen that Equation (11.63) becomes:

$$p_x = \frac{\mu_{f_a}}{\mu_{f_a} + \mu_{r_a}} \times \frac{\mu_{f_b}}{\mu_{f_b} + \mu_{r_b}} \quad (11.65)$$

or:

$$\mu_{A_x} = \mu_{A_a} \mu_{A_b} \quad (11.66)$$

and the mean fractional dead time is:

$$\mu_{D_x} = 1 - \mu_{A_a} \mu_{A_b} \quad (11.67)$$

or:

$$\mu_{D_x} = \mu_{D_a} + \mu_{D_b} - \mu_{D_a} \mu_{D_b} \quad (11.68)$$

So it is seen that for the alternating renewal process, the combinations of mean availabilities and fractional dead times for the elements of the system obey the same laws as the equivalent combinations of independent probabilities.

If the exponential distribution applies to the times to failure and repair such that θ_a and τ_a are the mean failure-rate and mean repair time of element A and θ_b and τ_b the corresponding values for element B, then:

$$\mu_{A_x} = \frac{1}{(1 + \theta_a \tau_a)(1 + \theta_b \tau_b)} \quad (11.69)$$

SYNTHESIS OF SYSTEM RELIABILITY

or:

$$\mu_{A_x} = \frac{1}{(1+\theta\tau)^2} \quad \text{if} \quad \theta = \theta_a = \theta_b \quad \text{and} \quad \tau = \tau_a = \tau_b \tag{11.70}$$

and:

$$\mu_{D_x} = \frac{\theta_a \tau_a}{1+\theta_a \tau_a} + \frac{\theta_b \tau_b}{1+\theta_b \tau_b} - \frac{\theta_a \theta_b \tau_a \tau_b}{(1+\theta_a \tau_a)(1+\theta_b \tau_b)} \tag{11.71}$$

or:

$$\mu_{D_x} = \frac{2\theta\tau}{(1+\theta\tau)} - \frac{\theta^2 \tau^2}{(1+\theta\tau)^2} \quad \text{if} \quad \theta = \theta_a = \theta_b \quad \text{and} \quad \tau = \tau_a = \tau_b \tag{11.72}$$

and for the condition $\theta\tau \ll 1$, Equation (11.72) becomes:

$$\mu_{D_x} = 2\theta\tau \tag{11.73}$$

In general, for an n-element non-redundant system subject to the alternating renewal process:

$$\mu_{A_x} = \prod_{j=1}^{n} \mu_{A_j} \tag{11.74}$$

and:

$$\mu_{D_x} = 1 - \prod_{j=1}^{n} (1 - \mu_{D_j}) \tag{11.75}$$

or, if all the elements are nominally identical, then:

$$\mu_{A_x} = \mu_A^n \tag{11.76}$$

and:

$$\mu_{D_x} = 1 - (1 - \mu_D)^n \tag{11.77}$$

and for the approximate form of the exponential distribution this latter equation becomes:

$$\mu_{D_x} = n\theta\tau \tag{11.78}$$

Example 11.5

A temperature-measuring system consists of three main elements which are a thermocouple, an electronic amplifier and a chart recorder. Each element is subject to an alternating renewal process such that when an element fails a new repair process is started and when the repair is complete the element is restored to the 'as new' condition. The means of the distributions of times to failure and repair of each element are as follows:

	Mean time to failure	Mean time to repair
Thermocouple	100,000 h	5 h
Amplifier	5000 h	10 h
Recorder	10,000 h	10 h

What is the limiting mean availability and fractional dead time of the system and how would these values be affected if the thermocouple was perfect?

For the thermocouple:

$$\mu_D = \frac{5}{100,005} \simeq 5 \times 10^{-5}$$

For the amplifier:

$$\mu_D = \frac{10}{5010} = 1.9960 \times 10^{-3}$$

For the recorder:

$$\mu_D = \frac{10}{20,010} = 4.9975 \times 10^{-4}$$

Hence, since these values are small, the total mean fractional dead time for the system may be obtained from their sum, that is:

$$\mu_{D_x} = 2.546 \times 10^{-3}$$

and, without the thermocouple, this figure would be:

$$\mu_{D_x} = 2.496 \times 10^{-3}$$

The corresponding mean availabilities for these two cases are:

$$\mu_{A_x} = 0.99745$$

and:

$$\mu_{A_x} = 0.99750$$

For an alternating renewal process, as with an irreversible process, the first failure of a non-redundant system still occurs when the first element fails. Hence the time to the first system failure is given, as before, by Equations (11.29) and (11.31).

The mean number of failures that are likely to occur in a system subject to the alternating renewal process was derived in Equation (10.172). This equation, therefore, applies to each element in the case of an n-element non-redundant system. If each element is independent and contributes separately

SYNTHESIS OF SYSTEM RELIABILITY

and exclusively to the total number of system failures, then the mean number of system failures over a long period of time T is:

$$\mu_{N_x} = \sum_{j=1}^{n} \mu_{N_j} \qquad (11.79)$$

where μ_{N_j} is the mean number of failures of the jth element given by:

$$\mu_{N_j} = \frac{T}{\mu_{f_j} + \mu_{r_j}} \qquad (11.80)$$

If $\mu_r \ll \mu_{f_j}$, then:

$$\mu_{N_x} = \sum_{j=1}^{n} \frac{T}{\mu_{f_j}} \qquad (11.81)$$

which, for the exponential distribution of times to failure becomes:

$$\mu_{N_x} = \sum_{j=1}^{n} T\theta_j \qquad (11.82)$$

A comparison between this equation and Equation (11.31) shows that the following approximation now holds:

$$\mu_f = \frac{T}{\mu_{N_x}} \qquad (11.83)$$

Example 11.6

A system is made up of 1000 nominally identical component parts. Each part is subject to an alternating renewal process with times to failure and repair which both follow an exponential distribution. The parameters of the distributions are $\theta = 10^{-5}$ faults per hour and $\tau = 10$ h. What is the mean number of system failures over a period of 10,000 h?

From Equations (11.79) and (10.173):

$$\mu_{N_x} = \frac{n\theta T}{1 + \theta\tau}$$

and substituting the example values:

$$\mu_{N_x} \simeq 100 \text{ failures}$$

Note that if all the components were renewed each time any one failed and the repair process to do this was still $\tau = 10$ h, then:

$$\mu_{N_x} = \frac{n\theta T}{1 + n\theta\tau}$$

$$\simeq 90.9 \text{ failures}$$

This latter note in Example 11.6 leads to considerations of complete system renewals, instead of element renewals, in multi-element systems.

System renewals

It may well arise in practice that a sub-system, or particularly an item of equipment in a sub-system, is completely renewed or replaced when one of its component parts fails. For instance if a resistor fails in an electronic amplifier, the complete amplifier may be replaced and the original amplifier taken away and repaired at leisure. As far as the overall system is concerned, of which this amplifier is a part, the effect of any element failure in the amplifier is to replace or renew all elements in the amplifier. For a two-element case, this means that the failure of either element leads to the renewal of both. Hence, the mean availability for the two-element system is simply:

$$\mu_{A_z} = \frac{\mu_f}{\mu_f + \mu_r} \tag{11.84}$$

but here, in the first place, μ_r is the mean of the distribution $f_r(t)$ which is the distribution of times to repair both elements or, simply, the distribution of system replacement times. In the second place, μ_f is the mean of the distribution $f_f(t)$ which is the distribution of times to the first failure of either element A or element B. This latter type of composite distribution was discussed in Chapter 8 in connexion with the derivation of Equations (8.119) to (8.128) inclusive. Hence:

$$f_f(t) = z_f(t) \exp\left[-\int z_f(t)\,dt\right] \tag{11.85}$$

where:

$$z_f(t) = z_{f_a}(t) + z_{f_b}(t) \tag{11.86}$$

and where $z_{f_a}(t)$ and $z_{f_b}(t)$ are the hazard functions associated with the distributions of times to failure, $f_{f_a}(t)$ and $f_{f_b}(t)$, of elements A and B respectively. If all distributions are of the exponential form with failure-rate parameters θ_a and θ_b for elements A and B and a mean time to repair, τ, for the system, then Equation (11.84) becomes:

$$\mu_{A_z} = \frac{1}{1 + (\theta_a + \theta_b)\tau} \tag{11.87}$$

Obviously, for an n-element system Equations (11.84) and (11.85) still apply and Equation (11.86) becomes:

$$z_f(t) = \sum_{j=1}^{n} z_{f_j}(t) \tag{11.88}$$

SYNTHESIS OF SYSTEM RELIABILITY

which, for a system with n nominally identical elements each with an exponential failure-rate parameter, θ, yields:

$$\mu_{A_z} = \frac{1}{1+n\theta\tau} \qquad (11.89)$$

The next possibility with complete system renewals is to consider the following case. Element A is subject to faults which are 'revealed' and which would, therefore, be normally subject to an alternating renewal process. Element B is subject to faults which are 'unrevealed' and which would normally lead to an irreversible process over some fixed period of time, τ. The system, made up of these two non-redundant elements, is then replaced or renewed when a failure of either element is discovered. This means that if element A fails somewhere in the period 0 to τ, then a repair process is started on both elements A and B. Element B may or may not be in a failed state at this time, but, if it is, then it is repaired along with A. If, on the other hand, element B fails in the period 0 to τ then no action is taken since this is an 'unrevealed' or 'irreversible' fault. Such a fault will remain in existence on element B until either element A subsequently fails or the end of the period τ is reached. At the end of the period τ both elements are renewed irrespectively as in an ordinary sequential irreversible process.

To analyse this situation, the following density functions need to be defined:

$f_{f_r}(t) =$ distribution of times to occurrence of revealed faults in element A

$f_{f_u}(t) =$ distribution of times to occurrence of unrevealed faults in element B

$f_f(t) =$ distribution of times to occurrence of the earliest of either revealed or unrevealed faults

$f_r(t) =$ distribution of times to renew or replace both elements A and B

Obviously, in line with Equations (11.85) and (11.86):

$$f_f(t) = z_f(t)\exp\left[-\int z_f(t)\,dt\right] \qquad (11.90)$$

where, here:

$$z_f(t) = z_{f_r}(t) + z_{f_u}(t) \qquad (11.91)$$

The system has three basic states. These may be denoted by 's' for the successful or no failure state, 'u' for the unrevealed fault state and 'r' for the repair state. A composite state, 'f', may also be defined which includes both types of failure state. The change of state probability, $p_{jk}(t)$, as defined in

Chapter 10 now has nine possible forms since either j or k can be represented by s, u or r. For the three particular ones of interest:

$$p_{ss}(t)+p_{su}(t)+p_{sr}(t) = 1 \qquad (11.92)$$

and:

$$p_{sf}(t) = p_{su}(t)+p_{sr}(t) \qquad (11.93)$$

The probability $p_{ss}(t)$ can be achieved either if no failure of either kind occurs in the interval 0 to t, which is:

$$[1-p_f(t)]$$

or, irrespective of whether an unrevealed failure occurs or not, a revealed failure occurs at some time u (for $0 \leq u \leq t$) and a change of state from repair, r, to success, s, occurs in the interval u to t. This probability is:

$$\int_0^t p_{rs}(t-u) f_{f_r}(u) \, du$$

If an unrevealed failure occurs without a revealed failure, then this condition cannot lead to $p_{ss}(t)$, hence:

$$p_{ss}(t) = [1-p_f(t)] + \int_0^t p_{rs}(t-u) f_{f_r}(u) \, du \qquad (11.94)$$

which, on taking the Laplace transformation, yields:

$$sP_{ss}(s) = 1 - F_f(s) + sP_{rs}(s) F_{f_r}(s) \qquad (11.95)$$

The change from the repair state to the successful state can only take place through the completion of the repair process. Therefore:

$$p_{rs}(t) = \int_0^t p_{ss}(t-u) f_r(u) \, du \qquad (11.96)$$

or:

$$P_{rs}(s) = P_{ss}(s) F_r(s) \qquad (11.97)$$

and combining Equations (11.95) and (11.97) gives:

$$sP_{ss}(s) = \frac{1-F_f(s)}{1-F_{f_r}(s) F_r(s)} \qquad (11.98)$$

and this is seen to be of a similar form to Equation (10.152) except that $F_f(s)$ now represents the composite distribution of the two types of failure. $F_f(s)$ may be obtained from Equations (11.90) and (11.91).

Provided that the overall time interval τ is sufficiently large, the usual limiting process may be carried out on Equation (11.98) which gives the

SYNTHESIS OF SYSTEM RELIABILITY

mean availability for the system as:

$$\mu_{A_x} = \frac{\mu_f}{\mu_{f_r} + \mu_r} \quad (11.99)$$

The corresponding mean fractional dead time is most simply found by subtracting Equation (11.99) from unity. However, any of the other failed state probabilities can be found by a similar process to that just discussed. In the case of $p_{sr}(t)$, this is dependent only on the revealed fault occurrence and the repair processes. Hence, straightaway it may be said, on lines with the derivation of Equation (10.151), that:

$$sP_{sr}(s) = \frac{F_{f_r}(s)[1 - F_r(s)]}{1 - F_{f_r}(s) F_r(s)} \quad (11.100)$$

and, then, combining Equations (11.92), (11.98) and (11.100):

$$sP_{su}(s) = \frac{F_f(s) - F_{f_r}(s)}{1 - F_{f_r}(s) F_r(s)} \quad (11.101)$$

Equations (11.98), (11.100) and (11.101) only lead to relatively simple analytical results in the case of the exponential distribution. In particular, for this case, the reader might like to verify that:

$$\mu_{A_x} = \frac{\theta_{f_r}}{(\theta_{f_r} + \theta_{f_u})(1 + \theta_{f_r} \tau_r)} \quad (11.102)$$

where:

$$\mu_{f_r} = 1/\theta_{f_r} \quad (11.103)$$

$$\mu_{f_u} = 1/\theta_{f_u} \quad (11.104)$$

$$\mu_r = \tau_r \quad (11.105)$$

Example 11.7

The elements of a two-element non-redundant system are each subject to an exponential distribution of times to failure. The failures of one element are revealed and occur at a mean rate of 10^{-2} faults per hour. The failures of the other element are unrevealed and occur at a mean rate of 10^{-4} faults per hour. Any repair process which is carried out has times to completion which are also exponentially distributed with a mean value of 10 h. What is the system's mean availability over a period of 5000 h if
(a) each element is treated separately and independently, and
(b) a repair of a revealed fault also implies a repair of the element subject to unrevealed faults?

For question (a), the mean availability of the 'unrevealed' fault element is given by Equation (10.96) which yields:

$$\mu_{A_u} = \frac{1}{\theta_{f_u}\tau}[1 - \exp(-\theta_{f_u}\tau)] \qquad (11.106)$$

which for the example values gives:

$$\mu_{A_u} = 0.787$$

The mean availability of the 'revealed' fault element is given by Equation (10.165), that is:

$$\mu_{A_r} = \frac{1}{1 + \theta_{f_r}\tau_r} \qquad (11.107)$$

or:

$$\mu_{A_r} = 0.909$$

Hence:

$$\mu_A = \mu_{A_u}\mu_{A_r} = 0.715$$

For the case of question (b), the system's mean availability is given by Equation (11.102). This yields:

$$\mu_{A_x} = 0.900$$

So, for this example, the policy of (b) gives a better mean availability than the policy of (a).

Variational aspects

So far, in the two- or multi-element non-redundant system only 'catastrophic' changes of state have been considered. This means that, as far as the synthesis of overall system reliability is concerned, the discussion has centred around the determination of the system p in Equation (10.63) or, more specifically, of the system μ_A in Equations (10.109) and (10.175).

To complete the system reliability synthesis, it is necessary to determine the overall $p_{x_h}(x)$ or $f_{x_h}(x)$ for the non-redundant system. This means, first of all, determining the way in which the elements functionally combine for any particular system performance parameter x. The only criterion for a non-redundant system is that all the elements must be in the working or successful state. Within this definition any functional combination of elements is possible. So, for example, the two system elements of a battery connected across a resistance combine functionally as a quotient to give the system current. The time delays of two system elements operating in series combine additively to give the overall system time delay. The methods of carrying out these functional combinations to give the resultant system density function for the overall function were fully described in Chapter 8

SYNTHESIS OF SYSTEM RELIABILITY

and so need not be repeated here. Having obtained this functional density function, the reliability expression of Equation (10.63) may be evaluated by the methods discussed under the heading of 'The Correlation Process' in Chapter 10.

This, then, has pointed the way to the overall reliability synthesis of a non-redundant system. The next step is to examine various types of redundant configurations that may occur in technological systems.

Parallel Redundant Elements

Change of state representation

The parallel redundant arrangement of two elements implies that they are both designed to be functioning but only one or the other of them is required to be working for the system to be successful. This may be expressed by the logical statement:

If:

(Element A in working state) OR (Element B in working state)

then:

(System X in working state)

Using similar Boolean notation to that employed in Equations (11.1) and (11.2), then for this case:

$$X = A + B \tag{11.108}$$

and:

$$\bar{X} = \bar{A}\bar{B} \tag{11.109}$$

and it is seen that these equations are, in a certain way, complementary to the corresponding equations for the two non-redundant elements given in Equations (11.1) and (11.2).

A functional representation of the two-element parallel redundant system could be given as shown in Figure 11.3(a), but this can lead to misinterpretations. The equivalent, and less ambiguous, Boolean diagrams are shown in Figures 11.3(b) and 11.3(c) which may be compared with Figures 11.1(b) and 11.1(c).

In a similar way to the derivations of Equations (11.3) and (11.4), it is seen that the equivalent probability state equations for the two-element redundant system are:

$$p_x = p_a + p_b - p_a p_b \tag{11.110}$$

and:

$$\bar{p}_x = \bar{p}_a \bar{p}_b \tag{11.111}$$

and these equations may also be obtained by summating the system substates 1, 2 and 3 in Table 11.2 for the successful state, p_x, and taking substate 4 for the system failed state, $\bar{p}_x$.

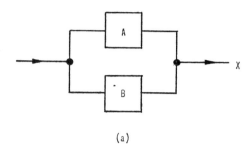

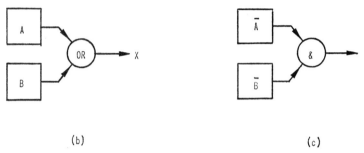

Figure 11.3 Diagrammatic representations of two-element parallel redundant systems

Alternative representations of Equations (11.110) and (11.111) are:

$$p_x = p_a + \bar{p}_a p_b \tag{11.112}$$

or:

$$p_x = 1 - \bar{p}_a \bar{p}_b \tag{11.113}$$

and:

$$\bar{p}_x = \bar{p}_a - \bar{p}_a p_b \tag{11.114}$$

or:

$$\bar{p}_x = 1 - p_a - p_b + p_a p_b \tag{11.115}$$

which may be compared with Equations (11.6) to (11.9).

The diagrammatic representations of probability states for the two-element parallel redundant system, corresponding with the non-redundant representation of Figure 11.2, are given in Figure 11.4. It should be remembered, as discussed previously, that an 'AND' function represents multiplication of probabilities and an 'OR' function the combination of probabilities as given, for example, in Equation (11.110).

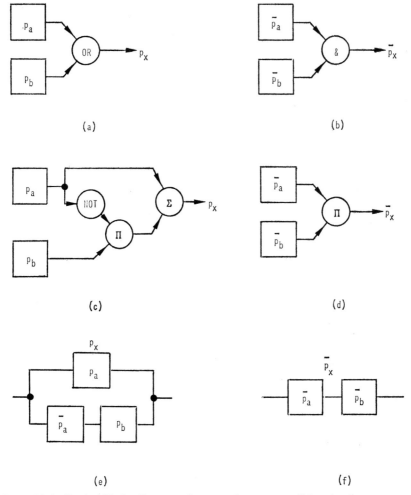

Figure 11.4 Probabilistic diagrams for two-element parallel redundant systems

In the more general case of an n-element system:

$$p_x = 1 - \prod_{j=1}^{n} \bar{p}_j \tag{11.116}$$

and:

$$\bar{p}_x = \prod_{j=1}^{n} \bar{p}_j \tag{11.117}$$

and the expression for p_x involving only sums and products becomes:

$$p_x = p_a + \bar{p}_a p_b + \bar{p}_a \bar{p}_b p_c + \bar{p}_a \bar{p}_b \bar{p}_c p_d + \cdots \qquad (11.118)$$

The reader will have noticed during this discussion that all the expressions and representations of the parallel redundant system are the same as those for the non-redundant system with each event or probability changed for its complement. That is, p changed to $\bar{p}$ and vice versa.

Irreversible changes of state

For this type of change of state, Equations (11.18) to (11.23) still apply to the individual elements of a redundant system. Hence, from these equations and Equation (11.111):

$$\bar{p}_x = \bar{p}_a \bar{p}_b$$

$$= \left\{1 - \exp\left[-\int z_a(t)\,dt\right]\right\}\left\{1 - \exp\left[-\int z_b(t)\,dt\right]\right\} \qquad (11.119)$$

but this does not lead to a simple relationship between the hazard functions as it did in the case of Equation (11.24). The general solutions of Equation (11.119) were, in fact, discussed in Chapter 8 in connexion with Equations (8.136) to (8.141) and need not be gone into further at this stage, except to say that the mean time to first system failure is given by the mean of the distribution function of the type of Equation (8.141).

However, in the particular case of the exponential distribution, Equation (11.119) yields:

$$\bar{p}_x = 1 - e^{-\theta_a t} - e^{-\theta_b t} + e^{-(\theta_a + \theta_b)t} \qquad (11.120)$$

and the mean time to first system failure, μ_f, becomes:

$$\mu_f = \frac{1}{\theta_a} + \frac{1}{\theta_b} - \frac{1}{\theta_a + \theta_b} \qquad (11.121)$$

or:

$$\mu_f = \frac{3}{2\theta} \quad \text{if} \quad \theta_a = \theta_b = \theta \qquad (11.122)$$

If $(\theta_a + \theta_b)t \ll 1$, then Equation (11.120) approximates to:

$$\bar{p}_x = \theta_a \theta_b t^2 \qquad (11.123)$$

Generally, then, for an n-element system under these conditions:

$$\bar{p}_x = t^n \prod_{j=1}^{n} \theta_j \qquad (11.124)$$

and:

$$p_x = 1 - t^n \prod_{j=1}^{n} \theta_j \qquad (11.125)$$

SYNTHESIS OF SYSTEM RELIABILITY

and these latter two equations may be compared with the equivalent values for a non-redundant system given in Equations (11.33) and (11.34).

Example 11.8

Two generators are running in parallel under such conditions that either can supply the required load. Once a generator has failed it remains irreversibly in the failed state. The times to failure of each generator follow a Weibull distribution with parameters $\lambda = 500$ h and $\beta = 2$. What is the probability that no load is being supplied at 250 h?

The probability of each generator being in the failed state, $\bar{p}$, at time t is:

$$\bar{p} = 1 - e^{-(t/\lambda)^\beta}$$

$$= 0.2212$$

Now, from Equation (11.119):

$$\bar{p}_x = (\bar{p})^2$$

$$= 0.0489$$

The next parameter to be considered for the parallel redundant system is that of availability or its complement, fractional dead time. For the irreversible changes of state in the elements of this system, the same assumptions are made as for the non-redundant system. This means that the distribution of system availabilities is given by Equation (11.35). In this case, though, the $p_x(\tau)$ in Equation (11.35) is derived from the relationship of Equation (11.111), hence:

$$p_x(\tau) = p_a(\tau) p_b(\tau) \qquad (11.126)$$

and the $f_x(t)$ of Equation (11.35) is obtained from the differential of Equation (11.126), that is:

$$f_x(t) = f_a(t) p_b(t) + f_b(t) p_a(t) \qquad (11.127)$$

A relationship which was, of course, previously deduced in Equation (8.138).

For any set of element failure distributions, the distribution of system availability may be obtained from Equations (11.35), (11.126) and (11.127). In general, however, this leads to a complex, though possibly soluble, function. For illustration, therefore, the case will only be considered of the approximate exponential distribution. In this instance, from Equation (11.123), the equivalent forms of Equations (11.127) and (11.126) become:

$$f_x(t) = 2\theta_a \theta_b t \qquad (11.128)$$

and:

$$p_x(\tau) = \theta_a \theta_b \tau^2 \qquad (11.129)$$

Substituting these expressions in the availability distribution equation of Equation (10.95) yields:

$$f_{A_x}(z) = \frac{2}{\theta_a \theta_b \tau^2} \left[\frac{z}{\theta_a \theta_b \tau^2} - \frac{(1 - \theta_a \theta_b \tau^2)}{\theta_a \theta_b \tau^2} \right] \quad \text{for} \quad 1 - \theta_a \theta_b \tau^2 \leq z \leq 1 \quad (11.130)$$

The mean availability may then be obtained from this distribution function in the usual way, or, more directly, from Equations (10.96) and (11.129). Whence:

$$\mu_{A_x} = 1 - \frac{\theta_a \theta_b \tau^2}{3} \quad (11.131)$$

In a similar way:

$$f_{D_x}(z) = \frac{2}{\theta_a \theta_b \tau^2} \left[1 - \frac{z}{\theta_a \theta_b \tau^2} \right] \quad \text{for} \quad 0 \leq z \leq \theta_a \theta_b \tau^2 \quad (11.132)$$

and:

$$\mu_{D_x} = \frac{\theta_a \theta_b \tau^2}{3} = \tfrac{1}{3} p_a(\tau) p_b(\tau) \quad (11.133)$$

Moreover, it should be noted, in a similar manner to the inequality of Equation (11.44) that:

$$\mu_{D_x} \neq \mu_{D_a} \times \mu_{D_b} \quad (11.134)$$

This can be checked with reference to Equation (10.108).

With redundant systems more than with non-redundant systems, it is pertinent to consider a number of elements of nominally the same type and characteristics since this is the way that practical redundancy is normally applied. For an n-element parallel redundant system it may be readily seen that Equations (11.130) to (11.133) generalize to the following forms:

$$f_{A_x}(z) = \frac{n}{\theta^n \tau^n} \left[\frac{z}{\theta^n \tau^n} - \frac{(1 - \theta^n \tau^n)}{\theta^n \tau^n} \right]^{n-1} \quad \text{for} \quad 1 - \theta^n \tau^n \leq z \leq 1 \quad (11.135)$$

$$\mu_{A_x} = 1 - \frac{\theta^n \tau^n}{n+1} \quad (11.136)$$

$$f_{D_x}(z) = \frac{n}{\theta^n \tau^n} \left(1 - \frac{z}{\theta^n \tau^n} \right)^{n-1} \quad \text{for} \quad 0 \leq z \leq \theta^n \tau^n \quad (11.137)$$

$$\mu_{D_x} = \frac{\theta^n \tau^n}{n+1} \quad (11.138)$$

where θ is the mean failure rate for each independent but nominally identical element, where each element has an exponential distribution of times to failure and where every τ is the time at which all elements are renewed such that $\theta^n \tau^n \ll 1$.

SYNTHESIS OF SYSTEM RELIABILITY

Since:
$$p_x(\tau) = \theta^n \tau^n = [p_j(\tau)]^n \qquad (11.139)$$

Equations (11.135) to (11.138) may be written in terms of $p_x(\tau)$. Also, using this same relationship, Equations (11.136) and (11.138) may be obtained directly from Equations (10.96) and (10.103) respectively.

By way of illustration, the general distribution curve for fractional dead time as given by Equation (11.137) is plotted in Figure 11.5.

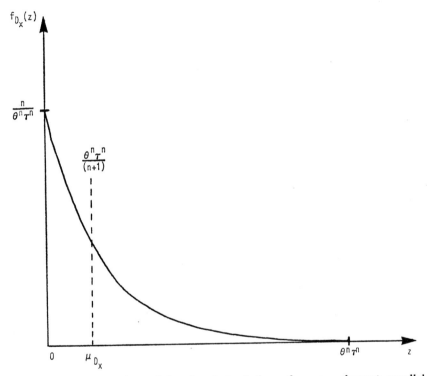

Figure 11.5 Distribution of fractional dead times for an n-element parallel redundant system subject to element failures which follow the approximate exponential distribution

If, at the end of each time interval τ, the elements of the system are out of action for a finite period of time τ_r during which time the renewal process takes place, then the values of μ_A and μ_D for the system are modified according to the relationships of Equations (10.114) and (10.116) respectively.

However, with an n-element parallel redundant system the testing of each element is unlikely to take place at the same time since this could defeat the purpose of the redundancy. The testing of the n-elements may, though, take place consecutively over a period $n\tau_r$ at the end of each interval τ. In this case, if μ_{A_y} and μ_{D_y} are the parameters of the $(n-1)$-element redundant system, then:

$$\mu_{A'_z} = \frac{\tau}{(\tau+n\tau_r)}\mu_{A_z} + \frac{n\tau_r}{(\tau+n\tau_r)}\mu_{A_y} \qquad (11.140)$$

and:

$$\mu_{D'_z} = \frac{n\tau_r}{(\tau+n\tau_r)}\mu_{D_y} + \frac{\tau}{(\tau+n\tau_r)}\mu_{D_z} \qquad (11.141)$$

These equations, however, are not strictly correct since at the end of each full test sequence the elements of the system are starting off at slightly different ages. Suppose, therefore, that the first element is tested in the interval τ to $(\tau+\tau_r)$, the second in the interval $(\tau+\tau_r)$ to $(\tau+2\tau_r)$ up to the nth element which is tested in the interval $[\tau+(n-1)\tau_r]$ to $(\tau+n\tau_r)$. A complete cycle of events then lasts for a total period of $(\tau+n\tau_r)$. Hence, applying Equation (10.103) for each interval, apart from the first:

$$\mu_{D_z} = \frac{1}{(\tau+n\tau_r)}\int_0^\tau p_n(t)p_{n-1}(t+\tau_r)\ldots p_1[t+(n-1)\tau_r]\,dt$$

$$+\frac{1}{(\tau+n\tau_r)}\int_\tau^{\tau+\tau_r} p_n(t)p_{n-1}(t+\tau_r)\ldots p_2[t+(n-2)\tau_r]p_1(\infty)\,dt$$

$$+\frac{1}{(\tau+n\tau_r)}\int_{\tau+\tau_r}^{\tau+2\tau_r} p_n(t)p_{n-1}(t+\tau_r)\ldots p_3[t+(n-3)\tau_r]$$

$$\times p_2(\infty)p_1[t-(\tau+\tau_r)]\,dt$$

$$+\ldots \qquad (11.142)$$

If now it is assumed that $p_j(t) = \theta t$ for all j and $n\tau_r \ll \tau$, then:

$$\mu_{D_z} = \frac{1}{\tau}\int_0^\tau \theta^n t^n\,dt + \frac{1}{\tau}\int_\tau^{\tau+\tau_r}\theta^{n-1}t^{n-1}\,dt$$

$$+\frac{1}{\tau}\int_{\tau+\tau_r}^{\tau+2\tau_r}\theta^{n-1}t^{n-2}[t-(\tau+\tau_r)]\,dt+\ldots \qquad (11.143)$$

Only the first two terms of Equation (11.143) are generally significant, therefore:

$$\mu_{D_z} = \frac{\theta^n \tau^n}{n+1} + \theta^{n-1}\tau^{n-2}\tau_r \qquad (11.144)$$

and putting $n = 2$, yields:

$$\mu_{D_z} = \frac{\theta^2 \tau^2}{3} + \theta\tau_r \qquad (11.145)$$

which can be seen to be the fractional dead time due to two parallel redundant elements over a period τ plus that due to two non-redundant elements over a period τ_r, and this is the approximate form that would be expected.

An alternative method of testing, as discussed in connexion with non-redundant systems, is to stagger the test of each element over the period τ. Suppose, then, that in a two-element redundant system, element A starts new at time zero, runs for a time τ and is then renewed over an interval τ_r. Element B, on the other hand, starts new at a time $k\tau + \tau_r$ and is then run and renewed in the same manner as element A. The mean fractional dead time is then given by:

$$\mu_{D_x} = \frac{1}{(\tau+\tau_r)} \int_0^{k\tau} \theta^2[t+(1-k)\tau]t\,dt + \frac{1}{(\tau+\tau_r)} \int_{k\tau}^{k\tau+\tau_r} \theta t\,dt$$
$$+ \frac{1}{(\tau+\tau_r)} \int_{k\tau+\tau_r}^{\tau} \theta^2[t-(k\tau+\tau_r)]t\,dt + \frac{1}{(\tau+\tau_r)} \int_{\tau}^{\tau+\tau_r} \theta[t-(k\tau+\tau_r)]\,dt \quad (11.146)$$

for two nominally identical elements following the approximate exponential distribution of times to failure.

If $\tau_r \ll \tau$, then Equation (11.146) reduces to:

$$\mu_{D_x} = \frac{\theta^2 \tau^2}{3} + \theta \tau_r - \frac{\theta^2 \tau^2}{2} k(1-k) \quad (11.147)$$

As k approaches unity, this equation reduces to Equation (11.145). To find the minimum value of Equation (11.147), μ_{D_x} may be differentiated with respect to k and equated to zero. This yields:

$$k = \tfrac{1}{2} \quad (11.148)$$

So that this value of k represents the optimum method of staggered testing under the various conditions assumed. The mean fractional dead time for this optimum is:

$$\mu_{D_x} = \tfrac{5}{24} \theta^2 \tau^2 + \theta \tau_r \quad (11.149)$$

or:

$$\mu_{D_x} = \tfrac{5}{24} \theta^2 \tau^2 \quad \text{if} \quad \tau_r = 0 \quad (11.150)$$

Proceeding along the same lines and assumptions as those from which Equation (11.149) was derived, it can be shown, for an n-element parallel redundant system which is subject to symmetrical staggered testing, that:

$$\mu_{D_x} = \frac{n!(n+3)(\theta\tau)^n}{4n^n(n+1)} + \frac{(n-1)!(\theta\tau)^{n-2}\theta\tau_r}{n^{n-2}} \quad (11.151)$$

where one element of the system is tested every τ/n and where $n \leq 5$. For

$\tau_r = 0$, this reduces to:

$$\mu_{D_z} = \frac{n!(n+3)(\theta\tau)^n}{4n^n(n+1)} \qquad (11.152)$$

The value of the mean fractional dead time for an n-element system where all the elements are tested together has already been given in Equation (11.138). If this expression is divided into Equation (11.152) then this gives a measure, I, of the approximate improvement obtained by symmetrical staggered testing over simultaneous testing:

$$I = \frac{4n^n}{n!(n+3)} \qquad (11.153)$$

Some typical values of I are tabulated in Table 11.3.

Table 11.3 Improvement factors due to staggered testing

n	I
1	1.00
2	1.60
3	3.00
4	6.12
5	13.17

Example 11.9

A twin-screw ship is capable of sailing satisfactorily on the thrust from either screw. Nominally identical machinery is used to drive both screws and each has times to failure which are exponentially distributed with a mean value of 1000 h. Each drive system is thoroughly overhauled and restored to the 'as new' condition every 200 h, but is not tested or repaired at any other time. If the ship is designed to be in continuous operation what is the mean fractional dead time for
(a) simultaneous testing and zero repair time,
(b) consecutive testing at the end of each interval and a repair time of 10 h per system, and
(c) symmetrical staggered testing with zero repair time?
For case (a), from Equation (11.138):

$$\mu_D = \frac{\theta^2 \tau^2}{3}$$

$$= 1.33 \times 10^{-2}$$

SYNTHESIS OF SYSTEM RELIABILITY

For case (b), from Equation (11.145):

$$\mu_D = \frac{\theta^2 \tau^2}{3} + \theta \tau_r$$

$$= 2.33 \times 10^{-2}$$

and, for case (c), from Equation (11.150):

$$\mu_D = \tfrac{5}{24} \theta^2 \tau^2$$

$$= 8.67 \times 10^{-3}$$

The number of failures that are likely to occur in a system subject to irreversible changes of state was derived in Equations (10.130) and (10.131). These formulae apply to an n-element parallel redundant system where the system cumulative probability function, $p_x(\tau)$ or $\bar{p}_x$, is given by Equation (11.119). In the case of the approximate exponential distribution applying to each of the n elements, the mean number of failures over some total time T made up of a number of intervals of length τ becomes:

$$\mu_{N_x} = T\tau^{n-1} \prod_{j=1}^{n} \theta_j \qquad (11.154)$$

and the reader will now be able to deduce corresponding formulae for other cases.

Reversible changes of state

As the reader was reminded when discussing this type of state change for non-redundant systems, Equation (10.141) applies for time-dependent failures and repairs and Equation (10.151) applies for the alternating renewal process. Either of these equations can be used to find the appropriate probability of an element of the system, such as $\bar{p}_a$, being in the failed state. The system probability of success or failure then follows from Equations (11.116) or (11.117).

Typically, for a two-element parallel redundant system, the chance of the system being in the failed state at some time t for elements subject to the alternating renewal process is:

$$\bar{p}_x = \left[L^{-1}\left\{ \frac{F_{f_a}(s)[1 - F_{r_a}(s)]}{s[1 - F_{f_a}(s) F_{r_a}(s)]} \right\} \right] \left[L^{-1}\left\{ \frac{F_{f_b}(s)[1 - F_{r_b}(s)]}{s[1 - F_{f_b}(s) F_{r_b}(s)]} \right\} \right] \qquad (11.155)$$

which may be compared with the chance of system success for the equivalent non-redundant case given in Equation (11.63). And again, in a similar way to before, the limiting mean fractional dead time from Equation (11.155) is:

$$\mu_{D_x} = \frac{\mu_{r_a}}{\mu_{f_a} + \mu_{r_a}} \times \frac{\mu_{r_b}}{\mu_{f_b} + \mu_{r_b}} \qquad (11.156)$$

or:
$$\mu_{D_x} = \mu_{D_a}\mu_{D_b} \qquad (11.157)$$

and the mean availability is:
$$\mu_{A_x} = 1 - \mu_{D_a}\mu_{D_b} \qquad (11.158)$$

or:
$$\mu_{A_x} = \mu_{A_a} + \mu_{A_b} - \mu_{A_a}\mu_{A_b} \qquad (11.159)$$

In the general case of an n-element system Equations (11.157) and (11.158) become:

$$\mu_{D_x} = \prod_{j=1}^{n} \mu_{D_j} \qquad (11.160)$$

$$\mu_{A_x} = 1 - \prod_{j=1}^{n}(1 - \mu_{A_j}) \qquad (11.161)$$

The values of Equations (11.160) and (11.161) for the exponential distribution of failures and repairs follow the same lines of derivation as Equations (11.69) to (11.73). For instance a two-element system gives:

$$\mu_{D_x} = \frac{\theta_a \theta_b \tau_a \tau_b}{(1 + \theta_a \tau_a)(1 + \theta_b \tau_b)} \qquad (11.162)$$

and an n-element system of nominally identical elements yields:

$$\mu_{D_x} = \frac{\theta^n \tau^n}{(1 + \theta\tau)^n} \qquad (11.163)$$

or:
$$\mu_{D_x} \simeq \theta^n \tau^n \quad \text{if} \quad \theta\tau \ll 1 \qquad (11.164)$$

Example 11.10

A pressure measurement system consists of three nominally identical chains of measurement any one of which is adequate to provide the required information. Each chain of measurement is subject to an alternating renewal process. The means of the distributions of times to failure and times to repair for each chain of measurement are 1000 h and 10 h respectively. What is the mean availability of the system?

For each element,
$$\mu_{D_j} = \frac{\mu_{r_j}}{\mu_{f_j} + \mu_r}$$
$$= 9.9 \times 10^{-3}$$

For the system:
$$\mu_{D_x} = [\mu_{D_j}]^3$$
$$= 9.7 \times 10^{-7}$$

SYNTHESIS OF SYSTEM RELIABILITY 467

Hence:
$$\mu_{A_z} = 1 - 9.7 \times 10^{-7}$$

The mean number of failures for an alternating renewal process was derived in Equation (10.172). However, for a parallel redundant system, all elements have to be in the failed state at the same time to produce a system failure. For a two-element system, the mean number of failures of element A in a total time T is:

$$\mu_{N_A} = \frac{T}{\mu_{f_a} + \mu_{r_a}} \tag{11.165}$$

but only a proportion of these which fall in the dead time of element B constitute a system failure. Hence, for a system failure, Equation (11.165) needs to be multiplied by the mean fractional dead time of element B. So:

$$(\mu_{N_z})_A = \frac{T}{\mu_{f_a} + \mu_{r_a}} \times \frac{\mu_{r_b}}{\mu_{f_b} + \mu_{r_b}} \tag{11.166}$$

and similarly:

$$(\mu_{N_z})_B = \frac{T}{\mu_{f_b} + \mu_{r_b}} \times \frac{\mu_{r_a}}{\mu_{f_a} + \mu_{r_a}} \tag{11.167}$$

Now:
$$\mu_{N_z} = (\mu_{N_z})_A + (\mu_{N_z})_B \tag{11.168}$$

So:
$$\mu_{N_z} = \frac{(\mu_{r_a} + \mu_{r_b}) T}{(\mu_{f_a} + \mu_{r_a})(\mu_{f_b} + \mu_{r_b})} \tag{11.169}$$

which in terms of the exponential distribution parameters becomes:

$$\mu_{N_z} = \frac{\theta_a \theta_b (\tau_a + \tau_b) T}{(1 + \theta_a \tau_a)(1 + \theta_b \tau_b)} \tag{11.170}$$

and when $\theta_a = \theta_b = \theta$ and $\tau_a = \tau_b = \tau$, then:

$$\mu_{N_z} = \frac{2 \theta^2 \tau T}{(1 + \theta \tau)^2} \tag{11.171}$$

The equivalent of Equation (11.171) for the n-element system is then seen to be:

$$\mu_{N_z} = \frac{n \theta^n \tau^{n-1} T}{(1 + \theta \tau)^n} \tag{11.172}$$

The mean time to first system failure for a redundant system subject to the alternating renewal process is not capable of being expressed in terms of simple analytical functions. However, for the particular case of exponential

16

distributions of times to failure and repair where the latter is very much smaller than the former, then a relationship of the type given in Equation (11.83) may be applied. Hence, from Equations (11.83) and (11.172), the mean time to first failure, μ_f, of an n-element redundant system becomes:

$$\mu_t = \frac{T}{\mu_{N_x}} = \frac{1}{n\theta^n \tau^{n-1}} \quad \text{for} \quad \theta\tau \ll 1 \tag{11.173}$$

and, typically, if $n = 2$ under these conditions, then:

$$\mu_t = \frac{1}{2\theta^2 \tau} \tag{11.174}$$

Example 11.11

Find the mean number of expected failures for the system of Example 11.10 over a total time of 2×10^6 h if each element of the system has an exponential distribution of times to failure and repair.

Putting the example values in Equation (11.172) yields:

$$\mu_{N_x} \simeq 0.6$$

System renewals

The discussion under this heading in connexion with non-redundant systems showed that a policy could be adopted whereby all the elements of a system were renewed every time one of the elements failed. Such a repair policy could also be adopted for a redundant system but in this case it is not practically very pertinent. A little reflexion will show that the policy would put all the redundant elements out of action for the relevant renewal time each time any redundant element failed. This would make the redundant system similar to the non-redundant one and hence defeat one of the main purposes of redundancy.

One aspect, though, which was considered in connexion with non-redundant systems, has a relevance with redundant systems. This is the aspect associated with the occurrence in the elements of unrevealed faults which lead to an irreversible process coupled with revealed faults which can lead to an alternating renewal process.

Consider, therefore, a two-element parallel redundant system where element A is subject to unrevealed faults and element B to revealed faults. Unlike the non-redundant system, though, the occurrence of these two types of faults in the two elements are considered entirely separately. Hence a revealed fault occurring in element B does not cause a renewal of element A, for the reasons just given. However, both elements of the system may be renewed at the end of the interval τ_a coupled with the irreversible process of element A.

SYNTHESIS OF SYSTEM RELIABILITY

For any time up to τ_a, then, the probability of element A being in the failed state at time t is given by Equation (10.71) and that of element B being in the failed state at time t is given by Equation (10.151). Hence, the system probability of being in the failed state, $\bar{p}_x$, is:

$$\bar{p}_x = p_a(t) \left[L^{-1} \left\{ \frac{F_{f_b}(s)\,[1 - F_{r_b}(s)]}{s[1 - F_{f_b}(s)\,F_{r_b}(s)]} \right\} \right] \qquad (11.175)$$

which in its limiting form becomes:

$$\bar{p}_x = \frac{\mu_{r_b} p_a(t)}{\mu_{f_b} + \mu_{r_b}} \qquad (11.176)$$

This is now a constant multiplied by $p_a(t)$ where $p_a(t)$ is the cumulative distribution function of the times to unrevealed (or irreversible) failures in element A. Hence, Equation (10.103) may be applied to give a mean system fractional dead time of:

$$\mu_D = \frac{\mu_{r_b}}{(\mu_{f_b} + \mu_{r_b})\tau_a} \int_0^{\tau_a} p_a(t)\,dt \qquad (11.177)$$

The reader may then like to verify that for the approximate exponential distribution this reduces to:

$$\mu_D = \tfrac{1}{2} \theta_b \tau_b \theta_a \tau_a \qquad (11.178)$$

Variational aspects

When discussing this subject in connexion with non-redundant systems, it was seen that any combination of variational distributions could apply to a non-redundant system. A redundant system, however, by definition implies that each element performs the same or equivalent function. When all elements are free from catastrophic failures, any one element will perform the system function. This means, for instance, that the element with the best performance or the quickest time of action will govern the system performance. If one element is in the catastrophic failed state, then the system function depends upon the best performance out of the remaining healthy elements. Obviously in each case, the relevant combination of variational aspects follows the methods discussed under 'Selected Combinations of Distributions' in Chapter 8.

The situation may be illustrated with reference to a two-element parallel redundant system where each element is subject to an alternating renewal process. For this case, and applying Equation (10.175), the overall reliability of the system is given by:

$$R = \mu_{A_a}\mu_{A_b}\{[p_{h_a}(x_{qu})-p_{h_a}(x_{ql})]+[p_{h_b}(x_{qu})-p_{h_b}(x_{ql})]$$
$$-[p_{h_a}(x_{qu})-p_{h_a}(x_{ql})][p_{h_b}(x_{qu})-p_{h_b}(x_{ql})]\}$$
$$+\mu_{A_a}\mu_{D_b}[p_{h_a}(x_{qu})-p_{h_a}(x_{ql})]+\mu_{D_a}\mu_{A_b}[p_{h_b}(x_{qu})-p_{h_b}(x_{ql})]$$
(11.179)

Making use of the relationship between μ_A and μ_D, Equation (11.179) can be rewritten as:

$$R = -\mu_{A_a}\mu_{A_b}\{[p_{h_a}(x_{qu})-p_{h_a}(x_{ql})][p_{h_b}(x_{qu})-p_{h_b}(x_{ql})]\}$$
$$+\mu_{A_a}[p_{h_a}(x_{qu})-p_{h_a}(x_{ql})]+\mu_{A_b}[p_{h_b}(x_{qu})-p_{h_b}(x_{ql})] \quad (11.180)$$

and where elements A and B are nominally identical:

$$R = -\mu_{A_j}^2[p_{h_j}(x_{qu})-p_{h_j}(x_{ql})]^2+2\mu_{A_j}[p_{h_j}(x_{qu})-p_{h_j}(x_{ql})] \quad (11.181)$$

Other similar equations to Equations (11.180) and (11.181) may be derived for alternative situations. This, then, has outlined the method for overall reliability synthesis of parallel redundant systems. The final step in this chapter is to undertake, as mentioned at the beginning of the chapter, a brief examination of standby redundant systems.

Standby Redundant Elements

Change of state representation

A standby redundant arrangement of two elements implies that they are each capable of performing the required system function on their own but only one of the elements is actively working at any one time. The system may start, for instance, with element A working. If element A then fails, a change-over is made to element B which is then assumed to take over the system function. A subsequent failure of element B results in a change back to element A and this alternating process is repeated for every failure event.

An alternative method of operation, which is perhaps more usual in practice, is for the main element to generally perform the system function and for the other element only to come into operation when the main element is in the failed state.

In both these schemes of operation, if the detection of an element failure and the change-over process are assumed to be perfect, then the logical and probabilistic representations of the system are the same as for a parallel redundant arrangement as given in Equations (11.108) to (11.115). In other words, the standby redundant system with perfect switch-over only fails if both elements are in the failed state at the same time.

However, the change-over process may not be perfect but have some probability, p_c, of being in the working state at the time when the active

SYNTHESIS OF SYSTEM RELIABILITY

element fails. Assume, for instance, that element A is the main element such that it always performs the system function whenever it is working. Then the system must always be successful if element A is working irrespective of the state of either element B or the change-over element, C. On the other hand, if element A fails, then both element B and element C must be working for the system to remain in the successful state. The Boolean notation for this situation is then:

$$X = A + \bar{A}BC \tag{11.182}$$

whence, by Boolean manipulation:

$$\bar{X} = \bar{A}(\bar{B} + \bar{C}) \tag{11.183}$$

and the equivalent probability state expressions to Equations (11.182) and (11.183) are respectively:

$$p_x = p_a + \bar{p}_a p_b p_c \tag{11.184}$$

$$\bar{p}_x = \bar{p}_a(\bar{p}_b + p_b \bar{p}_c) \tag{11.185}$$

If, of course, $p_c = 1$, then these last four equations become identical with those for a parallel redundant system given in Equations (11.108), (11.109), (11.112) and (11.111).

Irreversible changes of state

The considerations under this heading depend upon the method of operation of the standby system. For purposes of illustration, therefore, it is assumed that element A is the main element, element B is the standby element and element C performs the change-over from A to B. The system starts at time zero with all elements in the working state and proceeds for a total time τ when the time zero condition is restored. Element A performs the system function for the total time τ. However, if it fails before τ then it remains in the failed state until τ but at the time of failure element C is designed to switch the system function to element B which is then required to start from new and to work for the remaining part of the period. Obviously the system can proceed irreversibly into the failed state before τ, if either C or B is in the failed state when element A fails or element B subsequently fails in operation.

The state probability equation of (11.184) applies up to the moment of any change-over but requires modifying to include the probability of subsequent successful operation of element B. Each of the elements may be considered as having two failure characteristics, one for their active operational role, 'o', and one for their inactive or dormant role, 'd'. Element A starts in the active role but elements B and C start in the dormant role. When a switch-over takes place element B changes to the active role. In any role, a change of state is considered irreversible. Hence, Equation (11.184) becomes:

$$p_x = p_{a_o} + \bar{p}_{a_o} p_{b_d} p_{c_d} p_{b_o} \tag{11.186}$$

and writing this in terms of the appropriate failure distribution functions yields:

$$p_x = \overline{p_x(t)} = \overline{p_{a_o}(t)} + \int_0^t \overline{p_{b_d}(u)}\, \overline{p_{c_d}(u)}\, \overline{p_{b_o}(t-u)}\, f_{a_o}(u)\, du$$

for $0 \leqslant t \leqslant \tau$ and $0 \leqslant u \leqslant t$ (11.187)

and the corresponding probability of the system being in the failed state by time t is:

$$\bar{p}_x = p_x(t) = p_{a_o}(t) - \int_0^t \overline{p_{b_d}(u)}\, \overline{p_{c_d}(u)}\, \overline{p_{b_o}(t-u)}\, f_{a_o}(u)\, du \tag{11.188}$$

The integral Equations of (11.187) and (11.188) do not reduce to any much simpler forms. However, if all distributions are exponential, then Equation (11.187) becomes:

$$\overline{p_x(t)} = \exp(-\theta_{a_o} t) + \frac{\theta_{a_o} \exp(-\theta_{b_o} t)\{1 - \exp[-(\theta_{a_o} - \theta_{b_o} + \theta_{b_d} + \theta_{c_d}) t]\}}{\theta_{a_o} - \theta_{b_o} + \theta_{b_d} + \theta_{c_d}}$$

(11.189)

If $\theta_{c_d} = 0$ and $\theta_{b_d} = \theta_{b_o}$, then this equation reduces to the same form as that for a parallel redundant system as would be expected.

From Equations (7.67) and (11.189), the mean time to system failure, μ_f, is:

$$\mu_f = \frac{1}{\theta_{a_o}} + \frac{\theta_{a_o}}{\theta_{a_o} - \theta_{b_o} + \theta_{b_d} + \theta_{c_d}} \left(\frac{1}{\theta_{b_o}} - \frac{1}{\theta_{a_o} + \theta_{b_d} + \theta_{c_d}} \right) \tag{11.190}$$

If $(\theta_{a_o} + \theta_{b_d} + \theta_{c_d}) t \ll 1$ and $\theta_{b_o} t \ll 1$, then Equation (11.189) approximates to:

$$\overline{p_x(t)} = 1 - \tfrac{1}{2} \theta_{a_o} t^2 (\theta_{b_o} + \theta_{b_d} + \theta_{c_d}) \tag{11.191}$$

and correspondingly:

$$p_x(t) = \tfrac{1}{2} \theta_{a_o} t^2 (\theta_{b_o} + \theta_{b_d} + \theta_{c_d}) \tag{11.192}$$

which may be compared with Equation (11.123).

A variation in the above-discussed model of a standby redundant system is to assume that the elements never fail during a dormant mode but have a constant probability of failure to start operation when demanded. The cumulative distribution functions $p_{b_d}(t)$ and $p_{c_d}(t)$ then have constant probabilities of failure $\bar{p}_{b_d}$ and $\bar{p}_{c_d}$ respectively and Equation (11.188), for instance, becomes:

$$p_x(t) = p_{a_o}(t) - p_{b_d} p_{c_d} \int_0^t \overline{p_{b_o}(t-u)}\, f_{a_o}(u)\, du \tag{11.193}$$

SYNTHESIS OF SYSTEM RELIABILITY 473

The integral part of Equation (11.193) is now in the form of a convolution integral and hence a Laplace transformation of Equation (11.193) yields:

$$F_x(s) = F_{a_o}(s)\{1 - p_{b_d} p_{c_d}[1 - F_{b_o}(s)]\} \quad (11.194)$$

This leads, for exponential distributions, to:

$$f_x(t) = \theta_{a_o} \exp(-\theta_{a_o} t) - p_{b_d} p_{c_d} \theta_{a_o} \exp(-\theta_{a_o} t)$$
$$+ \frac{p_{b_d} p_{c_d} \theta_{a_o} \theta_{b_o}}{\theta_{b_o} - \theta_{a_o}} [\exp(-\theta_{a_o} t) - \exp(-\theta_{b_o} t)] \quad (11.195)$$

and:

$$\overline{p_x(t)} = \exp(-\theta_{a_o} t) + \frac{p_{b_d} p_{c_d} \theta_{a_o}}{\theta_{b_o} - \theta_{a_o}} [\exp(-\theta_{a_o} t) - \exp(-\theta_{b_o} t)] \quad (11.196)$$

and:

$$\mu_f = \frac{1}{\theta_{a_o}} + \frac{p_{b_d} p_{c_d}}{\theta_{b_o}} \quad (11.197)$$

If $\theta_{a_o} t \ll 1$ and $\theta_{b_o} t \ll 1$, then Equation (11.196) simplifies to:

$$\overline{p_x(t)} = 1 - \theta_{a_o} t(1 - p_{b_d} p_{c_d}) \quad (11.198)$$

It is now a simple matter to find the corresponding mean fractional dead times and mean availabilities by applying Equations (10.103) and (10.96) respectively to the various cumulative probabilities of failure derived for the standby redundant system. For example, under the appropriate assumed conditions, Equation (11.192) yields:

$$\mu_{D_x} = \tfrac{1}{6} \theta_{a_o} \tau^2 (\theta_{b_o} + \theta_{b_d} + \theta_{c_d}) \quad (11.199)$$

and, as another example, Equation (11.198) leads to:

$$\mu_{A_x} = 1 - \tfrac{1}{2} \theta_{a_o} \tau (1 - p_{b_d} p_{c_d}) \quad (11.200)$$

Still assuming that the approximate exponential distribution applies and that the cumulative probability of success is as given by Equation (11.198), then, by applying Equation (10.131), the mean number of system failures in a total time T made up of consecutive intervals of length τ is:

$$\mu_{N_x} \simeq \theta_{a_o} T(1 - p_{b_d} p_{c_d}) \quad (11.201)$$

Example 11.12

An electrical supply system has a standby generator set. All elements in the system have exponential distributions of times to failure and remain in the failed state once they have failed. The following mean failure-rates apply:

$$\theta_{a_o} = 5 \times 10^{-4} \text{ faults per hour}$$
$$\theta_{b_o} = 8 \times 10^{-4} \text{ faults per hour}$$
$$\theta_{b_d} = 1 \times 10^{-4} \text{ faults per hour}$$
$$\theta_{c_d} = 1 \times 10^{-4} \text{ faults per hour}$$

The complete system is renewed every 100 h. What is:
(a) the probability that the system is in the failed state at time of renewal,
(b) the mean availability of the system, and
(c) the mean availability if:

$$\theta_{b_d} = \theta_{c_d} = 0$$
$$p_{b_d} = 0.99$$
$$p_{c_d} = 0.99$$

The probability of the system being in the failed state at time t is the complement of Equation (11.191). Hence, for $t = 100$ h:

$$p_x(t) = 2.5 \times 10^{-3}$$

The corresponding mean availability may be derived from the mean fractional dead time expression of Equation (11.199), whence:

$$\mu_{A_z} = 0.99917$$

For constant probabilities of success for elements B and C, Equation (11.200) applies. With the example values this leads to:

$$\mu_{A_z} = 0.99950$$

Reversible changes of state

The basic model still considered under this heading is with element A acting as the main element and with the standby element B only operating when element A is in the failed state. Directly element A fails, element B is invoked via the change-over element C as before. If at any time, whilst element B is invoked, element B fails then the system is considered to have failed. For the moment, this final system-failed state is still considered irreversible. The first reversible process assumed is in connexion with element A. When element A fails it is taken that a repair process starts immediately. If this repair process is then completed before element B fails, then the system is immediately and perfectly switched back to element A. At this time, which represents the completion of element A repair, the whole system starts again from the time zero condition. It is also assumed, initially, that element C is perfect.

SYNTHESIS OF SYSTEM RELIABILITY

The system then starts in the condition of element A on line and working and the final state of interest is with element B on line and failed. Let the probability of this transition by a time t be denoted by $p_{a\bar{b}}(t)$. Other possible transitions are from a to b and from b to either a or $\bar{b}$, with a, b and $\bar{b}$ being the only possible states. Hence, for any value of t:

$$p_{aa}(t) + p_{ab}(t) + p_{a\bar{b}}(t) = 1 \qquad (11.202)$$

The initial conditions are:

$$p_{aa}(0) = 1, \quad p_{ab}(0) = 0, \quad p_{a\bar{b}}(0) = 0$$

and, since the failure of B on line is an irreversible state, the final conditions are:

$$p_{a\bar{b}}(\infty) = 1$$

The probability, $p_{a\bar{b}}(t)$, therefore, has the properties of a cumulative distribution function.

In this system model, the transition out of state b requires further investigation. Two transitions are possible which are mutually exclusive. Either a repair is completed on element A before element B fails and a transition takes place from b to a, or a failure of B occurs before A is repaired and the change is from b to $\bar{b}$. If the probability of the change b to a in the time increment t to $(t+\delta t)$ is $g_z(t)\,dt$, then:

$$g_z(t) = [1 - p_b(t)] f_r(t) \qquad (11.203)$$

where $f_r(t)$ and $f_b(t)$ are the distributions of times to repair element A and times to failure of element B respectively. Similarly, if $h_z(t)\,dt$ is the probability of the change from b to $\bar{b}$ in t to $(t+\delta t)$, then:

$$h_z(t) = [1 - p_r(t)] f_b(t) \qquad (11.204)$$

It should be noted that although $g_z(t)$ and $h_z(t)$ are distribution functions they are not complete density functions since their sums to infinity are only fractions of unity. However, they are mutually exclusive and exhaustive and so their partial sums make a whole such that:

$$f_z(t) = g_z(t) + h_z(t) \qquad (11.205)$$

or:

$$f_z(t) = \overline{p_b(t)} f_r(t) + \overline{p_r(t)} f_b(t) \qquad (11.206)$$

The function, $f_z(t)$, is now a complete density function and will be recognized as the function for the selection of the smaller of two variates as originally derived in Equation (8.122).

The complete system change of state equations may now be examined, for instance:

$$p_{a\bar{b}}(t) = \int_0^t p_{b\bar{b}}(t-u) f_a(u) \, du \tag{11.207}$$

or:

$$P_{a\bar{b}}(s) = P_{b\bar{b}}(s) F_a(s) \tag{11.208}$$

where $f_a(t)$ is the density function for times to failure of element A. Also:

$$p_{b\bar{b}}(t) = \int_0^t p_{\bar{b}\bar{b}}(t-u) h_z(u) \, du + \int_0^t p_{a\bar{b}}(t-u) g_z(u) \, du \tag{11.209}$$

or:

$$sP_{b\bar{b}}(s) = H_z(s) + sP_{a\bar{b}}(s) G_z(s) \tag{11.210}$$

since $p_{\bar{b}\bar{b}}(t) = 1$ for all t.

Combining Equations (11.208) and (11.210) yields:

$$sP_{a\bar{b}}(s) = \frac{F_a(s) H_z(s)}{1 - F_a(s) G_z(s)} \tag{11.211}$$

and the reader may easily verify that the other two pertinent probability states are given by:

$$sP_{aa}(s) = \frac{1 - F_a(s)}{1 - F_a(s) G_z(s)} \tag{11.212}$$

$$sP_{ab}(s) = \frac{F_a(s) [1 - F_z(s)]}{1 - F_a(s) G_z(s)} \tag{11.213}$$

If exponential distributions apply in all cases, then Equation (11.211), for instance, becomes:

$$sP_{a\bar{b}}(s) = \frac{\theta_a \theta_b}{s^2 + s(\theta_a + \theta_b + \theta_r) + \theta_a \theta_b} \tag{11.214}$$

This transform may be inverted using standard tables, but the normal interest in this function is to derive the corresponding mean time to entering the irreversible failed state of $\bar{b}$. The function $p_{a\bar{b}}(t)$ is, in fact, the cumulative distribution function of $f_f(t)$ which is the distribution of times to system failure. Hence:

$$F_f(s) = sP_{a\bar{b}}(s) \tag{11.215}$$

and, as is known:

$$\mu_f = -\left[\frac{d}{ds} F_f(s)\right]_{s=0} \tag{11.216}$$

SYNTHESIS OF SYSTEM RELIABILITY

So, the combination of these last three equations yields:

$$\mu_\mathrm{f} = \frac{\theta_a + \theta_b + \theta_\mathrm{r}}{\theta_a \theta_b} \qquad (11.217)$$

If the possible imperfection of element C or element B at the moment of demand is now taken into account by postulating a constant chance of success, p_c, for the change-over from A to B, then Equation (11.211) becomes:

$$sP_{a\bar{b}}(s) = \frac{p_c F_a(s) H_z(s) + \bar{p}_c F_a(s)}{1 - p_c F_a(s) G_z(s)} \qquad (11.218)$$

and similarly:

$$sP_{aa}(s) = \frac{1 - F_a(s)}{1 - p_c F_a(s) G_z(s)} \qquad (11.219)$$

$$sP_{ab}(s) = \frac{p_c F_a(s)[1 - F_z(s)]}{1 - p_c F_a(s) G_z(s)} \qquad (11.220)$$

and Equation (11.217) is modified to:

$$\mu_\mathrm{f} = \frac{p_c \theta_a + \theta_b + \theta_\mathrm{r}}{\theta_a(\theta_b + \bar{p}_c \theta_\mathrm{r})} \qquad (11.221)$$

Obviously, other types of imperfections in the change-over process can be dealt with in a similar way.

If once the complete system has failed a further renewal process is carried out whereby the whole system is restored to the time zero condition again then the final state of the system is no longer irreversible. The system will now settle down to a mean availability and mean fractional dead time. These values can be found by using the value of μ_f from such equations as (11.217) or (11.221) together with the mean time to complete renewal in Equations (10.158) and (10.159).

Example 11.13

An electrical supply and standby system, similar to that of Example 11.12, has elements which all have exponential distributions of event times. The mean time to failure of the main unit is 2000 h and the mean time to restore it when it fails is 50 h. A change-over element switches the supply to a standby unit when the main unit fails. The probability of the change-over element being operable at the time required is 0.99 and the probability that the standby unit is available at the instant required is also 0.99. When the main unit is repaired, the supply is perfectly switched back to the main unit, provided that the standby unit has not failed up to this time. The mean time to failure of the standby unit from the moment that it comes into operation is 1000 h.

If the whole system should fail at any time the mean time to complete restoration is 100 h. What is the mean availability of the system over a long period of time?

The mean time to complete system failure can be derived from Equation (11.221), which on inserting the example values, yields:

$$\mu_f = 3.01 \times 10^4 \text{ h}$$

and hence the limiting mean availability over a long period of time, from Equation (10.158), is:

$$\mu_A = 0.997$$

Variational aspects

With a standby redundant system, the system is generally functionally dependent on only one element at any one time. If variational aspects of performance are taken into account, this means that each successful state of the system is modified by the probability that variations in performance of the elements contributing to that state do not fall outside the required performance limits. Hence, the various probability states discussed so far in connexion with standby redundant systems require to be reviewed in the light of such relationships as given in Equation (10.63). Since there are so many possible configurations that may arise, it is not intended to deal with this method of additional analysis beyond what has already been said.

The discussion in this chapter, then, has shown how various typical system elements may be combined in non-redundant, parallel redundant and standby redundant arrangements. The methods described enable those reliability parameters of interest which were discussed in Chapter 10 to be synthesized at the system level from the corresponding parameters at the element level.

Although the element combinations dealt with cover most of the likely arrangements in typical technological systems, there are some particular combinations dealing with such aspects as majority-vote schemes and common cross-connexions which have not so far been covered. A range of these more complex aspects are studied in the next chapter.

Questions
1. A three-element system consists of elements which can either be in the working state or the failed state at any one time. Construct a truth table showing all the possible system sub-states and derive the following:
   (a) the total number of system sub-states,
   (b) the probability of being in the successful state if the system depends upon the working of either element A or the working of both element B and C,

(c) the Boolean expression for system failure under the conditions of question (b), and
(d) the probability failure diagram for the condition of question (b).

2. A measurement system consists of a sensor unit and an indicator unit. Starting from time zero, the times to failure for each unit are exponentially distributed. The mean time to failure for the sensor is 6000 h and that for the indicator is 3000 h. If either unit fails the system remains in the failed state. What is:
(a) the mean time to system failure,
(b) the probability of the system being in the failed state after 1000 h, and
(c) the probability of the system being in the successful state after 4000 h?

3. An electronic amplifier is made up of 50 resistors, 20 capacitors, 10 transistors, 10 diodes and 10 variable resistors. Every component must be in the working state for the amplifier to be working. Each component has times to failure which follow an exponential distribution with mean values of 5×10^6 h, 2×10^6 h, 10^6 h, 10^7 h and 5×10^5 h for resistors, capacitors, transistors, diodes and variable resistors respectively. What is:
(a) the mean time to failure of the amplifier, and
(b) the probability that the amplifier has failed by a time of 100 h?

4. If the electronic amplifier described in Question 3 is replaced by a new one every 200 h, what is:
(a) the mean availability over a long period of time,
(b) the probability that the availability is less than 0.99235, and
(c) the probability that the fractional dead time is less than 0.00765?

5. An engine unit and gearbox unit each have times to failure which follow a special Erlangian distribution with a parameter of 4000 h. The combined system fails if either unit fails. When a failure occurs no action is taken until the end of every 1000 h period. At the end of each of these periods the complete system is restored to the time zero condition. What is:
(a) the mean availability of the system, and
(b) the modified mean availability if the restoration process takes a constant time of 50 h each time it occurs?

6. Two generator units each have times to failure which are exponentially distributed. The mean time to failure of each unit is 1000 h. When a failure occurs on a unit no action is taken until the maintenance time is reached. One generator unit has its first maintenance after 250 h and thereafter the unit is renewed every 500 h. The other generator has its first maintenance at 500 h and thereafter is also renewed every 500 h. The system depends upon both generator units being available. What is the mean availability of the system after a long period of time?

7. A lamp system contains two lamps whose times to failure are normally distributed with a mean of 1000 h and a standard deviation of 100 h. Both lamps are renewed every 800 h. If the system depends upon both lamps operating what is the mean number of system failures in a total time of 80,000 h?

8. The lamp system of Question 7 is subject to an alternating renewal process instead of a complete system renewal every 800 h. In this process each lamp is replaced by a new one each time it fails. The mean time to detect a failure and replace a lamp is 10 h. What is the mean availability of the system?

9. A hydraulic control valve contains 20 components each of which is subject to an alternating renewal process such that when a component fails it is replaced by a new one. The mean time to failure of each component is 10,000 h and the mean time to replace it with a new one is 1 h. If the valve is out of action each time a component is being replaced, what is the mean fractional dead time?

10. A control system consists of three main parts which are a measuring device, a control amplifier and an actuator. Each part is subject to an alternating renewal process with times to failure and repair which follow an exponential distribution. The means of the distribution for each part are as follows:

	Mean time to failure	Mean time to repair
Measuring device	50,000 h	10 h
Control amplifier	5000 h	5 h
Actuator	2000 h	2 h

What is:
(a) the mean availability of the system,
(b) the mean time to the first system failure, and
(c) the mean number of system failures in a total time of 10,000 h?

11. An alarm system is subject to both unrevealed and revealed faults each of which have times to occurrence which are exponentially distributed with mean values of 200 h and 100 h respectively. If a revealed fault occurs, then the complete system is restored to the time zero condition by a repair process which has exponentially distributed times to completion. The mean value of this repair distribution is 10 h. If an unrevealed fault occurs then it remains in existence until a revealed fault occurs when it is repaired along with the revealed fault. What is the mean fractional dead time of the alarm system?

SYNTHESIS OF SYSTEM RELIABILITY 481

12. An aircraft flight panel is fitted with two types of artificial horizon indicators. The times to failure of each indicator from the start of a flight follow an exponential distribution with a mean value of 15 h for one and 30 h for the other. A flight lasts for a period of 3 h. What is:
    (a) the probability that the pilot will be without an artificial horizon indication by the end of a flight, and
    (b) the mean time to this event if the flight is of a long duration?

13. Three nominally identical shipboard generators run in parallel such that any one of the three is capable of supplying the required load. The generators are renewed at the beginning of a voyage but no repairs are possible during a voyage. The times to failure of each generator are exponentially distributed with a mean value of 4000 h. If a typical voyage lasts 8 days, what is the mean availability of the supply system over a large number of typical voyages?

14. A hydraulic system has two pumps running in parallel such that either pump can supply the required flow. Each pump has an exponential distribution of times to failure with a mean value of 2000 h. The pumps are normally unattended but at the end of every 100 h, one pump is taken out and replaced by an 'as new' pump followed by a similar replacement of the other pump. The individual outage time for each pump during replacement is 1 h. What is the approximate value of the mean fractional dead time for the system?

15. If, in the hydraulic system of Question 14, the individual pump replacements are symmetrically staggered so that one pump is renewed at 0, 100, 200, etc. h and the other at 50, 150, 250, etc. h, what is:
    (a) the new value of the mean fractional dead time with all other conditions being the same, and
    (b) the new value with the replacement time per pump assumed to be zero?

16. A direct reading temperature gauge and a temperature recorder are both connected to the same point in a process plant. Each instrument is capable of supplying the necessary information on its own. Each instrument is also independently subject to an alternating renewal process. The mean time to failure and the mean time to repair of the gauge are 5000 h and 2 h respectively. The equivalent figures for the recorder are 2000 h and 5 h. What is:
    (a) the mean fractional dead time of the combined temperature indicating system, and
    (b) the mean number of system failures in a total time of 10^5 h?

17. Three nominally identical control sub-systems are used to control an important aspect of a process plant. Each sub-system is capable of independently performing the required control function. The sub-systems are each subject to exponential distributions of times to failure

and repair with mean values of 500 h and 10 h respectively. The failure and repair process follows an alternating renewal process for each subsystem. For the overall control function, what is:
(a) the mean availability,
(b) the mean time to first system failure, and
(c) the mean number of failures in a total time of 10^6 h?

18. In the system of Question 16, both temperature indicators are subject, in addition to catastrophic failures, to continuous random variations in accuracy. These variations for each indicator follow a normal distribution with a mean value of 200°C and a standard deviation of 3°C. If the required range of accuracy for success is from 190.7°C to 209.3°C, what is the overall reliability of the system with this additional restriction?

19. A ship has a standby propulsion unit which is deemed to be adequate if the main propulsion unit fails. The ship starts a voyage with the main unit on line and in the working condition. The times to failure of the main unit are exponentially distributed with a mean value of 1000 h. If the main unit fails it remains failed to the end of the voyage. The standby unit also starts in working order but inactive. This unit may fail whilst inactive with times to failure which are also exponentially distributed but have a mean value of 2000 h. If the main unit fails there is perfect and automatic change-over to the standby unit. If the standby unit is available at this time then its failure characteristics become the same as the main unit at the start of the voyage. Any failure of the standby unit is also irrepairable during the voyage. If a typical voyage lasts for 100 h, what is:
(a) the chance of a successful voyage,
(b) the overall mean availability during voyages, and
(c) the mean time to complete propulsion failure for very long voyages?

20. For the same ship as described in Question 19, the main propulsion unit is now assumed to be repairable with an exponential distribution of times to repair which has a mean value of 20 h. Once repaired, this unit takes over perfectly from the standby unit and starts again to propel the ship in the 'as new' condition. The standby propulsion unit, when working, still has the failure characteristics of the main unit but is unrepairable. Also, the standby unit is assumed to be incapable of failure whilst in its standby or dormant role. What is:
(a) the probability that a voyage is unsuccessful, and
(b) the mean time to complete propulsion failure on a very long voyage?

CHAPTER 12

Synthesis of Complex Systems

Types of Complex Systems

The basis of the discussion on system synthesis in the previous chapter was that a system could be built up from various two-element combinations. Although this is very often true it does not always apply. Two-element combinations depend to a large extent on logical 'AND' and 'OR' functions. There are, however, as mentioned in Chapter 4, other derived functions such as the 'NOT', 'NOR' and 'NAND' which may arise in a system's functional arrangement. These additional logical functions may require a modification of the treatment previously described.

Another case where the simple two-element combination does not apply is where a number of system elements are inter-dependent or 'cross-linked' in some way. In this case, all the elements which are associated or linked with certain common elements may have to be treated as a whole. A simple example of this type is where a number of system sub-functions depend on the presence of some common supply. Alternatively, the same functional element may have a role to play in different parts of the overall system. This might apply, for example, in the case of a common piece of test equipment which is used to monitor several different aspects of system performance. Quite a common example of a cross-linked configuration arises in the case of majority-vote systems which have previously been mentioned in Chapters 1, 4 and 8.

Techniques are required, then, to deal with common elements or cross-linked configurations. These techniques are generally concerned with the synthesis of larger groups of elements than previously discussed. For this reason, the techniques may also be useful in synthesizing those simple systems where a detailed build-up of the system on a two-element basis may be tedious or laborious.

The following sections in this chapter illustrate some of the appropriate techniques which may apply by considering some typical cross-linked arrangements. In each case it is shown how the system may be reduced to standard forms and how the various reliability parameters of interest may be derived.

Exclusive Elements

In this case, one common element in the system dictates one of various alternative modes of system functioning. Only one mode of system functioning can exist at any one time and hence each mode is mutually exclusive of any other. If all modes are taken into account, then the modes which depend on the one common element are exhaustive as well as mutually exclusive. For instance, a process plant may have a pump which is driven by mains electricity when the latter is available. If, however, the electrical supply is not available

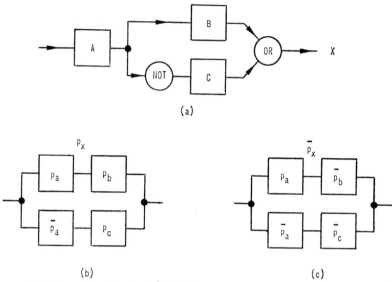

Figure 12.1 Functional and probabilistic representations of an exclusive element system

then another pump comes into operation driven by an internal combustion engine. The system, then, has two functional modes characterized by the two types of pumps and either pump assures the correct functioning of the system. The choice of the mode depends upon the presence or absence of the electrical supply and this supply, therefore, forms the common element.

The functional flow diagram for a system of the type just described can be drawn as shown in Figure 12.1(a). This diagram cannot be reduced into simple non-redundant and redundant elements, although, as will be seen, it is a more general case of standby redundancy.

In Boolean notation:
$$X = AB + \bar{A}C \tag{12.1}$$
and for system failure:
$$\bar{X} = A\bar{B} + \bar{A}\bar{C} \tag{12.2}$$

SYNTHESIS OF COMPLEX SYSTEMS

Converting Equations (12.1) and (12.2) into equivalent probabilities yields respectively:

$$p_x = p_a p_b + \bar{p}_a p_c \tag{12.3}$$

$$\bar{p}_x = p_a \bar{p}_b + \bar{p}_a \bar{p}_c \tag{12.4}$$

and diagrammatic representations of these latter equations are shown in Figures 12.1(b) and 12.1(c).

If $p_c = 0$, that is element C does not perform or does not exist, then, as can be seen by reference to the previous chapter, Equation (12.3) and Figure 12.1(b) reduce to the forms for a non-redundant system. Similarly, if $p_b = 1$, that is element B is perfect or a 'short-circuit', then Equation (12.4) and Figure 12.1(c) reduce to the forms for either a parallel or standby redundant system. An exclusive element representation can, therefore, be used to describe both non-redundant and redundant systems.

One advantage of an exclusive element representation is that the product terms in a probability 'OR' function such as Equation (11.4) become zero. This means that the Boolean notation and probability notation are virtually identical for both 'AND' and 'OR' type functions. Hence, under these conditions, a Boolean 'product' is a probability 'product' and a Boolean 'sum' is a probability 'sum'. This can be seen by comparing Equations (12.1) and (12.2) with Equations (12.3) and (12.4) respectively. The reader may have already noticed this aspect from some of the work done in the previous chapter.

The first probability term in Equation (12.3) is similar to Equation (11.3). Hence the combination of all reliability parameters from this term follow identical methods to those described under 'Non-redundant Elements' in Chapter 11. Similarly, the second probability term in Equation (12.4) is equivalent to Equation (11.111) and hence the combinational techniques for this term are the same as those dealt with under 'Parallel Redundant Elements' in Chapter 11. This latter similarity, however, only applies if element C is continually in an active role and not in a standby role.

Since the synthesis of both non-redundant and redundant elements involves the analysis of probability products such as $p_a p_b$ and $\bar{p}_a \bar{p}_c$, the reader will by now be familiar with the general method for tackling products such as $\bar{p}_a \bar{p}_c$ and $p_a \bar{p}_b$ provided again that element C is active. Hence, for irreversible changes of state it will be seen that the probability of the exclusive element system being in the successful state by a time t is:

$$p_x = \exp\left\{-\int [z_a(t)+z_b(t)]\,dt\right\} + \exp\left[-\int z_c(t)\,dt\right] \\ - \exp\left\{-\int [z_a(t)+z_c(t)]\,dt\right\} \tag{12.5}$$

where:

$$p_a p_b = \exp\left\{-\int [z_a(t) + z_b(t)] \, dt\right\} \quad (12.6)$$

as in Equation (11.24), and:

$$\bar{p}_a p_c = \exp\left[-\int z_c(t) \, dt\right] - \exp\left\{-\int [z_a(t) + z_c(t)] \, dt\right\} \quad (12.7)$$

For the approximate exponential distribution:

$$p_a p_b = 1 - (\theta_a + \theta_b) t + \tfrac{1}{2}(\theta_a + \theta_b)^2 t^2 \quad (12.8)$$

and:

$$\bar{p}_a p_c = \theta_a t - \tfrac{1}{2}(\theta_a^2 + 2\theta_a \theta_c) t^2 \quad (12.9)$$

whence:

$$p_x = 1 - \theta_b t + \tfrac{1}{2}[2\theta_a(\theta_b - \theta_c) + \theta_b^2] t^2 \quad (12.10)$$

and if $\theta_b = 0$, this reduces to the form of Equation (11.125).

The mean time to the first failure is:

$$\mu_f = \frac{1}{\theta_a + \theta_b} + \frac{1}{\theta_c} - \frac{1}{\theta_a + \theta_c} \quad (12.11)$$

which, if $\theta_c = \infty$, reduces to the form of Equation (11.31).

The mean fractional dead time for the approximate exponential distribution becomes:

$$\mu_D = \tfrac{1}{2}\theta_b \tau - \tfrac{1}{6}[2\theta_a(\theta_b - \theta_c) + \theta_b^2] \tau^2 \quad (12.12)$$

with similar modifications to this expression, as discussed in Chapter 11, for different test routines.

The mean number of failures over a time T made up of many periods of length τ where the system is repaired and renewed each τ is:

$$\mu_N = \theta_b T - \tfrac{1}{2}[2\theta_a(\theta_b - \theta_c) + \theta_b^2] \tau T \quad (12.13)$$

Example 12.1

Two generators, one diesel driven and the other hydraulically driven, are running continuously and either is capable of supplying the required electrical load. Normally only the diesel generator is on line provided that the required diesel oil supply is maintained. If it is not, then the hydraulically driven generator takes over the supply of the load. Both generators have exponentially distributed times to failure with a mean value of 2000 h. The times to failure of the oil supply are also exponentially distributed with a mean value of 4000 h. Any failure is irreversible in a time period of 500 h. What is the mean availability of the system over this time period?

From Equation (12.5), since $z_c(t) = z_b(t)$,

$$p_x = e^{-\theta_b t} \tag{12.14}$$

Hence:

$$\mu_A = \frac{1}{\theta_b \tau}(1 - e^{-\theta_b \tau}) \tag{12.15}$$

which, for the example values, gives:

$$\mu_A = 0.885$$

Notice that, for this particular set of circumstances, the system availability is independent of the time to failure of the oil supply.

If each element of the exclusive element system is subject to an alternating renewal process, then the long-term mean availability, from Equation (12.3), becomes:

$$\mu_{A_x} = \mu_{A_a}\mu_{A_b} + \mu_{D_a}\mu_{A_c} \tag{12.16}$$

which is general for any distribution of failure and repair. Also:

$$\mu_{D_x} = \mu_{A_a}\mu_{D_b} + \mu_{D_a}\mu_{D_c} \tag{12.17}$$

which, in the particular case of the approximate exponential distribution, reduces to:

$$\mu_{D_x} = \frac{\theta_b \tau_b}{(1+\theta_a \tau_a)(1+\theta_b \tau_b)} + \frac{\theta_a \theta_c \tau_a \tau_c}{(1+\theta_a \tau_a)(1+\theta_c \tau_c)} \tag{12.18}$$

or:

$$\mu_{D_x} \simeq \theta_b \tau_b + \theta_a \theta_c \tau_a \tau_c \tag{12.19}$$

Other relationships can be found in a similar manner to those derived in Chapter 11.

So far it has been taken that element C is continually in an active state as far as its failure characteristics are concerned. However, as with the standby systems discussed in the previous chapter, element C may be inactive up to the time of failure of element A and then if available, start operation in the 'as new' condition. This means, for irreversible changes of state, that the system probability of being in the successful state needs to be represented by an equation of the type of Equation (11.187). The cumulative probability function, $p_{a_0}(t)$, in this case, however, is derived from both elements A and B and if it is assumed that element C is perfect up to the time of failure of element A, then:

$$p_x = \overline{p_x(t)} = \overline{p_a(t)}\,\overline{p_b(t)} + \int_0^t \overline{p_c(t-u)} f_a(u)\,du \tag{12.20}$$

which, for the exponential distribution, yields:

$$\overline{p_x(t)} = \exp[-(\theta_a+\theta_b)t] + \frac{\theta_a}{\theta_c-\theta_a}[\exp(-\theta_a t)-\exp(-\theta_c t)] \quad (12.21)$$

and which may be compared with Equation (11.196). Also, the mean time to first system failure is:

$$\mu_f = \frac{1}{\theta_a+\theta_b} + \frac{1}{\theta_c} \quad (12.22)$$

For the approximate exponential distribution, Equation (12.21) becomes:

$$p_x = \overline{p_x(t)} = 1 - \theta_b t + \tfrac{1}{2}[\theta_a(2\theta_b-\theta_c)+\theta_b^2]t^2 \quad (12.23)$$

which may be compared with Equation (12.10). Under the same conditions:

$$\mu_{D_x} = \tfrac{1}{2}\theta_b\tau - \tfrac{1}{6}[\theta_a(2\theta_b-\theta_c)+\theta_b^2]\tau^2 \quad (12.24)$$

Superimposed upon any standby arrangement are various possibilities for repair routines as discussed in Chapter 11. One possibility, in the exclusive element system, is that element C perfectly takes over when element A fails and element A is subject to an alternating renewal process. Directly element A is restored, element C is no longer required for system functioning. On the other hand, if element C fails whilst it is in operation then this is irreversible and the system fails. This is one of the situations dealt with in Chapter 11 and for the present case, if element B is perfect, this means that Equations (11.211) to (11.213) apply. On the other hand, if element B is not perfect but is subject to an alternating renewal process, then, in the limit, element B has a fixed probability of being in the failed state. This is given by:

$$\mu_{D_b} = \frac{\theta_b\tau_b}{1+\theta_b\tau_b} \quad (12.25)$$

where θ_b is the mean time to failure of element B and τ_b is the mean time to repair element B assuming exponential distributions.

As far as the system is concerned, an equation of the form of Equation (11.211) still represents a system failed state by a time t. For exponential distributions this yields in the present instance:

$$sP_{a\bar{c}}(s) = \frac{\theta_a\theta_c}{s^2+s(\theta_a+\theta_c+1/\tau_a)+\theta_a\theta_c} \quad (12.26)$$

In addition, an equation of the form of Equation (11.212) can also represent a system failed state if, at the same time, element B is in the failed state. This means that the appropriate form of Equation (11.212) needs to be multiplied by μ_{D_b} in order to yield the other exclusive form of system failure:

$$\mu_{D_b}sP_{aa}(s) = \frac{\theta_b\tau_b[s^2+s(\theta_c+1/\tau_a)]}{(1+\theta_b\tau_b)[s^2+s(\theta_a+\theta_c+1/\tau_a)+\theta_a\theta_c]} \quad (12.27)$$

SYNTHESIS OF COMPLEX SYSTEMS

The total probability of the system being in the failed state at time t is then given by the sum of the expressions in Equations (12.26) and (12.27). The inverse transform of this combined expression may be obtained from Table A.2. The expression is rather complex, but it can be seen that on a short time scale Equation (12.27) predominates and approximates to the form of Equation (12.25) and on a long time scale Equation (12.26) predominates. So, if the system as a whole is regularly renewed, then the probability of it being in the failed state is given almost entirely by Equation (12.25). A similar facet will already have been noticed in previous equations where the characteristics of element B tend to determine the complete characteristics of the exclusive element system.

Example 12.2

The generating system of Example 12.1 now has an oil supply situation which is subject to an alternating renewal process. The mean time to the restoration of oil supply once it has failed is 40 h and the times are exponentially distributed. The diesel-driven generator is also subject to a renewal process with repair times which are exponentially distributed with a mean value of 20 h. Any failure of the hydraulically driven generator is still irreversible over the time period of interest. What is the probability of the system being in the failed state at:

(a) 5 h, and
(b) 500 h?

Since, in this example, the repair times are very much less than the corresponding times to failure, then for small values of t (or large s), Equations (12.26) and (12.27) can be approximated to:

$$P_{a\bar{c}}(s) = \frac{\theta_a \theta_c}{s^2(s + 1/\tau_a)} \qquad (12.28)$$

which yields:

$$p_{a\bar{c}}(t) = \theta_a \theta_c \tau_a^2 \left\{ \frac{t}{\tau_a} - [1 - \exp(-t/\tau_a)] \right\} \qquad (12.29)$$

and:

$$\mu_{D_b} P_{aa}(s) = \frac{\theta_b \tau_b}{s} \qquad (12.30)$$

which yields:

$$\mu_{D_b} p_{aa}(t) = \theta_b \tau_b \qquad (12.31)$$

At a time of 5 h, only Equation (12.31) is significant and this gives a value of:

$$\mu_{D_b} p_{aa}(t) = 0.01$$

at a time of 500 h:

$$\mu_{D_b} p_{aa}(t) + p_{a\bar{c}}(t) = 0.01 + 0.0025$$
$$= 0.0125$$

Variations in performance of the elements of an exclusive element system may be dealt with by the methods already described in Chapter 11. For instance, if elements A and B are working then the variational analysis involves the combined performance functions of elements A and B. If element A has failed, then it is only a question of whether element C remains within the required performance limits. The reader is referred to relationships of the type given in Equation (11.179).

Simple Common Elements

In this case, the information or supply from one common element jointly feeds two other elements either of which may be used for system functioning. The system is very similar to the exclusive element arrangement just discussed except that no 'NOT' function is involved and element A is continually connected to both elements B and C. The relevant flow diagram is shown in Figure 12.2(a).

The first thing to notice from the flow diagram is that elements B and C form, on their own, a standard parallel redundant arrangement. This part of the system may therefore be synthesized, independently of element A, on the lines described in Chapter 11 to form a combined element which may be called D. The synthesized element D is now in a non-redundant arrangement with element A and hence the complete solution of the system also follows the lines previously described. It is obvious then that any redundant–non-redundant arrangement, or series–parallel arrangement, of this type can be synthesized by a step-by-step process of reduction using the two basic methods of element combination.

The simple common element system, however, has some characteristics which it is useful to explore in connexion with the synthesis of more complex arrangements which will be discussed in later sections of this chapter. Examining first the Boolean notation it is seen that:

$$X = A(B+C) \qquad (12.32)$$

that is, element A is in 'series' with the 'parallel' combination of elements B and C. The system may also be looked upon as a 'parallel' combination of elements A and B in 'series' with elements A and C in 'series', that is:

$$X = AB + AC \qquad (12.33)$$

SYNTHESIS OF COMPLEX SYSTEMS

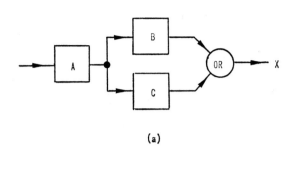

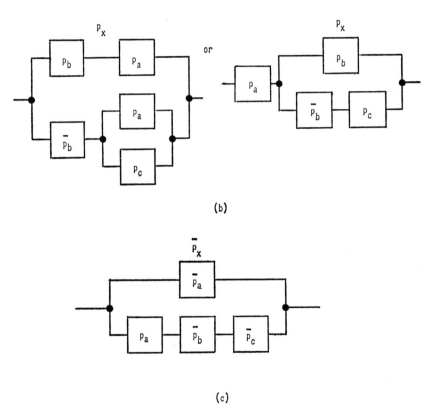

Figure 12.2 Functional and probabilistic representations of a simple common element system

which is just a simple expansion of Equation (12.32). Similarly:

$$\bar{X} = \bar{A} + \bar{B}\bar{C} \qquad (12.34)$$

or:

$$\bar{X} = (\bar{A} + \bar{B})(\bar{A} + \bar{C}) \qquad (12.35)$$

where the standard Boolean reductions of Equation (12.35) yield Equation (12.34).

If Equation (12.34) is converted into equivalent probabilities, then:

$$\bar{p}_x = \bar{p}_a + \bar{p}_b\bar{p}_c - \bar{p}_a\bar{p}_b\bar{p}_c \qquad (12.36)$$

applying the usual conversion rules for 'AND' and 'OR' functions.

On the other hand, converting Equation (12.35) yields:

$$\bar{p}_x = (\bar{p}_a + \bar{p}_b - \bar{p}_a\bar{p}_b)(\bar{p}_a + \bar{p}_c - \bar{p}_a\bar{p}_c) \qquad (12.37)$$

and the result:

$$\bar{p}_x = \bar{p}_a + \bar{p}_b\bar{p}_c - \bar{p}_a\bar{p}_b\bar{p}_c \qquad (12.38)$$

is only obtained by using the relationship that:

$$\bar{p}_a\bar{p}_a = \bar{p}_a \qquad (12.39)$$

This, of course, is similar to the Boolean law that:

$$\bar{A}\bar{A} = \bar{A} \qquad (12.40)$$

and it is apparent that it also applies to probabilities since a probability multiplied by itself still represents the single probability of the same event. On the same lines, it is obvious that:

$$p_a\bar{p}_a = 0 \qquad (12.41)$$

These simple probability laws generally form an important aspect of any system which contains common elements. For instance, if the two bracketed terms in Equation (12.37) were each evaluated separately to produce a probability number then these numbers would be correct in themselves. However, the product of these numbers would yield an incorrect value because of the common element associated with each one. This emphasizes that a system with a common element needs to be algebraically reduced to its simplest form before any numerical evaluation is carried out. The reader will remember that these aspects were also mentioned under the heading of 'Dependent correlations' in Chapter 10.

When discussing the exclusive element system it was seen that both non-redundant and redundant arrangements could be expressed in 'exclusive' form. For instance, putting $B = 1$ and $C = B$ in Equation (12.1) gives:

$$X = A + \bar{A}B \qquad (12.42)$$

which is identical to the redundant arrangement of elements A and B as given

SYNTHESIS OF COMPLEX SYSTEMS

in Equation (11.108), that is:
$$X = A + B \tag{12.43}$$
also of identical form is:
$$X = B + \bar{B}A \tag{12.44}$$
Changing Equation (12.32) to 'exclusive' form then yields:
$$X = A(B + \bar{B}C) \tag{12.45}$$
and a similar process on Equation (12.34) gives:
$$\bar{X} = \bar{A} + A\bar{B}\bar{C} \tag{12.46}$$
The conversion of 'exclusive' Boolean forms into the corresponding probabilities is direct as seen in the previous section. Hence the probability equivalent of Equation (12.45) is:
$$p_x = p_a(p_b + \bar{p}_b p_c) \tag{12.47}$$
and the equivalent of Equation (12.46) is:
$$\bar{p}_x = \bar{p}_a + p_a \bar{p}_b \bar{p}_c \tag{12.48}$$
which is seen to be identical with Equation (12.36). The 'exclusive' or probability series–parallel diagrams for Equations (12.47) and (12.48) are shown in Figures 12.2(b) and 12.2(c) respectively.

There is an alternative way in which Equations (12.47) and (12.48) may be obtained. The effects of the common or linked element being in either of its two states may be treated separately since they are exclusive events. This leads to a probabilistic statement of the following form:

(Probability of system in successful state)

= (probability of element A in successful state)

× (probability of system in successful state with element A successful)

+ (probability of element A in failed state)

× (probability of system in successful state with element A failed)

$$\tag{12.49}$$

which may be written symbolically as:
$$p_x = p(X|A)p_a + p(X|\bar{A})\bar{p}_a \tag{12.50}$$
The expressions such as $p(X|A)$ are the conditional probabilities of state X given that state A exists. For the present case it can be seen that:
$$p(X|A) = p_b + \bar{p}_b p_c \tag{12.51}$$
and:
$$p(X|\bar{A}) = 0 \tag{12.52}$$

Hence, from Equation (12.50):

$$p_x = p_a(p_b + \bar{p}_b p_c) \tag{12.53}$$

as before.
Similarly:

$$\bar{p}_x = p(\bar{X}|A)p_a + p(\bar{X}|\bar{A})\bar{p}_a \tag{12.54}$$

where:

$$p(\bar{X}|A) = \bar{p}_b \bar{p}_c \tag{12.55}$$

and:

$$p(\bar{X}|\bar{A}) = 1 \tag{12.56}$$

so that:

$$\bar{p}_x = p_a \bar{p}_b \bar{p}_c + \bar{p}_a \tag{12.57}$$

as before.

Equations of the type of Equations (12.50) and (12.54) represent particular examples of a general theorem and technique which will be discussed in more detail in the concluding section of this chapter.

The synthesis of reliability parameters for the simple common element system may be carried out by the non-redundant and redundant synthesis described at the beginning of this section. Alternatively, the system may be treated as a whole and the synthesis carried out on the basis of the sums and products of the probability states as given in Equations (12.53) and (12.57). The methods of dealing with such sums and products have already been covered earlier in this chapter and in Chapter 11.

Cross-linked Common Elements

An illustration of this type of system is given in Figure 12.3. The functional flow for system success is from left to right through elements A, B, C and D and either way through element E. Note that this system cannot be divided into simple non-redundant and redundant arrangements.

A method of finding the expression for success in Boolean notation is to trace each success path from node 1 to node 4. The Boolean products of each element in a success path are then added together. This technique was originally examined for a similar type of system in Chapter 4 and leads to the result:

$$X = AB + CD + AED + CEB \tag{12.58}$$

Any similar or higher-order terms disappear in the Boolean manipulation. Applying de Morgan's rule, then yields the corresponding expression for the system failed state:

$$\bar{X} = \bar{A}\bar{C} + \bar{B}\bar{D} + \bar{A}\bar{E}\bar{D} + \bar{C}\bar{E}\bar{B} \tag{12.59}$$

SYNTHESIS OF COMPLEX SYSTEMS   495

An alternative approach is to express the system network node connexions in a matrix form where each element of the matrix represents the Boolean connexion between two nodes. This representation was also mentioned in Chapter 4. For instance, the nodal connexion (1, 2) is represented by A since this is the successful notation for a passage from node 1 to node 2. The symbol A is then written in the first row and second column of the matrix. In general the symbol located in the jth row and kth column of the matrix is the Boolean symbol for success of the element connected between the jth and kth node in the direction of j to k. The nodes are allocated so that each element is bounded by a node on either side. The elements of the

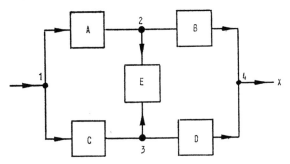

Figure 12.3 A cross-linked common element system

matrix (j, j) are equated to unity since this represents a node connected to itself. Any nodal element (j, k) which embraces more than one direct series element or represents an inappropriate reverse functional flow is equated to zero. The connexion matrix, M, for the system of Figure 12.3 illustrates the procedure:

$$M = \begin{bmatrix} 1 & A & C & 0 \\ 0 & 1 & E & B \\ 0 & E & 1 & D \\ 0 & 0 & 0 & 1 \end{bmatrix} \quad (12.60)$$

The matrix, M, may then be reduced to contain a single transmission term between the nodes 1 and 4 by a process of node removals as outlined in Chapter 4. An alternative approach is to create the power matrix M^n by multiplying the matrix, M, n times by itself. This power matrix, M^n, is then evaluated with a minimum value of n such that any higher powers of n leave the resultant power matrix unchanged. With the present example this is achieved with $n = 3$, and M^3 becomes:

$$M^3 = \begin{bmatrix} 1 & (A+CE) & (C+AE) & (AB+CD+AED+CEB) \\ 0 & 1 & E & (B+ED) \\ 0 & E & 1 & (D+EB) \\ 0 & 0 & 0 & 1 \end{bmatrix} \quad (12.61)$$

It is now seen that the matrix is complete and each element of the matrix represents the success, in Boolean notation, between all the relevant nodes. Of particular interest, of course, is the (1, 4) element since this represents the complete system success. It is seen that this element of the matrix is identical with Equation (12.58).

The connexion matrix method may be valuable in a complex system where all the possible success paths are not readily apparent or where there is a chance of human error in omitting to include some of the relevant paths.

Equation (12.59) may be converted into an 'exclusive' Boolean form. There are a number of possible and equally acceptable results obtainable by this conversion depending on the way in which the elements of the equation are grouped. One possible grouping can be carried out as follows:

$$\bar{X} = [\bar{A}\bar{C} + \overline{\bar{A}\bar{C}}\bar{B}\bar{D}] + [\overline{\bar{A}\bar{C}} + \overline{\bar{A}\bar{C}\bar{B}\bar{D}}][\bar{A}\bar{E}\bar{D} + \overline{\bar{A}\bar{E}\bar{D}}\bar{C}\bar{E}\bar{B}] \quad (12.62)$$

which by successive reductions yields:

$$\bar{X} = \bar{A}\bar{C} + A\bar{B}\bar{D} + A\bar{B}\bar{C}D\bar{E} + \bar{A}BC\bar{D} + \bar{A}BC\bar{D}\bar{E} \quad (12.63)$$

and the corresponding probability equation is:

$$\bar{p}_x = \bar{p}_a\bar{p}_c + p_a\bar{p}_b\bar{p}_d + p_a\bar{p}_b\bar{p}_c p_d \bar{p}_e + \bar{p}_a\bar{p}_b p_c \bar{p}_d + \bar{p}_a p_b p_c \bar{p}_d \bar{p}_e \quad (12.64)$$

which may then be used to synthesize the overall system reliability parameters using the methods for sums and products of probabilities previously described.

As another approach, use may be made of equations of the type of Equations (12.50) and (12.54). In this case the pertinent element to choose is element E, whence:

$$p_x = p(X|E)p_e + p(X|\bar{E})\bar{p}_e \quad (12.65)$$

With element E perfectly successful, the system reduces to a parallel redundant arrangement of elements A and C in a non-redundant configuration with a parallel redundant arrangement of elements B and D, hence:

$$p(X|E) = (p_a + \bar{p}_a p_c)(p_b + \bar{p}_b p_d) \quad (12.66)$$

Also, with element E perfectly failed, the system reduces to a non-redundant arrangement of elements A and B in a parallel redundant configuration with

SYNTHESIS OF COMPLEX SYSTEMS

a non-redundant arrangement of elements C and D, hence:

$$p(X|\bar{E}) = p_a p_b + \overline{p_a p_b} p_c p_d$$
$$= p_a p_b + (\bar{p}_a + \bar{p}_b) p_c p_d \qquad (12.67)$$

and substituting Equations (12.66) and (12.67) in Equation (12.65) yields:

$$p_x = p_e(p_a + \bar{p}_a p_c)(p_b + \bar{p}_b p_d) + \bar{p}_e[p_a p_b + (\bar{p}_a + \bar{p}_b) p_c p_d] \qquad (12.68)$$

and the reader may like to verify that this is the complement of Equation (12.64).

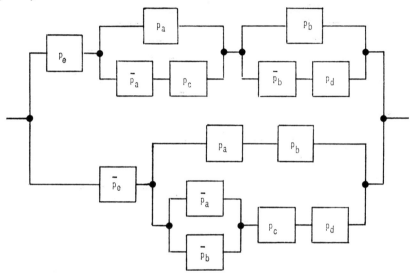

Figure 12.4 Probabilistic representation of the system of Figure 12.3

An advantage of this latter method is that Equations (12.66) and (12.67) represent combinations of standard non-redundant and redundant elements and Equation (12.68) represents an exclusive element system. So the system can be synthesized from these three standard forms. The corresponding exclusive probability diagram is shown in Figure 12.4 and a rearrangement in 'tree' form is shown in Figure 12.5.

Obviously, a wide variety of cross-linked configurations are possible but a typical type occurs in connexion with majority-vote schemes. These are therefore reviewed in more detail in the following section.

Majority-vote Elements

Change of state representation

A majority-vote element system is one consisting of n independent elements which are so arranged that any m or more ($m \leq n$) of the elements are required to be in the successful state for the system to be successful. Conversely, the system may be defined in that any r or more of the elements are required to be in the failed state in order to reduce the system to the overall failed state. It then follows that:

$$r = n - m + 1 \tag{12.69}$$

a relationship previously expressed in Equation (5.59).

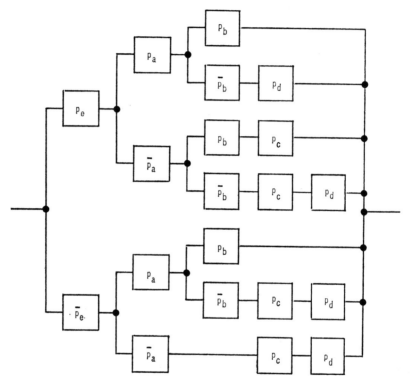

Figure 12.5 Rearrangement of Figure 12.4 in 'tree' form

The simplest majority-vote system consists of 3 elements of which any 2 or more are required to be successful for system success. Here:

$$\left. \begin{array}{l} n = 3 \\ m = 2 \\ r = 2 \end{array} \right\} \tag{12.70}$$

A possible system functional flow diagram is shown in Figure 12.6(a) and it is noticed that each element fulfils a functional position in two of the flow paths. There are thus cross-linked connexions between all the elements.

SYNTHESIS OF COMPLEX SYSTEMS

Figure 12.6(b) shows the same system written in the logical form of 'AND' and 'OR' functions and Figure 12.6(c) a shorthand notation for expressing this logical configuration on the lines originally introduced in Chapter 4.

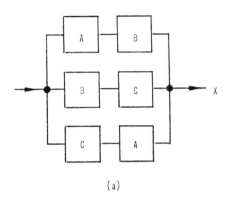

(a)

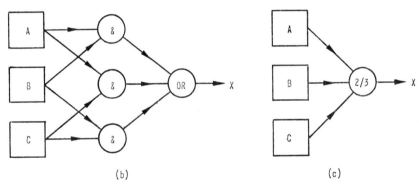

Figure 12.6 Functional and logical representations of a 2-out-of-3 majority-vote system

The Boolean expressions for the system are:

$$X = AB + BC + CA \qquad (12.71)$$

and:

$$\bar{X} = \bar{A}\bar{B} + \bar{B}\bar{C} + \bar{C}\bar{A} \qquad (12.72)$$

and the corresponding probability expressions:

$$p_x = p_a p_b + p_b p_c + p_c p_a - 2 p_a p_b p_c \qquad (12.73)$$

and:

$$\bar{p}_x = \bar{p}_a \bar{p}_b + \bar{p}_b \bar{p}_c + \bar{p}_c \bar{p}_a - 2 \bar{p}_a \bar{p}_b \bar{p} \qquad (12.74)$$

Re-expressing these last four equations in 'exclusive' terms yields:

$$X = \bar{A}BC + A(B + \bar{B}C) \qquad (12.75)$$

$$\bar{X} = A\bar{B}\bar{C} + \bar{A}(\bar{B} + B\bar{C}) \qquad (12.76)$$

$$P_x = \bar{p}_a p_b p_c + p_a(p_b + \bar{p}_b p_c) \qquad (12.77)$$

$$\bar{P}_x = p_a \bar{p}_b \bar{p}_c + \bar{p}_a(\bar{p}_b + p_b \bar{p}_c) \qquad (12.78)$$

and the corresponding probability diagrams are shown in Figure 12.7.

Normally, as with redundant element systems, all the elements of a majority-vote system are nominally the same. This means that all the elements have the same probability value but are still independent of each other and are not identical in the probabilistic sense. Hence, under these conditions, if:

$$p_a = p_b = p_c = p \qquad (12.79)$$

then:

$$P_x = 3p^2 - 2p^3 \qquad (12.80)$$

or:

$$P_x = p^2(1 + 2\bar{p}) \qquad (12.81)$$

and:

$$\bar{P}_x = 3\bar{p}^2 - 2\bar{p}^3 \qquad (12.82)$$

or:

$$\bar{P}_x = \bar{p}^2(1 + 2p) \qquad (12.83)$$

These latter equations will now be recognized as particular forms of the binomial cumulative distribution function for the probability of m or more outcomes out of a total of n events when each outcome has a probability of occurrence, p, per event. The reader is referred to the discussion on this distribution in Chapters 5 and 7. Obviously then, in the general case, and in line with Equation (5.54):

$$P_x = \sum_{j=m}^{n} \binom{n}{j} p^j (1-p)^{n-j} \qquad (12.84)$$

or:

$$P_x = \sum_{j=m}^{n} \binom{n}{j} p^j (\bar{p})^{n-j} \qquad (12.85)$$

and:

$$\bar{P}_x = \sum_{j=r}^{n} \binom{n}{j} (\bar{p})^j (1-\bar{p})^{n-j} \qquad (12.86)$$

or:

$$\bar{P}_x = \sum_{j=r}^{n} \binom{n}{j} (\bar{p})^j p^{n-j} \qquad (12.87)$$

SYNTHESIS OF COMPLEX SYSTEMS

The inter-relationships between these types of functions has already been illustrated in Equations (5.54) to (5.62).

Putting $n = 3$, $m = 2$ and $r = 2$ in Equations (12.84) to (12.87) brings them to the form of Equations (12.80) to (12.83) respectively.

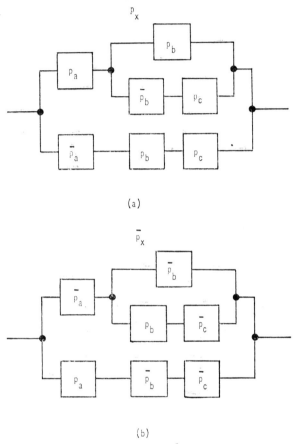

Figure 12.7 Probabilistic representations of a 2-out-of-3 majority-vote system

In practice, $\bar{p}$, the probability of an element being in the failed state, will be small. In this case, Equation (12.87) approximates to:

$$\bar{p}_x \simeq \binom{n}{r}(\bar{p})^r \qquad (12.88)$$

which is of the same form as the equivalent expression for parallel redundant elements [see Equation (11.117)] multiplied by the binomial coefficient $\binom{n}{r}$.

In the same way:

$$p_x \simeq 1 - \binom{n}{r}(\bar{p})^r \qquad (12.89)$$

Irreversible changes of state

For any distribution function relating to the failure characteristics of each element of an *n*-element majority-vote system, Equation (12.87) becomes:

$$p_x(t) = \bar{p}_x = \sum_{j=r}^{n} \binom{n}{j}\left\{1 - \exp\left[-\int z(t)\,dt\right]\right\}^j \exp\left[-\int (n-j)z(t)\,dt\right] \qquad (12.90)$$

or, for the exponential distribution:

$$p_x(t) = \bar{p}_x = \sum_{j=r}^{n} \binom{n}{j}[1 - \exp(-\theta t)]^j \exp[-(n-j)\theta t] \qquad (12.91)$$

or:

$$p_x(t) = \bar{p}_x \simeq \binom{n}{r}\theta^r t^r \quad \text{if} \quad \theta t \ll 1 \qquad (12.92)$$

For the 2-out-of-3 system, Equation (12.91) becomes:

$$p_x(t) = 1 - 3\exp(-2\theta t) + 2\exp(-3\theta t) \qquad (12.93)$$

or, if $\theta t \ll 1$, then:

$$p_x(t) = 3\theta^2 t^2 \qquad (12.94)$$

which may be compared with Equation (11.123) for a two-element redundant system. The mean time to first system failure is:

$$\mu_f = \int_0^\infty \overline{p_x(t)}\,dt$$

$$= \frac{5}{6\theta} \qquad (12.95)$$

which may be compared with Equation (11.122).

Example 12.3

Four nominally identical measurement channels are connected so that they operate in a 3-out-of-4 logic. This means that any three or more must be successful to achieve system success. Any failure of any element is irreversible in the time scale of interest. The times to failure of each element are exponentially distributed with a mean value of 2000 h. What is:
(a) the probability of the system being in the successful state at 500 h, and
(b) the mean time to first system failure?

SYNTHESIS OF COMPLEX SYSTEMS 503

From Equation (12.85):

$$\overline{p_x(t)} = 4\exp(-3\theta t) - 3\exp(-4\theta t) \tag{12.96}$$

which with the example values yields:

$$\overline{p_x(t)} = 0.788$$

Also, the mean time to first failure for this system is:

$$\mu_f = \frac{7}{12\theta} \tag{12.97}$$

or:

$$\mu_f = 1167 \text{ h}$$

The mean number of failures that are likely to occur with a majority-vote system which is subject to irreversible changes of state but which is also completely renewed every time interval τ is derived from Equation (10.131) together with the value of $p_x(\tau)$ from Equation (12.90). For the 2-out-of-3 system subject to exponential distributions of times to failure, this gives:

$$\mu_N = \frac{T}{\tau}[1 - 3\exp(-2\theta\tau) + 2\exp(-3\theta\tau)] \tag{12.98}$$

or, if $\theta\tau \ll 1$, then:

$$\mu_N = 3\theta^2 T\tau \tag{12.99}$$

For an n-element majority-vote system where the approximate exponential distribution of times to failure applies for each element:

$$\mu_N = \binom{n}{r} \theta^r \tau^{r-1} T \tag{12.100}$$

The next reliability parameter of interest for the majority-vote system is that of availability or fractional dead time. The distribution of availability is obtained, as before, from Equation (10.95) with $p_x(\tau)$, in this case, evaluated from Equation (12.90), that is:

$$p_x(\tau) = \sum_{j=r}^{n} \binom{n}{j} [p(\tau)]^j [\overline{p(\tau)}]^{n-j} \tag{12.101}$$

and $f_x(t)$ is obtained from the differential of Equation (12.101) which gives:

$$f_x(t) = n[p(t)]^{n-1}f(t) + \sum_{j=r}^{n-1} \binom{n}{j}[p(t)]^{j-1}[\overline{p(t)}]^{n-j-1}[j - np(t)]f(t) \tag{12.102}$$

which, if $n = 3$ and $r = 2$, reduces to:

$$f_x(t) = 6p(t)f(t)[1 - p(t)] \tag{12.103}$$

a formula which was previously obtained in Equation (8.150) in connexion with 2-out-of-3 selected combinations of distributions.

For the approximate exponential distribution, Equations (12.101) and (12.102) become:

$$p_x(\tau) = \binom{n}{r} \theta^r \tau^r \qquad (12.104)$$

$$f_x(t) = \binom{n}{r} r \theta^r t^{r-1} \qquad (12.105)$$

whence, from Equations (10.95) and (10.102):

$$f_A(z) = \left[r \bigg/ \binom{n}{r} \theta^r \tau^r \right] \left\{ \left[z \bigg/ \binom{n}{r} \theta^r \tau^r \right] - \left[1 - \binom{n}{r} \theta^r \tau^r \bigg/ \binom{n}{r} \theta^r \tau^r \right] \right\}^{r-1}$$

$$\text{for} \quad 1 - \binom{n}{r} \theta^r \tau^r \leqslant z \leqslant 1 \quad (12.106)$$

and:

$$f_D(z) = \left[r \bigg/ \binom{n}{r} \theta^r \tau^r \right] \left\{ 1 - \left[z \bigg/ \binom{n}{r} \theta^r \tau^r \right] \right\}^{r-1} \quad \text{for} \quad 0 \leqslant z \leqslant \binom{n}{r} \theta^r \tau^r \qquad (12.107)$$

and these latter two equations may be compared with the equivalent relationships for parallel redundant systems given in Equations (11.135) and (11.137) respectively.

The mean availability or fractional dead time may be obtained from Equations (12.106) and (12.107) for the approximate exponential distribution or more directly, in the general case, from Equation (12.101) since, for instance:

$$\mu_D = \frac{1}{\tau} \int_0^\tau p_x(t) \, dt \qquad (12.108)$$

For the full exponential distribution, this integral yields:

$$\mu_D = 1 - \frac{1}{\theta \tau} \sum_{j=0}^{r-1} \left\{ \frac{1}{(n-j)} \sum_{k=j+1}^{n} \binom{n}{k} [p(\tau)]^k [\overline{p(\tau)}]^{n-k} \right\} \qquad (12.109)$$

where:

$$p(\tau) = 1 - \exp(-\theta \tau) \qquad (12.110)$$

It is seen that the term in brackets in Equation (12.109) is still in the form of the binomial cumulative distribution. This term may, therefore, be evaluated numerically from binomial tables or, more approximately, from the appropriate curve in Figures A.1 to A.27 inclusive.

SYNTHESIS OF COMPLEX SYSTEMS

For the approximate exponential distribution, Equation (12.109) reduces to:

$$\mu_D = \binom{n}{r} \frac{\theta^r \tau^r}{(r+1)} \qquad (12.111)$$

a result which could also have been obtained from Equations (12.108) and (12.104) or by finding the mean value of Equation (12.107).

Similarly:

$$\mu_A = 1 - \binom{n}{r} \frac{\theta^r \tau^r}{(r+1)} \qquad (12.112)$$

The corresponding equations for a parallel redundant system were given in Equations (11.138) and (11.136) respectively. If $r = n$ then this implies complete redundancy and no majority-voting principle and hence the equations become identical. Similarly, if $r = 1$ then the majority-vote mean fractional dead time from Equation (12.111) becomes:

$$\mu_D = \frac{n}{2} \theta \tau \qquad (12.113)$$

which is the corresponding formula for a non-redundant system derived previously in Equation (11.53).

For the 2-out-of-3 system, Equation (12.109) becomes:

$$\mu_D = 1 - \frac{3}{2\theta\tau}[1 - \exp(-2\theta\tau)] + \frac{2}{3\theta\tau}[1 - \exp(-3\theta\tau)] \qquad (12.114)$$

which, in its approximate form, reduces to:

$$\mu_D = \theta^2 \tau^2 \qquad (12.115)$$

Example 12.4

An autopilot system contains three nominally identical but independent control units which are only effective if any two or more are functioning correctly. The units are checked and restored to the 'as new' condition at the start of every flight. A flight lasts 3 h and during this time the control units each have times to failure which are exponentially distributed with a mean value of 60 h. What is:
(a) the form of the distribution of availability,
(b) its mean value, and
(c) the probability that the availability is greater than 0.9975?

With the example values:

$$\theta\tau = 0.05$$

This value of $\theta\tau$ is sufficiently small to use the approximate relationship of Equation (12.106), hence:

$$f_A(z) = \frac{2 \times 10^8}{5625}(z - 0.9925) \quad \text{for} \quad 0.9925 \leqslant z \leqslant 1$$

This is a linear function of z rising from zero at a value of $z = 0.9925$ to $f_A(z) = 2/0.0075$ at $z = 1$.

The mean availability may be obtained from this distribution or Equation (12.112) and yields:

$$\mu_A = 0.9975$$

and the probability, p, that the availability is greater than this value is:

$$p = \int_{0.9975}^{1} f_A(z) \, dz$$
$$= \tfrac{5}{9}$$

So far it has been assumed that each of the elements in the majority-vote system is tested simultaneously and in zero time at the end of each interval of length τ, such a test restoring all elements to the time zero condition. If the test takes a finite time τ_r then the values of μ_A and μ_D for the system require modifying according to the relationships of Equations (10.114) and (10.116) respectively.

However, as with a parallel redundant system the more likely test sequence is either to test each element sequentially over a period $n\tau_r$ at the end of each interval τ or to stagger the testing of the elements over the whole period τ. These aspects will now be considered for a majority-vote system taking the 2-out-of-3 arrangement as an illustrative example.

Suppose, therefore, that each element of a 2-out-of-3 system is tested every τ with a test time of τ_r. However, element C is tested in the interval τ to $(\tau + \tau_r)$, element B in the interval $(\tau + \tau_r)$ to $(\tau + 2\tau_r)$ and element A in the interval $(\tau + 2\tau_r)$ to $(\tau + 3\tau_r)$. The complete cycle of events spans a period of $(\tau + 3\tau_r)$ with element A renewed at the beginning of the period and elements B and C at τ_r and $2\tau_r$ before the beginning. The probabilities of each element being in the failed state during the time that they are supposed to be operational is now slightly different. It is seen that:

$$\bar{p}_a = p(t)$$
$$\bar{p}_b = p(t + \tau_r)$$
$$\bar{p}_c = p(t + 2\tau_r)$$

SYNTHESIS OF COMPLEX SYSTEMS

Hence applying Equation (12.74) for any interval apart from the first:

$$p_x(t) = \bar{p}_x = p(t)p(t+\tau_r) + p(t+\tau_r)p(t+2\tau_r)$$
$$+ p(t)p(t+2\tau_r) - 2p(t)p(t+\tau_r)p(t+2\tau_r) \quad (12.116)$$

or, if $\bar{p}_a \bar{p}_b \bar{p}_c \ll 1$ and the approximate exponential distribution applies, then:

$$p_x(t) = \theta^2[3t^2 + 6t\tau_r + 2\tau_r^2] \quad (12.117)$$

Hence, for the interval 0 to τ:

$$\mu_D = \frac{1}{(\tau+3\tau_r)} \int_0^\tau p_x(t)\,dt$$

$$= \frac{\theta^2 \tau(\tau+\tau_r)(\tau+2\tau_r)}{(\tau+3\tau_r)} \quad (12.118)$$

or, if $\tau_r \ll \tau$, then:

$$\mu_D = \theta^2 \tau^2$$

as shown before in Equation (12.115).

In the interval τ to $(\tau+\tau_r)$ only elements A and B are functional and element C is being tested. If the elements of the system perform a dynamic function then, in general, the testing of element C will render it inoperable and the system depends on both elements A and B functioning during this test period. Hence, for this period:

$$p_x(t) = \bar{p}_x = \bar{p}_a + \bar{p}_b - \bar{p}_a \bar{p}_b \quad (12.119)$$

or:

$$p_x(t) \simeq \bar{p}_a + \bar{p}_b \quad (12.120)$$

However, if the elements perform a simple two-state 'switching' function, then the function of element C may be simply placed in the operable switched state during test. The system then depends on either element A or element B functioning. In this case, for the test period of element C:

$$p_x(t) = \bar{p}_a \bar{p}_b \quad (12.121)$$

There are thus two cases to consider. Applying Equation (12.120) yields a mean fractional dead time of:

$$\mu_D = \frac{1}{(\tau+3\tau_r)} \int_\tau^{\tau+\tau_r} p_x(t)\,dt$$

$$= \frac{2\theta\tau_r(\tau+\tau_r)}{(\tau+3\tau_r)} \quad (12.122)$$

or, if $\tau_r \ll \tau$, then:

$$\mu_D = 2\theta\tau_r \quad (12.123)$$

Secondly, the mean fractional dead time for the situation of Equation (12.121) is:

$$\mu_D = \frac{\theta^2 \tau_r(6\tau^2 + 12\tau\tau_r + 5\tau_r^2)}{6(\tau + 3\tau_r)} \qquad (12.124)$$

or, if $\tau_r \ll \tau$, then:

$$\mu_D = \theta^2 \tau_r \tau \qquad (12.125)$$

For the interval $(\tau + \tau_r)$ to $(\tau + 2\tau_r)$ if element B remains inoperable during test, the probability of the system being in the failed state is:

$$p_x(t) \simeq \bar{p}_a + \bar{p}_c \qquad (12.126)$$

where, since element C has been renewed at the beginning of this interval:

$$\bar{p}_c = p(t - \tau - \tau_r) \qquad (12.127)$$

Hence:

$$p_x(t) = \theta(2t - \tau - \tau_r) \qquad (12.128)$$

and:

$$\mu_D = \frac{\theta \tau_r(\tau + 2\tau_r)}{(\tau + 3\tau_r)} \qquad (12.129)$$

or, if $\tau_r \ll \tau$, then:

$$\mu_D = \theta \tau_r \qquad (12.130)$$

Similarly, if element B can be rendered perfectly operable during test, then:

$$p_x(t) = \bar{p}_a \bar{p}_c \qquad (12.131)$$

or:

$$p_x(t) = \theta^2 t(t - \tau - \tau_r) \qquad (12.132)$$

Whence, for this case:

$$\mu_D = \frac{\theta^2 \tau_r^2(3\tau + 5\tau_r)}{6(\tau + 3\tau_r)} \qquad (12.133)$$

or, if $\tau_r \ll \tau$, then:

$$\mu_D = \frac{\theta^2 \tau_r^2}{2} \qquad (12.134)$$

The reader may now like to verify, for the final interval from $(\tau + 2\tau_r)$ to $(\tau + 3\tau_r)$, that the fractional dead time in the case where element A is inoperable during test is:

$$\mu_D = \frac{2\theta\tau_r^2}{(\tau + 3\tau_r)} \qquad (12.135)$$

SYNTHESIS OF COMPLEX SYSTEMS

or:

$$\mu_D = \frac{2\theta \tau_r^2}{\tau} \quad \text{if} \quad \tau_r \ll \tau \tag{12.136}$$

and in the case where element A is perfectly operable during test:

$$\mu_D = \frac{5\theta^2 \tau_r^3}{6(\tau+3\tau_r)} \tag{12.137}$$

or:

$$\mu_D = \frac{5\theta^2 \tau_r^3}{6\tau} \quad \text{if} \quad \tau_r \ll \tau \tag{12.138}$$

The total system mean fractional dead time, μ_{D_z}, for the case where the element under test is made inoperable is given by the sum of Equations (12.118), (12.122), (12.129) and (12.135), namely:

$$\mu_{D_z} = \frac{\theta(\tau+2\tau_r)\{\theta\tau(\tau+\tau_r)+3\tau_r\}}{(\tau+3\tau_r)} \tag{12.139}$$

or, if $\tau_r \ll \tau$, then:

$$\mu_{D_z} = \theta^2 \tau^2 + 3\theta\tau_r \tag{12.140}$$

Similarly, the total mean system fractional dead time for the case where the element under test can be made perfectly operable during the test is given by the sum of Equations (12.118), (12.124), (12.133) and (12.137) namely:

$$\mu_{D_z} = \frac{\theta^2(2\tau^3 + 8\tau^2 \tau_r + 9\tau\tau_r^2 + 5\tau_r^3)}{2(\tau+3\tau_r)} \tag{12.141}$$

or, if $\tau_r \ll \tau$, then:

$$\mu_{D_z} = \theta^2 \tau^2 \tag{12.142}$$

Obviously, both Equations (12.140) and (12.142) approach the value of the mean fractional dead time for the perfectly tested system given in Equation (12.115).

Example 12.5

Three nominally identical but independent diesel generators are running in parallel such that any two or more are capable of supplying the required load. The times to generator failures are exponentially distributed with a mean value of 1000 h. When a failure occurs it is not corrected until the next routine maintenance operation. The maintenance takes a time of 10 h for each generator during which time the maintained diesel is out of operation. The maintenance of the system occurs every 100 h and the three units are then maintained one at a time and consecutively starting at this time. What is the mean availability of the system over a long period of time, and how would this be altered if the maintenance took zero time?

This is a case where the formula of Equation (12.139) applies. Substituting the example values, yields:

$$\mu_{D_x} = 0.0378$$

Hence:

$$\mu_{A_x} = 0.9622$$

If maintenance is perfect and takes zero time, then Equation (12.115) applies, whence:

$$\mu_{D_x} = 0.01$$

or:

$$\mu_{A_x} = 0.99$$

The alternative method of staggered testing has already been discussed in connexion with redundant elements in Chapter 11. There it was seen that the optimum method was to make each of the staggered intervals between testing of equal length. For the 2-out-of-3 system, this means that element A is run for a period τ and then renewed over a test period τ_r, element B is run for a period $2\tau/3$ and renewed over a period τ_r and element C is run over a period $\tau/3$ and renewed over a period τ_r. After this first cycle of events, each element is then run for a period τ before it is tested again. After the first sequence, and if each element is nominally identical, then any sub-interval of length $(\tau_r + \tau/3)$ is the same. Typically then, for such a sub-interval:

$$\mu_D = \frac{1}{[\tau_r + (\tau/3)]} \int_0^{\tau/3} \left[p_a(t) p_b\left(t + \frac{\tau}{3}\right) + p_b\left(t + \frac{\tau}{3}\right) p_c\left(t + \frac{2\tau}{3}\right) \right.$$

$$\left. + p_a(t) p_c\left(t + \frac{2\tau}{3}\right) - 2 p_a(t) p_b\left(t + \frac{\tau}{3}\right) p_c\left(t + \frac{2\tau}{3}\right) \right] dt$$

$$+ \frac{1}{[\tau_r + (\tau/3)]} \int_{\tau/3}^{\tau_r + (\tau/3)} \left[p_a(t) + p_b\left(t + \frac{\tau}{3}\right) - p_a(t) p_b\left(t + \frac{\tau}{3}\right) \right] dt \quad (12.143)$$

for the case where the element being tested is inoperable during the period of test.

Assuming the approximate exponential distribution, Equation (12.143) reduces to:

$$\mu_D = \frac{\theta[\theta(4\tau^3 - 4\tau^2 \tau_r - 9\tau \tau_r^2 - 6\tau_r^3) + 18\tau_r(\tau + \tau_r)]}{6(\tau + 3\tau_r)} \quad (12.144)$$

and, if $\tau_r \ll \tau$, then:

$$\mu_D = \frac{2\theta^2 \tau^2}{3} + 3\theta\tau_r \qquad (12.145)$$

which may be compared with the equivalent result for consecutive testing of the elements at the end of the interval τ given in Equation (12.140).

If $\tau_r = 0$, then Equations (12.140) and (12.145) become respectively:

$$\mu_D = \theta^2 \tau^2 \qquad (12.146)$$

and:

$$\mu_D = \frac{2\theta^2 \tau^2}{3} \qquad (12.147)$$

Hence, the improvement factor, I, for the ratio of mean fractional dead time due to perfect simultaneous testing to that due to perfect symmetrical staggered testing is:

$$I = 1.5 \qquad (12.148)$$

Proceeding along the same lines and assumptions as those from which Equation (12.147) was derived, the mean fractional dead time can be found for any m-out-of-n majority-vote system subject to perfect symmetrical staggered testing. The general formula is cumbersome, but particular values are tabulated in Table 12.1 for up to a five-element system. Some of

Table 12.1 Mean fractional dead times for majority-vote systems subject to perfect symmetrical staggered testing and approximate exponential distributions of times to failure

r \ n	1	2	3	4	5
1	$\frac{1}{2}\theta\tau$	$\theta\tau$	$\frac{3}{2}\theta\tau$	$2\theta\tau$	$\frac{5}{2}\theta\tau$
2	—	$\frac{5}{24}\theta^2 \tau^2$	$\frac{2}{3}\theta^2 \tau^2$	$\frac{11}{8}\theta^2 \tau^2$	$\frac{7}{3}\theta^2 \tau^2$
3	—	—	$\frac{1}{12}\theta^3 \tau^3$	$\frac{3}{8}\theta^3 \tau^3$	$\theta^3 \tau^3$
4	—	—	—	$\frac{251}{7680}\theta^4 \tau^4$	$\frac{24}{125}\theta^4 \tau^4$
5	—	—	—	—	$\frac{19}{1500}\theta^5 \tau^5$

the values in Table 12.1 which are equivalent to non-redundant and parallel redundant systems will be recognized from the work previously done in Chapter 11.

By dividing each value in Table 12.1 by the corresponding value for simultaneous testing as given by Equation (12.111), the various improvement factors for each system may be evaluated. These improvement factors are tabulated in Table 12.2, and this table may be compared with Table 11.3

Table 12.2 Improvement factors for majority-vote systems comparing perfect simultaneous testing with perfect symmetrical staggered testing

r \ n	1	2	3	4	5
1	1	1	1	1	1
2	–	1.60	1.50	1.45	1.43
3	–	–	3.00	2.67	2.50
4	–	–	–	6.12	5.21
5	–	–	–	–	13.17

which applied to parallel redundant systems only.

Example 12.6

With the system of Example 12.5 and assuming that the maintenance is perfect and symmetrically staggered for each of the three units, what is the overall mean availability and how does this compare with the values previously calculated?

For perfect staggered testing and provided that $\theta\tau \ll 1$ then Equation (12.147) applies. Hence the mean availability is:

$$\mu_A = 1 - \frac{2\theta^2\tau^2}{3}$$

which with the example values yields:

$$\mu_A = 0.9933$$

This is seen to be a better availability than either of the two previously calculated values for this system.

Reversible changes of state

This type of change of state was introduced in Chapter 10 and discussed in connexion with non-redundant and redundant systems in Chapter 11. The reader is referred to the previous discussions for details of how to apply time-dependent failures and repairs and also how to apply the alternating renewal process to any system. By way of illustration, only the limiting case of the alternating renewal process will be considered in connexion with majority-vote systems. This is very often the case which most nearly represents practical situations. For this situation, the limiting probability of an element of the system being in the successful state approaches a constant value given by Equation (10.156). Similarly, the probability of the element being in the failed state is given by Equation (10.157). Typically,

SYNTHESIS OF COMPLEX SYSTEMS

then, the probability of the jth element being in the successful state is:

$$p_j = \frac{\mu_{f_j}}{\mu_{f_j} + \mu_{r_j}} \qquad (12.149)$$

and:

$$\bar{p}_j = \frac{\mu_{r_j}}{\mu_{f_j} + \mu_{r_j}} \qquad (12.150)$$

where μ_{f_j} is the mean time to failure and μ_{r_j} is the mean time to repair of the jth element.

For a majority-vote system it is then a question of substituting the values of p_j and $\bar{p}_j$ as given by Equations (12.149) and (12.150) in the appropriate expression for system probability states as given, for instance, by Equations (12.85) and (12.87). For the exponential distribution and using the usual symbols of θ for mean failure-rate and τ for mean repair time, this gives:

$$\mu_{A_x} = \sum_{j=m}^{n} \binom{n}{j} \left(\frac{1}{1+\theta\tau}\right)^j \left(\frac{\theta\tau}{1+\theta\tau}\right)^{n-j}$$

$$= \frac{1}{(1+\theta\tau)^n} \sum_{j=m}^{n} \binom{n}{j} (\theta\tau)^{n-j} \qquad (12.151)$$

Similarly:

$$\mu_{D_x} = \frac{1}{(1+\theta\tau)^n} \sum_{j=r}^{n} \binom{n}{j} (\theta\tau)^j \qquad (12.152)$$

and these equations may, of course, be written in terms of the mean availabilities and mean fractional dead times of the elements for any distribution, namely:

$$\mu_{A_x} = \sum_{j=m}^{n} \binom{n}{j} \mu_A^j \mu_D^{n-j} \qquad (12.153)$$

$$\mu_{D_x} = \sum_{j=r}^{n} \binom{n}{j} \mu_D^j \mu_A^{n-j} \qquad (12.154)$$

If in Equations (12.151) and (12.152), $\theta\tau$ is very much less than unity, then:

$$\mu_{A_x} = 1 - \binom{n}{r} \theta^r \tau^r \qquad (12.155)$$

and:

$$\mu_{D_x} = \binom{n}{r} \theta^r \tau^r \qquad (12.156)$$

which may be compared with Equation (11.164) for a parallel redundant system and Equation (11.78) for a non-redundant system under the same conditions.

In the case of the 2-out-of-3 system, for example, Equation (12.154) becomes:

$$\mu_{D_z} = 3\mu_D^2 - 2\mu_D^3 \qquad (12.157)$$

or, in terms of the exponential distribution parameters from Equation (12.152):

$$\mu_{D_z} = \frac{3\theta^2 \tau^2}{(1+\theta\tau)^2} - \frac{2\theta^3 \tau^3}{(1+\theta\tau)^3} \qquad (12.158)$$

or:

$$\mu_{D_z} = 3\theta^2 \tau^2 \quad \text{if} \quad \theta\tau \ll 1 \qquad (12.159)$$

Example 12.7

The internal pressure of a boiler is monitored by three nominally identical but independent pressure sensors. The output from these sensors is processed on a 2-out-of-3 basis so that an alarm is given if any two or more pressure signals exceed a certain value. Each pressure sensing channel is subject to an alternating renewal process with a mean time to failure of 1000 h and a mean time to repair of 10 h. What is the mean fractional dead time of the system?

The fractional dead time of each element is:

$$\mu_D = \frac{\theta\tau}{1+\theta\tau}$$

$$= 9.9 \times 10^{-3}$$

Hence, from Equation (12.157):

$$\mu_{D_z} = 2.92 \times 10^{-4}$$

The mean number of failures in a total time T for an element subject to the alternating renewal process was derived in Equation (10.172). Hence:

$$\mu_N = \frac{T}{\mu_f + \mu_r}$$

$$= \frac{T\mu_D}{\mu_r} \qquad (12.160)$$

A similar argument to that used in connexion with the derivation of Equation (11.169) for a redundant system may now be used. Only a proportion of the failures of a single element in a majority-vote system will result in a system

SYNTHESIS OF COMPLEX SYSTEMS

failure. This proportion depends upon the mean fractional dead time due to the failure of $(r-1)$ or more elements out of the remaining $(n-1)$ elements. Hence, the number of system failures due to, say, the element A is:

$$(\mu_{N_z})_A = \mu_N \sum_{j=r-1}^{n-1} \binom{n-1}{j} \mu_D^j \mu_A^{n-1-j} \qquad (12.161)$$

A similar relationship applies for each of the n elements. Therefore the total number of system failures is:

$$\mu_{N_z} = n\mu_N \sum_{j=r-1}^{n-1} \binom{n-1}{j} \mu_D^j \mu_A^{n-1-j} \qquad (12.162)$$

where μ_N is given by Equation (12.160).

If $\mu_D \ll 1$, then Equation (12.162) reduces to:

$$\mu_{N_z} = \binom{n}{r} \frac{rT\mu_D^r}{\mu_r} \qquad (12.163)$$

which in terms of the exponential distribution parameters becomes:

$$\mu_{N_z} = \binom{n}{r} rT\theta^r \tau^{r-1} \qquad (12.164)$$

With the approximate exponential distribution, as discussed in connexion with Equation (11.173), the mean time to the first system failure may be approximately obtained by dividing the mean number of failures into the total time. Hence, in this case:

$$\mu_f = \frac{1}{\binom{n}{r} r\theta^r \tau^{r-1}} \qquad (12.165)$$

and for the 2-out-of-3 system this becomes:

$$\mu_f = \frac{1}{6\theta^2 \tau} \qquad (12.166)$$

which may be compared with Equation (11.174).

Example 12.8

For the system of Example 12.7 find the mean number of failures that are likely to occur over a total time of 100,000 h and also the approximate time to the first system failure.

For this system, Equation (12.163) becomes:

$$\mu_{N_z} = \frac{6T\mu_D^2}{\mu_r}$$

$$= 5.88$$

and, approximately, the mean time to first system failure is this number divided into the total time:

$$\mu_f = 17{,}020 \text{ h}$$

Variational aspects

These may be dealt with in a similar fashion to that applied to parallel redundant systems and the reader is referred to the appropriate discussion under this heading in Chapter 11. However, by way of example in this case, the 2-out-of-3 system with nominally identical elements may be considered. Here, it is seen that:

$$R = \mu_A^3 \{3[p_h(x_{qu}) - p_h(x_{ql})]^2 - 2[p_h(x_{qu}) - p_h(x_{ql})]^3\}$$
$$+ 3\mu_A^2 \mu_D [p_h(x_{qu}) - p_h(x_{ql})]^2 \quad (12.167)$$

or:

$$R = 3\mu_A^2 [p_h(x_{qu}) - p_h(x_{ql})]^2 - 2\mu_A^3 [p_h(x_{qu}) - p_h(x_{ql})]^3 \quad (12.168)$$

which may be compared with Equation (11.181) for the two-element parallel redundant case.

It will now have become apparent to the reader throughout the discussion on majority-vote systems that these systems represent a more general case of both non-redundant and parallel redundant systems. Hence, many of the formulae originally derived in Chapter 11 can be obtained from the corresponding formulae in this present section by putting $r = n$ for a parallel redundant system and by putting $r = 1$ for a non-redundant system. The majority-vote system, therefore, represents a single and general element of synthesis which encompasses both non-redundant and parallel redundant arrangements.

General Procedures for Synthesis

By applying the methods outlined in this chapter and in the previous chapter, it is now possible to synthesize almost any system configuration in terms of its constituent elements. Relationships can be obtained and evaluated for the various reliability parameters at the system level from a knowledge of the failure and restoration characteristics at the element or component part level. Running through the discussions in this chapter and Chapter 11 have been some general procedures by which any systematic synthesis of a system may be performed. It is pertinent, therefore, at this stage to summarize these general methods of approach and at the same time to mention some ways of simplifying the evaluation of the reliability functions that are generated by large and complex systems.

The first step is to obtain a complete definition of the system and its configuration together with its functional method of performance. This step

SYNTHESIS OF COMPLEX SYSTEMS 517

was introduced in the early chapters of the book and, in particular, in Chapter 4. In addition, Chapter 11 and this present chapter have illustrated the need, for the purposes of synthesis, to define the system functional arrangement. This may be specified in terms of words or by suitably defined functional diagrams such as Figures 11.1(a), 11.3(a), 12.1(a), 12.2(a), 12.3 and 12.6(a).

Once the system has been functionally defined, the second step is to express the overall function in logical form. Again this may be in the form of words, an appropriate logical symbolism or an equivalent logic diagram. In terms of pure logic, the Boolean notation represents a convenient although by no means exclusive form of logical representation. It is restricted, of course, to a system where all the elements can be modelled on a two-state principle carrying the Boolean notations of 0 and 1. As pointed out on a number of previous occasions, an n-element system composed of two-state elements has 2^n possible system states. There are, therefore, 2^n Boolean expressions corresponding to each of these states. These may be all written down separately or tabulated in the form of a truth table. For large systems, this is generally impracticable and, in any case, interest normally lies in a combination of these separate states rather than in each one individually. If there are just two states of interest at the system level, say success and failure, then these can be represented by two complementary Boolean expressions which are derived from the two complementary partial sums of all the 2^n individual system states. These partial sums can be obtained, of course, from a truth table, but there are other less tedious and more direct methods of obtaining them.

One method is to construct the characteristic connexion matrix for the system as illustrated with reference to Equations (12.60) and (12.61). Another method is to trace all the successful paths through the system functional diagram and sum the Boolean logic for each path as shown in connexion with Equation (12.58). This latter method, however, needs care in order to ensure that no successful paths that exist have been inadvertently omitted. Once the Boolean expression for system success or failure has been obtained, this may be written down in notational form as in Equations (11.1), (11.108), (11.182), (12.1), (12.32), (12.58) and (12.71). Alternatively, the same system logic may be represented in diagrammatic form taking, for example, Figures 11.1(b), 11.3(b) and 12.6(b).

The third step is to convert the system functional logic, which may be expressed as a Boolean function, into the corresponding probability state equations. In this process, as has been seen, Boolean products represent a direct conversion but Boolean sums do not. The reader is referred, for instance, to the conversion of Equation (11.1) to (11.3) and of Equation (11.2) to (11.4). Alternatively, the Boolean expression may be written in 'exclusive' form, as instanced when discussing 'Exclusive Elements' in this chapter and also Equations (12.42) to (12.46). In this case the conversion to equivalent probabilities is direct for both sums and products. It is possible, of course, knowing the equivalent logical and probabilistic relationships to combine steps two and three into one step and write down the probability state equations directly from the system functional arrangement. In the same way, a combined probabilistic and logical diagram may be drawn encompassing both the second and third step in the procedure. If this is done, the Boolean notation as such is superfluous but the process may be more prone to error.

An alternative way of proceeding directly to the probability state equations is to make use of the conditional probability relationships which were introduced in connexion with Equations (12.49) to (12.57). This method now warrants some further examination. The symbol $p(X|A)$ has already been defined as the conditional probability of state X given that state A exists. Hence, if $p(AB)$ is the probability of both states A and B existing then:

$$p(AB) = p(A)p(B|A) \qquad (12.169)$$

or, alternatively:

$$p(AB) = p(B)p(A|B) \qquad (12.170)$$

If A and B represent independent events, then both Equations (12.169) and (12.170) reduce to:

$$p(AB) = p(A)p(B) \qquad (12.171)$$

So, given this condition, Equations (12.169) and (12.170) are alternative ways of expressing the product rule of probability calculus originally discussed in Chapter 1.

Suppose now that some state X is dependent upon a number of mutually exclusive and exhaustive states A_j, then:

$$p(XA_j) = p(X)p(A_j|X) \qquad (12.172)$$

and:

$$p(XA_j) = p(A_j)p(X|A_j) \qquad (12.173)$$

whence:

$$p(A_j|X) = \frac{p(A_j)p(X|A_j)}{p(X)} \qquad (12.174)$$

Also, summing Equation (12.173) for all j:

$$p(X) = \sum_{j=1}^{n} p(A_j)p(X|A_j) \qquad (12.175)$$

since:

$$\sum_{j=1}^{n} p(XA_j) = p(X) \qquad (12.176)$$

SYNTHESIS OF COMPLEX SYSTEMS

Therefore, from Equations (12.174) and (12.175):

$$p(A_j|X) = p(A_j)p(X|A_j) \Big/ \sum_{j=1}^{n} p(A_j)p(X|A_j) \qquad (12.177)$$

The expression derived in Equation (12.177) is a particular case of Bayes' Theorem and, as can be seen, is a direct consequence of the product laws of probability.

Of particular interest in system analysis is the expression of Equation (12.175) where X may represent the overall system successful state and A_j the various mutually exclusive and exhaustive states of one of the system elements. The successive application of Equation (12.175) can yield the complete probability state equation for the system in 'exclusive' form. This has already been illustrated in connexion with the system of Figure 12.3 and the derivation of Equation (12.68). The method is, in fact, equivalent to summing all the relevant terms of the truth table and the 'tree' diagram of Figure 12.5 is an example of a diagrammatic representation of such a truth table partial sum.

The system is now described in the form of a probability state equation containing sums and products, in exclusive form, of the probability states of each element. The fourth step is to express the probability state of each element, as it appears in the overall system equation, in terms of the failure and restoration characteristics of the element. This involves the determination of the type of failure such as whether it is partial or catastrophic and whether it is 'unrevealed' or 'revealed'. The determination of the distribution of times to failure for each type of failure and the ways in which any restoration process may be carried out. For unrevealed faults, this involves the knowledge of the detailed test and maintenance procedures, the times between these procedures and the action taken at each procedure. For revealed faults, a knowledge is required of the type of repair or renewal process, the effect of this process on the element and the distribution of times to completion of renewal or repair. In addition, the distribution of the element's performance variation may be required to be known coupled with the appropriate requirement limits associated with this aspect of performance. The reader is referred, in this context, to Chapters 3, 5 and 8. Methods of deriving the overall probability state of an element from all these element characteristics were fully discussed in Chapter 10.

The fifth step is to combine the expressions for the probability states of the elements according to the relationships in the derived equation for the probability state of the overall system. This may be done by taking the various product and sum terms and combining them according to the techniques which have been illustrated throughout Chapter 11. Alternatively, the system's probability-logic diagram and the corresponding overall probability state equation may be arranged to form a group of 'standard' forms or combinations. It was shown earlier in this chapter that any system which does not contain cross-linked elements can be arranged in the form of non-redundant and redundant element groupings. This means that the non-redundant and redundant methods of synthesis described in Chapter 11 may be applied alternatively on a step-by-step basis until the complete system is solved. The system solution is thus obtained from a knowledge of the techniques for only two types of standard arrangements. This may, in fact, be reduced to one standard type since, as was seen in this chapter, the majority-vote grouping is a general case of both the non-redundant and redundant arrangements.

Where cross-linked elements are involved, the system probability-logic may be first changed into an exclusive form based on each of the common elements in the cross-linked configuration. This can be done by applying the particular form of Bayes' Theorem just discussed and was illustrated in connexion with Figure 12.3 and Equation (12.68). This process can be repeated until each part of the resulting exclusive equation is in the majority-vote standard form or particular cases of such a standard form. Again, this aspect was shown in connexion with Equation (12.68). The complementary exclusive terms in the overall system equation can then be dealt with as another 'standard' form as described under the heading of 'Exclusive Elements' in this chapter. This latter standard form can also be used as a general case for standby-redundant arrangements as illustrated in the section of this chapter just referred to.

It is now possible, as a sixth step, to arrive at an overall analytical expression for the complete system in the form of any particular reliability parameter of interest. This may be the probability of the system being in a particular state at a particular time or of being in a particular state at the end of some defined period of time. It may also be in the form of an availability or fractional dead-time distribution or the mean values of these respective distributions. Other possible derived system parameters are the mean time to first system failure and the mean number of likely failures over a given period of time. In a large system, the complete analytical expression for any of these reliability parameters in terms of the element characteristics may be very complex and only the adoption of certain simplifying assumptions may yield an expression in readily expressible analytical terms.

The seventh step is to insert into the complete analytical expression for the system's reliability the appropriate numerical values for each of the element characteristics and work out the overall numerical result. Since, as has just been pointed out, the analytical expression may be complex, this calculation could be very long and tedious. It is in cases of this sort where computer techniques for numerical solution may be very valuable.

Going back one or two steps in the procedure, a computer may be used to provide a system solution by means of a truth table approach provided that the individual probability states of each element are known. Obviously, a computer can scan and select from 2^n states in a very much shorter time than a human operator. However, if n is of any reasonably large size, this process may become uneconomical even for a computer. In any case, the probability states for each element may involve calculations which are not amenable to a truth table approach.

Knowing the relevant parameters, any probability may be synthesized by a process of random selection from an artificially created population of events. This is often known as a Monte Carlo process and many computer programmes have been written employing this technique for numerous applications. The probability state of any element at any time may be derived in this way by selecting a sufficient random sample· from artifically created populations of failure and restoration processes. The technique may then be extended to derive a numerical solution of the complete system probability-logic function and, if necessary, its time–space distribution. The method tends to have limitations in those cases where the overall probability of the system being in a particular state is very low. The reason for this is that the number of random Monte Carlo selections that have to be made to achieve a reasonably accurate probability value may be extremely large.

A third way of using a computer is to employ it as a calculator to solve all the appropriate analytical expressions. This is, in fact, the same as the manual process already described in these system analysis procedures but using a computer to take the time and effort out of the complex processes of probability combinations and integrations. This method has been used effectively to calculate certain reliability parameters for systems containing up to 1000 different elements.

Having obtained by manual or computer means a reliability value for a technological system, the eighth and final step is to state the accuracy of this value in the light of the accuracy of the input data and any assumptions made in the system mathematical model. Methods of expressing the accuracy of or confidence in various types of distribution parameters have already been discussed in Chapter 9 and some of these methods may be used to give an indication of the validity of the final result. Complex combinations of confidence limits, however, is an intricate subject and outside the scope of this present book. The reader is referred, therefore, to other publications on this particular subject.

The accuracy of the input data, however, is also related to the methods of data collection and the form in which relevant data is normally obtained. Some of these aspects are discussed in the next chapter. In addition, any assumptions made in the mathematical model may need careful review in order to substantiate the validity of any calculated reliability value. Some of the implications which may arise in this connexion are also discussed in the following chapter.

This, then, completes the summary of the procedures to be followed in order to arrive at an overall reliability evaluation of a complete technological system by a process of synthesis. It may be finally useful to the reader to re-state each of the eight steps just described in the following more succinct form:

(1) Obtain a complete definition of the technological system and its configuration together with its functional method of performance;
(2) Express the complete functioning of the system in logical form either as a logic equation or as a logic diagram;
(3) Convert the logic equation into the equivalent probabilistic system state equation (in some cases this may be done directly and step (2) omitted);
(4) Express the probability state of each element, as it appears in the overall system state equation, in terms of the failure and restoration characteristics of the element;
(5) Combine the probability states for each element according to the relationships defined by the overall system equation and the characteristics of the elements;
(6) Derive the complete analytical expression for the system reliability parameters of interest in terms of the element characteristics;
(7) Insert the appropriate numerical values for the element characteristics and calculate the overall system reliability (this may be carried out by a computer in which case steps (5) and (6) may be omitted);
(8) Stipulate the accuracy of the final result in the light of the accuracy of the input data and any assumptions made in the mathematical model.

Questions

1. A system has the following functional flow diagram:

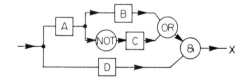

Write down:
(a) the Boolean expression for success,
(b) the equivalent probability state equation, and
(c) state what standard form the system reduces to if element C is permanently unsuccessful.

SYNTHESIS OF COMPLEX SYSTEMS

2. In a hospital operating theatre the lighting is normally achieved by an array of lamps supplied from mains electricity. If the mains fail, an emergency battery and lamp unit which is continuously in operation supplies the required illumination. The times to failure of the mains electricity, the main lamp unit and the emergency lamp system are all exponentially distributed with means of 5000 h, 2000 h and 100 h respectively. If any failure is considered irreversible over the time period of interest, what is:
   (a) the probability of no lighting after 20 h, and
   (b) the mean time to the first failure of all lighting?

3. If the system of Question 2 is operated continuously but restored to the time zero condition every 100 h, what are the approximate mean number of system failures that are likely to occur in an overall period of 10,000 h?

4. The braking system of a vehicle normally works via a hydraulic servo on to a set of primary brakes. Whenever the hydraulic servo is in the failed state, then the braking effort is obtained mechanically using a secondary set of brakes. The hydraulic servo, the primary brake system and the secondary brake system are each subject to an alternating renewal process such that when they fail a repair process is started which, on completion, restores the failed element to the time zero condition. The mean time to failure and for repair of each of the system elements when the vehicle is in use are as follows:

	Mean time to failure (h)	Mean time for repair (h)
Hydraulic servo	1000	5
Primary brake system	500	1
Secondary brake system	200	1

Over the period of usage of the vehicle what is the mean fractional dead time of the complete braking system?

5. If in the braking system of Question 4, all the distributions of events are exponential but the repair time of the secondary brake system is taken as infinite, what is the probability of the system being in the failed state at:
   (a) 5 h from start of usage, and
   (b) 500 h from start of usage?

6. A system has the following functional flow diagram:

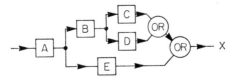

Express the probability of the system being in the failed state, $\bar{p}_x$, in terms of:
(a) the probability of each element being in the failed state, and
(b) the simplest combination of element probabilities which contains only sums and products.

7. If each of the elements in the system of Question 6 are subject to an alternating renewal process with a mean time to failure of 100 h what is the mean fractional dead time of the system if the mean time to repair for each element is:
(a) 1 h,
(b) 10 h, and
(c) 100 h?

8. A system has the following functional flow diagram:

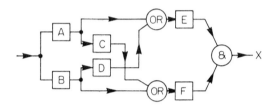

What is:
(a) the Boolean expression for system success expressed in terms of the successful notation for each element, and
(b) the probability expression for the system being in the failed state expressed in terms of exclusive probabilities?

9. Four elements of a system each have a constant probability of 0.1 of being in the failed state at any time. What is the system probability of being in the failed state if the elements are so connected that system success is achieved when:
(a) any 1 or more of the 4 elements are successful,
(b) any 2 or more of the 4 elements are successful,
(c) any 3 or more of the 4 elements are successful, and
(d) only all 4 elements are successful?

10. Three elements of a system, A, B and C, have constant probabilities of 0.9, 0.8 and 0.7 respectively of being in the successful state at any time. What is the probability of the system being in the successful state if:
(a) the elements are connected in a 2-out-of-3 configuration, and
(b) all the elements must be successful for system success?

11. Three nominally identical temperature sensing elements are connected to nominally the same point on a process plant. An alarm is designed to be given if any two or more of these temperature sensors record a

SYNTHESIS OF COMPLEX SYSTEMS 525

temperature above a certain prescribed level. The times to failure of each element are exponentially distributed with a mean value of 5000 h and any failure is considered to be irreversible in the time scale of interest. What is:
 (a) the probability of the alarm system not working if an excessive plant temperature rise takes place at 500 h, or secondly, at 2000 h, and
 (b) the mean time to the complete failure of the alarm system?

12. How many system failures are likely to occur in a total time of 10,000 h, if the temperature alarm system of Question 11 is restored to the time zero condition every:
 (a) 100 h,
 (b) 500 h, and
 (c) 2000 h?

13. A shut-down signal for a process plant actuates three nominally identical flow control valves. The plant is safely shut down if any two or more of these valves close in response to the shut-down signal. The valves are checked for failures and restored to the time zero condition every 1000 h. Within this period valve failures are irreversible and have times to occurrence which follow an exponential distribution with a mean value of 20,000 h. What is:
 (a) the mean fractional dead time, and
 (b) the probability that the fractional dead time is greater than this mean value in any 1000-h period?

14. What is the mean fractional dead time for the system described in Question 13 if the restoration of the valves is:
 (a) carried out consecutively at the end of each 1000-h interval, and
 (b) carried out on a symmetrical staggered basis still giving 1000 h between tests on any one valve but allowing a restoration of one valve to take place every 1000/3 h?
 Each valve is taken to be inoperable for a time of 20 h during its restoration process under either scheme of maintenance.

15. Four nominally identical but independent diesel generators are running in parallel and synchronized to supply a common electrical load. The times to generator failure for each generator are exponentially distributed with a mean value of 5000 h. When a failure occurs it is not corrected until the next maintenance operation which occurs at intervals of 200 h for each generator. The actual maintenance time is taken to be negligible. What is the mean fractional dead time of the system supply if:
 (a) any three or more generators are required to supply the load and they are all maintained together,
 (b) any three or more generators are required to supply the load and the maintenance is symmetrically staggered, and

(c) any two or more generators are required to supply the load and the maintenance is symmetrically staggered?

16. The position of a ship at sea may be obtained by three different and independent methods of measurement. Confidence in the accuracy of overall measurement is only obtained if any two or more of the different measurement methods produce the same result. Each measurement method is subject to an alternating renewal process with mean times to failure of 400 h, 500 h and 800 h. The corresponding mean times to repair are 20 h, 40 h and 40 h respectively. What is the mean availability for acceptable accuracy in the overall measurement of the ship's position?

17. A fire alarm signal is produced if any two or more out of four nominally identical but independent detectors sense a certain rise in temperature. Each detector is subject to an alternating renewal process with a mean time to failure of 2000 h and a mean time to repair of 5 h. What is the mean fractional dead time of the system?

18. A shut-down system for a nuclear reactor depends upon a signal from any two or more out of three excessive temperature measurements. The temperature measurement channels are nominally identical and subject to an alternating renewal process. The times to failure and repair of each channel are exponentially distributed with mean values of 2000 h and 10 h respectively. What is:
    (a) the mean fractional dead time of the system,
    (b) the mean number of failures in a total time of 100,000 h, and
    (c) the mean time to first system failure?

19. In the system described in Question 18, each of the temperature measurements is subject to variations about a mean value which follow a normal distribution. If these variations exceed plus or minus three standard deviations, then this also constitutes a failure of the element. What is the new mean fractional dead time of the system under these conditions?

20. A pump element within a process plant may be in one of two possible states which may be termed success or failure. The probability of the pump being in the successful state is 0.99. When the pump is in the successful state, the plant has a probability of 0.001 of being in the failed state. If the pump is in the failed state, however, the probability of the plant being in the failed state is increased to 0.1. Given that the plant is in the failed state, what is the probability that the pump is in the failed state?

CHAPTER 13
The Application of Reliability Assessment

General Review

As pointed out in Chapter 1, the characteristic of reliability is of interest to the management of any technological project or system. With the interest arises the need for a scientific evaluation of reliability in quantitative terms so that the overall costs of reliability achievement can be balanced against the costs of unreliable behaviour. This need becomes more paramount as the risk associated with the technological system becomes higher. The same applies whether the risk and cost is purely financial or whether it encompasses such aspects as danger to human life or loss of national prestige.

Very often a reliability assessment of a technological system is required before the system is put into active operation or even before it has emerged from the design stage. There is always an element of prediction, therefore, associated with the assessment process. This again highlights the need for a scientific approach in order to impart validity to the predicted behaviour.

In quantitative form, reliability is a measure of the success of a technological system in meeting its stated objectives. Unsuccessful behaviour arises generally from unpredictable or uncontrollable variations in the system's performance. These variations, although individually unpredictable, may be described on mass by probability functions or statistical distributions. This leads to a definition of reliability, as given in Chapter 1, in a probabilistic form.

The definition gives rise to probabilistic expressions for reliability at specific times or over specific time intervals. These expressions may take the form of discrete probabilities, cumulative probability functions or availability functions. The most suitable function or parameter of interest depends upon the reliability requirements and the failure and restoration characteristics of the system as discussed in Chapter 1 and, more particularly, in Chapter 10. Other related reliability parameters that may be of interest in particular cases are the mean values of the times to failure, the times to repair, the number of failures and the availability or fractional dead time.

The degree of success, or the reliability, of a system can only be obtained by a correlation between the system requirement and achievement. This correlation process was extensively discussed at the beginnings of Chapters 2, 3, 4 and 10. Here, the requirement was described in terms of a Q function and the achievement in terms of an H function. The importance of first determining a complete and accurate definition of the requirement function, Q, was emphasized in Chapter 3. The questions such as 'What fundamentally is the system required to do' and 'Under what conditions is it required to do it' need to be answered with exactitude.

The next stage in the process of reliability assessment, as indicated in Chapter 4, is to examine the actual system to ensure that it has the inherent capability of functioning in the intended manner. In other words, if both Q and H are free from variations can the requirement be met? This is done by ascribing singular or mean values to Q and H and in a completely deterministic way arriving at the value for the reliability, R. Such a value of R, in this context, is either 0 or 1 since a singular value of H either does or does not meet singular limits imposed by Q. This step is also very important since a detailed probabilistic evaluation of R would be a complete waste of time if some element in the system, due to its inherent lack of capability, rendered the overall system completely incapable.

The requirement Q and the achievement H are both performance functions which require to be expressed in the same form so that they may be correlated. This aspect was discussed in Chapters 4 and 5. In some cases, a convenient way of describing Q and H is to express them in the form of a transfer matrix where each element of the matrix represents the appropriate transfer function between the corresponding system output and input functions. This method of representation was used in Chapters 3, 4 and 5 and the properties of the transfer function and the transfer matrix described in Chapter 6. It was seen how the transfer matrices for Q and H could be made of the same form by appropriate selection of the values for the coefficients of the transfer functions.

The probabilistic nature of system reliability arises when both the Q and H functions are recognized as possessing systematic and random variations in performance values. The different types of variation that may exist were introduced to the reader in the latter half of Chapter 3 in connexion with the requirement and in Chapter 5 in connexion with the achievement function. This involved the concept of discrete and continuous statistical distributions and the properties of such distributions were enlarged upon in Chapter 7. It was seen in Chapters 3 and 5 that the variational aspect in performance could be ascribed completely to the variations of the coefficients in the transfer functions of the transfer matrix.

THE APPLICATION OF RELIABILITY ASSESSMENT

Whatever notational method is used to describe complete performance, the overall picture for achievement is that of an n-dimensional surface of performance which is changing systematically and randomly in space and time. This surface then nominally lies between two similar types of changing surface which represent the lower and upper limits of required performance. The chance of the achievement surface 'breaking through' the requirement surfaces is then the unreliability of the overall situation.

A practical distinction was made in Chapter 2 and enlarged upon in Chapters 5 and 10 between catastrophic and discrete changes of state in performance value on the one hand and relatively small variations in performance value on the other hand. This gave rise to the development of a two-state probabilistic model where a system performance aspect is either completely in the failed state or subject to variations in value in a nominal working state. The reliability of the system with such a model then depends not only on the probability of the system being in the nominal working state but also on the probability, in this state, that the variations do not exceed the required performance limits. The usefulness of this type of model in the overall correlation process was discussed in Chapter 10.

It was then seen that the concept of change of state brought into consideration the ways by which a system might change from one state to another and back again. Various processes of the irreversible, reversible and renewable type were discussed together with their practical implications. Chapter 10 concluded with a review of the various reliability parameters of interest which arose out of the possible system failure and restoration characteristics that might exist.

This then is the overall picture for the complete determination of R, in the form of interest, at the system level. Unfortunately, as pointed out in Chapters 5, 8 and 11, the necessary data for complete evaluation of R are very rarely available in connexion with a complete system. Such data as normally exist apply more often to the elements or component parts of which a system is comprised. A need very often exists, therefore, to be able to synthesize system reliability from a knowledge of the reliability parameters of its component parts. In the two-state model, two distinct though basically similar types of synthesis are required. One associated with the combinations of continuous performance variations of each element in the working state and the other associated with the combinations of elements being in the working state at all. The former techniques were discussed in Chapter 8 and the latter methods of synthesis in Chapters 11 and 12.

From Chapters 11 and 12 arose techniques, originally introduced in a deterministic form in Chapter 4, for the logical representation of a system's functional behaviour in terms of the behaviour of individual elements. It was also seen that most systems could be broken down into a small number of 'standard' combinations of elements depending upon the degree and type of redundancy employed. The various mathematical models for system synthesis on the basis of these standard forms were described and developed and procedures suggested whereby the whole process could be expedited.

Finally, whether the system reliability is evaluated directly from the correlation between the characteristics of Q and H at the system level or whether such characteristics have to be synthesized from a knowledge of the corresponding characteristics of the system elements, the calculated result is dependent upon the validity of any mathematical models used and the data fed into such models. It is now the intention in this final chapter to examine some of the principal aspects which may affect the accuracy of reliability data and mathematical models in any practical application of reliability assessment.

Reliability Data

General aspects

Inaccuracies in reliability data may arise both in connexion with their acquisition and in connexion with their use. In both cases, possible errors must be judged in the light of what sort and form of data are required. The requirements for data which are pertinent to the present discussion are those related to the needs of reliability assessment as discussed throughout this book. The availability of data in the required form then determines the depth of synthesis which needs to be carried out.

Obviously, if the required data exist at the system level then no synthesis is necessary. Any system should, therefore, be examined from this point of view as a first step. If the required data are not available, then the system should be broken down according to its hierarchic levels until a level or range of levels is reached where all the necessary data are available. Beyond this no further subdivisions of the system is necessary or even, in many instances, desirable. The amount and degree of system synthesis, therefore, are dependent on the levels at which the data obtainable meet the requirements in form and accuracy. In some instances the depth of synthesis dictated by these considerations may extend down to the level of the smallest component part.

The type of data required at any level in a system may be deduced, in most practical cases, from the failed-state/working-state model which has been extensively discussed previously and which is illustrated in Figure 5.4. First of all, then, data are needed in order to calculate the probability at any time of a transition from the working state to the failed state and back again. For the first type of transition this involves, as has been seen, a knowledge of the probability density function for the distribution of times to catastrophic failure related to the starting time or times of interest. For the second type

THE APPLICATION OF RELIABILITY ASSESSMENT

of transition back to the working state the data required depend upon the manner in which the failed state is detected and rectified. In those cases where some repair or restoration process starts immediately the fault occurs, then a similar p.d.f. to that just mentioned for failures is required for the distribution of times to repair orientated to an origin dependent upon the time of failure. In the case of an 'unrevealed' type of fault, the additional information is required of the time or distribution of times between the actual occurrence of the fault and the commencement of the restoration process.

Secondly, data are needed in order to calculate the probability at any time of an element in the nominal working state moving into an unreliable situation due to variations in its performance parameters. This leads to a requirement for the distribution of variations in each performance parameter relative to the appropriate time and space conditions.

So, in general, the type of data required for reliability assessment purposes is basically in the form of statistical distributions. The distributions are those related to the failure and restoration characteristics of the system or element and those related to the variations in performance parameters of the element. It is now relevant to see whether these characteristics are generally available and in what form they may be practically obtained.

Acquisition of data

Data on reliability characteristics, like that on any other performance attribute, may be obtained by two main methods, namely:

Field experience

Sample testing

By field experience is meant the monitoring and recording of the elements' reliability characteristics over a period of practical usage. Sample testing, on the other hand, refers to an experiment deliberately set up to record the required information from a representative sample of the elements of interest. Both methods, of course, involve sampling techniques since even extensive field experience of an element's reliability performance can never embrace the whole population of its characteristics. Techniques for taking, analysing and interpreting sample data were discussed in Chapter 9. These techniques apply, in general, to all methods of data acquisition. Tests need to be carried out to determine the forms of the appropriate distributions and estimates made of the distribution parameters, such as the mean and the variance, at certain levels of confidence. Since these types of procedures and their interpretation were dealt with in Chapter 9 they will not be discussed any further at this stage. Some more general comments may be made, however, on the factors involved in the acquisition of data from past experience and from sample experiments.

When data are required on an element in a particular application, the first step is to see whether a nominally identical element has been used in a nominally identical application on a previous occasion. If it has, then it may well be that some of the characteristics of interest have been recorded. These characteristics then require to be analysed according to the sampling techniques previously described and a statement made of the confidence associated with the data. At this stage it has to be judged whether this degree of confidence is adequate in light of the overall accuracy requirements for the analysis. If it is, then the necessary data are now available and no further steps need to be taken. Care is needed in this process, however, particularly in ensuring that the past data are relevant to both the present element and the present application and environment of the element. Even a nominally identical element may be subject to considerable changes in reliability characteristics in different applications and environments. This aspect will be considered further when discussing the practical use of reliability data.

When the confidence in the data is judged to be too low a number of further steps are open for consideration. First, it may be investigated as to whether the appropriate data exist for a generally similar element. For instance, if data are required for a particular transistorized electronic amplifier of a certain complexity, then a review of data may be made of all generic forms of this type of amplifier. If this can be extended to show a spread of likely values, then an estimate may be made of boundary values within which the characteristics of the particular amplifier can be assumed to lie. In some cases, where the effects of differences in design, manufacture and application are known, correction factors may be applied to the generic estimates in order to arrive at a more realistic set of data for the particular case. The use and limitations of such correction factors are also discussed further in the next section. Obviously, the use of generic data, whether corrected or not, is not likely to be as accurate as data which are directly relevant to the element of interest.

Secondly, as mentioned earlier, it may be possible to sub-divide the element under consideration into further component parts and extend the depth of synthesis. An examination of the data available at this new and lower hierarchic level is then carried out in the manner just described. Such a process can obviously be continued until no further sub-division is practically feasible. This, of course, increases the complexity of the system synthesis and also the total amount of data required. Both accuracy and economic considerations may now also dictate the level to which this technique can be applied.

Thirdly, and at any stage in the sub-division process, the possibility of setting up a practical experiment to obtain the required data may be reviewed.

THE APPLICATION OF RELIABILITY ASSESSMENT

This is the process of sample testing as distinct from field experience. It involves the controlled operational running of a sufficiently large number of the elements for a sufficiently long period of time under the correct conditions of application and environment. The interpretation of the word 'sufficiently' depending upon the degree of confidence which can be ascribed to the analysis of the test results. As shown in Chapter 9, confidence increases with increase in sample size. This generally precludes, from practical and economic considerations, the application of sample testing to the higher hierarchic levels of any complex system. However, for smaller component parts the technique may often be a viable proposition. As with any experimental technique, care needs to be taken to ensure that the conditions of the experiment are directly relevant to the element in its intended application.

The confidence in a sample test of, say, the distribution of times to failure of an element is governed by the number of separate times to failure that can be recorded. This concept and its implications were fully discussed in Chapter 9. One corollary of this fact is that the more reliable an element is the longer a sample test needs to be run for any stated confidence level. This means that proof of low reliability may be relatively easily obtained but that proof of high reliability may be much more difficult. One method of increasing the sample size for relatively small component parts is to run a large number of elements for a short period of time instead of a single or small number of elements for a long period of time. This implies, of course, that mass in numbers and mass in time can be validly interchanged. For elements whose reliability characteristics are likely to change radically with time, such a method may not be possible. However, if in this instance, the mechanism of change with time is known, then such an approach may still have some value. Another method of speeding up the time scale of a sample test experiment is to run the elements under much higher than normal stress conditions. Again, this may be a feasible approach if there is prior and adequate knowledge of the effect of varying stress conditions on the reliability characteristics.

Fourthly, if it is possible to wait for a reliability analysis until the system has gained some actual operational experience, then the reliability characteristics of the system and its various elements may be monitored and recorded during its initial or subsequent operating life. This is obviously the most direct and can be the most accurate approach to the acquisition of the required data. The accuracy arises because the method is not dependent upon past experience of doubtfully similar elements under doubtfully similar conditions, nor is it dependent upon the validity of a sample test experiment, but is, in effect, allowing the system to speak for itself. There is still the limitation, with a highly reliable system, that an unacceptably long period of time may have to be covered before sufficient confidence is obtained in such directly recorded data.

In monitoring actual system performance, the data collection process needs to be accurately and conscientiously carried out otherwise mistakes or omissions in the recording procedures may lead to inaccuracies. With a large plant or system it is generally the local operator or maintenance man upon whose shoulders the responsibility of initially recording the relevant data rests. With these people the collection of data becomes an adjunct to their normal tasks. Therefore, the procedures and methods of collection need to be reasonably simple and straightforward if the required accuracy is to be maintained. Very often existing operational logs and maintenance/job cards, which are required for normal system operation, may be slightly modified to include the information and data which are of direct interest to the reliability analyst.

Typical plant or system aspects which may be included in this way are the times and dates when elements of the system are installed, put on line, changed, modified, replaced or repaired and also the man effort and materials that are needed for these processes. This can lead to an analysis of the running hours, storage times, repair times and costs associated with each element of the system. Additionally, any change in the system or element performance characteristics, whether it be a catastrophic change or a small variation, needs to be recorded together with the time, date, type of change and identification of the elements concerned. From this sort of information, the variational characteristics and the distribution of times to complete failure can be computed.

Various approaches to obtaining an adequate indication of the required reliability data have been outlined in the preceding remarks. Accepting the fact that full knowledge always supplants any form of estimation, it is most unlikely that any one method of data acquisition is ever adequate in its own right. Generally, data acquisition needs to be tackled from the initial system design concept through the various stages of development and manufacture and only concludes when the system has reached the end of its useful life in the field. In this way, reliability data begin life as a tentative estimation, proceed through a practical sampling sequence with certain levels of confidence and finish with the confirmed results of actual experience.

As mentioned at the beginning of this section, whatever method is used to acquire data it nearly always represents a sample from some possible infinite population. Hence, the interpretation of such data and the confidence that may be ascribed to such an interpretation depend upon the sampling theory discussed in Chapter 9. Sometimes this confidence may turn out to be unacceptably low in spite of resort to all possible methods of collecting data in the required forms. In cases of this sort it may be prudent to examine whether the requirement itself may be degraded. The basic requirements for reliability data, as stated earlier, are for the distributional forms of the failure,

THE APPLICATION OF RELIABILITY ASSESSMENT   535

restoration and performance variational characteristics of the relevant elements. It may be, in some instances, that these complete distributional forms are not necessarily required and an adequate reliability analysis can be performed on the basis of, say, just the mean values of the distributions. This, together with some of the other aspects mentioned previously, will be enlarged upon in the next section.

The use of data

As just pointed out, the precise form of the relevant statistical distributions for the appropriate reliability parameters may be difficult to acquire. Under certain boundary conditions, any statistical distribution may be characterized by its hazard function as demonstrated in Chapter 7 and, in particular, in Equation (7.107). The general form of the hazard function is fairly well established for some of the reliability characteristics of many engineering elements or component parts. This may be illustrated with reference to the distributions for times to failure which apply over the lifetimes of various types of elements.

For distributions representing times to failure, the hazard function is often known as the failure-rate function. Experience suggests that this failure-rate function follows a fairly standard pattern with respect to time for most technological devices. A typical pattern of this sort is illustrated in Figure 13.1. Here the failure-rate function is plotted against time and the resulting characteristic can be seen to fall into three distinct phases.

The first phase represents a pattern of failure events which typically arises from initial production, test or assembly faults. This is sometimes called the 'burn-in' phase or the 'infant mortality' phase and reflects the early 'teething' troubles which often arise in practice with engineering devices. The last phase illustrates the effects of ageing when the device or element is beginning to wear out and the rate of failures follows a rising characteristic. This is often known as the 'wear-out' phase or 'end-of-life' phase. In between the first and last phase is a phase which may be termed the 'useful life' where the failure-rate function remains either sensibly constant or follows a relatively slow change in value.

This overall pattern of events is typical of a wide range of engineering elements and also has a basic similarity with the corresponding death-rate characteristic for human beings. A typical characteristic of this latter type, based on the male population in England and Wales for the years 1960 to 1962, is illustrated in Figure 13.2 for comparison purposes. It is this similarity which has led to the adoption in technology of human characteristic terms such as 'infant mortality' and 'end of life'. Also, for the same reason, the failure-rate function is sometimes termed the 'force of mortality' as previously mentioned in Chapter 7.

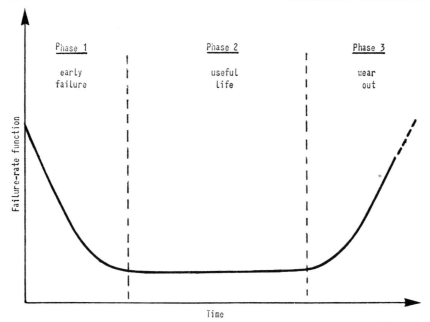

Figure 13.1 Typical failure-rate characteristic for engineering devices

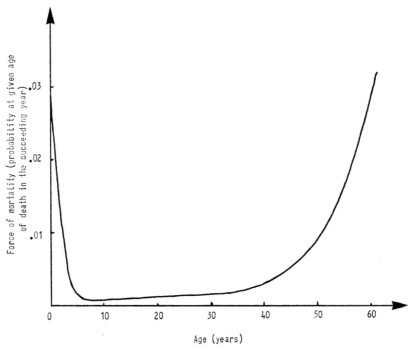

Figure 13.2 Death-rate characteristic for males living in England and Wales for the years 1960 to 1962

It will be seen, from a comparison between the shapes of the failure-rate characteristic in Figure 13.1 and the hazard function curves of Figure 7.1, that certain standard types of distribution can be used to represent element failure characteristics. For instance, phase 1 might be represented by a gamma distribution, phase 2 by an exponential or Weibull distribution and phase 3 by a normal distribution. Most technological systems, however, are designed and used so that their elements or component parts operate in phase 2, that is, the useful-life phase. This is normally achieved by the application of soak testing and certain commissioning procedures to weed out elements with a high initial failure-rate. At the other end of the scale, arrangements are generally made to replace or refurbish elements before they enter the wear-out phase.

If these procedures are adopted and substantiated then many systems can be considered as operating in phase 2 where the failure-rate function has a fairly constant value. This leads to perhaps a Weibull or, more approximately, to an exponential distribution for times to element failure. In many instances the exponential distribution may be a sufficient approximation. It is for this reason that the characteristics and the application of the exponential distribution have been given a certain degree of prominence throughout this book.

As has been seen in the previous chapters, the exponential distribution is completely determined by the single parameter θ which represents the constant value of the hazard function or failure-rate function, $z(t)$. This parameter θ is also the reciprocal of the mean of the exponential distribution or the mean time to first failure.

On the assumption of a constant failure-rate for elements working in their useful life phase, many authorities over the past decade or so have collected and summarized generic data in terms of representative values for the parameter θ. On this basis, the bar chart shown in Figure 13.3 gives an indication of failure-rate values that might be expected for a few typical component parts, equipments and systems. Each bar line on the chart represents the likely failure-rate range for each element specified. The midpoint of the range is equivalent to the mean failure-rate value for the element when working at design stress in a general-purpose ground-based environment. Such an environment being typified, for example, by many light engineering or industrial installations. The lower limit of the range generally represents the most reliable type of the element when lightly stressed and operating in a controlled and favourable environment. This could apply, for instance, in an air-conditioned and temperature-controlled situation which is typical of many computer installations. The upper limit of the range normally represents the least reliable type of the element when heavily stressed and working in a relatively severe environment. This would be equivalent to some heavy industrial installations or some types of transportation systems.

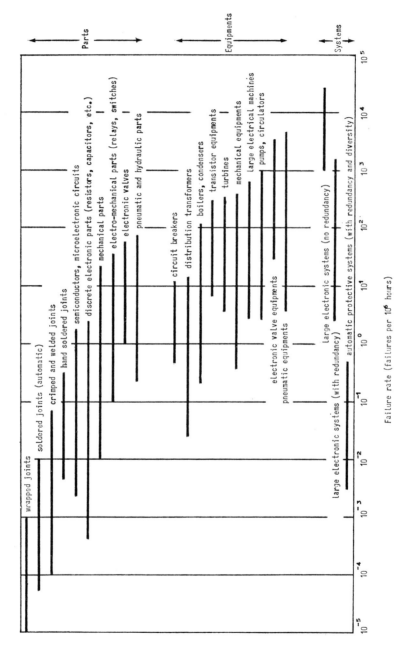

Figure 13.3 Typical ranges of failure-rates for parts, equipments and systems

THE APPLICATION OF RELIABILITY ASSESSMENT   539

The data on which the bar chart of Figure 13.3 is based were taken from a large number of different sources. Generally, the data on failure-rates were based on population sizes ranging from 10^9 element-hours for small component parts to 10^5 element-hours for large and specialized items. The chart, of course, gives only a very general indication of failure-rates since each element depicted covers a wide variety of types and may be operating in a range of different environments. However, it can be useful in substantiating boundary arguments. For instance, if a system required a circuit breaker with a mean failure-rate of 10^{-2} failures per 10^6 h, then reference to the chart would show that such a figure was virtually unattainable on the basis of past generic experience. It would therefore require some break-through in design or manufacturing techniques with reference to circuit breakers or some new system configuration in order to stand a chance of meeting the postulated requirement. As another example, if an electronic system contained a million wrapped joints, then it could be reasonably assumed that the overall failure-rate due to wrapped joints would be about 10^2 failures per 10^6 h to within an order of magnitude.

In many cases, however, the reliability analyst needs data of a more precise and specific nature than that given in Figure 13.3. Such data are available for a wide selection of typical component parts which are used in electronic, electrical and mechanical systems. The reader is referred to the specialized literature on this subject, but a selection of the type of failure-rate figures for such component parts that have been derived from past experience in a large number of applications is given in Table A.7. The figures are average values taken from large samples of the appropriate components over a number of years. They also apply or have been corrected to apply to components working at or near to their normal design stress in a general-purpose ground-based environment, but some further comment will be made about this aspect later on.

Although the figures in Table A.7 are average values based on reasonably large sample sizes, they still represent only the best estimate of the true population mean and this fact must always be borne in mind. However, in many practical cases this estimated mean may be equated to the true mean with sufficient accuracy for analytical purposes. For instance, reference to Table A.7 shows that the assumed failure-rate for tantalum foil electrolytic capacitors is 1.0 fault per 10^6 h. This means that if a large number of such capacitors were run under the appropriate conditions for a long time then, on average, one failure would take place for each 10^6 capacitor-hours of operation in the useful-life phase. It is important, however, to remember that this statement does not say anything about the behaviour of individual tantalum foil capacitors. There may, in fact, be large discrepancies between the failure-rate of an individual component and the estimated mean for the whole population. However, if a system or item of equipment contains a large number of non-redundant components then the use of mean values of failure-rate for each component tends to be a justifiable assumption when computing the overall failure-rate. This is so since the variance of the overall mean value reduces as the number of individual values is increased. The reader is referred again to the discussion on 'Sample means' in Chapter 9 in connexion with this point. Care may have to be taken, though, where a system contains only one sample of a particular component type and this individual component has a significant effect upon the overall failure pattern. In cases of this sort, resort may have to be made to obtaining a boundary value or set of boundary values for the failure-rate of the components and use made of these values rather than of the mean value.

Mention was made earlier of the fact that component failure-rate figures apply only under a particular set of conditions. It can be readily appreciated that failure-rates are bound to be affected, for instance, by environmental conditions. A 1 W resistor dissipating only 1 mW in a temperature-controlled, air-conditioned computer is not likely to exhibit such a high failure-rate as the same type of resistor running at maximum rating in the confines of a missile during its launching phase. In fact, field experience suggests that the ratio of component failure-rates appertaining to these two extremes of environment may be as high as 1000 : 1. The figures in Table A.7, as stated earlier, are 'standardized' to a particular set of conditions. As an approximate approach, the figures may be multiplied by various K factors in order to relate them to other conditions of environment and stress. Typical values of these K factors are given in Table A.8 where K_1 relates to the general environment of operation, K_2 to the specific rating or stress of the component and K_3 to the general effect of temperature. The standard figures of failure-rate in Table A.7 only apply, therefore, under the conditions given by:

$$K_1 = K_2 = K_3 = 1$$

The use of simple K factors, in the way just described, will only lead to an approximate estimate of failure-rate under any particular set of conditions. For a more exact approach, the reader is referred to standard published manuals on this subject where detailed curves are given for the relationship between failure-rate and stress for a range of particular component parts. In the ideal, of course, the failure-rate data used should be those which have been collected for the exact element of interest under the exact conditions of environment and stress that apply.

Many component parts have more than one significant mode of failure. Resistors, for instance, may fail in the extreme and catastrophic case to either an open-circuit or short-circuit condition. The figures for resistors

THE APPLICATION OF RELIABILITY ASSESSMENT   541

given in Table A.7 relate to the total failure-rate which includes all modes of failure. Only a proportion of this total, for example, applies to the short-circuit mode of failure. This proportion is generally about 10% in the case of resistors. For capacitors, the open-circuit and short-circuit modes of failure tend to be equally likely and hence each is taken as 50% of the total. This also applies to other electronic components such as contacts and diodes. In the case of three-terminal components, six principal modes of failure are taken into account. These are three open circuits of the equivalent tee network and three short circuits of the equivalent delta network. For transistors these six modes are taken generally as equally likely with 16.7% of the total failure-rate allocated to each.

The open-circuit and short-circuit type of classification does not apply, of course, to all types of components. An electrical component, like a circuit breaker for example, may have a failure mode of failing to close when it is normally open or a mode of failing to open when it is normally closed. A pneumatic or hydraulic component may fail due to a leakage or fail due to a blockage. Some of the principal modes of failure for some of the component parts listed in Table A.7 are given in Table A.9.

Only extreme modes of failure have been commented on in the preceding discussion. However, with the two-state failure model described in earlier chapters, a component may be considered to have failed in one of its extreme modes if its characteristic parameter has moved outside the laid down specification limits. On this basis, the failure-rate figures given in Table A.7 may be taken to cover the full range of failures for components which are not within specification. This only leaves those variations in performance which take place within the specification limits. These can only be conveniently described by the appropriate distributions which require to be specifically obtained for each component part of interest. Any review of this latter type of information would be extremely extensive and is beyond the scope of this present book. In certain instances, however, these distributions may be assumed to take a normal form where the specification limits are equivalent to a certain number of standard deviations from the mean. The actual number of deviations then depends upon the particular component part and its application.

The Mathematical Model

Characteristics and type

As pointed out in the general review at the beginning of this chapter, a complete and accurate reliability model would represent the system of interest in all its variational characteristics for all aspects of performance. Most variational characteristics of a practical system are random in form or can be represented by an appropriate random distribution even when they are associated with some systematic cause or effect. Therefore, the reliability model of a system is essentially a model based upon the concepts of probability and statistics. This basic nature of the model should be continually borne in mind during its use and in any subsequent process of interpretation.

Probability and statistics are concerned with the description of the mass behaviour of chance events. The two key words are 'chance' and 'mass'. The result of any reliability calculation for a system only represents a likelihood or chance of certain things being achieved under certain conditions. There is no certainty connected with a reliability number except in the context of its statistical interpretation. This is where the connotation of 'mass' is important for a probability number implies the average result or outcome from a very large or 'massive' sample of the total population. The reader is reminded of the definition of probability originally given in Chapter 1. This means, of course, that the exact reliability of a single system can never be predicted but only the average outcome that would be expected from a large number of nominally identical systems operating over nominally identical periods or upon nominally identical occasions.

The interpretation of the reliability modelled and calculated for a system may be illustrated by the following simple example. Suppose that the reliability parameter of interest for a particular system is that of its availability. The appropriate reliability model for the system may then yield, for instance, the distribution of availability. Beyond this, there is now no definite indication as to what particular value of availability on this distribution curve may apply to the system in question. The mean value of the distribution may be quoted but this only indicates the availability, on average, that would be expected from an infinite number of such systems. Alternatively, any specific value of availability may be quoted together with, say, the probability derived from the distribution function of this value being exceeded. Here, once again, the statement contains a probability which must be interpreted in terms of a 'mass' of systems rather than a single system. So, in whatever form the availability is derived or quoted there is a need to interpret the result in terms of the mass behaviour of chance events.

The actual form and type of reliability model depend upon a large number of factors which the reliability analyst should carefully examine during the formulation process. Some of these factors are considered and summarized in the following paragraphs.

The first influence on the reliability model is the system functional arrangement. This functional arrangement, together with the appropriate specifications, forms the starting point of any reliability analysis. From this arises the initial setting up of the reliability model in logical form and the checking in a deterministic way that the system has, in fact, the capability of performing in the manner required. The type of system logic then dictates the forms of analytical expressions which are required to represent such logical configurations as non-redundant elements, redundant elements, standby elements and majority-voting elements.

The second influence on the character of the model arises from the types of variation which may take place in the performance aspects of the various elements of the system. Each performance aspect of each element may be represented by a complete set of distributional characteristics or, more simply, by a two-state failure/success model as previously discussed. Either form of representation, or any other form, will dictate the particular characteristics of the reliability model. In addition, the way in which any form of variation, such as a complete element failure, is detected or revealed will influence the form of the model. Considerations, in this respect, are those associated with irreversible and reversible processes as discussed in the previous chapters.

Leading from the method of revelation of the various system modes of behaviour are the procedures adopted for restoration, maintenance and testing. These form the third and final main influence on the type of model that has to be formulated. Any of the procedures for restoration may be manual, automatic or a combination of the two. Manual procedures may involve the modelling of human behaviour as well as that of equipment behaviour, but this aspect will be commented upon later. In the case of automatic procedures, such as automatic testing, the equipment which carries out this function may need to be represented in the overall reliability model as well as the system itself. The testing may be partial or complete. Partial testing may not substantiate continued capability and may not reveal all possible fault conditions or other system modes of behaviour. The reliability model needs to be designed in recognition of these possibilities. Restoration, testing or maintenance is almost certain to involve some disturbances of the system. Such disturbances may lead to mistakes, for example, in adjustments, calibrations, replacement parts and reassembly. This means that other fault modes may be introduced apart from those which may be described as inherent in the system elements themselves. These aspects, too, need to be considered in the setting up of the logic and form of the reliability model.

Where human influences are involved in testing and maintenance there is generally some form of portable test equipment or other 'tools' which are not an integral part of the basic system. As with automatic testing and restoration, these items of test equipment or ancillary elements need to be allowed for in the modelling process. The level of testing or maintenance may be important in connexion with whether it is carried out at the system level, the equipment level or the component-part level. At the lower hierarchic levels, the testing and maintenance of the elements or group of elements may be simultaneous, consecutive or staggered. Various procedures of this type will also influence the form of the model as indicated in the previous chapters. The time taken to carry out a test or maintenance procedure may be pertinent if it is significant in length compared with other times of interest. Here the reliability model may have to take account of times between tests or maintenance, the mean length of time involved in the test or the distribution of times for the different types of restoration procedures. In addition, the state of the element or system during such procedures can have a direct bearing on the form of model. For example, the element's normal performance functions may be completely or, perhaps, only partly inhibited for the duration of the test or maintenance activity.

This then has been a brief review of the ways in which the character and forms of the reliability model are influenced by the logical functioning of the system, by the modes of system behaviour and by the various types of restoration procedures. A comprehensive model may generally be built up which takes all known factors into account. For any reasonably sized technological system, this is likely to be a very complex model both in its logical form and in its analytical representation. Even with computer techniques, considerable time and effort may be involved in solving the probability logic and the various multiple integrations that may be involved. In many cases, however, such an extensive modelling procedure may not be expedient or even necessary. In addition the availability and accuracy of basic data to feed into the model may limit the degree of sophistication to which it is prudent to take the modelling process. For these reasons, most practical reliability models are simplified to a form which is an adequate representation of the system under certain assumed conditions. Because of the assumptions involved, simplified models tend to invoke some dangers and these are outlined in the next section.

Typical assumptions

In order to illustrate some of the assumptions which may have to be taken into account, the following simplified model which was derived in Equation (12.111) of Chapter 12 is used as an example:

$$\mu_D = \binom{n}{r} \frac{\theta^r \tau^r}{(r+1)} \tag{13.1}$$

This, it will be remembered, is an approximate or simplified form of the mean fractional dead time for a system made up of n nominally identical elements arranged to operate in a majority-vote logic so that r or more elements are required to fail to produce a system failure. In addition, it applies to the case of unrevealed faults which are irreversible except at the point in time where the elements of the system are proof tested and maintained.

The parameters in Equation (13.1), therefore, carry the following meanings:
- θ = the mean failure-rate for unrevealed faults,
- τ = the time between maintenance and testing procedures,
- n = the total number of nominally identical elements in the system,
- r = the minimum number of elements required to fail in order to inhibit the system

Some of the assumptions involved in the use of a simple model of the type given by Equation (13.1) are now as follows:

(a) The occurrence of unrevealed faults in any of the elements, irrespective of their cause, appear on mass to follow the laws of chance and their times to occurrence are capable of being represented by a statistical distribution.

(b) The elements of the system are each operating in their 'useful-life' phase where the mean failure-rate is constant over a large number of failures. This means that the distribution of times to failure can be characterized by the exponential distribution.

(c) The value of the mean failure-rate, θ, which is the characteristic parameter of the exponential distribution, is known within the limits of accuracy required.

(d) Each of the elements is nominally identical so that the same numerical value of θ may be legitimately ascribed to each one.

(e) The elements are statistically and completely independent of each other so that a failure, for instance, of one element has no connexion with and cannot cause a failure of another element.

(f) There are no two or more coincident failures in the elements which can compensate for one another.

(g) All elements and the corresponding overall system have the deterministic capability of functioning in the required manner under the relevant conditions.

(h) The testing, maintenance and any necessary repair of the elements takes place simultaneously on all elements at the end of every time interval of length τ.

(j) Faults on the elements are only corrected at the point in time when the testing, maintenance and repair procedures are carried out.

(k) The term $(\theta\tau)^r$ in Equation (13.1) is very much less than unity.

(l) The system and its constituent elements are correct in every way at the start of system operation or at the time origin. The installation, commissioning procedures and initial test routines, for example, are perfect and introduce no faults.

(m) The testing, maintenance and repair procedures, which are carried out at the end of every time interval τ, are perfect on each occasion. This

means, for example, that testing time is negligible, all possible faults which may exist are detected, all faults that exist are repaired or corrected in zero time, all repairs of such faults are perfect and no additional faults or errors are introduced by any of the procedures.
(n) Any faults which occur in any of the elements are permanent, static and irreversible until they are discovered and corrected by the laid-down procedures. In other words, there are no transient failures or self-correcting faults.
(o) The arithmetic involved in the manipulation of the mathematical model and the subsequent calculations of the mean fractional dead time are perfect. This implies that the user and interpreter of the model cannot make a mistake.

The reader will appreciate that this is a fairly imposing list of assumptions and yet each one of them is pertinent to the simple model of Equation (13.1). Many of the assumptions referred to in (a) to (o) inclusive may, of course, be taken into account by refining the mathematical model. This fact has already been illustrated by the range of more comprehensive formulae which have been developed in the previous chapters. However, as pointed out at the end of the last section of this chapter, a simplified model may have certain advantages and therefore the assumptions associated with it require careful consideration. A general criterion is that a simplified model of the type of Equation (13.1) may be used provided that the order of the probability number it yields is significantly greater than the probability of all the relevant assumptions being incorrect. This raises the pertinent question as to the level at which the typical assumptions discussed are no longer valid. There is no general answer to this question as the level obviously depends upon the degree of refinement in the model. However, it is possible to deduce one or two guiding principles which are applicable to many of the functional arrangements of typical technological systems. These are enlarged upon in the following paragraphs.

With many majority-voting or redundant element systems it is generally fairly easy to substantiate the use of formulae such as Equation (13.1) provided that the mean fractional dead-time numbers yielded are no lower than about 10^{-3}. Below this, any one of the assumptions made in (a) to (o) inclusive may begin to predominate and substantiation that they can be justifiably applied becomes increasingly difficult. As one illustration of this, it can be seen from Equation (13.1) that theoretically μ_D can be made smaller by reducing τ. However, shortening the interval between testing and maintenance implies more tests and more disturbances on the system in a given length of time. Assumption (m) may break down if the time taken to test becomes significant compared with the total time or if the effect of the increased testing is to increase the chance of additional faults being introduced into the system. Since most test and maintenance routines depend

THE APPLICATION OF RELIABILITY ASSESSMENT

upon human intervention, the increase in testing means that the system is becoming more sensitive to the chances of human error and faulty test equipment. General evidence suggests that the chance of a fault being put on an element due to human test or maintenance activity lies in the range 10^{-2} to 10^{-4} per operation depending upon the complexity of the task. Hence, for a reasonably complex test routine, the rate of introduction of faults due to the test process may well become significant compared with the 'basic' value of θ.

Imperfections in test and maintenance procedures may also have another insiduous effect particularly in redundant systems. It is quite likely that the same methods and items of equipment are used to test every nominally identical element in a redundant arrangement. This means, for some types of faults, that a fault may be introduced not only into one element but into all the elements at the same time. The elements are, therefore, no longer independent as far as this 'common' fault is concerned and assumption (e) will almost certainly break down. The effect of such a common fault is to defeat the intended redundancy which has been built into the system.

Apart from the possibility of introducing faults into a system, the test and maintenance procedures may fail to discover and rectify faults which are already in existence. This could arise because of mistakes in the procedures or because the 'design' of the procedures is such that they are basically incapable of recognizing the existence of certain types of faults. This is the aspect of partial testing which was commented upon in the previous section. Ideally, any test or maintenance should be completely thorough and check all functional aspects of the system. Such an ideal, of course, is not always practical or expedient and some compromise partial test procedure has to be adopted. When a test is only partial, some inferences have to be drawn about the complete behaviour of the system on the basis of the incomplete results which are yielded by the partial test. A partial test, then, is not only prone to leave faults undetected but may also be subject to errors of simulation and errors of inductive reasoning. Undetected faults and simulation errors are another common factor which may mitigate against the assumed independence of the elements as expressed in assumption (e).

As has been seen then, testing and maintenance may introduce common faults into a system, but this is not the only danger. There may be common faults due to errors in design, manufacture, installation or operating environment of all the nominally identical elements. If the probability of a common fault, from whatever cause, is significant compared with the probability number yielded by Equation (13.1) then assumption (e) has broken down and the use of Equation (13.1) becomes invalid.

Practical experience indicates that it is generally the common fault which limits the attainability of high reliability with a purely redundant system as represented by Equation (13.1). The same experience suggests that the common fault may begin to have a significant effect upon the fractional dead times of purely redundant systems when the estimated value of this mean fractional dead time falls below about 10^{-3}. At or below this level, therefore, the assumptions involved in (a) to (o) inclusive should be very carefully examined if a simplified model is used for reliability calculations.

One of the difficulties associated with a purely redundant system made up of nominally identical elements is that common faults may arise due to the very fact that the elements are nominally identical. It is obvious in this case that any design error, operational error or maintenance error on one element is very likely to affect all elements because of their basic similarities. A possible extension of the redundancy principle, therefore, would be to employ elements which still perform the same basic overall function required but are otherwise totally different in design, concept and method of operation. So, for instance, a two-element redundant system required to produce a d.c. supply might consist of a battery as one of the elements and a diesel generator as the other element. As another example, a redundant braking system on a vehicle might have a purely mechanical linkage system as one element and a hydraulically actuated system as the other element. The reader will be able to postulate many other similar examples. The term 'diversity' is very often used in the place of 'redundancy' to describe such a redundant arrangement where the elements are functionally similar but different in type and concept.

With a diverse type of redundancy, the simple model of Equation (13.1) no longer applies since the θ and τ values for each element may be different. However, the modification to the model to take this fact into account does not involve any undue additional complexity. At the same time, most of the assumptions listed in (a) to (o) still apply even for the modified model. The main practical difference is that the probability of the assumptions being incorrect, particularly those reflecting common fault possibilities, will generally be smaller than with a purely redundant system. This is because a common fault is generally less likely in a system employing different types of elements than in one employing nominally identical elements.

However, any practical system of diversity only reduces the chance of a common fault occurrence and does not eliminate it altogether. The elements still have common functional properties and, probably, common functional inputs or outputs. For instance, in the case of the diverse vehicle-braking system, if the driver's foot slips on the brake pedal or the brake linings are badly worn then both systems of braking are defeated by either this common input or common output fault. Once again, therefore, there is some limit beyond which the simplified mathematical model of even a diverse system cannot be used due to the breakdown of, say, assumption (e). Practical experience with diverse systems suggests that such an assumption may require

THE APPLICATION OF RELIABILITY ASSESSMENT

very careful substantiation at estimated system mean fractional dead times below about 10^{-4}. The actual usage of any one particular system may, of course, show that the typical assumptions discussed are valid to lower orders of probability than this figure. However, both this figure and that quoted previously in connexion with purely redundant systems are only guide figures to show those areas in which care in the use and interpretation of a simplified model may need to be taken. In general, the reader is reminded of the obvious fact that low orders of probability for estimated system behaviour and the possible breakdown of the relevant assumptions are always very difficult to prove or substantiate.

Any simplified model for reliability analysis can, of course, be extended or refined to take any or all of the relevant assumptions into account. In practice, therefore, a balance needs to be struck between the degree of simplification adopted on the one hand and the probability of assumptions being incorrect on the other hand. This is perhaps something which the reliability analyst learns by experience rather than something which can be adequately expressed in a book of this nature. However, the reader will have noticed that a certain emphasis has been given to effects of common faults and their influence on the assumption of statistical independence. This is one of the main factors which, in the authors' experience, requires special consideration in the analysis of any system employing a simplified reliability model. This is particularly true in those cases where a high order of reliability is expected or estimated.

If the general character of any common fault possibility is known, of course, it can be inserted into the reliability model in the form of an additional element or range of elements. This does not normally present any difficulties from the modelling point of view but may present difficulties from the aspect of obtaining the necessary reliability data. The guide figures suggested in the previous paragraphs may be of some general help in connexion with this latter point. However, common fault modes and their probability of occurrence are very often tied up with human influence factors and human errors. This is a problem associated with human behaviour and the man–machine interface and as such is beyond the scope of the present book. The reader is therefore referred to other works on the statistics of the behaviour of human beings in connexion with the design, operation and maintenance of technological systems when he becomes confronted with this problem. Provided that the data on such behaviour exist in the required form, the man–machine element can very often be modelled in a similar way to that of inanimate elements discussed throughout this book.

The Perspective of Reliability Assessment

The needs for the reliability analysis of technological systems and the methods and techniques by which such an analysis may be carried out have now been described in the various chapters of this book. In particular, a summary of the relevant arguments and procedures involved in a reliability analysis was given in the first section of this present chapter. It has been seen that any system may be theoretically modelled to almost any desired depth of sophistication. However, in most practical applications it will generally be expedient to use some simplified model which is determined by the accuracy of the available data and the accuracy required in the final analytical result. In the application of simplified models care has to be taken to understand all the assumptions that may be made and to check that such assumptions are adequately valid. Used in this way, the methods described can represent useful evaluation tools in the theoretical reliability analysis of any technological system.

It should be emphasized, however, that any theoretical analysis is only an adjunct, though a valuable adjunct, to any evidence which can be acrued from the direct operational experience of the system in question. Nor would it be suggested that the analytical techniques can replace experienced engineering judgement in the overall assessment and understanding of system reliability. Rather the techniques are complementary to the sound engineering practices which are usually employed in design, manufacture, quality control and efficient operation. On the other hand, reliability analysis can be of particular value in the early stages of system design and even used when the knowledge of actual reliability behaviour is limited or unknown.

In any application, the advantages of reliability assessment tend to fall under two main headings. These are the increased understanding of the system which can acrue from the method of approach and the increased degree of communication which can result in expressing reliability in quantitative terms.

The increased understanding can arise due to the discipline involved in carrying out a reliability analysis. As has been seen, the techniques lead the analyst, of necessity, through a detailed and exact understanding of the system's requirement to a corresponding detailed and correct understanding of the system's configuration and functional behaviour. Furthermore, the very process of logical analysis on which any reliability model must be based contributes to a knowledge of the system which may be difficult to acquire by any other means. Even before any probabilistic or statistical theory is applied to the analysis, the process of assessment may have highlighted any fundamental weaknesses in the system concept or any complete lack of functional capability. This in itself can be a valuable tool to the designer or user of any system.

Once the logical expression of the system is converted into some probabilistic measure of reliability, the second main advantage of the overall technique becomes apparent. There is no longer any need to speak in such

THE APPLICATION OF RELIABILITY ASSESSMENT

subjective terms as 'highly reliable', 'good quality', 'value for money' or 'no risk'. The strengths and weaknesses of the system, from whatever point of view, may be expressed in quantitative terms and numerical quantities. This almost always leads to improved understanding and communication between the various parties concerned in the overall system concept. Even if the various numbers are not exact, they are invariably more precise than the use of descriptive adjectives.

The precision required of reliability numbers very often depends on the use to which they are to be put. Less precision may be required, for example, if the numbers are used for comparative purposes rather than for some indication of absolute performance. Comparatively, quantitative reliability can be used to determine the degrees of the strength or weakness in a system or, alternatively, to determine the degree by which one overall system design meets the various requirements better than another. In the absolute sense, a reliability number may need to be treated with more care and interpreted on the basis of all the assumptions involved in the data acquisition and modelling processes. Used with this care the absolute value of any reliability parameter of interest can be of great value to the designer and particularly to the management of any technological system or project. It can form the basis of overall risk assessment, of cost effectiveness, of full life-cycle costing, of the optimization of running and maintenance procedures and of stock-level control. Since such processes as these obviously depend upon the accuracy of the reliability numbers it may be useful, in conclusion, to see what orders of precision are typically obtainable using the techniques of analysis that have been described.

To obtain an index of precision the theoretically predicted values of the reliability parameters for a particular system need to be compared with the reliability actually achieved by the same system over a long period of operation. In the past a number of systems or elements of systems have been analysed at the design stage or production stage and the data on their reliability performance subsequently collected during a reasonable sample period of practical usage. The ratio, r, of the value of the particular observed reliability parameter to the corresponding predicted reliability parameter has then been obtained for each system so studied. A ratio of unity means, of course, that the prediction is apparently exact, a ratio greater than unity that the prediction is an underestimate and a ratio less than unity that the prediction is an overestimate. In a typical example of about fifty different system elements examined in this way the value of r was found to be randomly variable as might be expected. This variance arises from two main causes:

(a) Variability in prediction due to the method of analysis, assumptions in the mathematical model and errors in the component part data.

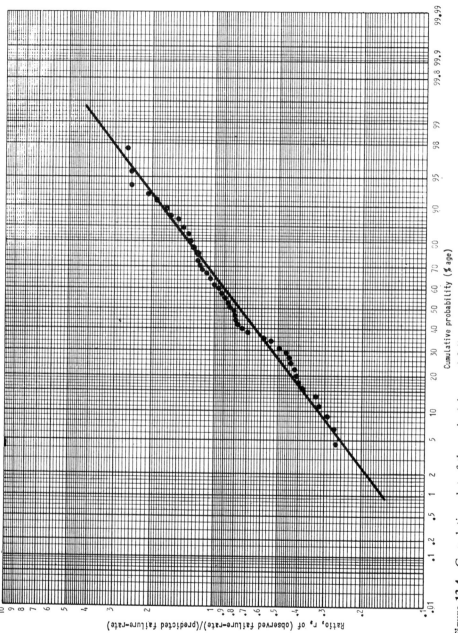

Figure 13.4 Cumulative plot of the ratio (observed failure-rate)/(predicted failure-rate) on lognormal probability paper

THE APPLICATION OF RELIABILITY ASSESSMENT

(b) Variability in observation due to the method of recording system performance. Typically, for instance, errors arise in practice due to difficulties in detecting, classifying, timing and recording all changes and fault modes of an actual system behaviour.

The distribution of the ratio, r, arising from these causes of variation has been found to be approximately lognormal. Figure 13.4 shows the cumulative distribution of r which is based on the sample of system elements previously mentioned and which is plotted on lognormal probability paper. The nearness of the sample points to a straight line indicating the degree of correspondence with a lognormal distribution. From the plot it can be seen that the median of r is 0.76, that the chance of the ratio r being within a factor of 2 of the median value is 70% and that the chance of the ratio being within a factor of 4 is 96%.

This means, with due care and attention, that the results of a reliability analysis of a system carried out along the lines indicated in this book can be reasonably expected to yield numerical values to within a factor of 2 of the actual values and can be taken to be within a factor of 4 with quite a high degree of confidence. Generally, the limitations are not in the mathematical model but in the precision to which adequate and valid reliability data can be collected and recorded. The discussions in this book have mainly been concerned with the techniques involved in the modelling process and it is hoped that the subject has now been covered in sufficient detail to enable the reader to set up for himself an adequate reliability model for most types of technological systems and, having set it up, to apply it with confidence and understanding.

Appendix

Table A.1 The cumulative normal distribution function

$$p(t) = \frac{1}{\sqrt{(2)\pi}} \int_0^t e^{-\frac{1}{2}t^2} \, dt$$

t	0.00	0.01	0.02	0.03	0.04	0.05	0.06	0.07	0.08	0.09
0.0	0.0000	0.0040	0.0080	0.0120	0.0159	0.0199	0.0239	0.0279	0.0319	0.0359
0.1	0.0398	0.0438	0.0478	0.0517	0.0557	0.0596	0.0636	0.0675	0.0714	0.0753
0.2	0.0793	0.0832	0.0871	0.0909	0.0948	0.0987	0.1026	0.1064	0.1103	0.1141
0.3	0.1179	0.1217	0.1255	0.1293	0.1331	0.1368	0.1406	0.1443	0.1480	0.1517
0.4	0.1555	0.1591	0.1628	0.1664	0.1700	0.1736	0.1772	0.1808	0.1844	0.1879
0.5	0.1915	0.1950	0.1985	0.2019	0.2054	0.2088	0.2123	0.2157	0.2190	0.2224
0.6	0.2257	0.2291	0.2324	0.2356	0.2389	0.2421	0.2454	0.2486	0.2517	0.2549
0.7	0.2580	0.2611	0.2642	0.2673	0.2703	0.2734	0.2764	0.2793	0.2823	0.2852
0.8	0.2881	0.2910	0.2939	0.2967	0.2995	0.3023	0.3051	0.3078	0.3108	0.3133
0.9	0.3159	0.3186	0.3212	0.3238	0.3264	0.3289	0.3315	0.3340	0.3365	0.3389
1.0	0.3413	0.3437	0.3461	0.3485	0.3508	0.3531	0.3554	0.3577	0.3599	0.3621
1.1	0.3643	0.3665	0.3686	0.3708	0.3729	0.3749	0.3770	0.3790	0.3810	0.3830
1.2	0.3849	0.3869	0.3888	0.3906	0.3925	0.3943	0.3962	0.3980	0.3997	0.4015
1.3	0.4032	0.4049	0.4066	0.4082	0.4099	0.4115	0.4131	0.4147	0.4162	0.4177
1.4	0.4192	0.4207	0.4222	0.4236	0.4251	0.4265	0.4279	0.4292	0.4306	0.4319
1.5	0.4332	0.4345	0.4357	0.4370	0.4382	0.4394	0.4406	0.4418	0.4429	0.4441
1.6	0.4452	0.4463	0.4474	0.4484	0.4495	0.4505	0.4515	0.4525	0.4535	0.4545
1.7	0.4554	0.4564	0.4573	0.4582	0.4591	0.4599	0.4608	0.4616	0.4625	0.4633
1.8	0.4641	0.4648	0.4656	0.4664	0.4671	0.4678	0.4686	0.4693	0.4699	0.4706
1.9	0.4713	0.4719	0.4726	0.4732	0.4738	0.4744	0.4750	0.4756	0.4761	0.4767
2.0	0.4772	0.4778	0.4783	0.4788	0.4793	0.4798	0.4803	0.4808	0.4812	0.4817
2.1	0.4821	0.4826	0.4830	0.4834	0.4838	0.4842	0.4846	0.4850	0.4854	0.4857
2.2	0.4861	0.4864	0.4868	0.4871	0.4874	0.4878	0.4881	0.4884	0.4887	0.4890
2.3	0.4893	0.4896	0.4898	0.4901	0.4904	0.4906	0.4909	0.4911	0.4913	0.4916
2.4	0.4918	0.4920	0.4922	0.4924	0.4927	0.4929	0.4930	0.4932	0.4934	0.4936
2.5	0.4938	0.4940	0.4941	0.4943	0.4945	0.4946	0.4948	0.4949	0.4951	0.4952
2.6	0.4953	0.4955	0.4956	0.4957	0.4958	0.4960	0.4961	0.4962	0.4963	0.4964
2.7	0.4965	0.4966	0.4967	0.4968	0.4969	0.4970	0.4971	0.4972	0.4973	0.4974
2.8	0.4974	0.4975	0.4976	0.4977	0.4977	0.4978	0.4979	0.4979	0.4980	0.4981
2.9	0.4981	0.4982	0.4982	0.4983	0.4984	0.4984	0.4985	0.4985	0.4986	0.4986
3.0	0.4986	0.4987	0.4987	0.4988	0.4988	0.4989	0.4989	0.4989	0.4990	0.4990
3.1	0.4990	0.4991	0.4991	0.4991	0.4991	0.4992	0.4992	0.4992	0.4993	0.4993
3.2	0.4993	0.4993	0.4994	0.4994	0.4994	0.4994	0.4994	0.4995	0.4995	0.4995
3.3	0.4995	0.4995	0.4995	0.4996	0.4996	0.4996	0.4996	0.4996	0.4996	0.4996
3.4	0.4997	0.4997	0.4997	0.4997	0.4997	0.4997	0.4997	0.4997	0.4997	0.4998

APPENDIX

Table A.2 Transform pairs of functions

No.	$F(s) = L[f(x)]$	$f(x) = L^{-1}[F(s)]$
1	1	$\delta(x)$
2	$\dfrac{1}{s}$	$U(x)$
3	$\dfrac{a}{s}$	$aU(x)$
4	$\dfrac{1}{s^2}$	x
5	$\dfrac{1}{s^{\frac{1}{2}}}$	$\dfrac{1}{\sqrt{(\pi x)}}$
6	$\dfrac{1}{s^{\frac{3}{2}}}$	$\sqrt{(4x/\pi)}$
7	$\dfrac{1}{s^n}\quad (n>-1)$	$\dfrac{x^{n-1}}{\Gamma(n)}$
8	$\dfrac{1}{s+a}$	e^{-ax}
9	$\dfrac{1}{(s+a)^n}\quad (n>-1)$	$\dfrac{x^{n-1}e^{-ax}}{\Gamma(n)}$
10	$\dfrac{s}{(s+a)^2}$	$(1-ax)e^{-ax}$
11	$\dfrac{1}{s(s+a)}$	$\dfrac{1}{a}(1-e^{-ax})$
12	$\dfrac{1}{s^2(s+a)}$	$\dfrac{1}{a^2}[ax-(1-e^{-ax})]$
13	$\dfrac{1}{s(s+a)^2}$	$\dfrac{1}{a^2}[1-(1+ax)e^{-ax}]$
14	$\dfrac{1}{s^2(s+a)^2}$	$\dfrac{1}{a^3}[(2+ax)e^{-ax}-(2-ax)]$
15	$\dfrac{n!}{s^{n+1}}\sum_{j=0}^{n}\dfrac{(-1)^j(as)^j}{j!}$ $(n=1,2,\text{etc.})$	$(x-a)^n U(x)$
16	$\dfrac{1}{(s+a)(s+b)}\quad (a\neq b)$	$\dfrac{1}{(b-a)}(e^{-ax}-e^{-bx})$
17	$\dfrac{s}{(s+a)(s+b)}\quad (a\neq b)$	$\dfrac{1}{(b-a)}(b\,e^{-bx}-a\,e^{-ax})$
18	$\dfrac{1}{(s+a)(s+b)^2}\quad (a\neq b)$	$\dfrac{e^{-ax}-e^{-bx}[1-(a-b)x]}{(a-b)^2}$

Table A.2 (cont.)   Transform pairs of functions

No.	$F(s) = L[f(x)]$	$f(x) = L^{-1}[F(s)]$
19	$\dfrac{1}{(s+a)(s+b)(s+c)} \quad (a \neq b \neq c)$	$\dfrac{(c-b)\,e^{-ax}+(a-c)\,e^{-bx}+(b-a)\,e^{-cx}}{(a-b)(b-c)(c-a)}$
20	$\dfrac{s}{(s+a)(s+b)(s+c)} \quad (a \neq b \neq c)$	$\dfrac{a(b-c)\,e^{-ax}+b(c-a)\,e^{-bx}+c(a-b)\,e^{-cx}}{(a-b)(b-c)(c-a)}$
21	$\dfrac{1}{as^2+bs+c}$ $(b^2-4ac>0,\ a\neq 0)$	$\dfrac{e^{-(b-k)x/2a}-e^{-(b+k)x/2a}}{k}$ $k=\sqrt{(b^2-4ac)}$
22	$\dfrac{1}{as^2+bs+c}$ $(b^2-4ac=0,\ a\neq 0)$	$\dfrac{x\,e^{-bx/2a}}{a}$
23	$\dfrac{1}{as^2+bs+c}$ $(b^2-4ac<0,\ a\neq 0)$	$\dfrac{2\,e^{-bx/2a}}{k}\sin\dfrac{kx}{2a}$ $k=\sqrt{(4ac-b^2)}$
24	$\dfrac{1}{(s+d)(as^2+bs+c)}$ $(k=ad^2-bd+c\neq 0,\ a\neq 0)$	$\dfrac{e^{-dx}}{k}+\dfrac{1}{k}L^{-1}\!\left(\dfrac{-as+ad-b}{as^2+bs+c}\right)$
25	$\log\!\left(\dfrac{s+a}{s+b}\right)$	$\dfrac{1}{x}(e^{-bx}-e^{-ax})$
26	e^{-as}	$\delta(x-a)$
27	$\dfrac{e^{-as}}{s}$	$U(x-a)$
28	$\dfrac{e^{-as}}{s^2}$	$(x-a)\,U(x-a)$
29	$\dfrac{e^{-as}}{s^n} \quad (n>-1)$	$\dfrac{(x-a)^{n-1}U(x-a)}{\Gamma(n)}$
30	$\dfrac{n!\,e^{-as}}{s^{n+1}}\displaystyle\sum_{j=0}^{n}\dfrac{(as)^j}{j!}$ $(n=1,2,\text{etc.})$	$x^n\,U(x-a)$
31	$\dfrac{(1+as)\,e^{-as}}{s^2}$	$xU(x-a)$
32	$\dfrac{1-e^{-as}}{s}$	$U(x)-U(x-a)$
33	$\dfrac{e^{-as}-e^{-bs}}{s}$	$U(x-a)-U(x-b)$

APPENDIX

Table A.2 (*cont.*) Transform pairs of functions

No.	$F(s) = L[f(x)]$	$f(x) = L^{-1}[F(s)]$
34	$\dfrac{e^{-a(s-c)} - e^{-b(s-c)}}{s-c}$	$e^{cx}[U(x-a) - U(x-b)]$
35	$\dfrac{e^{-as}}{s+b}$	$e^{-b(x-a)}\, U(x-a)$
36	$\dfrac{e^{-as}}{(s+b)^n}\quad (n>-1)$	$\dfrac{(x-a)^{n-1}\, e^{-b(x-a)}\, U(x-a)}{\Gamma(n)}$
37	$\dfrac{e^{-ab}\, e^{-as}}{s+b}$	$e^{-bx}\, U(x-a)$
38	$\dfrac{\omega}{s^2+\omega^2}$	$\sin \omega x$
39	$\dfrac{s}{s^2+\omega^2}$	$\cos \omega x$
40	$\dfrac{\omega}{(s+a)^2+\omega^2}$	$e^{-ax} \sin \omega x$
41	$\dfrac{s+a}{(s+a)^2+\omega^2}$	$e^{-ax} \cos \omega x$
42	$\dfrac{2\omega s}{(s^2+\omega^2)^2}$	$x \sin \omega x$
43	$\dfrac{s^2-\omega^2}{(s^2+\omega^2)^2}$	$x \cos \omega x$
44	$\tan^{-1}\left(\dfrac{\omega}{s}\right)$	$\dfrac{1}{x}\sin \omega x$
45	$\log\left(\dfrac{s^2+\omega^2}{s^2}\right)$	$\dfrac{2}{x}(1-\cos \omega x)$
46	$\dfrac{1}{s^2(s^2+\omega^2)}$	$\dfrac{1}{\omega^3}(\omega x - \sin \omega x)$
47	$\dfrac{1}{s(s^2+\omega^2)}$	$\dfrac{1}{\omega^2}(1-\cos \omega x)$
48	$\dfrac{s^2}{(s^2+\omega^2)^2}$	$\dfrac{1}{2\omega}(\sin \omega x + \omega x \cos \omega x)$
49	$\dfrac{1}{(s^2+\omega^2)^2}$	$\dfrac{1}{2\omega^3}(\sin \omega x - \omega x \cos \omega x)$
50	$\dfrac{\omega}{s^2-\omega^2}$	$\sinh \omega x$
51	$\dfrac{s}{s^2-\omega^2}$	$\cosh \omega x$

Table A.3 Transform pairs of operations

No.	$F(s) = L[f(x)]$	$f(x) = L^{-1}[F(s)]$
1	$\lim_{s \to \infty} sF(s)$	$\lim_{x \to 0} f(x)$
2	$\lim_{s \to 0} sF(s)$	$\lim_{x \to \infty} f(x)$
3	$aF(s)$	$af(x)$
4	$F_1(s) \pm F_2(s)$	$f_1(x) \pm f_2(x)$
5	$e^{-as} F(s)$	$f(x-a)$
6	$F(s-a)$	$e^{ax} f(x)$
7	$aF(as)$	$f\left(\dfrac{x}{a}\right)$
8	$\dfrac{dF(s)}{ds}$	$-xf(x)$
9	$\dfrac{d^n F(s)}{ds^n}$	$(-1)^n x^n f(x)$
10	$\int_s^\infty F(s)\, ds$	$\dfrac{f(x)}{x}$
11	$[F^{-n}(s)]_s^\infty$	$\dfrac{f(x)}{x^n}$
12	$sF(s) - f(0+)$	$\dfrac{df(x)}{dx}$
13	$s^n F(s) - s^{n-1} f(0+) - s^{n-2} f'(0+) \ldots$	$\dfrac{d^n f(x)}{dx^n}$
14	$\dfrac{\partial}{\partial a} F(s, a)$	$\dfrac{\partial}{\partial a} f(x, a)$
15	$\dfrac{F(s)}{s} + \dfrac{f^{-1}(0+)}{s}$	$\int f(x)\, dx$
16	$\dfrac{F(s)}{s^2} + \dfrac{f^{-1}(0+)}{s^2} + \dfrac{f^{-2}(0+)}{s}$	$\int \left[\int f(x)\, dx\right] dx$
17	$F_1(s) F_2(s)$	$\int_0^x f_1(x-u)\, f_2(u)\, du$

APPENDIX

Table A.4 The cumulative exponential distribution function, $p(x) = 1 - e^{-x/\lambda}$

x/λ	$p(x)$	x/λ	$p(x)$	x/λ	$p(x)$	x/λ	$p(x)$	x/λ	$p(x)$
0.01	0.010	0.44	0.356	0.87	0.581	1.30	0.728	1.96	0.859
0.02	0.020	0.45	0.362	0.88	0.585	1.31	0.730	1.98	0.862
0.03	0.030	0.46	0.369	0.89	0.589	1.32	0.733	2.00	0.865
0.04	0.039	0.47	0.375	0.90	0.593	1.33	0.736	2.05	0.871
0.05	0.049	0.48	0.381	0.91	0.598	1.34	0.738	2.10	0.878
0.06	0.058	0.49	0.387	0.92	0.602	1.35	0.741	2.15	0.884
0.07	0.068	0.50	0.394	0.93	0.606	1.36	0.743	2.20	0.889
0.08	0.077	0.51	0.400	0.94	0.609	1.37	0.746	2.25	0.895
0.09	0.086	0.52	0.406	0.95	0.613	1.38	0.748	2.30	0.900
0.10	0.095	0.53	0.411	0.96	0.617	1.39	0.751	2.35	0.905
0.11	0.104	0.54	0.417	0.97	0.621	1.40	0.753	2.40	0.909
0.12	0.113	0.55	0.423	0.98	0.625	1.41	0.756	2.45	0.914
0.13	0.122	0.56	0.429	0.99	0.628	1.42	0.758	2.50	0.918
0.14	0.131	0.57	0.435	1.00	0.632	1.43	0.761	2.55	0.922
0.15	0.139	0.58	0.440	1.01	0.636	1.44	0.763	2.60	0.926
0.16	0.148	0.59	0.446	1.02	0.639	1.45	0.765	2.65	0.929
0.17	0.156	0.60	0.451	1.03	0.643	1.46	0.768	2.70	0.933
0.18	0.165	0.61	0.457	1.04	0.647	1.47	0.770	2.75	0.936
0.19	0.173	0.62	0.462	1.05	0.650	1.48	0.772	2.80	0.939
0.20	0.181	0.63	0.467	1.06	0.654	1.49	0.775	2.85	0.942
0.21	0.189	0.64	0.473	1.07	0.657	1.50	0.777	2.90	0.945
0.22	0.198	0.65	0.478	1.08	0.660	1.52	0.781	2.95	0.948
0.23	0.206	0.66	0.483	1.09	0.664	1.54	0.786	3.00	0.950
0.24	0.213	0.67	0.488	1.10	0.667	1.56	0.790	3.10	0.955
0.25	0.221	0.68	0.493	1.11	0.670	1.58	0.794	3.20	0.959
0.26	0.229	0.69	0.498	1.12	0.674	1.60	0.798	3.30	0.963
0.27	0.237	0.70	0.503	1.13	0.677	1.62	0.802	3.40	0.967
0.28	0.244	0.71	0.508	1.14	0.680	1.64	0.806	3.50	0.970
0.29	0.252	0.72	0.513	1.15	0.683	1.66	0.810	3.60	0.973
0.30	0.259	0.73	0.518	1.16	0.687	1.68	0.814	3.70	0.975
0.31	0.267	0.74	0.523	1.17	0.690	1.70	0.817	3.80	0.978
0.32	0.274	0.75	0.528	1.18	0.693	1.72	0.821	3.90	0.980
0.33	0.281	0.76	0.532	1.19	0.696	1.74	0.825	4.00	0.982
0.34	0.288	0.77	0.537	1.20	0.699	1.76	0.828	4.20	0.985
0.35	0.295	0.78	0.542	1.21	0.702	1.78	0.831	4.40	0.988
0.36	0.302	0.79	0.546	1.22	0.705	1.80	0.835	4.60	0.990
0.37	0.309	0.80	0.551	1.23	0.708	1.82	0.838	4.80	0.992
0.38	0.316	0.81	0.555	1.24	0.711	1.84	0.841	5.00	0.993
0.39	0.323	0.82	0.560	1.25	0.714	1.86	0.844	5.50	0.996
0.40	0.330	0.83	0.564	1.26	0.716	1.88	0.847	6.00	0.998
0.41	0.336	0.84	0.568	1.27	0.719	1.90	0.850	6.50	0.999
0.42	0.343	0.85	0.573	1.28	0.722	1.92	0.853	7.00	0.999
0.43	0.350	0.86	0.577	1.29	0.725	1.94	0.856	7.50	0.999

Table A.5 The cumulative χ^2 distribution (P in %)—see Equation (9.112)

f \ P	0.5	1.0	2.5	5.0	10	20	25	30	50
1	0.000	0.000	0.001	0.004	0.016	0.064	0.102	0.148	0.455
2	0.010	0.020	0.051	0.103	0.211	0.446	0.575	0.713	1.386
3	0.072	0.115	0.216	0.352	0.584	1.005	1.213	1.424	2.366
4	0.207	0.297	0.484	0.711	1.064	1.649	1.923	2.195	3.357
5	0.412	0.554	0.831	1.145	1.610	2.343	2.675	3.000	4.351
6	0.676	0.872	1.237	1.635	2.204	3.070	3.455	3.828	5.348
7	0.989	1.239	1.690	2.167	2.833	3.822	4.255	4.671	6.346
8	1.344	1.646	2.180	2.733	3.490	4.594	5.071	5.527	7.344
9	1.735	2.088	2.700	3.325	4.168	5.380	5.899	6.393	8.343
10	2.156	2.558	3.247	3.940	4.865	6.179	6.737	7.267	9.342
11	2.603	3.053	3.816	4.575	5.578	6.989	7.584	8.148	10.341
12	3.074	3.571	4.404	5.226	6.304	7.807	8.438	9.034	11.340
13	3.565	4.107	5.009	5.892	7.042	8.634	9.299	9.926	12.340
14	4.075	4.660	5.629	6.571	7.790	9.467	10.165	10.821	13.339
15	4.601	5.229	6.262	7.261	8.547	10.307	11.036	11.721	14.339
16	5.142	5.812	6.908	7.962	9.312	11.152	11.912	12.624	15.338
17	5.697	6.408	7.564	8.672	10.085	12.002	12.792	13.531	16.338
18	6.265	7.015	8.231	9.390	10.865	12.857	13.675	14.440	17.338
19	6.844	7.633	8.907	10.117	11.651	13.716	14.562	15.352	18.338
20	7.434	8.260	9.591	10.851	12.443	14.578	15.452	16.266	19.337
21	8.034	8.897	10.283	11.591	13.240	15.445	16.344	17.182	20.337
22	8.643	9.542	10.982	12.338	14.041	16.314	17.240	18.101	21.337
23	9.260	10.196	11.688	13.091	14.848	17.187	18.137	19.021	22.337
24	9.886	10.856	12.401	13.848	15.659	18.062	19.037	19.943	23.337
25	10.520	11.524	13.120	14.611	16.473	18.940	19.939	20.867	24.337
26	11.160	12.198	13.844	15.379	17.292	19.820	20.843	21.792	25.336
27	11.808	12.879	14.573	16.151	18.114	20.703	21.749	22.719	26.336
28	12.461	13.565	15.308	16.928	18.939	21.588	22.657	23.647	27.336
29	13.121	14.256	16.047	17.708	19.768	22.475	23.567	24.577	28.336
30	13.787	14.953	16.791	18.493	20.599	23.364	24.478	25.508	29.336

Values of χ^2

Table shows area to left of f value

Table A.5 (*cont.*)   The cumulative χ^2 distribution (P in %)—see Equation (9.112)

P \ f	70	75	80	90	95	97.5	99	99.5	99.9
1	1.074	1.323	1.642	2.706	3.841	5.024	6.635	7.879	10.827
2	2.408	2.773	3.219	4.605	5.991	7.378	9.210	10.597	13.815
3	3.665	4.108	4.642	6.251	7.815	9.348	11.345	12.838	16.268
4	4.878	5.385	5.989	7.779	9.488	11.143	13.277	14.860	18.465
5	6.064	6.626	7.289	9.236	11.070	12.832	15.086	16.750	20.517
6	7.231	7.841	8.558	10.645	12.592	14.449	16.812	18.548	22.457
7	8.383	9.037	9.803	12.017	14.067	16.013	18.475	20.278	24.322
8	9.524	10.219	11.030	13.362	15.507	17.535	20.090	21.955	26.125
9	10.656	11.389	12.242	14.684	16.919	19.023	21.666	23.589	27.877
10	11.781	12.549	13.442	15.987	18.307	20.483	23.209	25.188	29.588
11	12.899	13.701	14.631	17.275	19.675	21.920	24.725	26.757	31.264
12	14.011	14.845	15.812	18.549	21.026	23.337	26.217	28.300	32.909
13	15.119	15.984	16.985	19.812	22.362	24.736	27.688	29.819	34.528
14	16.222	17.117	18.151	21.064	23.685	26.119	29.141	31.319	36.123
15	17.322	18.245	19.311	22.307	24.996	27.488	30.578	32.801	37.697
16	18.418	19.369	20.465	23.542	26.296	28.845	32.000	34.267	39.252
17	19.511	20.489	21.615	24.769	27.587	30.191	33.409	35.718	40.790
18	20.601	21.605	22.760	25.989	28.869	31.526	34.805	37.156	42.312
19	21.689	22.718	23.900	27.204	30.144	32.852	36.191	38.582	43.820
20	22.775	23.828	25.038	28.412	31.410	34.170	37.566	39.997	45.315
21	23.858	24.935	26.171	29.615	32.671	35.479	38.932	41.401	46.797
22	24.937	26.039	27.301	30.813	33.924	36.781	40.289	42.796	48.268
23	26.018	27.141	28.429	32.007	35.172	38.076	41.638	44.181	49.728
24	27.096	28.241	29.553	33.196	36.415	39.364	42.980	45.558	51.179
25	28.172	29.339	30.675	34.382	37.652	40.646	44.314	46.928	52.620
26	29.246	30.434	31.795	35.563	38.885	41.923	45.642	48.290	54.052
27	30.319	31.528	32.912	36.741	40.113	43.194	46.963	49.645	55.476
28	31.391	32.620	34.027	37.916	41.437	44.461	48.278	50.993	56.893
29	32.461	33.711	35.139	39.087	42.557	45.722	49.588	52.336	58.302
30	33.530	34.800	36.250	40.256	43.773	46.979	50.892	53.672	59.703

Abridged Version of Table IV from R. A. Fisher and F. Yates: *Statistical Tables for Biological, Agricultural and Medical Research* published by Oliver & Boyd Ltd., Edinburgh and by permission of the publishers and authors.

Table A.6 The cumulative t distribution (P in %)—see Equation (9.123)

P \ f	55	60	70	80	90	95	97.5	99	99.5	99.9
1	0.158	0.325	0.727	1.376	3.078	6.314	12.71	31.82	63.66	318.0
2	0.142	0.289	0.617	1.061	1.886	2.920	4.303	6.965	9.925	22.30
3	0.137	0.277	0.584	0.978	1.638	2.353	3.182	4.541	5.841	10.20
4	0.134	0.271	0.569	0.941	1.533	2.132	2.776	3.747	4.604	7.173
5	0.132	0.267	0.559	0.920	1.476	2.015	2.571	3.365	4.032	5.893
6	0.131	0.265	0.553	0.906	1.440	1.943	2.447	3.143	3.707	5.208
7	0.130	0.263	0.549	0.896	1.415	1.895	2.365	2.998	3.499	4.785
8	0.130	0.262	0.546	0.889	1.397	1.860	2.306	2.896	3.355	4.501
9	0.129	0.261	0.543	0.883	1.383	1.833	2.262	2.821	3.250	4.297
10	0.129	0.260	0.542	0.879	1.372	1.812	2.228	2.764	3.169	4.144
11	0.129	0.260	0.540	0.876	1.363	1.796	2.201	2.718	3.106	4.025
12	0.128	0.259	0.539	0.873	1.356	1.782	2.179	2.681	3.055	3.930
13	0.128	0.259	0.538	0.870	1.350	1.771	2.160	2.650	3.012	3.852
14	0.128	0.258	0.537	0.868	1.345	1.761	2.145	2.624	2.977	3.787
15	0.128	0.258	0.536	0.866	1.341	1.753	2.131	2.602	2.947	3.733
16	0.128	0.258	0.535	0.865	1.337	1.746	2.120	2.583	2.921	3.686
17	0.128	0.257	0.534	0.863	1.333	1.740	2.110	2.567	2.898	3.646
18	0.127	0.257	0.534	0.862	1.330	1.734	2.101	2.552	2.878	3.610
19	0.127	0.257	0.533	0.861	1.328	1.729	2.093	2.539	2.861	3.579
20	0.127	0.257	0.533	0.860	1.325	1.725	2.086	2.528	2.845	3.552
21	0.127	0.257	0.532	0.859	1.323	1.721	2.080	2.518	2.831	3.527
22	0.127	0.256	0.532	0.858	1.321	1.717	2.074	2.508	2.819	3.505
23	0.127	0.256	0.532	0.858	1.319	1.714	2.069	2.500	2.807	3.485
24	0.127	0.256	0.531	0.857	1.318	1.711	2.064	2.492	2.797	3.467
25	0.127	0.256	0.531	0.856	1.316	1.708	2.060	2.485	2.787	3.450
26	0.127	0.256	0.531	0.856	1.315	1.706	2.056	2.479	2.779	3.435
27	0.127	0.256	0.531	0.855	1.314	1.703	2.052	2.473	2.771	3.421
28	0.127	0.256	0.530	0.855	1.313	1.701	2.048	2.467	2.763	3.408
29	0.127	0.256	0.530	0.854	1.311	1.699	2.045	2.462	2.756	3.396
30	0.127	0.256	0.530	0.854	1.310	1.697	2.042	2.457	2.750	3.885
40	0.126	0.255	0.529	0.851	1.303	1.684	2.021	2.423	2.704	3.307
60	0.126	0.254	0.527	0.848	1.296	1.671	2.000	2.390	2.660	3.232
120	0.126	0.254	0.526	0.845	1.289	1.658	1.980	2.358	2.617	3.160
∞	0.126	0.253	0.524	0.842	1.282	1.645	1.960	2.326	2.576	3.090

This table is reproduced from Table III of R. A. Fisher and F. Yates: *Statistical Tables for Biological, Agricultural and Medical Research* published by Oliver & Boyd Ltd., Edinburgh and by permission of the publishers and authors.

Table A.7 Average component failure-rates—Electronic components

Type of component	Failure-rate $f/10^6$ h
RESISTORS (fixed)	
High stability, carbon	0.5
Wirewound, general	0.5
Wirewound, precision	1.0
Thyrite (silicon carbide)	1.0
Composition, Grade 2	0.1
Metal film	0.1
Oxide film	0.05
Thermistor	1.0
RESISTORS (variable)	
Miniature wirewound	5.0
Precision wirewound	6.0
Carbon track	2.0
General wirewound	3.0
CAPACITORS (fixed)	
Oil filled	3.0
Paper	1.0
Metallized paper	0.5
Synthetic film	0.5
Mica	0.3
Ceramic	0.1
Polystyrene	0.1
CAPACITORS (electrolytic)	
Aluminium foil	2.0
Tantalum foil	1.0
Tantalum pellet	0.5
Tantalum solid	0.2
CAPACITORS (variable)	
Air spaced	1.0
Ceramic	0.5
INDUCTORS	
Single coils, general	0.3
Saturable reactors	1.5
Tuning coils	1.5
R.F. coils	0.5
Solenoids	0.4
A.F. coils	0.3

Table A.7 (*cont.*) Average component failure-rates—Electronic components

Type of component	Failure-rate $f/10^6$ h
TRANSFORMERS	
Each winding, general	0.3
High voltage	9.0
Mains, encapsulated	5.0
Mains, oil filled	1.0
Pulse	1.5
Variable	1.0
I.F.	1.0
A.F.	0.3
Control and low voltage	0.3
RELAYS	
Each coil, general	0.3
Each contact pair, general	0.2
Miniature high speed	7.0
General high speed	5.0
Heavy duty	5.0
General control	2.0
P.O. type, general	2.0
P.O. type, fully tropicalized	1.0
Hermetically sealed	0.5
Carpenter, polarized, each coil	0.3
each contact pair	0.5
Time delay	4.0
Thermal	4.0
Protective	5.0
SWITCHES	
Each contact, general	0.2
Rotary, wafer	2.0
Micro	0.5
Thermal, heater	1.0
contacts	1.0
Toggle, general	1.0
each contact pair	0.2
Push button	0.5
CONNECTORS	
Each pin, general	0.05
Printed circuit, each contact	0.1
Valve holders, each pin	0.1
WIRING	
Terminals	0.5
Soldered joints, auto	0.001
Soldered joints, hand	0.05
Wrapped joints	0.0001
Crimped and welded joints	0.005

Table A.7 (*cont.*)   Average component failure-rates—
Electronic components

Type of component	Failure-rate $f/10^6$ h
VALVES	
Loss of emission, diodes	5.0
others	10.0
Each electrode, o/c fault	1.0
Each adjacent electrode pair, s/c fault	1.0
Diodes	12.0
Double diodes	16.0
Triodes	19.0
Double triodes	28.0
Tetrodes	21.0
Pentodes	23.0
Stabilizers, diode	13.0
triode	16.0
Rectifiers	20.0
Thyratrons	50.0
Thermal relay, intermittent	2.0
continuous	15.0
TRANSISTORS	
Germanium, high power	1.0
Germanium, low power	0.1
Silicon, high power	0.5
Silicon, low power	0.05
SEMI-CONDUCTOR DIODES	
Germanium, point contact	0.5
Germanium, junction	0.2
Germanium, gold bonded	0.2
Silicon, zener	0.1
Silicon, high power	0.5
Silicon, low power	0.02
METAL RECTIFIERS	
Selenium	5.0
Copper oxide	5.0
LAMPS	
Neon (indicator)	2.0
Filament, high wattage	50.0
Indicator	5.0
Fluorescent	10.0
MISCELLANEOUS	
Fuses	5.0
Meters (moving coil)	3.0
Recorders	25.0
Batteries	1.0
Vibrators	5.0

Table A.7 (*cont.*) Average component failure-rates—Electronic components

Type of component	Failure-rate $f/10^6$ h
SENSORS	
Ion chambers and leads	5.0
Thermocouples	10.0
Strain gauges	25.0
Photo-electric cells	15.0
MICROELECTRONICS	
Silicon integrated circuits	0.2

Table A.7 (*cont.*) Average component failure-rates—Electrical components

Type of component	Failure-rate $f/10^6$ h
GENERATORS	
a.c., general	7.0
d.c., general	9.0
Tachometers	5.0
Synchros	8.0
MOTORS	
General	10.0
Induction (above 200 kW)	10.0
Induction (below 200 kW)	5.0
Synchronous	7.0
Small, general	4.0
Stepper	5.0
CONTACTORS (general, less than 3.3 kV)	2.5
CIRCUIT BREAKERS	
General (less than 33 kV)	2.0
415 V to 11 kV	1.5
33 kV	3.0
132 kV	4.0
275 kV	7.0
400 kV	10.0
TRANSFORMERS	
Less than 15 kV	0.6
15 kV to 33 kV	2.0
33 kV to 132 kV	4.0
132 kV to 400 kV	7.0
CABLES	
Less than 1 kV (per km)	0.6
1 to 33 kV (per km)	4.4
33 kV to 275 kV (per km)	7.5
OVERHEAD LINES	
10 kV to 33 kV (per km)	12.5
110 kV to 400 kV (per km)	3.1

Table A.7 (*cont.*)   Average component failure-rates—Mechanical components

Type of component	Failure-rate /10^6 h
Ball bearings, heavy duty	20.0
Ball bearings, light duty	10.0
Roller bearings	5.0
Sleeve bearings	5.0
Shafts, heavily stressed	0.2
lightly stressed	0.02
Pins	15.0
Pivots	1.0
Couplings	5.0
Belt drives	40.0
Spur gears	10.0
Helical gears	1.0
Friction clutches	3.0
Magnetic clutches	6.0
Springs, heavily stressed	1.0
lightly stressed	0.2
Hair springs	1.0
Calibration springs, creep	2.0
breakage	0.2
Vibration mounts	9.0
Mechanical joints	0.2
Grub screws	0.5
Nuts	0.02
Bolts	0.02
Rack-and-pinion assemblies	2.0
Knife-edge fulcrums, wear	10.0

Table A.7 (*cont.*) Average component failure-rates—
Pneumatic and hydraulic components

Type of component	Failure-rate $f/10^6$ h
Bellows	5.0
Diaphragms, metal	5.0
rubber	8.0
Gaskets	0.5
Rotating seals	7.0
Sliding seals	3.0
'O' ring seals	0.2
Filters, blockage	1.0
leakage	1.0
Fixed orifices	1.0
Variable orifices	5.0
Restrictors	5.0
Pipes	0.2
Pipe joints	0.5
Unions and junctions	0.4
Hoses, heavily stressed	40.0
lightly stressed	4.0
Ducts	1.0
Pressure vessels, general	3.0
high standard	0.3
Relief valves, leakage	2.0
blockage	0.5
Hand-operated valves	15.0
Ball valves	0.5
Solenoid valves	30.0
Control valves	30.0
Pistons	1.0
Cylinders	0.1
Jacks	0.5
Pressure gauges	10.0
Pressure switches	15.0
Bourdon tubes, creep	0.2
leakage	0.05
Nozzle and flapper assemblies, blockage	6.0
breakage	0.2

Table A.8 Component stress levels—Overall environment

General environmental condition	K_1
Ideal, static conditions	0.1
Vibration free, controlled environment	0.5
General-purpose ground-based	1.0
Ship	2.0
Road	3.0
Rail	4.0
Air	10.0
Missile	100.0

Rating

Percentage of component nominal rating	K_2
140	4.0
120	2.0
100	1.0
80	0.6
60	0.3
40	0.2
20	0.1

Temperature

Component temperature (degrees C)	K_3
0	1.0
20	1.0
40	1.3
60	2.0
80	4.0
100	10.0
120	30.0

Table A.9 Allocation of component fault modes

Type of component	Percentage of total fault-rate	No. of fault modes
RESISTORS (fixed)		2
Open circuit	90.0	
Short circuit	10.0	
RESISTORS (variable, 3 pin)		6
Open circuit, each pin	30.0	
Short circuit, each adjacent pair	3.3	
CAPACITORS		2
Open circuit	50.0	
Short circuit	50.0	
COILS (all types)		2
Open circuit	80.0	
Short circuit	20.0	
CONTACTS (all types)		2
Open circuit	50.0	
Short circuit	50.0	
TRANSISTORS (3 pin)		6
Open circuit, each pin	16.7	
Short circuit, each adjacent pair	16.7	
DIODES (semi-conductor)		2
Open circuit	50.0	
Short circuit	50.0	
PNEUMATIC AND HYDRAULIC COMPONENTS		2
Leaks	80.0	
Blockages	20.0	

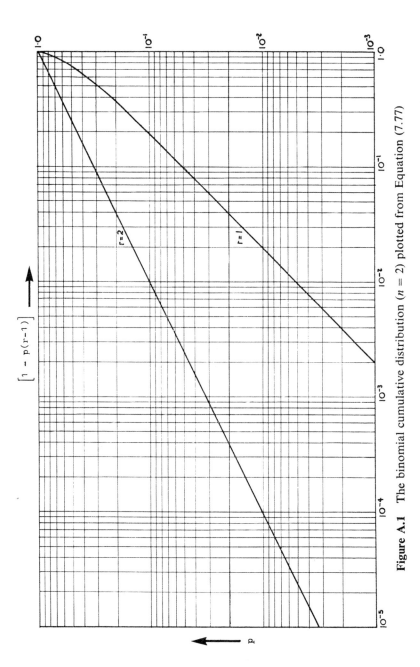

Figure A.1 The binomial cumulative distribution ($n = 2$) plotted from Equation (7.77)

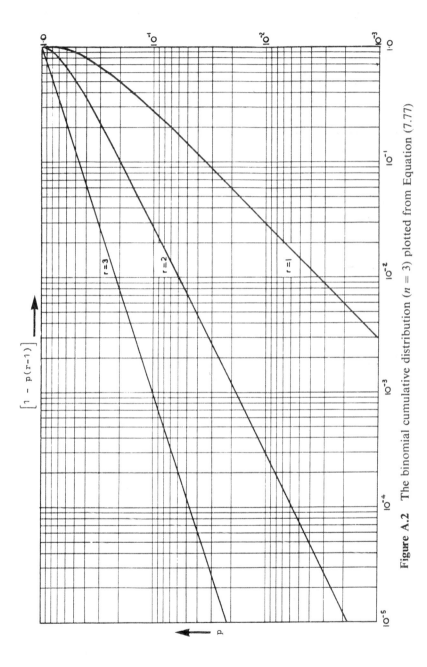

Figure A.2 The binomial cumulative distribution ($n = 3$) plotted from Equation (7.77)

APPENDIX

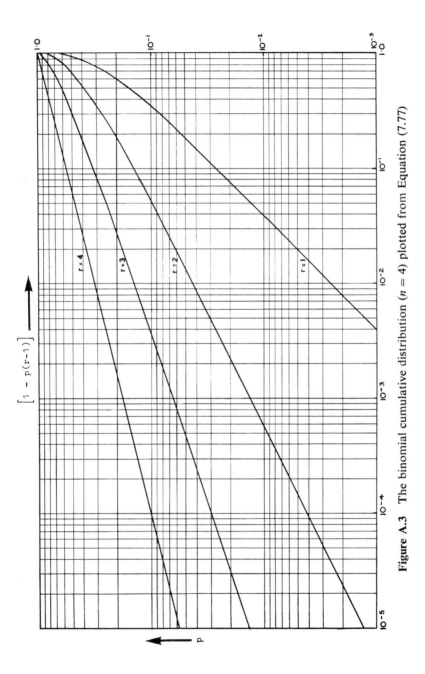

Figure A.3 The binomial cumulative distribution ($n = 4$) plotted from Equation (7.77)

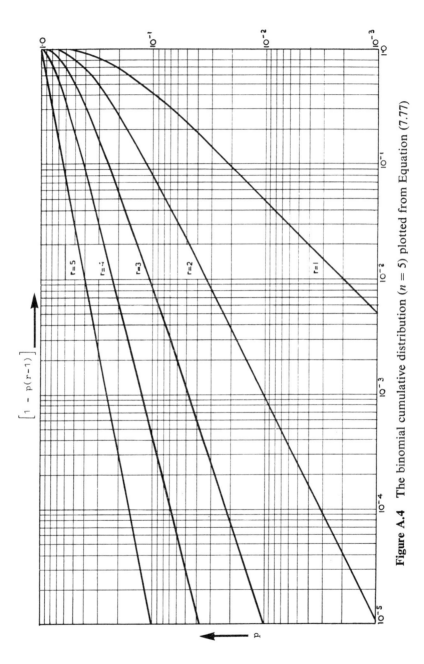

Figure A.4 The binomial cumulative distribution ($n = 5$) plotted from Equation (7.77)

APPENDIX

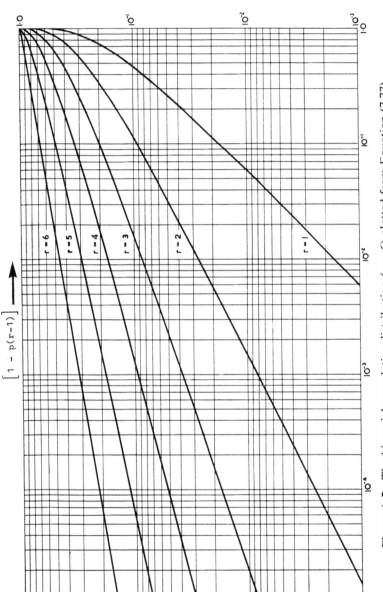

Figure A.5 The binomial cumulative distribution ($n = 6$) plotted from Equation (7.77)

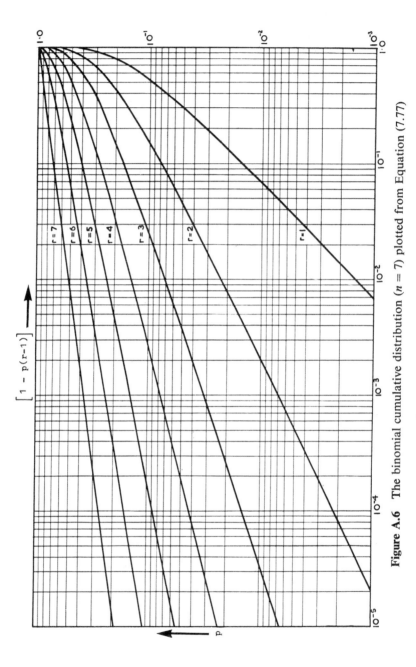

Figure A.6 The binomial cumulative distribution ($n = 7$) plotted from Equation (7.77)

APPENDIX

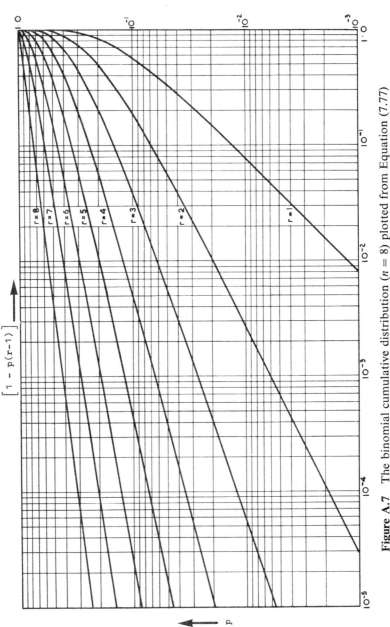

Figure A.7 The binomial cumulative distribution ($n = 8$) plotted from Equation (7.77)

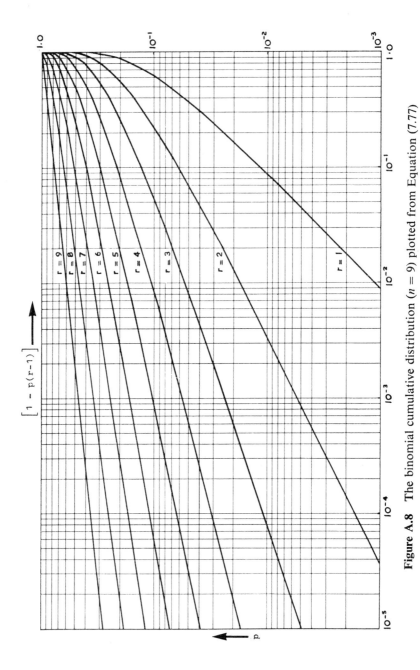

Figure A.8 The binomial cumulative distribution ($n = 9$) plotted from Equation (7.77)

APPENDIX

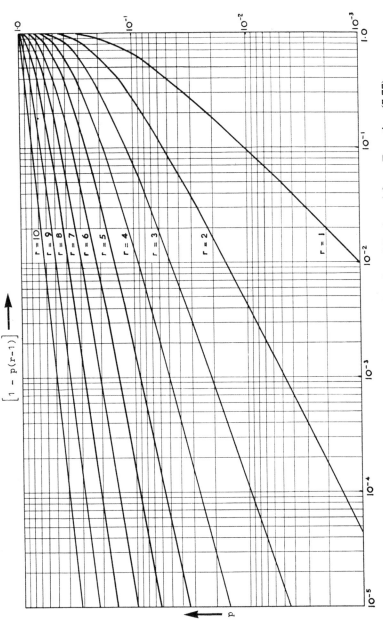

Figure A.9 The binomial cumulative distribution ($n = 10$) plotted from Equation (7.77)

580 RELIABILITY TECHNOLOGY

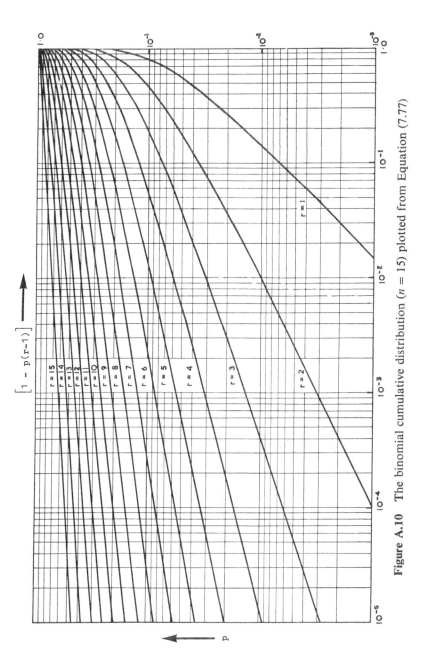

Figure A.10 The binomial cumulative distribution ($n = 15$) plotted from Equation (7.77)

APPENDIX 581

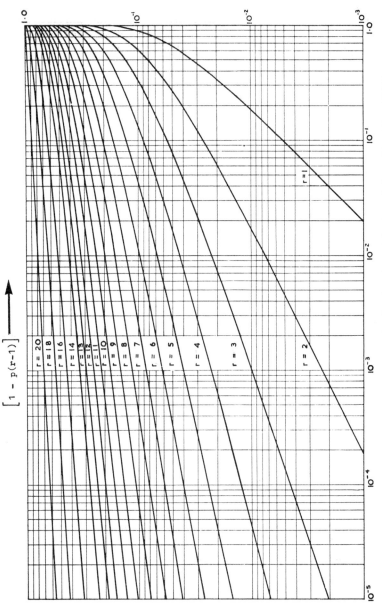

Figure A.11 The binomial cumulative distribution ($n = 20$) plotted from Equation (7.77)

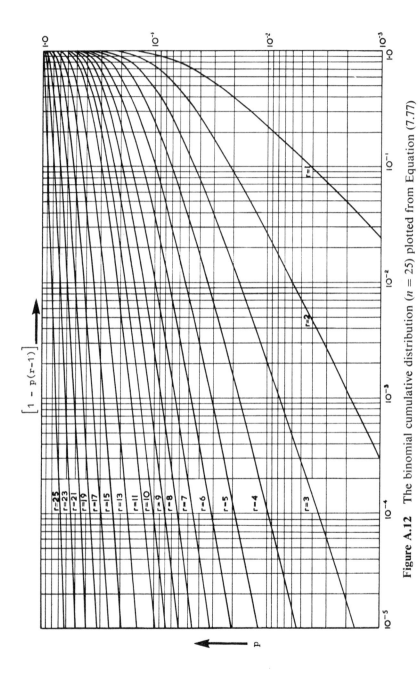

Figure A.12 The binomial cumulative distribution ($n = 25$) plotted from Equation (7.77)

APPENDIX

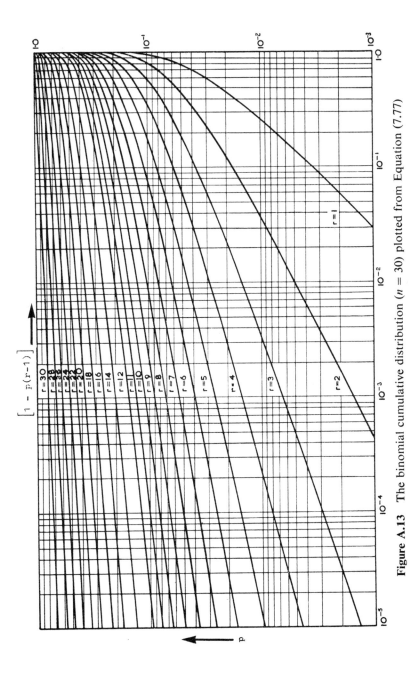

Figure A.13 The binomial cumulative distribution ($n = 30$) plotted from Equation (7.77)

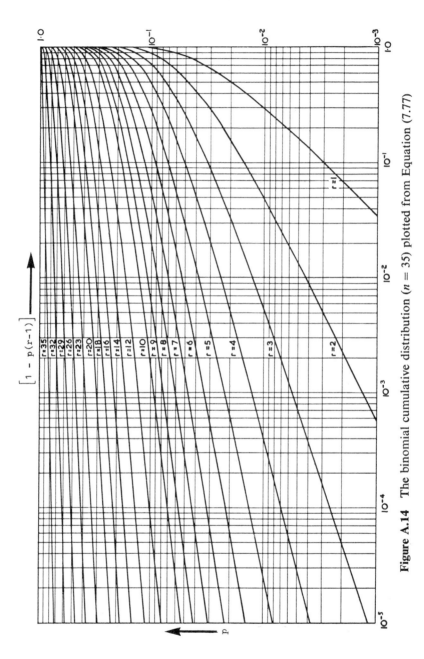

Figure A.14 The binomial cumulative distribution ($n = 35$) plotted from Equation (7.77)

APPENDIX

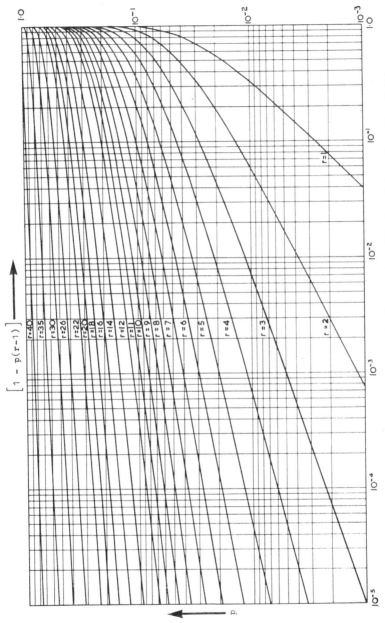

Figure A.15 The binomial cumulative distribution ($n = 40$) plotted from Equation (7.77)

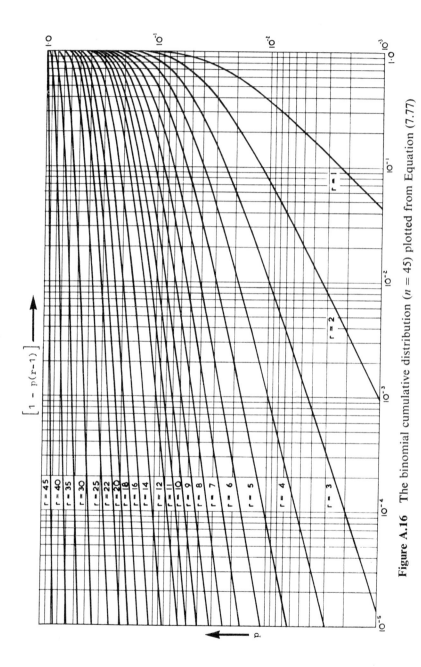

Figure A.16 The binomial cumulative distribution ($n = 45$) plotted from Equation (7.77)

APPENDIX 587

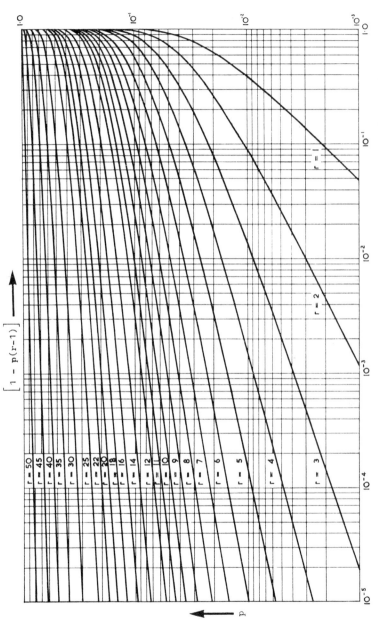

Figure A.17 The binomial cumulative distribution ($n = 50$) plotted from Equation (7.77)

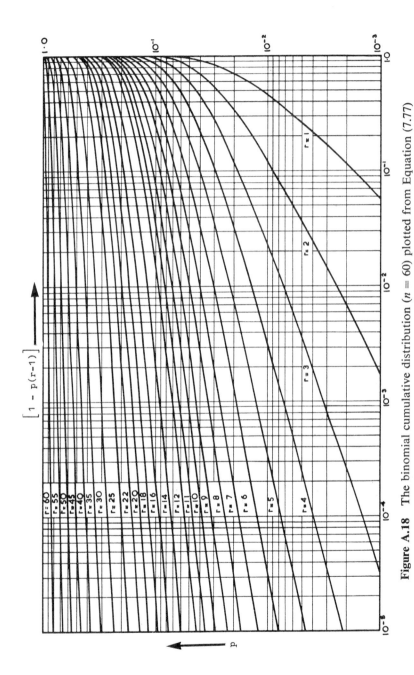

Figure A.18 The binomial cumulative distribution ($n = 60$) plotted from Equation (7.77)

APPENDIX

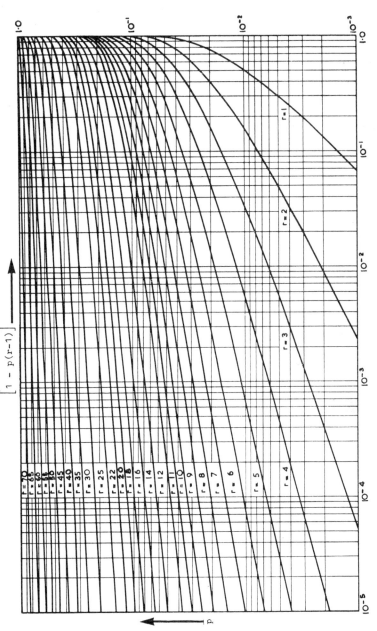

Figure A.19 The binomial cumulative distribution ($n = 70$) plotted from Equation (7.77)

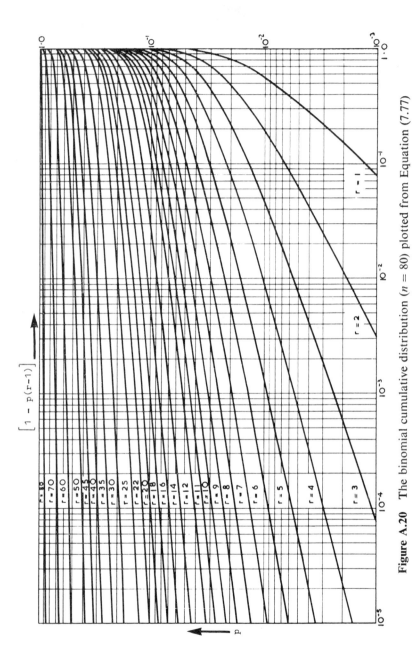

Figure A.20 The binomial cumulative distribution ($n = 80$) plotted from Equation (7.77)

APPENDIX

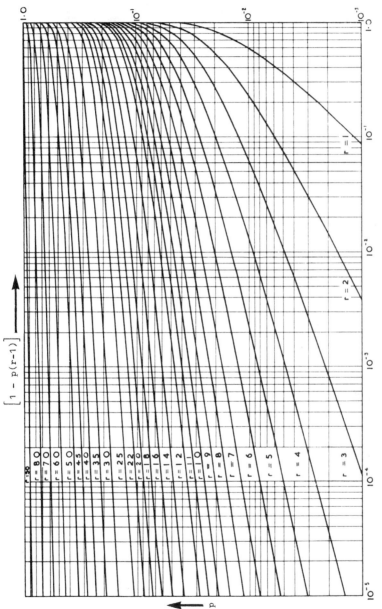

Figure A.21 The binomial cumulative distribution ($n = 90$) plotted from Equation (7.77)

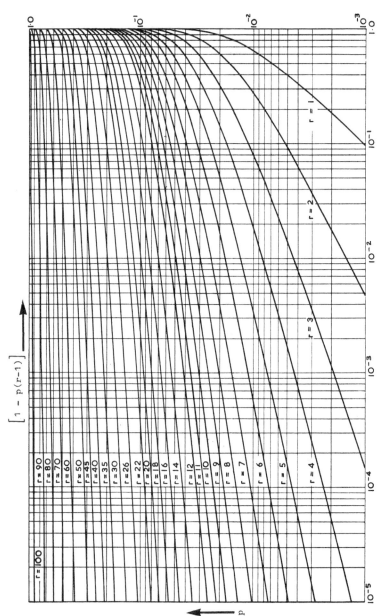

Figure A.22 The binomial cumulative distribution ($n = 100$) plotted from Equation (7.77)

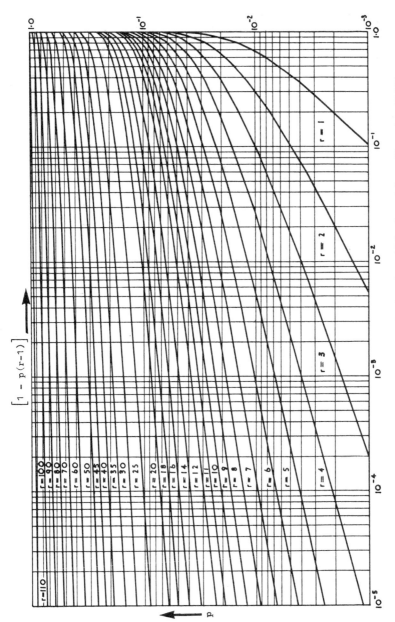

Figure A.23 The binomial cumulative distribution ($n = 110$) plotted from Equation (7.77)

594 RELIABILITY TECHNOLOGY

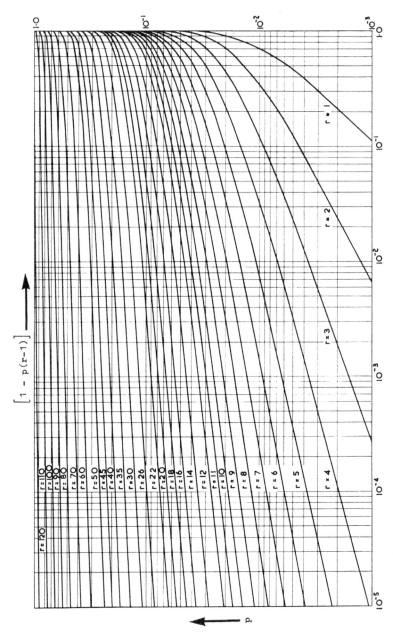

Figure A.24 The binomial cumulative distribution ($n = 120$) plotted from Equation (7.77)

APPENDIX

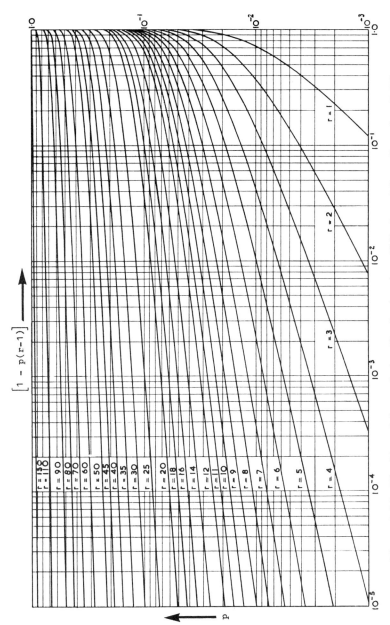

Figure A.25 The binomial cumulative distribution ($n = 130$) plotted from Equation (7.77)

596   RELIABILITY TECHNOLOGY

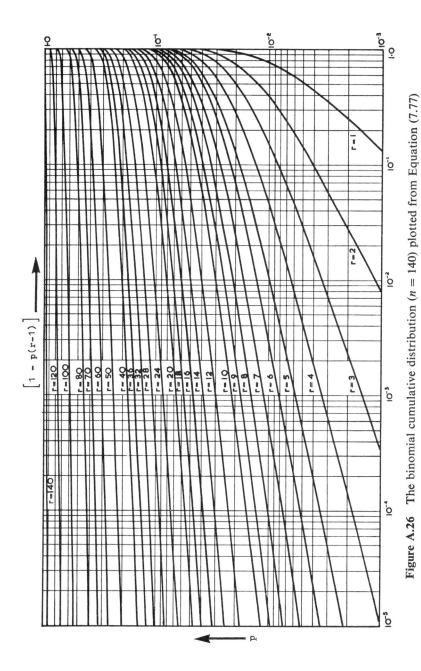

Figure A.26 The binomial cumulative distribution ($n = 140$) plotted from Equation (7.77)

APPENDIX 597

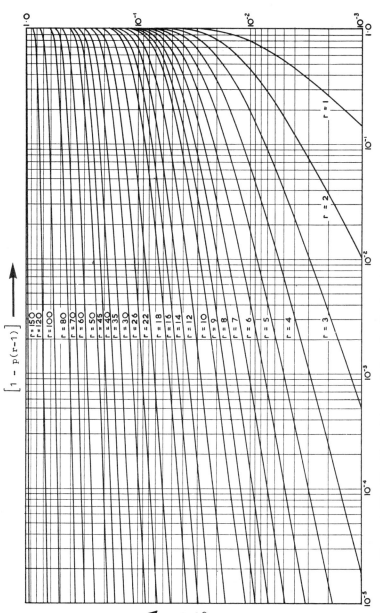

Figure A.27 The binomial cumulative distribution ($n = 150$) plotted from Equation (7.77)

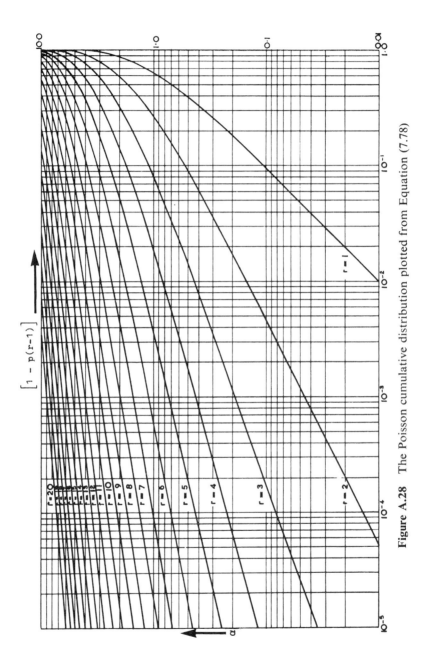

Figure A.28 The Poisson cumulative distribution plotted from Equation (7.78)

APPENDIX 599

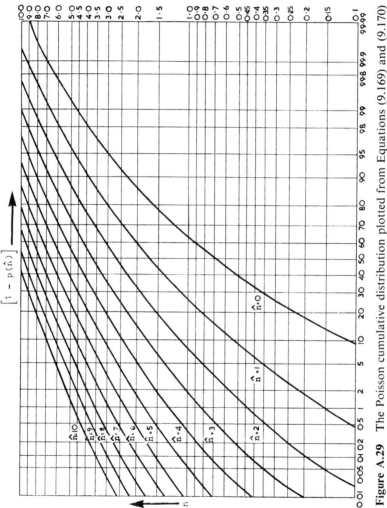

Figure A.29 The Poisson cumulative distribution plotted from Equations (9.169) and (9.170)

Answers to Questions

Chapter 1
4.   0.35
5.   Equations (1.17) and (1.18)
6.   140,400
7. (a) 0.003375
   (b) 0.385875
8.   54
9. (a) 0.7577
   (b) 0.1267
   (c) 0.0746
10. (a) 0.181
    (b) 0.632
    (c) 0.865
11. 5.76×10^{-4}
12.   99.865%
13. 10^{-4}
14. 2.1×10^{-7}

Chapter 2
1.   0.945
2.   0.824
3. (a) 0.2706
   (b) 0.6765
4. (a) 0.0952
   (b) 0.6321
5. (a) $f(x) = 0.1$
   (b) 0.6
7. (a) $f_w(w) = 0$ for $0 \leqslant w \leqslant 30$
        $f_w(w) = 0.05$ for $30 \leqslant w \leqslant 50$
        $f_w(w) = 0$ for $50 \leqslant w \leqslant \infty$
   (b) 0.7
8.   0.75
9. (a) 0.896
   (b) 0.992
   (c) 0.900

ANSWERS TO QUESTIONS   601

10.   0.9413
11. (a) 0.8550
    (b) 0.6412
    (c) 0.4275
12.   0.827

Chapter 4

3.

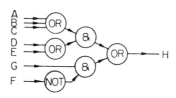

4.

(a)

(b)

5.

(a)

(b)

(c)

(d)

6. $AC+B+D$

7. (a) Property of circuit is equivalent to one contact x which in the state shown is an open-circuit.
   (b) Property of circuit is that it is an open-circuit.
8. $XYZ + XY\bar{Z} + \bar{X}Y\bar{Z}$
9. (a)

A	B	$A+\bar{B}$
1	1	1
1	0	1
0	1	0
0	0	1

   (b)

A	B	$A\bar{B}$
1	1	0
1	0	1
0	1	0
0	0	0

10. (a) Circuit is:

   [Circuit diagram with A—B in parallel with $\bar{A}$—$\bar{B}$, connected to L]

   or, using transfer contacts:

   [Circuit diagram with transfer contacts A/$\bar{A}$ and B/$\bar{B}$, connected to L]

   (b)

Row	Switch A	Switch B	Function for a row
1	1	1	$AB = 1$
2	1	0	$A\bar{B} = 0$
3	0	1	$\bar{A}B = 0$
4	0	0	$\bar{A}\bar{B} = 1$

   (c) Function $= AB + \bar{A}\bar{B}$
11. $T = XR + Y\bar{R} + Z\bar{Y} + \bar{R}Z\bar{X}$

Chapter 5

5. (a) $+0.5\%$
   (b) $+0.5\%$ for half the time and -3.5% for the other half
   (c) 100%, 50%
6. 71.3%
7. (a) 10.57%
   (b) 6.68%
   (c) 1.22%
   (d) 1.22%
   (e) 38.30%
   (f) 61.47%
8. 81.85%

ANSWERS TO QUESTIONS

9. 97.116%
10. (a) 1.98%
    (b) 97.02%
    (c) 97.884%
11. (a) 99.5328%
    (b) 97.2%
    (c) 35.2%
12. (a) 94.01%
    (b) 99.5%

Chapter 6

1. (a) $\dfrac{\Gamma(n+1)}{s^{n+1}}$ for any $n > -1$

   (b) $\dfrac{(b-a)}{(s+a)(s+b)}$

   (c) $\dfrac{s}{(s+a)^2}$

   (d) $\dfrac{a^2}{s(s+a)^2}$

   (e) $\dfrac{2\omega^2}{s(s^2+4\omega^2)}$

   (f) $\dfrac{\omega}{s^2-\omega^2}$

2. (a) $\dfrac{1}{s+a}$

   (b) $\dfrac{s^2}{(s+a)(s+b)}$

   (c) $\dfrac{1}{s(s^2+\omega^2)}$

3. (a) $\dfrac{2\omega s^2}{(s^2+\omega^2)^2}$

   (b) $\dfrac{2\omega s^3}{(s^2+\omega^2)^2}$

   (c) $\dfrac{2\omega^3(2s^2+\omega^2)}{(s^2+\omega^2)^2}$

4. (a) $\dfrac{2}{(s+a)^3}$

   (b) $\dfrac{6}{(s+a)^4}$

(c) $\dfrac{1}{s+a}$

5. (a) $e^{-bx}(1-e^{-ax})$
   (b) $e^{ax}-1$
   (c) $1-e^{-ax/b}$
   (d) $1-e^{-a(x-a)}$
   (e) $1-e^{-(x-a^2)}$

6. (a) $\dfrac{1}{a}[1+(a-1)e^{-ax}]$
   (b) $1-e^{-ax}-2bxe^{-bx}$
   (c) $\dfrac{be^{-ax}}{a(b-a)}+\dfrac{ae^{-bx}}{b(a-b)}+x-\dfrac{a+b}{ab}$

7. (a) $\dfrac{a}{(s+a)(s^2+a^2)}$
   (b) $\dfrac{2}{s^4(s+a)^3}$
   (c) $\dfrac{2}{s^3}\tan^{-1}\left(\dfrac{a}{s}\right)$

8. $\tfrac{1}{2}[(1+ax)e^{-ax}-\cos ax]$

9. (a) $\dfrac{(1+C_1 R_1 s)(1+C_2 R_2 s)}{C_1 R_1 C_2 R_2 s^2+(C_1 R_1+C_2 R_2+C_2 R_1)s+1}$
   (b) $\dfrac{(1+a^2\tau^2 s^2)}{a^2\tau^2 s^2+2(1+a^2)\tau s+1}$ where $\tau = CR$

10. (a) $\Phi(s) = \dfrac{2\zeta\omega_0 s}{s^2+2\zeta\omega_0 s+\omega_0^2}$ where $\omega_0 = \sqrt{\dfrac{K}{M}}$ and $\zeta = \dfrac{D}{2}\sqrt{\dfrac{1}{MK}}$
    (b)

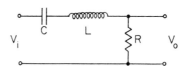

11. (a) $\Phi(s) = \dfrac{\tau s}{1+\tau s}$ where $\tau = \dfrac{D}{K}$
    (b) $x_0 = e^{-t/\tau}$
    (c) x_0 (initial value) $= 1$
    (d) x_0 (final value) $= 0$
    (e) x_0 (1/e of initial value) $= 5.0$ s

12. (a) $\omega_0 = \sqrt{\dfrac{K}{M}} = 50$ radians per second

   (b) $\zeta = \dfrac{D}{2}\sqrt{\dfrac{1}{MK}} = 0.5$

   (c) $\theta = \tan^{-1}\dfrac{1-(\omega/\omega_0)^2}{2\zeta(\omega/\omega_0)} = 56°\,18'$

   (d) $A = \dfrac{2\zeta(\omega/\omega_0)}{\sqrt{\{[1-(\omega/\omega_0)^2]^2+4\zeta^2(\omega/\omega_0)^2\}}} = 0.555$

13. (a) $\dfrac{\tau_2 s}{(1+\tau_1 s)(1+\tau_2 s)}$

   (b) $\dfrac{\tau_2 s}{(1+\tau_1 s)}$

   (c) $\dfrac{\tau_2 s}{[1+(\tau_1+\tau_2)s]}$

14. $[\Phi_{ij}(s)] = \begin{bmatrix} \dfrac{s}{1+\tau s} & 0 \\ ks & \dfrac{k}{s} \end{bmatrix}$

Chapter 7

2. (a) $f(x) = \tfrac{1}{6}[U(x-6) - U(x-12)]$

   (b) $L[f(x)] = \dfrac{1}{6s}[e^{-6s} - e^{-12s}]$

   (c) $\mu = 9$ s

3. (a) $M_2 = 2\sqrt{\pi},\quad M_4 = 16$
   (b) $M_2 = 3,\quad M_4 = 105$
   (c) $M_2 = 2,\quad M_4 = 24$

4. (a) 18
   (b) 90
   (c) 50

5. (a) 0.1353
   (b) 0.2706
   (c) 0.2706
   (d) 0.0000382

6. (a) 3
   (b) $\sqrt{3}$
   (c) 0
   (d) 1.8

7. (a) 1
   (b) $\frac{1}{6}$
   (c) 0
   (d) 2.4
8. (a) 0.324
   (b) 0.282
9. (a) 10 s
   (b) 0.913
10. (a) 10^{-5}
    (b) 2.5×10^{-4}
    (c) 10^{-3}
    (d) 5×10^{-3}
11. (a) 0.729
    (b) 0.493
    (c) 0.368
    (d) 20 years
12. (a) 0.018
    (b) 0.935
    (c) 0.047
13. (a) 0.019
    (b) 0.934
    (c) 0.047
    (d) 0.0000098
    (e) 0.9999721
    (f) 0.0000181
14. (a) 0.1306
    (b) 0.1371
    (c) 0.2612
15. (a) 0.8187
    (b) 0.8187
    (c) 0.8187
    (d) 0.9551
    (e) 0.9006
    (f) 0.8733
16. (a) $f(x) = \dfrac{e^{-\sqrt{x}}}{2\sqrt{x}}$
    (b) $p(x) = 1 - e^{-\sqrt{x}}$
    (c) $\mu = 2$
17. $f(x) = x e^{-x}$

Chapter 8

1. (a) $f_z(x) = \dfrac{1}{32{,}000}[(x-170)^2 U(x-170) - (x-190)^2 U(x-190)$
   $- (x-210)(x-130) U(x-210)$
   $+ (x-230)(x-150) U(x-230)]$
   (b) $\mu_z = 206\tfrac{2}{3}$ N
   (c) $\sigma_z = 11.055$ N
2. 0.023
3. (a) $f_z(z) = \tfrac{1}{7}(e^{-z/4} - e^{+z/3})$
   (b) $\mu_z = 1$ unit
   (c) $\sigma_z = 5$ units
4. (a) 0.1755
   (b) 0.1755
   (c) 0.6164
   (d) 0.3836
5. (a) 81
   (b) 0.5444
6. (a) 0.082
   (b) 0.515
7. (a) 8000 N
   (b) 113 N
8. (a) 110 V
   (b) 8.14 V
9. (a) $f_z(z) = \dfrac{1}{40}\left(55^2 - \dfrac{39^2}{z^2}\right)$ for $\dfrac{39}{55} \leqslant z \leqslant \dfrac{41}{55}$
   $f_z(z) = \dfrac{1}{40}\left(\dfrac{41^2 - 39^2}{z^2}\right)$ for $\dfrac{41}{55} \leqslant z \leqslant \dfrac{39}{45}$
   $f_z(z) = \dfrac{1}{40}\left(\dfrac{41^2}{z^2} - 45^2\right)$ for $\dfrac{39}{45} \leqslant z \leqslant \dfrac{41}{45}$
   (b) 0.8 m s^{-2}
   (c) 0.5
   (d) 0.1342
10. (a) 150 N m
    (b) 2.5 N m
    (c) 0.9758
11. (a) 50 Ω
    (b) 0.6494 Ω

12. (a) 250 W
    (b) 7.5 W
13. (a) 88.298 h
    (b) Weibull
14. (a) 10 s
    (b) $83\frac{1}{3}$ s
15. (a) $11,333\frac{1}{3}$ h
    (b) 11,750 h
16. (a) $f_z(z) = e^{-z/6}(1-e^{-z/12})^2$
    (b) 13 s

Chapter 9

1. $9.75 \text{ mm} \leqslant x \leqslant 14.25 \text{ mm}$
2. $f(x) = 0 \quad \text{for} \quad 0 \leqslant x \leqslant 100$

   $f(x) = \dfrac{1}{100}(x-100) \quad \text{for} \quad 100 \leqslant x \leqslant 110$

   $f(x) = \dfrac{1}{100}(120-x) \quad \text{for} \quad 110 \leqslant x \leqslant 120$

   $f(x) = 0 \quad \text{for} \quad 120 \leqslant x \leqslant \infty$

3. (a) Normal
   (b) 110 kW
   (c) 1.29 kW
4.   41.36
5. $56.4 \ \mu s \leqslant \tau \leqslant 67.6 \ \mu s$
6. (a) $0.228 \ \Omega$
   (b) $0.216 \ \Omega$
7. $1.15 \text{ mm} \leqslant \sigma \leqslant 3.54 \text{ mm}$
8. $18.55 \text{ mm} \leqslant \mu \leqslant 21.45 \text{ mm}$
9. $7190 \text{ h} \leqslant \lambda \leqslant 40{,}800 \text{ h}$
10. $14{,}270 \text{ h} \leqslant \lambda \leqslant 76{,}140 \text{ h}$
11.   10,010 h
12. (a) 3.00
    (b) 4.74
    (c) 9.15
13. $0.606 \leqslant p \leqslant 0.995$
14. (a) 394 kW
    (b) 1.5 kW
16. (a) $\chi^2 = 1.72$
    (b) Yes, χ^2 at $\alpha = 0.1$ is 10.645

ANSWERS TO QUESTIONS   609

17. (a) $D_{max} = 0.0135$
    (b) $D = 0.122$ at $\alpha = 0.1$
18. (a) $\lambda_t \simeq 397$ h
    (b) $n \simeq 5$ failures

Chapter 10
1. (a) 0.9
   (b) 0.9
2. (a) 0.66
   (b) 0.52
   (c) 0.37
3.   0.9938
4.   0.875
5.   0.8786
6.   0.977256
7.   0.897
8.   0.677 or $\frac{65}{96}$
9. (a) 0.6232
   (b) 0.9122
10.   0.742
11. (a) $f_{t_{sf}}(t) = \dfrac{\theta e^{-\theta t}}{1 - e^{-\theta \tau}}$

    (b) $\mu_{t_{sf}} = \dfrac{1}{\theta} - \dfrac{\tau e^{-\theta \tau}}{1 - e^{-\theta \tau}} = 0.494$ years

12. (a) $f_A(z) = \dfrac{\theta \tau}{[1 - \exp(-\theta \tau)]^2} \exp\left\{\dfrac{-\theta \tau[z - \exp(-\theta \tau)]}{[1 - \exp(-\theta \tau)]}\right\}$
    $= 28.58 \exp(-1.051 z)$

    (b) $\mu_A = \dfrac{1}{\theta \tau}(1 - e^{-\theta \tau}) = 0.952$

    (c) $f_D(z) = \dfrac{\theta \tau}{[1 - \exp(-\theta \tau)]^2} \exp\left\{\dfrac{-\theta \tau[1 - z - \exp(-\theta \tau)]}{[1 - \exp(-\theta \tau)]}\right\}$
    $= 9.986 \exp(+1.051 z)$

    (d) $\mu_D = 1 - \dfrac{1}{\theta \tau}(1 - e^{-\theta \tau}) = 0.048$

13. 1.485×10^{-2}
14. (a) 0.9174
    (b) 0.5235
15. (a) 0.0195
    (b) £834,285

16. (a) 0.9901
    (b) 990 h
17. (a) 0.943
    (b) 16.0

Chapter 11

1. (a) 8
   (b) $p_a + p_b p_c - p_a p_b p_c$
   (c) $\bar{A}(\bar{C} + \bar{B})$
   (d)

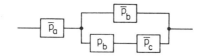

2. (a) 2000 h
   (b) 0.3935
   (c) 0.1353
3. (a) 19,610 h
   (b) 0.0051
4. (a) 0.9949
   (b) 0.25
   (c) 0.75
5. (a) 0.982
   (b) 0.935
6.   0.6128
7.   4.51
8.   0.98
9. 2×10^{-3}
10. (a) 0.9978
    (b) 1389 h
    (c) 7.2
11.   0.394
12. (a) 0.0173
    (b) 35 h
13.   0.9999723
14. 1.33×10^{-3}
15. (a) 1.02×10^{-3}
    (b) 5.21×10^{-4}

ANSWERS TO QUESTIONS   611

16. (a) 10^{-6}
    (b) 0.07
17. (a) $1 - 7.55 \times 10^{-6}$
    (b) 4.17×10^5 h
    (c) 2.4
18. $1 - 1.08 \times 10^{-5}$
19. (a) 0.9925
    (b) 0.9975
    (c) 1667 h
20. (a) 1.56×10^{-3}
    (b) 52,000 h

Chapter 12
1. (a) $X = D(AB + \bar{A}C)$
   (b) $p_x = p_d(p_a p_b + \bar{p}_a p_c)$
   (c) A simple non-redundant arrangement of elements A, B and D
2. (a) 0.0107
   (b) 1430.5 h
3.   6.8
4. 2.01×10^{-3}
5. (a) 0.002
   (b) 0.0144
6. (a) $\bar{p}_x = \bar{p}_a + \bar{p}_b \bar{p}_e - \bar{p}_a \bar{p}_b \bar{p}_e + \bar{p}_c \bar{p}_d \bar{p}_e - \bar{p}_a \bar{p}_c \bar{p}_d \bar{p}_e - \bar{p}_b \bar{p}_c \bar{p}_d \bar{p}_e + \bar{p}_a \bar{p}_b \bar{p}_c \bar{p}_d \bar{p}_e$
   (b) $\bar{p}_x = \bar{p}_a + p_a \bar{p}_e(\bar{p}_b + p_b \bar{p}_c \bar{p}_d)$
7. (a) 0.01
   (b) 0.099
   (c) 0.656
8. (a) $X = EF[A(B+C) + BD]$
   (b) $\bar{p}_x = \bar{p}_e + p_e\{\bar{p}_f + p_f[\bar{p}_a(\bar{p}_b + p_b \bar{p}_d) + p_a \bar{p}_b \bar{p}_c]\}$
9. (a) 0.0001
   (b) 0.0037
   (c) 0.0523
   (d) 0.3439
10. (a) 0.902
    (b) 0.504
11. (a) 0.0255
    (b) 0.2545
    (c) 4167 h

12. (a) 0.12
    (b) 0.51
    (c) 1.27
13. (a) 2.5×10^{-3}
    (b) $\frac{4}{9}$
14. (a) 5.44×10^{-3}
    (b) 4.43×10^{-3}
15. (a) 3.2×10^{-3}
    (b) 2.2×10^{-3}
    (c) 2.4×10^{-5}
16. 0.991
17. 6.2×10^{-8}
18. (a) 7.5×10^{-5}
    (b) 1.5
    (c) 6.67×10^{4} h
19. 1.8×10^{-4}
20. 0.5025

LIST OF SYMBOLS

Notes

* these symbols carry similar meanings to those given if the variates x and/or y in the symbols are replaced by other variates such as:
A, A_a, A_j, A_x, A_f, A_s, A_r, A_u, a, C, D, D_a, D_j, D_x, D_f, D_s, D_r, D_u, f, I, k, L, N, N_a, N_j, N_x, n, n_j, $\hat{n}$, p, $\hat{p}$, q, R, R_j, R_T, r, t, t_j, t_f, t_r, t_{ff}, t_{sf}, t_{ss}, u, u_j, V, v, W, w, w_j, X, X_j, x, x_j, x_h, x_q, x^2, $\sqrt{x}$, y, y_j, y^2, z, z_j, α, $\hat{\alpha}$, α_j, θ, $\hat{\theta}$, θ_j, λ, $\hat{\lambda}$, λ_j, μ, $\hat{\mu}$, μ_j, $\hat{\mu}^2$, σ, $\hat{\sigma}$, σ_c, $\hat{\sigma}^2$, $\hat{\sigma}_c^2$, τ, τ_h, τ_q, χ^2

† these symbols have a similar meaning in relation to an upper limit if the subscript l is replaced by the subscript u, e.g. taking x_l as the lower limit of x then x_u becomes the upper limit of x

‡ these symbols have a similar meaning in relation to an output function if the subscript i is replaced by the subscript o, e.g. taking W_i as input power then W_o becomes output power

^ the carat (^) over a symbol means an estimated value of the particular quantity, e.g. $\hat{p}$ is the estimated value of the probability p

⁻ the bar (⁻) over a symbol means the probabilistic complement or logical NOT of the quantity, e.g. $\bar{A}$ is the logical NOT or complement of A

A, B, C, ...	element or item designations
A, B, C, ...	outcome, event, state or successful state associated with elements A, B, C, ...
A	availability variate or amplitude variate
A_f	availability for an interval containing a failure
A_s	availability for a successful interval
A_r	availability associated with revealed faults
A_u	availability associated with unrevealed faults
$A_a, A_b, A_c, \ldots$	availability associated with elements A, B, C, ...
A_x	availability associated with system X
A_j	jth value of A, availability associated with jth element or particular values where $j = 1, 2, 3, \ldots$
a	parameter, area, or attenuation
$a, b, c, \ldots$	particular values of a variate, limits of a variate or distribution, constants, coefficients, contact states associated with elements A, B, C, ... or general designations for association with elements A, B, C, ...
$a_o, b_o, c_o, \ldots$	designations associated with elements A, B, C, ... in an operative role
$a_d, b_d, c_d, \ldots$	designations associated with elements A, B, C, ... in a dormant role
$a_j, b_j, c_j, \ldots$	jth value of $a, b, c, \ldots$ or particular values where $j = 1, 2, 3, \ldots$

	C	capacitance value or variate
	c	designation of a catastrophically failed state
	c	designation of a conditional aspect of a function or confidence level expressed as a probability
	D	fractional-dead-time variate, difference variate or diode resistance value
	D_f	fractional dead time for an interval containing a failure
	D_s	fractional dead time for a successful interval
	D_r	fractional dead time associated with revealed faults
	D_u	fractional dead time associated with unrevealed faults
	$D_a, D_b, D_c, \ldots$	fractional dead time associated with elements A, B, C, ...
	D_x	fractional dead time associated with system X
	D_j	jth value of D, fractional dead time associated with jth element or particular values where $j = 1, 2, 3, \ldots$
	d	designation for a demand or for a dormant state
	F	designation of a failed state
*	$F(s)$	Laplace transform of $f(x)$
*	$F_j(s)$	Laplace transform of $f_j(x)$, with similar meanings if j is replaced by other appropriate characters or symbols
	$F(ks)$	$F(s)$ with $s = ks$ where k is a constant or some subsidiary function of s
*	$F_y(s)$	Laplace transform of $f_y(x)$
	f	designation of a failed state
	f_r	designation of a revealed failed state
	f_u	designation of an unrevealed failed state
	$f_a, f_b, f_c, \ldots$	failed states associated with elements A, B, C, ...
	f_j	jth value of f, failed state associated with jth element or particular values where $j = 1, 2, 3, \ldots$
	f	function, force, number of degrees of freedom or number of faults
	f_j	jth value of f or particular values where $j = 1, 2, 3, \ldots$
*	$f(x)$	function, p.d.f. or distribution function in terms of x
*	$f_j(x)$	jth type of $f(x)$ or particular types where $j = 1, 2, 3, \ldots$
*	$f(0+)$	value of $f(x)$ as x approaches zero from positive values
*	$f(kx)$	$f(x)$ with $x = kx$ where k is a constant or some subsidiary function of x
*	$f_y(x)$	p.d.f. or distribution function of y in terms of x
*	$f_a(x)$	p.d.f. of hardware characteristics in terms of x
*	$f_o(x)$	p.d.f. of software characteristics in terms of x
*	$f_c(x)$	conditional p.d.f. in terms of x
*	$f_h(x)$	p.d.f. of achievement characteristics in terms of x
*	$f_q(x)$	p.d.f. of requirement characteristics in terms of x
†*	$f_{ql}(x)$	p.d.f. of lower limit values of requirement characteristics
*	$f_{hc}(x)$	p.d.f. of achievement characteristics in a catastrophically failed state
‡*	$f_i(x)$	input function in terms of x
‡*	$f_i(x\,|\,t, z)$	$f_i(x)$ under time, t, and space, z, conditions
	$f_a(t), f_b(t), \ldots$	p.d.f. of times to failure associated with elements A, B, ...
	$f_{a_o}(t), f_{b_o}(t), \ldots$	p.d.f. of times to failure associated with elements A, B, ... in an operative role
	$f_{a_d}(t), f_{b_d}(t), \ldots$	p.d.f. of times to failure associated with elements A, B, ... in a dormant role
	$f_x(t)$	p.d.f. of times to failure associated with system X
	$f_f(t)$	p.d.f. of times to failure
	$f_r(t)$	p.d.f. of times to repair

LIST OF SYMBOLS

$f_{f_r}(t)$	p.d.f. of times to occurrence of revealed faults
$f_{f_u}(t)$	p.d.f. of times to occurrence of unrevealed faults
$f_{f_a}(t), f_{f_b}(t), \ldots$	p.d.f. of times to failure associated with elements A, B, ...
$f_{r_a}(t), f_{r_b}(t), \ldots$	p.d.f. of times to repair associated with elements A, B, ...
* $G(s)$	Laplace transform of $g(x)$
* $G_j(s)$	Laplace transform of $g_j(x)$, with similar meanings if j is replaced by other appropriate characters or symbols
* $G_y(s)$	Laplace transform of $g_y(x)$
* $g(x)$	function or output function in terms of x
* $g_j(x)$	jth type of $g(x)$ or particular types where $j = 1, 2, 3, \ldots$
* $g_i(x)$	ith type of $g(x)$ or particular types where $i = 1, 2, 3, \ldots$
* $g_y(x)$	function of y in terms of x
‡* $g_i(x)$	input function in terms of x
H	designation of performance achievement or performance-achievement function
* $H(s)$	Laplace transform of $h(x)$
* $H_y(s)$	Laplace transform of $h_y(x)$
‡* $H_i(s)$	Laplace transform of $h_i(x)$
h	designation of performance achievement
$h_a, h_b, h_c, \ldots$	designation of performance achievement associated with elements A, B, C, ...
h_j	jth value of h, performance achievement associated with jth element or particular values where $j = 1, 2, 3, \ldots$
* $h(x)$	function or output function in terms of x
* $h_y(x)$	function of y in terms of x
‡* $h_i(x)$	input function in terms of x
I	current value or variate, inertia value, improvement factor or input conditions
I_q	required input conditions or characteristics
I_0	mean current value
† I_1	lower limit of current requirement
‡ I_i	input current
$I(s)$	Laplace transform of $I(t)$
$I(t)$	function of current variation with respect to time
i	designation of an input characteristic
i	designation of the ith member of a series or imaginary operator where $i = \sqrt{-1}$
J	inertia value
j	designation of the jth member of a series or population or jth row of a matrix
K	constant, spring restraint value or number of class intervals
K_1, K_2, K_3	correction factors for failure-rate data
K_m	mth coefficient of a distribution or particular coefficients where $m = 1, 2, 3, \ldots$
K_3	coefficient of skewness
K_4	coefficient of kurtosis
* K_{3x}	coefficient of skewness of distribution of x
* K_{4x}	coefficient of kurtosis of distribution of x
k	constant, variate, designation of the kth member of a series or population or kth column of a matrix
k_j	jth value of k or particular values where $j = 1, 2, 3, \ldots$

L	designation of Laplace transformation or inductance value or variate
L^{-1}	designation of inverse Laplace transformation
* $L[x]$	Laplace transformation of x
* $L[f(x)]$	Laplace transformation of $f(x)$
* $L^{-1}[F(s)]$	inverse Laplace transformation of $F(s)$
l	designation of a lower limit
M	viscous damping coefficient, moment of a distribution or a matrix
M_m	mth moment of a distribution about the origin or particular moments where $m = 1, 2, 3, \ldots$
$\overline{M}_m$	mth moment of a distribution about the mean or particular moments where $m = 1, 2, 3, \ldots$
* $M_m(x)$	mth moment of distribution of x about the origin or particular moments where $m = 1, 2, 3, \ldots$
m	number or range of items selected from some total number or range, number of output functions, number of rows of a matrix, an mth value of a parameter or minimum number of successful elements required for the success of a system
N	viscous damping coefficient, total number in a population, total number of states or total number of failures
$N_a, N_b, N_c, \ldots$	number of failures or occurrences associated with elements A, B, C, … or associated with events $A, B, C, \ldots$
N_x	number of failures associated with system X
N_j	jth value of N, number of failures associated with jth element or particular values where $j = 1, 2, 3, \ldots$
n	total number in a population, total number of events, total number of elements, number of measurements, number of input functions, number of dimensions, number of columns of a matrix or an nth value of a parameter
n_j	jth value of n or particular values where $j = 1, 2, 3, \ldots$
n_t	test criterion value of n
n_{mk}	kth sample of size n from the mth population
O_h	achieved output conditions or characteristics
O_q	required output conditions or characteristics
o	designation of an output characteristic or of an operative state
0	designation of zero, mean or standard value
P	probability or probability of occurrence of an event
$P(A), P(B), \ldots$	probability of occurrence of $A, B, \ldots$
$P(AB)$	probability of occurrence of A and of occurrence of B
$P(A+B)$	probability of occurrence of A, of occurrence of B or of occurrence of both
P_j	jth value of P or particular values where $j = 1, 2, 3, \ldots$
P_f	probability of occurrence of failure event or failed state
P_s	probability of occurrence of successful event or successful state
$P(T)$	probability of at least one event or failure in a total time T
* $P(s)$	Laplace transform of $p(x)$
* $P_j(s)$	Laplace transform of $p_j(x)$, with similar meanings if j is replaced by other appropriate characters or symbols
* $P_y(s)$	Laplace transform of $p_y(x)$
p	pressure value, parameter of binomial distribution, probability or probability of occurrence of an event, successful outcome or successful state

LIST OF SYMBOLS

p_j	jth value of p, probability of occurrence of the jth outcome or state or particular values where $j = 1, 2, 3, \ldots$
$p_a, p_b, p_c, \ldots$	probability of occurrence of an outcome, event or successful state associated with elements A, B, C, ...
p_x	probability of occurrence of an outcome, event or successful state associated with system X
$p_{a_o}, p_{b_o}, \ldots$	probability of successful state associated with elements A, B, ... in an operative role
$p_{a_d}, p_{b_d}, \ldots$	probability of successful state associated with elements A, B, ... in a dormant role
p_d	probability of occurrence of a demand
p_f	probability of a failed state
p_r	probability of a repair state
p_s	probability of a successful state
* $p(x)$	cumulative distribution function in terms of x
* $p_j(x)$	jth type of $p(x)$ or particular types where $j = 1, 2, 3, \ldots$
* $p(kx)$	$p(x)$ with $x = kx$ where k is a constant or some subsidiary function of x
* $p_y(x)$	cumulative distribution function of y in terms of x
* $p_h(x)$	cumulative distribution function of achievement characteristics in terms of x
* $p_q(x)$	cumulative distribution function of requirement characteristics in terms of x
* $p_{h_a}(x), p_{h_b}(x), \ldots$	function $p_h(x)$ associated with elements A, B, ...
* $p_{h_j}(x)$	jth type of $p_h(x)$, function $p_h(x)$ associated with jth element or particular types where $j = 1, 2, 3, \ldots$
†* $p_{ql}(x)$	cumulative distribution function of lower limit values of requirement characteristics
* $p_c(x)$	conditional cumulative distribution function in terms of x
* $p(a)$	$p(x)$ evaluated at $x = a$ with similar meanings if the value a is replaced by other values such as $b, c, \ldots$
†* $p_h(x_{ql})$	$p_h(x)$ evaluated at $x = x_{ql}$ and similarly for $p_{x_h}(x_{ql})$
$p_c(a)$	$p_c(x)$ evaluated at $x = a$ with similar meanings if the value a is replaced by other values such as $b, c, \ldots$
$p(t)$	cumulative distribution function in terms of time or cumulative probability of failure with respect to time
$p_a(t), p_b(t), \ldots$	cumulative probability of failure with respect to time associated with elements A, B, ...
$p_j(t)$	jth type of $p(t)$ or cumulative probability of failure with respect to time associated with jth element
$p_x(t)$	cumulative probability of failure with respect to time associated with system X
$p_{a_o}(t), p_{b_o}(t), \ldots$	cumulative probability of failure with respect to time associated with elements A, B, ... in an operative role
$p_{a_d}(t), p_{b_d}(t), \ldots$	cumulative probability of failure with respect to time associated with elements A, B, ... in a dormant role
$p(\tau)$	$p(t)$ evaluated at $t = \tau$
$p_a(\tau), p_b(\tau), \ldots$	$p_a(t), p_b(t), \ldots$ evaluated at $t = \tau$
$p_j(\tau)$	$p_j(t)$ evaluated at $t = \tau$
$p_x(\tau)$	$p_x(t)$ evaluated at $t = \tau$
$p_f(t)$	probability of being in failed state at time t
$p_r(t)$	probability of being in repair state at time t
$p_s(t)$	probability of being in successful state at time t
$p_{jk}(t)$	probability of being in state k at time t when in state j at time zero
$p_{ab}(t)$	probability of being in state b at time t when in state a at time zero with similar meanings if the symbols a and/or b are replaced by the symbols $a, \bar{a}, b, \bar{b}, c, \bar{c}, \ldots$

$p_{sf}(t)$	probability of being in state f at time t when in state s at time zero with similar meanings if the symbols f and/or s are replaced by the symbols f, s, r, u, ...
$p(X\|A)$	conditional probability of system X being successful when element A is successful with similar meanings if the symbols X and/or A are replaced by $A, \bar{A}, B, \bar{B}, C, \bar{C}, ..., X, \bar{X}$
Q	designation of performance requirement or performance-requirement function
$Q(s)$	polynomial in s
q	designation of performance requirement, discrete variate or an equivalent to $\bar{p}$
R	reliability, reliability function or resistance value or variate
R_i	ith value of R or particular values where $i = 1, 2, 3, ...$
R_j	jth value of R or particular values where $j = 1, 2, 3, ...$
R_0	mean resistance value
R_L	load resistance value or variate
R_T	total resistance value or variate
† R_l	lower limit of resistance variate
r	designation of a repair state
$r_a, r_b, r_c,$	repair states associated with elements A, B, C, ...
r_j	jth value of r, repair state associated with jth element or particular values where $j = 1, 2, 3, ...$
r	discrete variate, ratio, or minimum number of failed elements required for system failure
r_0	fixed or mean value of r
† r_1	lower limit of r
S	designation of a successful state
s	designation of a successful state
s	Laplace operator or complex variable
T	designation of total value
T	total time of interest, singular or representative value of time conditions or transmission function of a network
T_a	total available time
T_d	total dead time
T_{sf}	mean successful time in an interval containing a failure
T_{ff}	mean failed time in an interval containing a failure
t	time variable or variate, time condition, variate of the standardized normal distribution, variate of the Student-t distribution or discrete variate
t_j	jth value of t or particular values where $j = 1, 2, 3, ...$
t_f	variate of time to failure
t_r	variate of time to repair
t_{ff}	variate of failed times in intervals containing failures
t_{sf}	variate of successful times in intervals containing failures
t_{ss}	variate of successful times in successful intervals
* $U(x)$	unit step function in terms of x
u	designation of an unrevealed fault state or designation of an upper limit
u	variable or variate
u_j	jth value of u or particular values where $j = 1, 2, 3, ...$
V	voltage value or variate

LIST OF SYMBOLS 619

†	V_1	lower limit of voltage variate
‡	V_i	input voltage
	$V(t)$	function of voltage variation with respect to time
	$V_C(t)$	$V(t)$ across a capacitor
	$V_L(t)$	$V(t)$ across an inductor
	$V_R(t)$	$V(t)$ across a resistor
‡	$V_i(t)$	input $V(t)$
	$V(s)$	Laplace transform of $V(t)$ with similar meanings in the case of $V_C(s)$, $V_L(s)$, $V_R(s)$, $V_i(s)$ and $V_o(s)$
	v	variable or variate
	W	designation of a working state
	W	power value or variate
‡	W_i	input power
	w	variable, variate or weighting factor
	w_j	jth value of w or particular values where $j = 1, 2, 3, \ldots$
	X	system designation
	X	outcome, event, state or successful state associated with system X or a statistic
	X_j	jth value of X or particular values where $j = 1, 2, 3, \ldots$
	x	variable or variate, member of a population, designation associated with system X or outcome, event or state associated with system X
	x_j	jth value of x or particular values where $j = 1, 2, 3, \ldots$
	x_0	fixed or mean value of x
	x_h	achieved x
	x_q	required x
†	x_1	lower limit of x
†	x_{q1}	lower limit of x_q
	Y	system designation
	Y	outcome, event, successful state or characteristic associated with system Y
	y	variable, variate or member of a population
	y_j	jth value of y or particular values where $j = 1, 2, 3, \ldots$
	y_i	ith value of y or particular values where $i = 1, 2, 3, \ldots$
	y_0	fixed or mean value of y
†	y_1	lower limit of y
	Z	singular or representative value of space conditions
	z	variable, variate, space condition or member of a population
	z_j	jth value of z or particular values where $j = 1, 2, 3, \ldots$
†	z_1	lower limit of z
*	$z(x)$	hazard function in terms of x
*	$z_y(x)$	hazard function of y in terms of x
	$z(a)$	$z(x)$ evaluated at $x = a$ with similar meanings if the value a is replaced by other values such as $b, c, \ldots$
	$z(t)$	hazard or failure-rate function in terms of time or with respect to time
	$z_a(t), z_b(t), \ldots$	hazard or failure-rate function with respect to time associated with elements A, B, ...
	$z_x(t)$	hazard or failure-rate function with respect to time associated with system X
	$z_j(t)$	jth type of $z(t)$, $z(t)$ associated with jth element or particular types where $j = 1, 2, 3, \ldots$
	$z_f(t)$	hazard function of a p.d.f. of times to failure

$z_r(t)$		hazard function of a p.d.f. of times to repair
$z_{f_r}(t)$		hazard function of a p.d.f. of times to revealed failure
$z_{f_u}(t)$		hazard function of a p.d.f. of times to unrevealed failure
α		constant, level of significance, producer's risk or parameter of Poisson distribution
α_j		jth value of α or particular values where $j = 1, 2, 3, \ldots$
α_r		Poisson parameter for discrete variate r and similarly for α_q and α_t
β		consumer's risk or parameter of a distribution
* $\Gamma(x)$		gamma function in terms of x
* δx		small increment of x
* $\delta(x)$		unit impulse function or Dirac delta function in terms of x
ε		fractional error value
ζ		damping factor
Θ		summated mean failure-rate or mean failure-rate of a system
θ		parameter of a distribution, failure-rate parameter of the exponential distribution, mean failure-rate of an element, angular displacement or phase angle
θ_j		jth value of θ or particular values where $j = 1, 2, 3, \ldots$
$\theta_a, \theta_b, \theta_c, \ldots$		mean failure-rate associated with elements A, B, C, ...
θ_x		mean failure-rate associated with system X
θ_h		parameter of exponential achievement function
θ_q		parameter of exponential requirement function
† θ_l		lower limit of θ or lower confidence limit of θ
† θ_{ql}		lower limit of θ_q
‡ θ_i		input angular displacement
‡ $\theta_i(t)$		θ_i with respect to time
‡ $\theta_i(s)$		Laplace transform of $\theta_i(t)$
$\theta_{a_o}, \theta_{b_o}, \ldots$		mean failure-rate associated with elements A, B, ... in an operative state
$\theta_{a_d}, \theta_{b_d}, \ldots$		mean failure-rate associated with elements A, B, ... in a dormant state
θ_{f_r}		mean revealed failure-rate
θ_{f_u}		mean unrevealed failure-rate
λ		parameter of a distribution or mean time-to-failure (m.t.t.f.) parameter of the exponential distribution
λ_j		jth value of λ or particular values where $j = 1, 2, 3, \ldots$
$\lambda_a, \lambda_b, \lambda_c, \ldots$		mean time to failure associated with elements A, B, C, ...
* λ_x		parameter of the exponential distribution of x
λ_h		parameter of exponential achievement function
λ_q		parameter of exponential requirement function
λ_t		test criterion value of λ
† λ_l		lower limit of λ
† λ_{ql}		lower limit of λ_q
μ		mean value or mean value of a p.d.f. or distribution
μ_j		jth value of μ or particular values where $j = 1, 2, 3, \ldots$
μ_c		mean value of conditional p.d.f.
μ_h		mean value of achievement function

LIST OF SYMBOLS

μ_q	mean value of requirement function
† μ_l	lower limit of μ
† μ_{ql}	lower limit of μ_q
μ_f	mean of p.d.f. of times to failure
μ_r	mean of p.d.f. of times to repair
μ_{f_r}	mean of p.d.f. of times to revealed failure
μ_{f_u}	mean of p.d.f. of times to unrevealed failure
$\mu_{f_a}, \mu_{f_b}, \ldots$	mean of p.d.f. of times to failure associated with elements A, B, ...
$\mu_{r_a}, \mu_{r_b}, \ldots$	mean of p.d.f. of times to repair associated with elements A, B, ...
* μ_x	mean of p.d.f. of distribution of x
σ	standard deviation or standard deviation of a p.d.f. or distribution
σ_j	jth value of σ or particular values where $j = 1, 2, 3, \ldots$
$\hat{\sigma}_c$	corrected estimate of σ or Bessel correction of σ
σ_h	standard deviation of achievement function
σ_q	standard deviation of requirement function
† σ_{ql}	lower limit of σ_q
* σ_x	standard deviation of p.d.f. of distribution of x
τ	time constant, repair time or maintenance interval
τ_j	jth value of τ or particular values where $j = 1, 2, 3, \ldots$
$\tau_a, \tau_b, \tau_c, \ldots$	repair time associated with elements A, B, C, ...
τ_r	mean repair or restoration time
τ_h	achieved time constant
τ_q	required time constant
† τ_{ql}	lower limit of τ_q
$\Phi(s)$	transfer function
$\Phi_j(s)$	jth type of $\Phi(s)$ or particular types where $j = 1, 2, 3, \ldots$
$\Phi_{ij}(s)$	transfer function or Laplace transform of conversion characteristic ϕ_{ij}
ϕ	conversion or transfer characteristic of a system relating the output function to the input function
ϕ_{ij}	conversion or transfer characteristic of a system relating the ith output function to the jth input function
† $(\phi_{ij})_l$	lower required limit of ϕ_{ij}
$[\phi_{ij}\|t, z]_h$	achieved conversion or transfer characteristic, ϕ_{ij}, under time, t, and space, z, conditions
$[\phi_{ij}\|t, z]_q$	required conversion or transfer characteristic, ϕ_{ij}, under time, t, and space, z, conditions
χ^2	variate of the χ^2 distribution
χ_j^2	jth value of χ^2 or particular values where $j = 1, 2, 3, \ldots$
$\Psi(s)$	characteristic function
$\Psi_{ij}(s)$	Laplace transform of $\psi_{ij}(x)$
* $\psi_{ij}(x)$	integrodifferential operator acting on the jth output function of x in an equation for the ith input function of x
ω	constant or angular frequency
ω_0	natural frequency

GLOSSARY OF TERMS

Accuracy

The measure of exactness of a component's, equipment's or system's performance compared with some true or standard performance. This measure includes those deviations in performance which may be due to both systematic and non-systematic effects.

Availability

The proportion of the total time that a component, equipment or system is performing in the desired manner.

Average

(*see* Mean)

Best Estimate

(*see* Estimation)

Boundary Value

The limiting value of some performance characteristic which may be used for assessment purposes.

Capability

The fundamental ability of a component, equipment or system to perform in the desired manner.

Catastrophic Failure

The failure of a component, equipment or system in which its particular performance characteristic moves completely to one or the other of the extreme limits outside the normal specification range.

Confidence Interval

The estimated range or interval for a distribution parameter which is given at a certain confidence level.

Confidence Level

The probability that the assertion made about the value of the distribution parameter, from the estimates of that parameter, is true.

Confidence Limits

The end points of the confidence interval.

Consecutive Testing

The testing of a series of components or equipments one after the other. This may or may not involve a time interval between the completion of the test on one device and the starting of the test on the next device.

GLOSSARY OF TERMS

Consistent Estimator

The estimated value of a parameter such that the probability of this estimated value being the true value tends to unity as the sample size from which the estimate is taken approaches the size of the whole population.

Consumer's Risk

The probability of the user of a component, equipment or system accepting from the manufacturer of such devices an item whose failure-rate, or other performance characteristic, is worse than some agreed or stated value.

Cumulative Probability

The probability that a random variable, x, lies between some lower limit, which might be 0 or $-\infty$, and some upper limit determined by a particular value of x. For continuous random variables, the cumulative probability is given by the integration of the density function between the limits required, e.g.

$$p(x) = \int_{-\infty}^{x} f(x)\,dx$$

for discrete variables, the cumulative probability is given by the sum of the appropriate probability function over the range of values required, e.g.

$$p(r) = \sum_{j=0}^{r} f(j)$$

Dead Time

(*see* Fractional Dead Time)

Density Function (Probability Density Function)

If x is a continuous random variable, then the density function of x is defined as the first derivative of the cumulative probability function with respect to x, i.e.

$$f(x) = \frac{d}{dx} p(x)$$

Diversity

The performance of the same overall function by a number of independent and different means.

Early Failure

The failures of components, equipments and systems which occur during the initial life phase of such devices and which are generally caused by initial production, assembly, test, installation or commissioning errors.

Error

The deviation that can exist between the actual performance characteristic of a component, equipment or system and the true or required value of such performance.

Estimation

The evaluation of a statistical parameter from a set of sample data in such a way as to give the best indication as to the true value of such a parameter. Estimation generally involves the use of 'consistent' and 'unbiased' estimators (*q.v.*).

Failed State
The condition of a component, equipment or system during the time when it is subject to a failure.

Failed Time
The time for which a component, equipment or system remains in the failed state.

Failure
The condition of a component, equipment or system whereby a particular performance characteristic, or number of performance characteristics, of such a device has moved outside the assessed specification range for that characteristic in such a way that the component, equipment or system can no longer perform adequately in the desired manner.

Field Experience
The collection of data and information about the performance characteristics of components, equipments and systems from their actual use in appropriate practical installations.

Fractional Dead Time
The proportion of the total relevant time that a component, equipment or system is in the failed state.

Ideogram
The symbolic representation of ideas by means of patterns, diagrams or graphical techniques.

Infant Mortality
A term sometimes used to describe 'Early Failure' (*q.v.*).

Logic Diagram
A diagram showing in symbolic way the flow and/or processing of information.

Maintenance
The art of ensuring that the performance of a component, equipment or system is kept within a set of predetermined limits.

Majority-vote System
A system made up of n identical and independent elements which are so connected that any m or more ($m \leq n$) of the elements are required to give a correct output before a correct system output is obtained. The system may be considered as being partly redundant. In the particular cases where $m = 1$ the majority-vote system becomes completely redundant (*see* Redundant Elements) and where $m = n$ the system becomes completely non-redundant (*see* Non-redundant Elements).

Mean
The mean, μ, is defined as the sum of all values of a variate, x, where each value is weighted by its associated probability of occurrence. For discrete variables:

$$\mu = \sum_{r=0}^{\infty} rf(r)$$

GLOSSARY OF TERMS

and for continuous variables:

$$\mu = \int_{-\infty}^{\infty} xf(x)\,dx$$

Mean Time Between Failures (m.t.b.f.)

The total measured operating time of a population of components, equipments or systems divided by the total number of failures. The term is generally used in connexion with the exponential distribution of failures where the m.t.b.f. is a time-independent constant.

Mean Time to Failure (m.t.t.f.)

The mean of the distribution of the times to the first failure.

Non-redundant Elements

A configuration of elements which can only produce a correct output when each element in the configuration is functioning correctly.

Non-systematic Errors (or Random Errors)

Errors which do not represent a fixed deviation from the true value and have the implication of chance. It is not possible to evaluate a particular error of this sort but such errors in the mass can be understood and calculated using the laws of probability.

Population

The total collection of any items or events related to any particular common consideration. In statistics, the terms 'population', 'universe' or 'parent population' are generally synonymous and are used to describe an exceedingly large or infinite number of items or events from which samples may be taken for statistical analysis.

Precision

The degree to which the distribution of observations, items or events cluster upon themselves. The standard deviation of a distribution may be used as a measure of precision.

Probability

If the assertion is made that the probability of an event occurring as an outcome of a series of random and independent experiments is given by some value P, then the value of the probability, P, is defined as the limit of the ratio of the number of occurrences of the event to the total number of experiments (i.e. the relative frequency of occurrence) as the total number of experiments tends to infinity.

Probability Density Function (p.d.f.)

(*see* Density Function)

Producer's Risk

The probability of the manufacturer of a component, equipment or system rejecting such devices when their failure-rate, or other performance characteristic, is better than some agreed or stated value.

Proof testing

A method of ensuring that a component, equipment or system possesses all the required performance characteristics and is capable of responding to input conditions in the manner desired.

Random

A term used to describe the unpredictable occurrence of events in space, in time or in both space and time.

Redundancy

The performance of the same overall function by a number of independent but identical means.

Redundant Elements

A configuration of a number of identical and independent elements connected in such a way that the correct functioning of any one element, irrespective of the state of the others, produces a correct output.

Reliability

That characteristic of an item expressed by the probability that it will perform its required function in the desired manner under all the relevant conditions and on the occasions or during the time intervals when it is required so to perform.

Reliability Technology

The scientific study of the trustworthy nature of devices or systems in practical and industrial situations.

Revealed Failure

A failure of a component, equipment or system which is automatically brought to light on its occurrence.

Sample

The selection of a number of observations, items or events from the total population of such observations, items or events. The term is often construed to represent a 'random sample' where each item has an equal chance of being selected and the resultant sample is representative of the whole population.

Sampling Distribution

The distribution of a parameter of a parent population which is obtained from the estimated values of such a parameter from the results of the total possible number of identical and independent samples.

Simultaneous Testing

The testing of a number of components, equipments or systems at the same time.

Staggered Testing

A special case of consecutive testing (*q.v.*) where there is normally a time interval between the completion of a test on one device and the starting of a test on the next device.

Standard Deviation

The positive square root of the variance (*q.v.*).

Systematic Errors

An error which represents a predictable deviation from the true or standard value.

GLOSSARY OF TERMS

Technological System

That engineered complex which converts a given quantity of a certain range of facets such as wealth, materials, resources, facilities and information with specified characteristics into a given quantity of a certain range of information, power, products, services or wealth with specified characteristics under a certain set of space and time conditions.

Unbiased Estimator

The estimated value of a parameter such that the mean value of such estimates over a large number of estimations should be equal to the true value of the parameter.

Unrevealed Failure

A failure of a component, equipment or system which remains hidden until revealed by some thorough proof-testing procedure.

Useful Life Phase

That part of a component's, equipment's or system's life which lies between the phase of early failure and the phase of wear-out failure. In some instances the useful life phase is characterized by a constant average failure-rate.

Variance

The variance, σ^2, is the sum of all squared deviations from the mean where each such deviation is weighted by its associated probability of occurrence. As such, the variance gives a measure of the amount of spread that the variate of the distribution possesses about the mean. For discrete variates:

$$\sigma^2 = \sum_{r=0}^{\infty} (r-\mu)^2 f(r)$$

and for continuous variates:

$$\sigma^2 = \int_{-\infty}^{\infty} (x-\mu)^2 f(x)\, dx$$

Wear-out Failure

The failures of components, equipments and systems which occur after the end of their useful life.

SUGGESTED REFERENCES FOR FURTHER READING

Logic

Black, M., *Critical Thinking*, Prentice-Hall, New York, 1954.
Boole, G., *The Mathematical Analysis of Logic*, Blackwell, Oxford, 1948.
Boole, G., *Studies in Logic and Probability*, Watts, London, 1952.
Brinton, W. C., *Graphical Methods for presenting Facts*, Engineering Manufacturing Company, New York, 1914.
Cohen, M. R. and Nagel, E., *An Introduction to Logic and Scientific Method*, Routledge, London, 1934.
Johnson, W. E., *Logic*, Cambridge University Press, 1921.
Joyce, G. H., *Principles of Logic*, Longmans, London, 1916.
Lewis, P. I. and Langford, C. H., *Symbolic Logic*, Appleton, London, 1932.
Mace, C. A., *The Principles of Logic*, Longmans, London, 1933.
Polya, G., *How to Solve It*, Princeton University Press, 1945.
Polya, G., *Mathematics and Plausible Reasoning*, Oxford University Press, 1955.
Whitesitt, J. E., *Boolean Algebra and its Applications*, Addison-Wesley, Reading, Massachusetts, 1961.

Probability and Statistics

Beers, Y., *Introduction to the Theory of Error*, Addison-Wesley, Reading, Massachusetts, 1957.
Brownlee, K. A., *Statistical Theory and Methodology in Science and Engineering*, John Wiley, New York, 1961.
Cox, D. R., *Renewal Theory*, John Wiley, New York, 1962.
Cramer, H., *The Elements of Probability Theory and some of its Applications*, John Wiley, New York, 1955.
Feller, W., *An Introduction to Probability Theory and its Applications*, John Wiley, New York, 1957.
Freund, J., *Mathematical Statistics*, Prentice-Hall, Englewood Cliffs, N.J., 1962.
Hald, A., *Statistical Theory with Engineering Applications*, John Wiley, New York, 1952.
Hoel, P. G., *Introduction to Mathematical Statistics*, John Wiley, New York, 1954.
Kendal, M. G. and Buckland, W. R., *A Dictionary of Statistical Terms*, Oliver & Boyd, Edinburgh, 1957.
Kendal, M. G. and Stuart, A., *The Advanced Theory of Statistics*, Vols. 1 and 2, Charles Griffin, London, 1958 and 1961.
Papoulis, A., *Probability, Random Variables and Stochastic Processes*, McGraw-Hill, New York, 1965.
Paradine, C. G. and Rivett, B. H. P., *Statistical Methods for Technologists*, English University Press, 1960.
Schenck, H., *Theories of Engineering Experimentation*, McGraw-Hill, New York, 1961.
Sturges, H. A., The Choice of a Class Interval, *Journal of American Statistical Association*, Vol. 21, pp. 65–66, 1926.
Topping, J., *Errors of Observation and their Treatment*, Chapman & Hall, New York, 1962.
Wald, A., *Sequential Analysis*, John Wiley, New York, 1947.

Reliability

Aeronautical Radio, Inc., *Reliability Engineering*, Prentice-Hall, Englewood Cliffs, N.J., 1964.
Amstadter, B. L., *Reliability Mathematics: Fundamentals: Practices: Procedures*, McGraw-Hill, New York, 1971.
Barlow, R. E. and Proschan, F., *Mathematical Theory of Reliability*, John Wiley, New York, 1965.
Bazovsky, I., *Reliability Theory and Practice*, Prentice-Hall, Englewood Cliffs, N.J., 1961.
Calabro, S. R., *Reliability Principles and Practices*, McGraw-Hill, New York, 1962.
Chorafas, D. N., *Statistical Processes and Reliability Engineering*, D. Van Nostrand, Princeton, N.J., 1960.
Dummer, G. W. A. and Griffin, N., *Electronics Reliability—Calculation and Design*, Pergamon Press, Oxford, 1966.
Gnedenko, B. V., Belyayev, Yu. K. and Solovyev, A. D., *Mathematical Methods of Reliability Theory*, Academic Press, London, 1970.
Ireson, W. G., *Reliability Handbook*, McGraw-Hill, New York, 1966.
Myers, R. H., Wong, K. L. and Gordy, H. M., *Reliability Engineering for Electronic Systems*, John Wiley, New York, 1964.
Pieruschka, E., *Principles of Reliability*, Prentice-Hall, Englewood Cliffs, N.J., 1963.
Polovko, A. M., *Fundamentals of Reliability Theory*, Academic Press, London, 1968.
Roberts, N. H., *Mathematical Methods in Reliability Engineering*, McGraw-Hill, New York, 1964.
Sandler, G. H., *System Reliability Engineering*, Prentice-Hall, Englewood Cliffs, N.J., 1963.
Shooman, M. L., *Probabilistic Reliability: An Engineering Approach*, McGraw-Hill, New York, 1968.
Zelen, M., *Statistical Theory of Reliability*, The University of Wisconsin Press, Madison, Wisconsin, 1963.

System Analysis

Chestnut, H., *System Engineering Tools*, John Wiley, New York, 1965.
Chestnut, H. and Mayer, R. W., *Servomechanisms and Regulating System Design*, Vol. 1, John Wiley, New York, 1955.
Deutsch, R., *System Analysis Techniques*, Prentice-Hall, Englewood Cliffs, N.J., 1969.
Holbrook, J. G., *Laplace Transforms for Electronic Engineers*, Pergamon Press, London, 1959.
Kaplan, W., *Operational Methods for Linear Systems*, Addison-Wesley, Reading, Massachusetts, 1962.
Lewis, L. J., *Linear System Analysis*, McGraw-Hill, New York, 1969.
Truxal, J. G., *Control Engineer's Handbook*, McGraw-Hill, New York, 1958.

Tables

Burlington, R. and May, D., *Handbook of Probability and Statistics with Tables*, McGraw-Hill, New York, 1953.
Eilon, S., *Industrial Engineering Tables*, Van Nostrand, London, 1962.
Failure Rate Data Handbook (FARADA), Bureau of Naval Weapons, U.S. Naval Missile Systems Analysis and Evaluation Group, Corona, California.
Owen, D., *Handbook of Statistical Tables*, Addison-Wesley, Reading, Massachusetts, 1962.
Reliability Stress and Failure Rate Data, Department of Defense, MIL-HDBK-217, August, 1962.
Standard Mathematical Tables, Chemical Rubber Publishing Company, Cleveland, Ohio, 1954.

Tables of Normal Probability Functions, National Bureau of Standards, U.S. Government Printing Office, Washington, D.C., 1953.

Tables of the Binomial Probability Distribution, National Bureau of Standards, U.S. Government Printing Office, Washington, D.C., 1949.

Woodcock, E. R. and Eames, A. R., *Confidence Limits for Numbers from 0 to 1200 based on the Poisson Distribution*, Report AHSB(S) R179, United Kingdon Atomic Energy Authority or Her Majesty's Stationery Office, London, 1970.

Index

Achievement
  concepts 122–130
  function 71, 83, 122
  variations 162–165, 175–181, 190–191
Alternating renewal process (*see* Repair)
Assumptions (*see* Reliability model)
Auctioneering systems 301
Availability 21, 33
  distribution of 410
  due to change of state 407
  including repair 413
  mean 34

Basic requirement 86–91
  conditions 87–88
  hierarchies 89–90
Bayes' theorem (particular case) 519
Best estimate 355 (*see also* Estimation)
Binomial distribution 186
  cumulative 187
  cumulative plots 571–597
  general expressions 244–249
  graphs 265
  relationship with Poisson 240
  sampling from 350–353
Boolean algebra 136–145
  application to change of state 432
  exclusive form 496
  laws of 140
Boundary value approach 85
  applied to failure-rates 540

Capability
  and variability 129–130
  of performance 60–62
  (*see also* Functional capability)
Catastrophic failures
  two-state model 404
  types of 70
  (*see also* Discrete distributions)
Change of state 404–424
  alternating renewal process 419

Change of state (*cont.*)
  irreversible 404, 405–417
  majority-vote systems 497–502
  parallel redundant elements 455–458
  practical implications 404–405, 529
  reversible 417–424
  two-state catastrophic degradation 404
Change-over process (*see* Redundant)
Characteristic
  conversion 109–110 (*see also* Transfer characteristic)
  function 229
  matrix 230
  roots 230
Chi-square distribution 336
  cumulative table 560
  degree of freedom 336
  sample testing 378
Combinations of distributions 188–189
  and redundant systems 300
  need for 276–277 (*see also* Distributions)
  weighted estimates (*see* Estimation)
Common elements
  cross-linked type 494–497
  simple type 490–494
Common mode failure 11
  common fault 547
Complex systems 483
Conditional
  distribution functions 262–269
  expressions 133
  probability 263, 518
  relationships 114
Confidence
  in estimate 314 (*see also* Estimation)
  interval 316
  limits 316, 318–319
Consumer's risk 372 (*see also* Sample testing)

Continuous distributions of achievement variations 175–181
Continuous monitoring (*see* Repair)
Convolution
   Duhamel's integral 207
   relationships 206
Correlation process 387–404
   dependent 397–403
   discrete states 402–403
   in one dimension 387
   multiple 394–397
   simple 387–394
   upper and lower case 392
Cumulative distribution function 55, 257–262

Data
   accuracy of 521
   acquisition of 531–535
   monitoring system performance 534
   reliability 530–541
   use of 535–541
   variability in observation 553
Dependence
   interdependence and cross-linking 483
Density function (*see* Probability density function)
Deterministic
   approach to functional capability 129
   approximation to single-valued quantity 388
   evaluation 62
Dirac function 198 (*see also* Repair time)
Discrete distributions
   catastrophic failures 182
   of performance achievement variations 181–187
Distribution(s) (*see also* Distribution by name)
   central limit theorem 281, 323
   combinations of, need for 276–277
   continuous and discrete combinations (*see* Combinations of distributions)
   function 238–242
   of sample means (*see* Sampling)
   products of 290–296
   quotients of 296–299
   relationships when combined 285

Distribution(s) (*cont.*)
   sum and differences of 277–290
Diversity 11
   automatic protective systems 11
   bearing assemblies, example of 19
   use of 548

Environment average component failure-rates table 569
Erlangian special distribution
   general expressions 244–249
   probability density function 239
Estimation
   and confidence 315–359
   consistent 314, 316
   from sample 314
   general procedure for 319
   of parameter 315
   unbiased 314, 316
   weighted combinations 353–359
Evaluation process for system reliability 79
Exclusive elements 484–490
   Boolean form 517
   representation of 484
Exclusive events 28, 518
Exponential distribution 31
   and failure-rate characteristic 537
   cumulative table 559
   general expressions 244–249
   graphs 265
   probability density function 239
   sampling from 340–346

Fail-dangerous 13
Fail-safe 6, 13
Failure
   catastrophic 70, 519
   modes for components table 570
   partial 519
   revealed 451, 519
   unrevealed 451, 519
Failure-rate(s)
   and environment 537–539
   bar chart 538
   constant 272
   changing with time 272
   for components table 563
   function against time 535
   K factors and stress 540
   K factors table 569

INDEX 633

Failure probability (*see* Probability)
Field experience (*see also* Data)
   acquisition of data 531–534
   comparison with prediction 551
Financial
   and technological systems 116–117
   requirements 115–118
   system 115
Fractional dead time
   distribution of 461
   evaluation of mean due to change of state 409
   majority-vote systems tables of 511, 512
   mean 33
Frequency
   natural 214
   polygon 52
   proportionate 53
   table 52
Functional
   capability 130–136
   diagrams 152–155

Gamma distribution
   and failure-rate characteristic 537
   general expressions 244–249
   graphs 264
   probability density function 239
Gaussian distribution (*see* Normal distribution)
Goodness of fit 359–370
   analytical methods 363–370
   chi-square distribution test 366
   graphical methods 359–363
   Kolmogorov–Smirnov test 369
   level of significance 367
   normal probability paper 360
   Sturges' rule 359

Hardware variability 66
   and software performance 67
Hazard
   distribution functions 269–272
   function (*see* Failure-rate function)
   on a plant 14
High reliability systems, examples of 5–18
   aircraft landing systems 5–10
   chemical plants 14

High reliability systems, examples of (*cont.*)
   electrical supply systems 16
   power plants, protective systems 11
High risk situations 16
Histogram 52
Historical review 2–5
   aircraft 3
   electronic equipment 4
   missiles 3
   nuclear industry 4
Human influence
   factors and common mode faults 549
   in testing and maintenance 543, 546–547

Ideogrammatic representation 131, 133
Interface
   of system 123
   problem 122–124

Kolmogorov–Smirnov distribution 369
   (*see also* Goodness of fit)
Kurtosis coefficient of 254

Laplace transform(s) 111
   definition 194
   of a probability density function 241
   of particular functions 195
   of particular operations 201
Least squares principle 357
Logic 133–136
   diagrams 145–149
   operations 133, 135, 485
   symbols 146
   two-out-of-three 135
   two-state type 134
Logistics 131–132
Lognormal distribution
   general expressions 244–249
   graphs 264
   probability density function 239

Maintenance procedures 543
Majority vote system(s) 301, 307
   changes of state 497
   cross linking 483
   irreversible changes of state 502–512
   mean fractional dead times 511
   reversible changes of state 512–516
   variational aspects 516

Markov process and change of state 414 (*see also* Stochastic process)
Matching required system conditions 102
Matrix representation of a network 149–151, 495
  node removal technique 150
  primitive connexion matrix 149
  transmission function 151
Mean 243
  weighted 357
Mean time between failures (m.t.b.f.) 625 (*see also* Failure rate constant and Time to failure)
Moments of distributions 242–257

Non-redundant elements 431–455
  irreversible changes of state 437–445
  probabilistic diagrams 436
  representations of 433
  reversible changes of state 445–450
  system renewals 450
Normal distribution 178
  and failure-rate characteristic 537
  cumulative table 554
  general expressions 244–249
  graphs 264
  probability density function 239
  sampling from 334–340

Performance
  achieved 47, 57
  capability 60
  correlation procedure of 48
  large variations 70 (*see also* Catastrophic failure)
  n-dimensional surface of 529
  reliability diagram, example of 72
  required 47, 57, 72–75
  variability 62, 73
Poisson distribution
  cumulative plots 598–599
  probability function 240
  general expressions 244–249
  graphs 265
  relationship with binomial 240
  sampling from 346–350
Population 54
  estimates of 320

Prediction
  precision of results 551
  variability in prediction 551
Probabilities
  combination of 26
Probability
  definition 23
  distribution functions 272
  graphs 264–265
  of failure 23
  of success 24
Probability density function 54
  multivariate 70
Probability distributions general expressions 244–249
Problem
  artificially closed type of 85
  closed type of 85
  open type of 85
Producer's risk 372 (*see also* Sample testing)
Proportionate frequency histogram 54 (*see also* Probability density function)

Random variations in performance achievement 174–188
Rectangular distribution 55
  density function 282
  general expressions 244–249
  graphs 265
  sampling from 323
Redundancy 6
  auctioneering 301
  combination of probability distributions 300
  functional variations 301
  majority-vote system variations 307
  monitored duplicate system 6
  parallel 18
  standby 18
  triplex system 6
  voting 301 (*see also* Majority vote system)
Redundant
  change of state representation for standby elements 470–471
  change-over process 477
  irreversible changes of state 458–465
  irreversible change of state of standby system 471–474

INDEX 635

Redundant (*cont.*)
 parallel redundant elements 455–470
 parallel systems 516
 probabilistic diagrams for parallel systems 457
 representation of parallel systems 456
 reversible change of state of standby system 474–478
 reversible changes of state 465–468
 standby redundant elements 470–478
 structures 19
 systems and common fault 547
Reliability
 assessment general review 527–530
 assessment perspective 549–553
 correlation process 387–404 (*see also* Correlation process)
 definition 25
 deterministic correlation 388 (*see also* Deterministic)
 function 74
 model 541–549
 one-dimensional upper and lower case 392
 overall concept 47
 overall evaluation, summary of procedures 522
 parameters of interest 83, 424–425
 required and achieved performance 83
 time domain, in the 31–36
Reliability technology definition 21
Repair
 alternating renewal process 419 (*see also* Change of state)
 and availability evaluation 413
 continuous monitoring 417
 policy with redundant elements 468
 reversible change of state (*see* Restoration)
Repair time 33
 Dirac function representation 423
Required performance function 83
Requirement
 basic requirement 86–91
 concepts 83–86
 conditions 91–108
 degradation of 123
 limits 112–115
 lower limits 114

Requirement (*cont.*)
 overall 108–112
 single-valued 113
 upper limits 114
 variations 118–120
Response
 amplitude and frequency 219
 first-order system 217
 phase angle and frequency 220
 polar plot 221, 224
 ramp function 218
 second-order system 222
Restoration process of 405, 543 (*see also* Repair)

Sample testing 370–381
 acceptance probability curve 377
 chi-square distribution 378 (*see also* Chi-square distribution)
 consumer's risk 372
 data acquisition 531
 producer's risk 372
 truncated sequential test 380
Sampling 313–315
 Bessel correction 331
 chi-square distribution 336
 definition of random sample 315
 distribution 316
 distribution of sample means 321
 means 310–328
 sample variances 328
 Student t distribution 339
 variance of sample means 322
Skewness
 coefficient for a distribution 254, 323
Software
 combined variability with hardware performance 67
 variability 65
Standard deviation (*see* Variance)
Standby redundancy (*see* Redundancy standby)
Stochastic process 405
Stress levels for components table 569
Student t distribution 339
Substantiation of performance
 direct method 125
 indirect method 125
 points for assurance 126

Synthesis
  general procedures for 516–522
  need for synthesis 430–431
System(s)
  complex type 483
  configuration 125
  definition and specification 125
  functional arrangement 542
  general specification requirements 127
  generalized 155–158
  hardware 125
  logical analysis of 126
  non-redundant type 432, 516
  renewal of item of equipment 450–454
  synthesis of 155, 516–522
  two-element type 431–432
  two-state type 155
  types of complex systems 483
  variational aspects and synthesis 454–455
Systematic
  errors and their combination 169
  variations in performance achievement 169–174

Technological system
  applications 36, 100
  conversion process 105
  input variables 102
  typical input function 102
  typical output function 106
Testing
  improvement factors for majority-vote systems 511
  optimum method of staggered testing 463
  procedures 543
  staggered 510
Time to failure
  hazard function relationship 272
  mean (m.t.t.f.) 272, 371, 424
Tolerances 280
Transfer characteristic concepts 193–194
Transfer characteristic matrix 111, 229–234
  elements 120
  operational form 129
  singular value 130

Transfer function 111
  algebra 225–229
  closed-loop connexion 228
  effects of transfer functions 216–225
  electrical analogue 166
  first-order system 216
  integrodifferential equations 210
  parallel connexion 227
  second-order mechanical system application 213
  series connexion 226
  transform pairs of functions table 555
Tree diagram 149
Truth tables 143
  logical derivation 143, 185
  two-element system 431 (*see also* System)

Variance 252
  sample 328–334
Variate
  continuous 51, 175
  discrete 52
Variability 50
  function 75
  in prediction and observation 553
  of performance 62–72
Variational aspects
  non-redundant system 454–455, 469–470
  standby redundant system 478
Variations
  functional variations in redundant systems 301 (*see also* Redundancy)
  general combinations of small 299–300
  in a system performance parameter 189
  in hardware 62
  in software 62
  in system performance 62
  random (in requirement) 119
  systematic (in requirement) 119
  two-out-of-three system 516
Venn diagrams 141

Weibull distribution
  and failure-rate characteristic 537
  general expressions 244–249
  graphs 264
  probability density function 239
Weighted mean 357